PHYSICAL CHEMISTRY OF MACROMOLECULES

PHYSICAL CHEMISTRY OF MACROMOLECULES

BASIC PRINCIPLES AND ISSUES

S. F. SUN
St. John's University
Jamaica, New York

A Wiley-Interscience Publication
JOHN WILEY & SONS, INC.
New York · Chichester · Brisbane · Toronto · Singapore

This text is printed on acid-free paper.

Library of Congress Cataloging in Publication Data:

Sun, S. F.,
Physical chemistry of macromolecules: basic principles and issues / S. F. Sun.
p. cm.
"A Wiley-Interscience Publication."
Includes index.
ISBN 0-471-59788-0 (alk. paper)
1. Macromolcules. 2. Chemistry, Physical organic. I. Title.
QD381.8.S86 1994 93-28871
547.7'045—dc20

Printed in the United States of America
10 9 8 7 6 5 4 3 2 1

CONTENTS

PREFACE

Physical chemistry of macromolecules is a course that is frequently offered in the biochemistry curriculum of a college or university. Occasionally, it is also offered in the chemistry curriculum. When it is offered in the biochemistry curriculum, the subject matter is usually limited to biological topics and is identical to biophysical chemistry. When it is offered in the chemistry curriculum, the subject matter is often centered around synthetic polymers and the course is identical to physical polymer chemistry. Since the two disciplines are so closely related, students almost universally feel that something is missing when they take only biophysical chemistry or only physical polymer chemistry. This book emerges from the desire to combine the two courses into one by providing readers with the basic knowledge of both biophysical chemistry and physical polymer chemistry. It also serves a bridge between the academia and industry. The subject matter is basically academic, but its application is directly related to industry, particularly polymers and biotechnology.

This book contains seventeen chapters, which may be classified into three units, even though not explicitly stated. Unit 1 covers Chapters 1 through 5, unit 2 covers Chapters 6 through 12, and unit 3 covers Chapters 13 through 17. Since the materials are integrated, it is difficult to distinguish which chapters belong to biophysical chemistry and which chapters belong to polymer chemistry. Roughly speaking, unit 1 may be considered to consist of the core materials of polymer chemistry. Unit 2 contains materials belonging both to polymer chemistry and biophysical chemistry. Unit 3, which covers the structure of macromolecules and their separations, is relatively independent of units 1 and 2. These materials are important in advancing our knowledge of macromolecules, even though their use is not limited to macromolecules alone.

The book begins with terms commonly used in polymer chemistry and biochemistry with respect to various substances, such as homopolymers, copolymers, condensation polymers, addition polymers, proteins, nucleic acids, and polysaccharides (Chapter 1), followed by descriptions of the methods used to create these substances (Chapter 2). On the basis of classroom experience, Chapter 2 is a welcome introduction to students who have never been exposed to the basic methods of polymer and biopolymer syntheses. The first two chapters together comprise the essential background materials for this book.

Chapter 3 introduces statistical methods used to deal with a variety of distribution of molecular weight. The problem of the distribution of molecular weight is characteristic of macromolecules, particularly the synthetic polymers, and the statistical methods are the tools used to solve the problem. Originally Chapter 4 covered chain configurations and Chapter 5 covered macromolecular thermodynamics. Upon further reflection, the order was reversed. Now Chapter 4 on macromolecular thermodynamics is followed by Chapter 5 on chain configurations. This change was based on both pedagogical and chronological reasons. For over a generation (1940s to 1970s), Flory's contributions have been considered the standard work in physical polymer chemistry. His work together with that of other investigators laid the foundations of our way of thinking about the behavior of polymers, particularly in solutions. It was not until the 1970s that Flory's theories were challenged by research workers such as de Gennes. Currently, it is fair to say that de Gennes' theory plays the dominant role in research. In Chapter 4 the basic thermodynamic concepts such as χ, θ, ψ, and κ that have made Flory's name well known are introduced. Without some familiarity with these concepts, it would not be easy to follow the current thought as expounded by de Gennes in Chapter 5 (and later in Chapters 6 and 7). For both chapters sufficient background materials are provided either in the form of introductory remarks, such as the first section in Chapter 4 (a review of general thermodynamics), or in appendices, such as those on scaling concepts and correlation function in Chapter 5.

In Chapters 6 through 17, the subjects discussed are primarily experimental studies of macromolecules. Each chapter begins with a brief description of the experimental method, which, though by no means detailed, is sufficient for the reader to have a pertinent background. Each chapter ends with various theories that underlie the experimental work.

For example, in Chapter 6, to begin with three parameters, σ (shear stress), ε (shear strain), and E (modulus or rigidity), are introduced to define viscosity and viscoelasticity. With respect to viscosity, after the definition of Newtonian viscosity is given, a detailed description of the capillary viscometer to measure the quantity η follows. Theories that interpret viscosity behavior are then presented in three different categories. The first category is concerned with the treatment of experimental data. This includes the Mark-Houwink equation, which is used to calculate the molecular weight, the Flory-Fox equation, which is used to estimate thermodynamic quantities, and the Stockmayer-Fixman equation, which is used to supplement the intrinsic viscosity treatment. The second category describes the purely theoretical approaches to viscosity. These approaches include the Kirkwood-Riseman model and the Debye-Buche model. It also includes chain entanglement. Before presenting the third category, which deals with the theories about viscosity in relation to biological polymers, a short section discussing Stokes' law of frictional coefficient is included. The third category lists the theories proposed by Einstein, Peterlin, Kuhn and Kuhn, Simha, Scheraga and Mendelkern. With respect to viscoelasticity, Maxwell's model is adopted as a basis. Attention is focused on two theories that are very much in current thought, particularly in connection with the dynamic scaling law: the Rouse model and the Zimm model. These models are reminiscent of the Kirkwood-Riseman theory and the Debye-Buche theory in viscosity but are much more stimulating to the present way of thinking in the formulation of universal laws to characterize polymer behavior.

Chapter 7, on osmotic pressure, provides another example of my approach to the subject matter in this book. After a detailed description of the experimental determination of molecular weight and the second virial coefficient, a variety of models are introduced each of which focuses on the inquiry into inter- and intramolecular interactions of polymers in solution. The reader will realize that the thermodynamic function μ (chemical potential) introduced in Chapter 4 has now become the key term in our language. The physical insight that is expressed by theoreticians is unusually inspiring. For those who are primarily interested in experimental study, Chapter 7 provides some guidelines for data analysis. For those who are interested in theoretical inquiry, this chapter provides a starting point to pursue further research. Upon realizing the difficulties involved in understanding mathematical terms, several appendices are added to the end of the chapter to give some background information.

Chapters 8 through 12, are so intermingled in content that they are hardly independent from each other, yet they are so important that each deserves to be an independent chapter. Both Chapters 8 and 9 are about light scattering. Chapter 8 describes general principles and applications, while Chapter 9 discusses advanced techniques in exploring detailed information about the interactions between polymer molecules in solutions. Chapters 10 and 11 are both about diffusion. Chapter 10 deals with the general principles and applications of diffusion, while Chapter 11 describes advanced techniques in measurement. However, diffusion is only part of the domain in Chapter 11, for Chapter 11 is also directly related to light scattering. As a matter of fact, Chapters 8, 9, and 11 can be grouped together. In parallel, Chapters 10 and 12, one about diffusion and the other about sedimentation, are closely related. They describe similar principles and similar experimental techniques. Knowledge of diffusion is often complementary to knowledge of sedimentation and vice versa.

It should be pointed out that all the chapters in unit 2 (Chapters 6 through 12) so far deal with methods for determining molecular weight and the configuration of macromolecules. They are standard chapters for both a course of polymer chemistry and a course of biophysical chemistry. Chapters 13 through 17 describes some of the important experimental techniques that were not covered in Chapters 6 through 12. Briefly, Chapter 13, on optical rotatory dispersion (ORD) and circular dichroism (CD), describes the content of helices in a biological polymer under various conditions, that is, in its native as well as in its denatured states. The relationship between ORD and CD is discussed in detail. Chapter 14 provides basic knowledge of nuclear magnetic resonance phenomena and uses illustrations of several well-known synthetic polymers and proteins. Chapter 15, on x-ray crystallography, introduces the foundations of x-ray diffractions, such as Miller indices, Bravais lattices, seven crystals, 32 symmetries, and some relevant space groups. It then focuses on the study of a single crystal: the structure factor, the density map, and the phase problem. Chapter 16, on electron and infrared spectroscopy, provides the background for the three most extensively used spectroscopic methods in macromolecular chemistry, particularly with respect to biological polymers. These methods are ultraviolet absorption, fluorimetry, and infrared spectra. Chapter 17 belongs to the realm of separation science or analytical chemistry. It is included because no modern research in polymer chemistry or biophysical chemistry can completely neglect the techniques used in this area. This chapter is split into two parts. The first part, high-performance liquid chromatography (HPLC), describes key parameters of chromatograms and the four types of chromatography with an emphasis on size-exclusion chromatography, which enables us to determine the molecular weight, molecular weight distribution, and binding of small molecules to macromolecules. The second part, electrophoresis, describes the classical theory of ionic mobility and various types of modern techniques used for the separation and characterization of biological materials. Chapter 17 ends with an additional section, field-flow fractionation, which describes the combined methods of HPLC and electrophoresis.

In conclusion, the organization of this book covers the basic ideas and issues of the physical chemistry of macromolecules including molecular structure, physical properties, and modern experimental techniques.

Mathematical equations are used frequently in this book, because they are a part of physical chemistry. We use mathematics as a language in a way that is not different from our other language, English. In English, we have words and sentences; in mathematics, we use symbols (equivalent to words) and equations (equivalent to sentences). The only difference between the two is that mathematics, as a symbolic language, is simple, clear, and above all operative, meaning that we can manipulate symbols as we wish. The level of mathematics used in this text is not beyond elementary calculus, which most readers are assumed to have learned or are learning in college.

In this book, derivations, though important, are minimized. Derivations such as Flory's lattice theory on the entropy of mixing and Rayleigh's equation of light scattering are given only because they are simple, instructive, and, above all, they provide some sense of how an idea is translated from the English language to a mathematical language. The reader's understanding will not be affected if he or she skips the derivation and moves directly to the concluding equations. Furthermore, the presentation of the materials in this book has been tested on my classes for many years. No one has ever complained.

The selection of mathematical symbols (notations) used to designate a physical property (or a physical quantity) poses a serious problem. The same letter, for example, α or c, often conveys different meanings (that is, different designations). The Greek letter α can represent a carbon in a linear chain (α atom, β atom,...), one of the angles of a three-dimensional coordinate system (related to types of crystals), the expansion factor of polymer molecules in solutions (for example, $\alpha^5 - \alpha^3$), the polarizability with respect to the polarization of a molecule, and so on. The English letter c can represent the concentration of a solution (for example, g/mL, mol/L), the unit of coordinates (such as a, b, c), and so on. To avoid confusion, some authors use different symbols to represent different kinds of quantities and provide a glossary at the end of the book. The advantage of changing standard notation is the maintenance of consistency within a book. The disadvantage is that changing the well-known standard notation in literature (for example, S for expansion factor, T for polarizability, instead of α for both; or d for a unit coordinate, j for the concentration of a solution, instead of c for both), is awkward, and may confuse readers. In addressing this problem, the standard notations are kept intact. Sometimes the same letters are used to represent different properties in the same chapter. But I have tried to use a symbol to designate a specific property as clearly as possible in context by repeatedly defining the term immediately after the equation. I also add a prime on the familiar notations, for example, R' for gas constant and c' for the velocity of light. Readers need not worry about confusion.

At the end of each chapter are references and homework problems. The references are usually the source materials for the chapters. Some are original papers in literature, such as those by Flory, Kirkwood, Debye, Rouse, Des Cloizeau, deGennes, Luzzati, and Zimm, among others; and some are well-known books, such as those of Yamakawa and Hill, in which the original papers were cited in a rephrased form. Equations are usually given in their original forms from the original papers with occasional modifications to avoid confusion among symbols. It is hoped that this will familiarize readers with the leading literature. Homework problems are designed to help readers clarify certain points in the text.

A comment should be made on the title of the book, *Physical Chemistry of Macromolecules: Basic Principles and Issues.* The word "basic" refers to "fundamental," meaning "relatively timeless." In the selection of experimental methods and theories for each topic, the guideline was to include only those materials that do not change rapidly over time, for example, Fick's first law and second law in diffusion, Patterson's synthesis and direct method in x-ray crystallography, or those materials, though current, that are well established and frequently cited in the literature, such as the scaling concept of polymer and DNA sequencing by electrophoresis. The book is, therefore, meant to be "a course of study."

I wish to thank Professor Emily Sun for general discussion and specific advice. Throughout the years she has offered suggestions for improving the writing in this book. Chapters 1 through 12 were read by Patricia Sun, Esq., 13 through 17 by Caroline Sun, Esq., and an overall consultation was provided by Dr. Diana Sun. I am greatly indebted to them for their assistance. A special note of thanks goes to Mr. Christopher Frank who drew the figures in chapter 11 and provided comments on the appendix, and to Mr. Anthony DeLuca and Professor Andrew Taslitz, for improving portions of this writing. Most parts of the manuscript were painstakingly typed by Ms. Terry Cognard. For many years, students and faculty members of the Department of Chemistry of Liberal Arts and Sciences and the Department of Industrial Pharmacy of the College of Health Science at St. John's University have encouraged and stimulated me in writing this book. I am grateful to all of them.

S. F. SUN

Jamaica, New York
February 1994

PHYSICAL CHEMISTRY OF MACROMOLECULES

1

INTRODUCTION

Macromolecules are closely related to colloids, and historically the two are almost inseparable. Colloids were known first, having been recognized for over a century. Macromolecules were recognized only after much fierce struggle among chemists in the early 1900s. Today, we realize that while colloids and macromolecules are different entities, many of the same laws that govern colloids also govern macromolecules. For this reason, the study of the physical chemistry of macromolecules often extends to the study of colloids. Although the main topic of this book is macromolecules, we are also interested in colloids. Since colloids were known first, we will describe them first.

1.1 COLLOIDS

When small molecules with a large surface region are dispersed in a medium to form two phases, they are in a colloidal state and they form colloids. The two phases are liquid–liquid, solid–liquid, and so on. This is not a true solution (that is, not a homogeneous mixture of solute and solvent), but rather one type of material dispersed on another type of material, The large surface region is responsible for surface activity, the capacity to reduce the surface or interfaces tensions.

There are two kinds of colloids: lyophobic and lyophilic. Lyophobic colloids are solvent hating (that is, not easily miscible with the solvent) and thermodynamically unstable, whereas the lyophilic colloids are solvent loving (that is, easily miscible with the solvent) and thermodynamically stable. If the liquid medium is water, the lyophobic colloids are called hydrophobic colloids and the lyophilic colloids are called hydrophillic colloids. Examples of lyophobic colloids are foam, which is the dispersion of gas on liquid; emulsion, which is the dispersion of liquid on liquid; and sol, which is the dispersion of solid on liquid.

An example of lyophilic colloids is a micelle. A micelle is a temporary union of many small molecules or ions. It comes in shapes such as sphere or rods:

Sphere Rod

Typical micelles are soaps, detergents, bile salts, dyes, and drugs. A characteristic feature of the micelle is the abrupt change in physical properties at a certain concentration, as shown in Figure 1.1. The particular concentration is called the critical micelle concentration (CMC). It is at this concentration that the surface-active materials form micelles. Below the CMC, the small molecules exist as individuals. They do not aggregate.

Two micelle systems of current interest in biochemistry and pharmacology are sodium dodecylsulfate (SDS) and liposome. SDS is a detergent whose chemical formula is

$$\sim\!-\mathrm{O}-\overset{\overset{\displaystyle \mathrm{O}}{\|}}{\underset{\underset{\displaystyle \mathrm{O}}{\|}}{\mathrm{S}}}-\mathrm{O}^{-}\,\mathrm{Na}^{+}$$

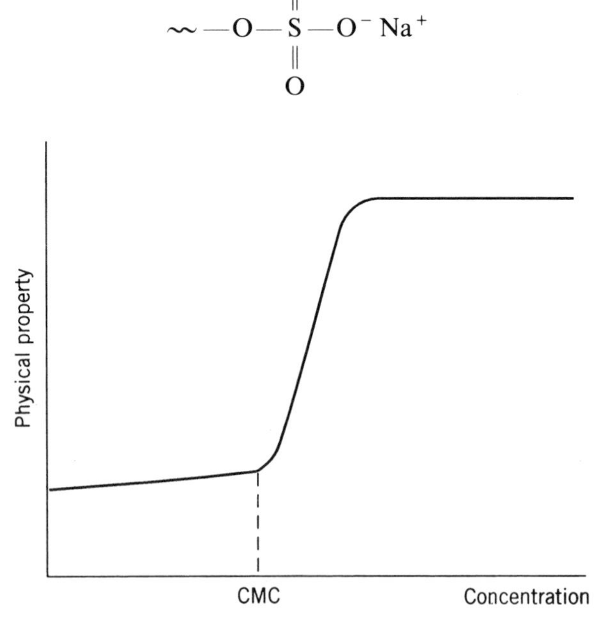

FIGURE 1.1 Critical micelle concentration.

The surface activity of this detergent causes a protein to be unfolded to a linear polypeptide. It destroys the shape of the protein molecule, for example, rendering a spherical molecule to a random coil. SDS binds to many proteins. The binding is saturated at the well-known 1.4-g/g level, that is, at the concentration of SDS exceeding 0.5 mM. Above this level SDS starts self-association and binding is reduced. At 8.2 mM, SDS forms micelles, with an aggregation number of 62 and a micellar molecular weight of 18 000.

Liposome is believed to be one of the best devices for the controlled release of drugs. There are three kinds of lipsomes:

1. Uncharged (ingredients: egg lecithin–cholesterol, weight ratio 33:4.64 mg).
2. Negatively charged (ingredients: egg lecithin–cholesterol–phosphatidic acid–dicetyl phosphate, ratio 33:4.46:10:3.24 mg).
3. Positively charged (ingredients: egg lecithin–cholestorol–stearylamine, ratio 33:4.46:1.6 mg)

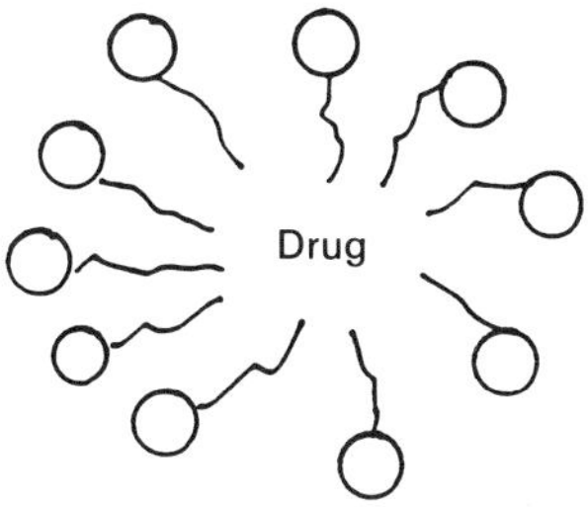

The surface of the micelle liposome is similar to that of membrane lipids; it does no harm to the body when administered.

1.2 MACROMOLECULES

The physical properties of macromolecules, such as sedimentation, diffusion, and light scattering, are very similar to those of colloids. For generations macromolecules have been regarded as associated colloids or lyophilic colloidal systems. But macromolecules are not colloids. Colloids are aggregations of small molecules, due to the delicate balance of weak attractive forces (such as the van der Waals force) and repulsive forces. The aggregation depends on the physical environment, particularly the solvent. When the solvent changes, the aggregation may collapse. Macromolecules are formed from many repeating small molecules which are connected by covalent bonds. Each macromolecule is an entity or a unit, not an aggregation. As the solvent changes, the properties of a macromolecule may change, but the macromolecule remains a macromolecule unless its covalent bands are broken.

Basically there are two types of macromolecules: synthetic polymers and biological polymers. Synthetic polymers are those that do not exist in nature; they are man-made molecules. Biological polymers do exist in nature, but they can also be synthesized in the laboratory. Synthetic polymers have a very small number of identical repeating units, usually one or two in a chain, whereas biological polymers have more identical repeating units in a chain, particularly proteins and enzymes, which have a variety of combinations (that is, amino acids). Synthetic polymers carry flexible chains; the molecules are usually not rigid. Biological polymer chains are more ordered; the molecules are, in general, rigid. The rigidity depends on the nature of the chains and their environment. Relatively speaking, nucleic acids are more rigid than proteins.

Recently, more similarity has been observed between the two types of macromolecules. For example, synthetic polymers, which are usually considered to be in the form of flexible random coils, can now be synthesized with the Ziegler-Natta catalysts to have stereoregularity. Furthermore, synthetic polymers can be designed to have helices, just like proteins and nucleic acids. As our knowledge of macromolecules increases, the sharp distinction between synthetic polymers and biological polymers becomes more and more arbitrary.

1.2.1 Synthetic Polymers

In 1929, Carothers classified synthetic polymers into two classes according to the method of preparation used: condensation polymers and addition polymers. For condensation (or step-reaction) polymers, the reaction occurs between two polyfunctional molecules by eliminating a small molecule, for example, water. The following are examples of condensation polymers:

$$\left[-O-\overset{\displaystyle O}{\overset{\|}{C}}-C_6H_4-\overset{\displaystyle O}{\overset{\|}{C}}-O-CH_2CH_2- \right]_x$$

$$HO\left[(CH_2)_xCOO\right]_y$$

$$\left[O-\overset{\displaystyle O}{\overset{\|}{C}}-(CH_2)_4\overset{\displaystyle O}{\overset{\|}{C}}-O-CH_2CH_2 \right]_x$$

Polyester (fiber)

$$\left[O(CH_2)_6OCONH(CH_2)_6NHCO \right]_x$$

Polyurethane (fiber)

$$\left[NH(CH_2)_6NHCONH(CH_2)_6NHCO \right]_x$$

Polyurea

Addition (or chain-reaction) polymers are formed in a chain reaction of monomers with double bonds. The following are examples of addition polymers:

$$+CH_2—CH_2+_n$$

Polyethylene

$$\left(CH_2—\underset{\underset{Cl}{|}}{CH}\right)_n$$

Poly(vinyl chloride)

$$\left(CH_2—\underset{\underset{C_6H_5}{|}}{CH}\right)_n$$

Polystyrene

$$\left(CH_2—\overset{\overset{CH_3}{|}}{\underset{\underset{COOCH_3}{|}}{CH}}\right)_n$$

Poly(methyl methacrylate)

Polymers may be classified into two structural categories: linear polymers and branched polymers. Linear polymers are in the form

$$A'—A—A\cdots A''$$

or

$$A'—(A)_{x-2}—A''$$

where A is the structural unit, x is the degree of polymerization, and A′, A″ are end groups of A. An example of a linear polymer is linear polystyrene. Branched polymers are in the form

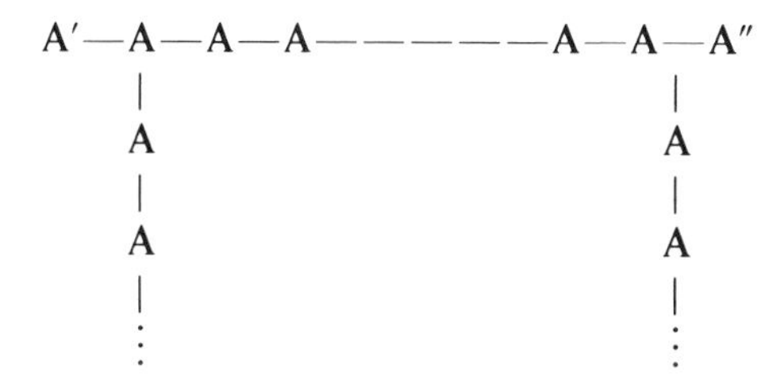

Two of the most well-known branched polymers are the star-shaped polymer

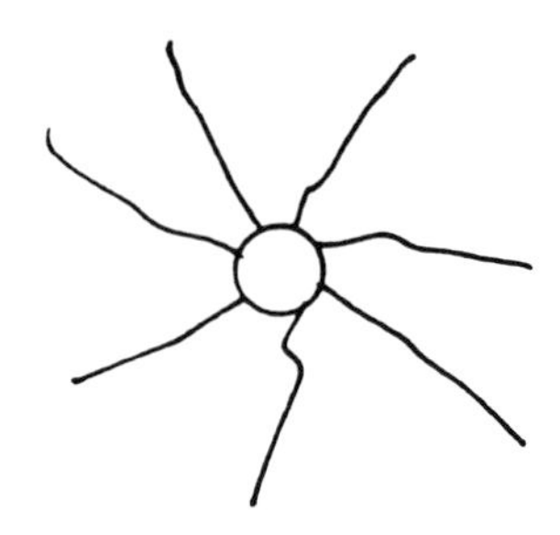

and the comb-shaped polymer

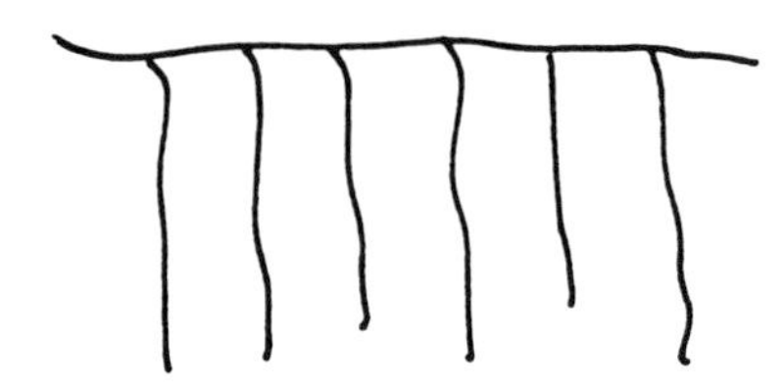

which both have various numbers of arms. Examples are star-shaped polystyrene and comb-shaped polystyrene.

In terms of repeating units there are two types of polymers: homopolymers and copolymers. A homopolymer is one in which only one monomer constitutes the repeating units, for example, polystyrene and poly(methyl methacrylate). A copolymer consists of two or more different monomers as repeating units, such as the diblock copolymer

$$\mathrm{A{-}A{-}A \cdots A{-}B{-}B{-}B \cdots B}$$

and the random or static copolymer

$$\mathrm{A{-}B{-}B{-}A{-}A{-}B{-}B{-}B{-}A{-}B{-}A{-}B{-}A{-}A \cdots}$$

An example is the polystyrene–poly(methyl methacrylate) copolymer.

In terms of stereoregularity synthetic polymers may have *trans* and *gauche* forms, similar to some small molecules (for example, ethane). Because of the steric position of substituents along the chain, the heterogeneity of the chain structure may be classified into three forms:

1. *Atactic polymers*—no regularity of R groups; for example,

$$\sim \mathrm{CH_2}{-}\underset{\text{Ⓡ}}{\overset{\mathrm{H}}{\mathrm{C}}}{-}\mathrm{CH_2}{-}\underset{\text{Ⓡ}}{\overset{\mathrm{H}}{\mathrm{C}}}{-}\mathrm{CH_2}{-}\underset{\mathrm{H}}{\overset{\text{Ⓡ}}{\mathrm{C}}}{-}\mathrm{CH_2}{-}\underset{\text{Ⓡ}}{\overset{\mathrm{H}}{\mathrm{C}}}{-} \sim$$

2. *Isotactic polymers*—regularity of R groups; for example,

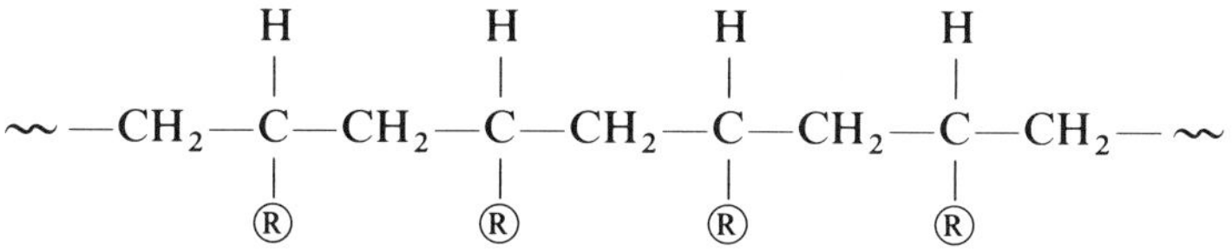

$$\sim\!\!-CH_2-\underset{\text{Ⓡ}}{\overset{H}{\underset{|}{\overset{|}{C}}}}-CH_2-\underset{\text{Ⓡ}}{\overset{H}{\underset{|}{\overset{|}{C}}}}-CH_2-\underset{\text{Ⓡ}}{\overset{H}{\underset{|}{\overset{|}{C}}}}-CH_2-\underset{\text{Ⓡ}}{\overset{H}{\underset{|}{\overset{|}{C}}}}-CH_2-\!\!\sim$$

3. *Syndiotactic polymers*—regularity involves *trans* and *gauche* forms in a uniform manner; for example,

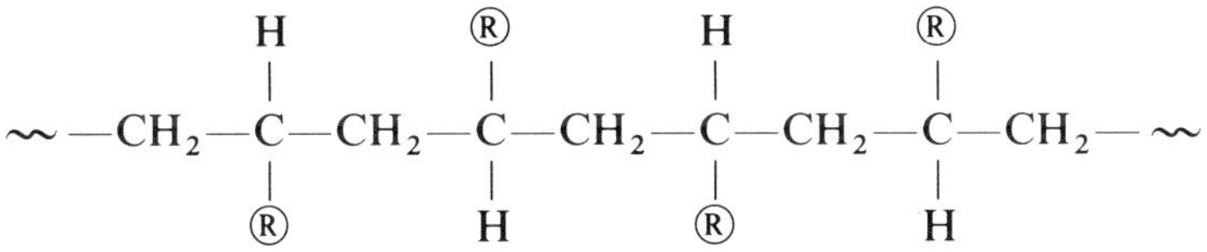

$$\sim\!\!-CH_2-\underset{\text{Ⓡ}}{\overset{H}{\underset{|}{\overset{|}{C}}}}-CH_2-\underset{H}{\overset{\text{Ⓡ}}{\underset{|}{\overset{|}{C}}}}-CH_2-\underset{\text{Ⓡ}}{\overset{H}{\underset{|}{\overset{|}{C}}}}-CH_2-\underset{H}{\overset{\text{Ⓡ}}{\underset{|}{\overset{|}{C}}}}-CH_2-\!\!\sim$$

The isotactic and syndiotactic polymers can be synthesized using Ziegler-Natta catalyst.

Synthetic polymers that are commercially manufactured in the quantity of billions of pounds may be classified in three categories: (1) plastics, which include thermosetting resins (for example, urea resins, polyesters, epoxides) and thermoplastic resins (for example, low-density as well as high-density polyethylene, polystyrene, polypropylene); (2) synthetic fibers, which include cellulosics (such as rayon and acetate) and noncellulose (such as polyester and nylon); and (3) synthetic rubber (for example, styrene–butadiene copolymer, polybutadiene, ethylene–propylene copolymer).

1.2.2 Biological Polymers

Biological polymers are composed of amino acids, nucleotides, or sugars. Here we describe three types of biological polymers: proteins and polypeptides, nucleic acids, and polymers of sugars.

Proteins and Polypeptides Amino acids are bound by a peptide bond which is an amide linkage between the amino group of one molecule and the carboxyl group of another. It is in the form

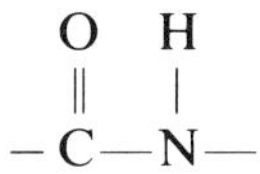

$$-\overset{O}{\overset{\|}{C}}-\overset{H}{\overset{|}{N}}-$$

For example,

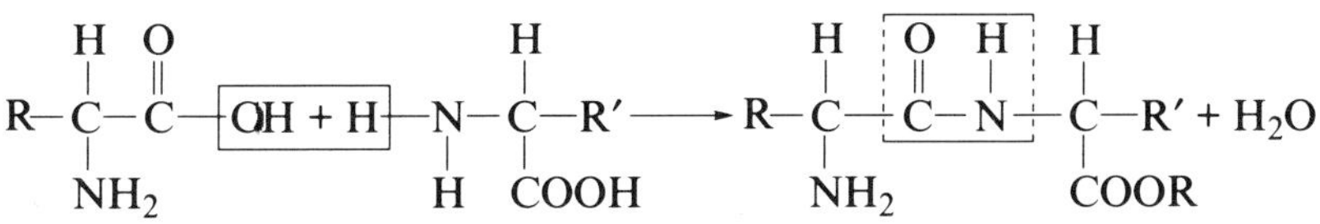

$$\underset{\text{Amino acid}}{R-\underset{NH_2}{\underset{|}{\overset{H}{\overset{|}{C}}}}-\overset{O}{\overset{\|}{C}}-\boxed{OH + H}}-\underset{\text{Amino acid}}{\underset{H}{\underset{|}{N}}-\underset{COOH}{\underset{|}{\overset{H}{\overset{|}{C}}}}-R'} \longrightarrow \underset{\text{A dipeptide}}{R-\underset{NH_2}{\underset{|}{\overset{H}{\overset{|}{C}}}}-\boxed{\overset{O}{\overset{\|}{C}}-\overset{H}{\overset{|}{N}}}-\underset{COOR}{\underset{|}{\overset{H}{\overset{|}{C}}}}-R'} + H_2O$$

A polypeptide is in the form

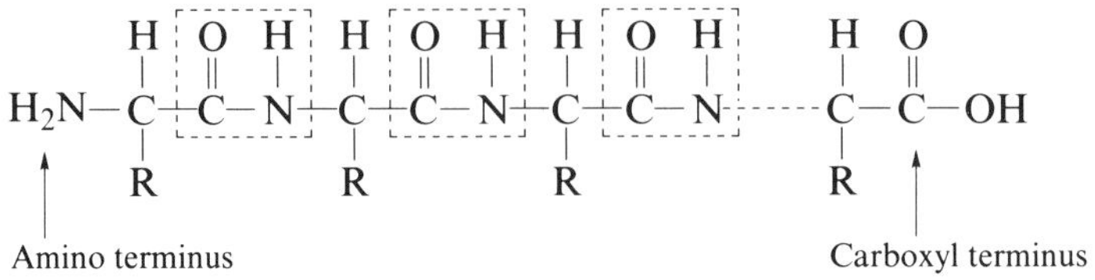

$$\underset{\substack{\uparrow \\ \text{Amino terminus}}}{H_2N}-\underset{R}{\overset{H}{C}}-\overset{O}{\overset{\|}{C}}-\overset{H}{N}-\underset{R}{\overset{H}{C}}-\overset{O}{\overset{\|}{C}}-\overset{H}{N}-\underset{R}{\overset{H}{C}}-\overset{O}{\overset{\|}{C}}-\overset{H}{N}\cdots\cdots\underset{R}{\overset{H}{C}}-\overset{O}{\overset{\|}{C}}-\underset{\substack{\uparrow \\ \text{Carboxyl terminus}}}{OH}$$

A protein is a polypeptide consisting of many amino acids (Table 1.1). A protein with catalytic activities is called an enzyme. All enzymes are proteins, but not all proteins are enzymes. A hormone is also a polypeptide, (for example, insulin), and is closely related to proteins.

There are two types of proteins: simple and conjugated. Simple proteins are classified in terms of their solubility in water into five groups (old classification):

1. *Albumins*—soluble in water and in dilute neutral salt solutions
2. *Globins*—soluble in water (for example, hemoglobins)
3. *Globulins*—insoluble in water, but soluble in dilute neutral salt solutions (for example, γ-globulins)
4. *Prolamines*—soluble in 70% ethyl alcohol, insoluble in water
5. *Histones*—strongly basic solutions, soluble in water

Conjugated proteins are classified by the nonprotein groups:

1. *Nucleoproteins*—a basic protein such as histones or prolamines combined with nucleic acid
2. *Phosphoproteins*—proteins linked to phosphoric acid (for example, casein in milk and vitellin in egg yolk)
3. *Glycoproteins*—a protein and a carbohydrate (for example, mucin in saliva, mucoids in tendon and cartilage, interferron, which is a human gene product made in bacteria using recombinant DNA technology)
4. *Chromoproteins*—a protein combined with a colored compound (for example, hemoglobin and cytochromes)
5. *Lipoproteins*—proteins combined with lipids (such as fatty acids, fat, and lecithin)
6. *Membrane proteins*—proteins embedded in the lipid core of membranes (for example, glycohorin A)

Proteins may be found in three shapes:

1. Thin length (for example, collagen, keratin, myosin, fibrinogen)
2. Sphere (for example, serum albumin, myoglobin, lysozyme, carboxypeptidase, chymotrypsin)

TABLE 1.1 Amino Acids

ALIPHATIC AMINO ACIDS (major amino acids contributed to a hydrophobic region)

Glycine (Gly) $H_2N—CH_2—COOH$

Alanine (Ala) $CH_3—CH(NH_2)—COOH$

Valine (Val) $CH_3—CH(CH_3)—CH(NH_2)—COOH$

Leucine (Leu) $CH_3—CH(CH_3)—CH_2—CH(NH_2)—COOH$

Isoleucine (Ileu) $CH_3—CH_2—CH(CH_3)—CH(NH_2)—COOH$

HYDROXY ACIDS

Serine (Ser) $CH_2(OH)—CH(NH_2)—COOH$

Threonine (Thr) $CH_3—CH(OH)—CH(NH_2)—COOH$

AROMATIC AMINO ACIDS (UV region)

Phenylalanine (Phe) $C_6H_5—CH_2—CH(NH_2)—COOH$

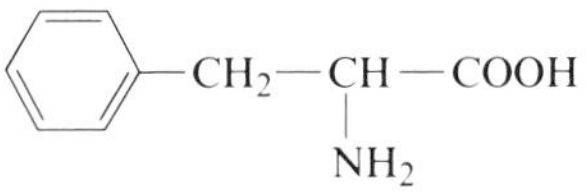

Tyrosine (Tyr) $HO—C_6H_4—CH_2—CH(NH_2)—COOH$

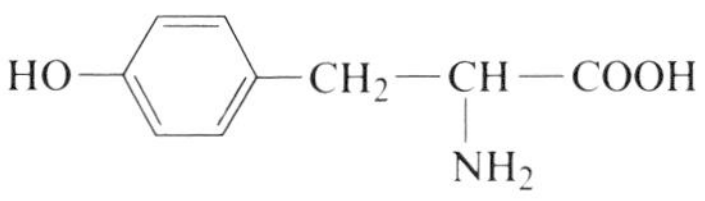

A HETEROCYCLIC GROUP

Tryptophan (Try) indolyl(N—H)$—CH_2—CH(NH_2)—COOH$

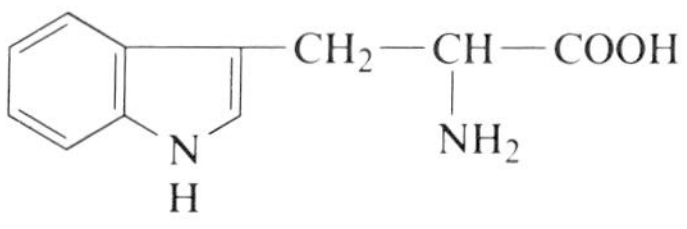

TABLE 1.1 (*Continued*)

SULFUR-CONTAINING AMINO ACIDS (cross-linkage)	
Cysteine (Cys)	$CH_2(SH)—CH(NH_2)—COOH$
Methionine (Met)	$CH_3—S—CH_2—CH_2—CH(NH_2)—COOH$
ACIDIC AMINO ACIDS (potentiometric titration)	
Aspartic acid (Asp)	$HOOC—CH_2—CH(NH_2)—COOH$
Glutamic acid (Glu)	$HOOC—CH_2—CH_2—CH(NH_2)—COOH$
BASIC AMINO ACIDS (potentiometric titration)	
Lysine (Lys)	$CH_2(NH_2)—CH_2—CH_2—CH_2—CH(NH_2)—COOH$
Arginine (Arg)	$H_2N—C(=NH)—NH—CH_2—CH_2—CH_2—CH(NH_2)—COOH$
Histidine (His)	$HC{=}C—CH_2—CH(NH_2)—COOH$ (imidazole ring: HC, N, C–H, NH)
IMINO ACIDS	
Proline (Pro)	$H_2C—CH_2$ / H_2C, $CH—COOH$, ring closed through N–H

TABLE 1.1 (*Continued*)

Hydroxyproline (Hyp)	$HO—HC——CH_2$ $\quad\;\; \vert \qquad\;\; \vert$ $\;\; H_2C \quad CH—COOH$ $\qquad\; N$ $\qquad\; H$
CARBOXAMIDE	
Asparagine	$NH_2 \qquad NH_2$ $\vert \qquad\qquad \vert$ $O{=}C—CH_2—C—COOH$ $\vert$ H
Glutamine	$NH_2 \qquad\qquad NH_2$ $\vert \qquad\qquad\qquad \vert$ $O{=}C—CH_2—CH_2—C—COOH$ $\vert$ H

3. Elastic (for example, elastin in the main constituent of ligament, aortic tissue, and the walls of blood vessels)

Nucleic Acids Nucleic acids consist of nucleotides, which in turn consist of nucleosides:

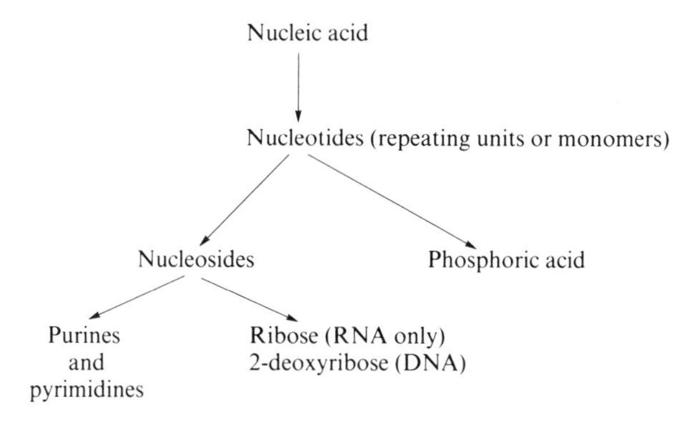

The major repeating units (nucleotides) are shown in Table 1.2. Each nucleotide consists of a base, a sugar, and a phosphate. There are only five bases, two sugars, and one phosphate from which to form a nucleotide. These are shown in Table 1.3. A nucleoside is a nucleotide minus the phosphate.

TABLE 1.2 Major Nucleotides

Nucleotide
Uridylic acid (uridine-3′-phosphate) 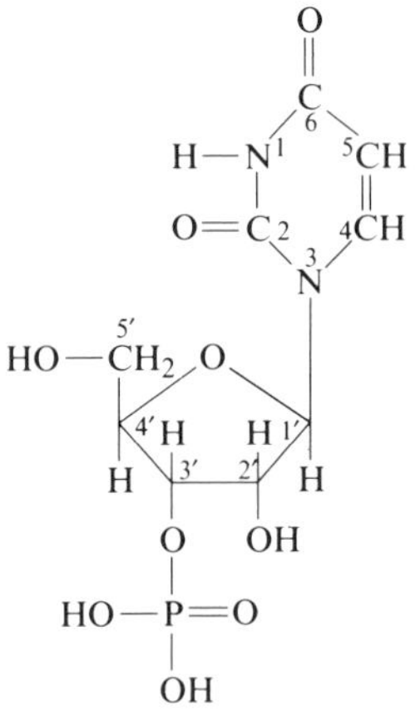
Cytidylic acid (cytidine-5′-phosphate) 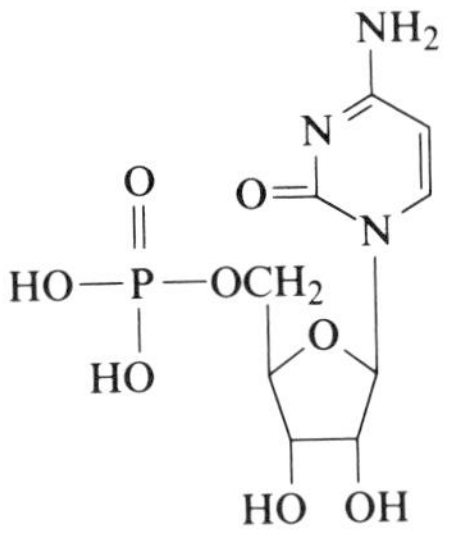
Deoxythymidylic acid (deoxythymidine-5′-phosphate) 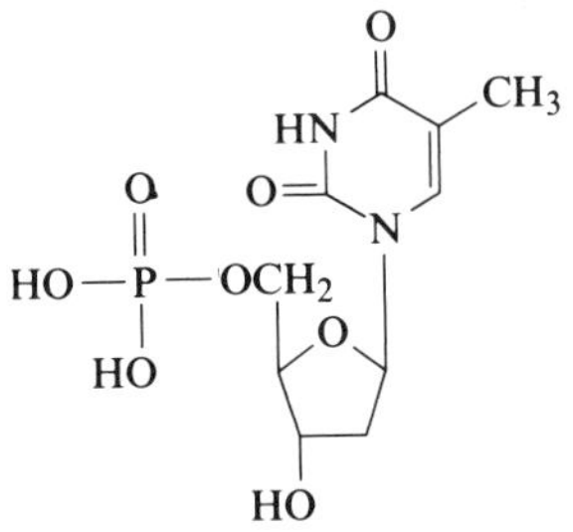
Adenylic acid (adenosine-5′-phosphate) 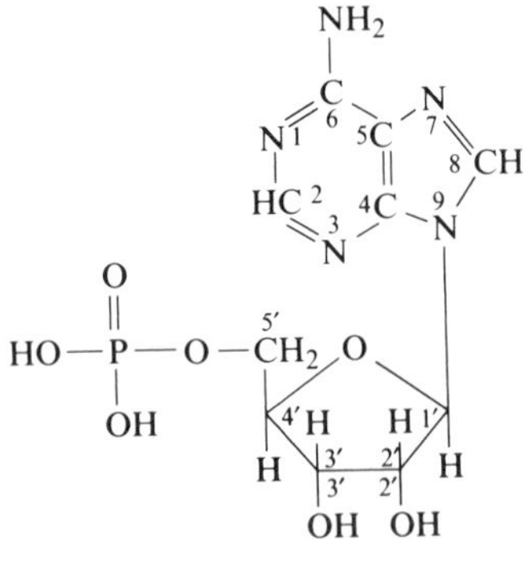

TABLE 1.2 (*Continued*)

Guanylic acid (guanosine-5′-phosphate)	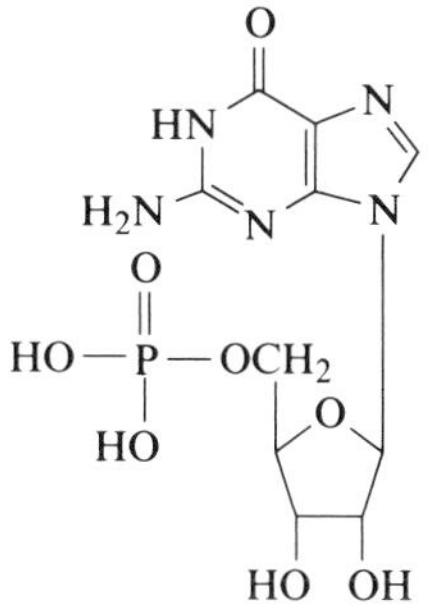

For illustrative purpose, we give two chemical reactions for the formation of nucleoside and one chemical reaction for the formation of a nucleotide:

Formation of nucleosides

Base + sugar ⟶ nucleoside

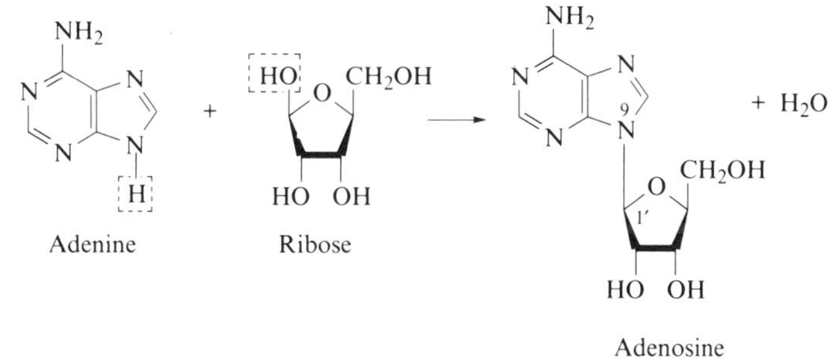

NH_2 N N N N H + HO O CH_2OH HO OH ⟶ NH_2 N N 9 N N O CH_2OH 1′ HO OH + H_2O

Adenine Ribose Adenosine

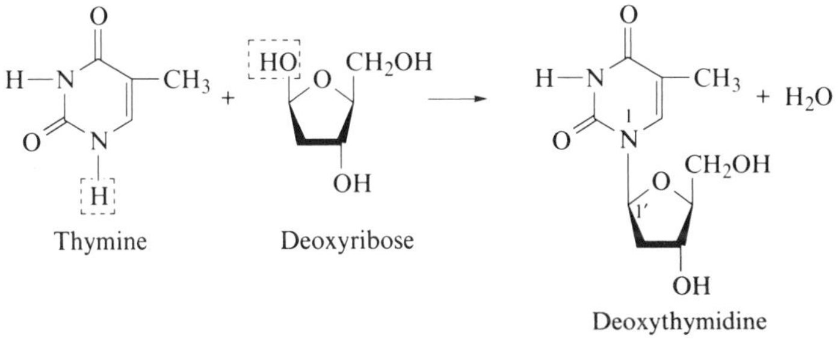

O H—N CH_3 O N H + HO O CH_2OH OH ⟶ O H—N CH_3 O 1 N O CH_2OH 1′ OH + H_2O

Thymine Deoxyribose Deoxythymidine

TABLE 1.3 Repeating Units of Nucleic Acids

FIVE BASES	
Adenine (A)	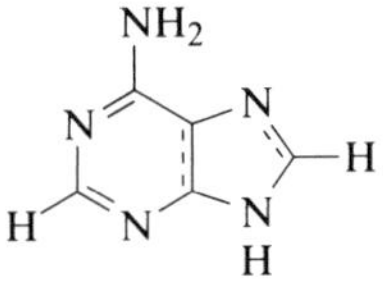
Guanine (G)	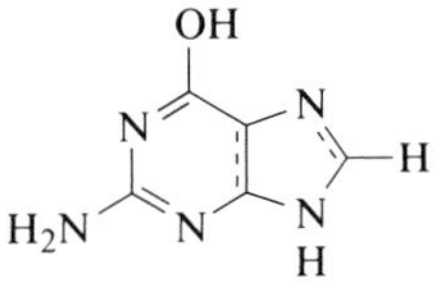
Thymine (T) (DNA only)	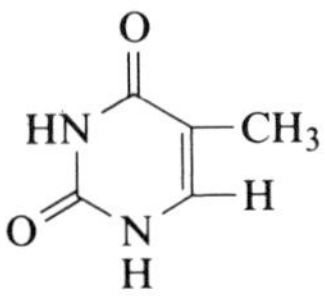
Cytosine (C)	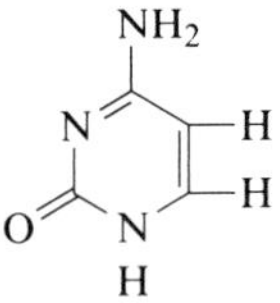
Uracil (U) (RNA only)	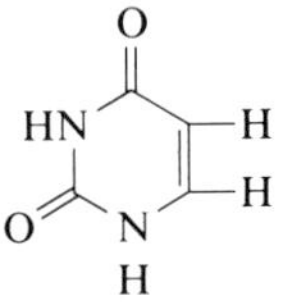
TWO SUGARS (a pentose in furanose form)	
Ribose (RNA only)	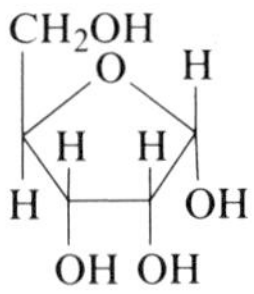
2-Deoxyribose (DNA only)	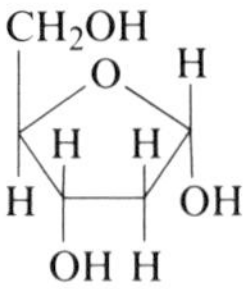
ONE PHOSPHATE	
Phosphate	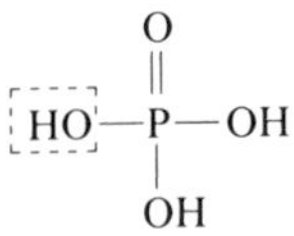

Formation of a nucleotide

$$\text{Nucleoside} + \text{phosphoric acid} \longrightarrow \text{nucleotide}$$

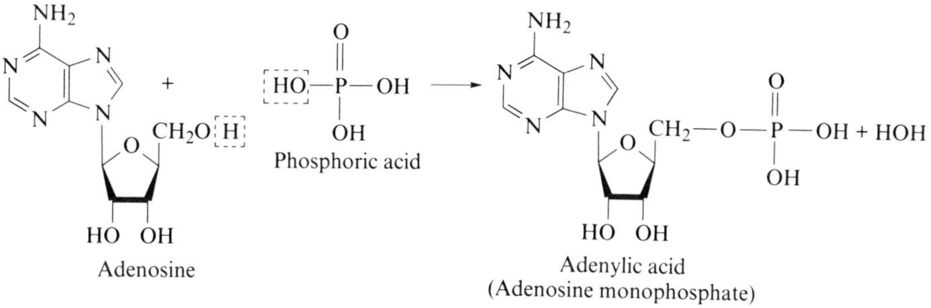

Phosphoric acid

Adenosine

Adenylic acid
(Adenosine monophosphate)

DNA Nucleotides are sequentially arranged to form a DNA molecule through 3′-5′ or 5′-3′ sugar–phosphate bonds:

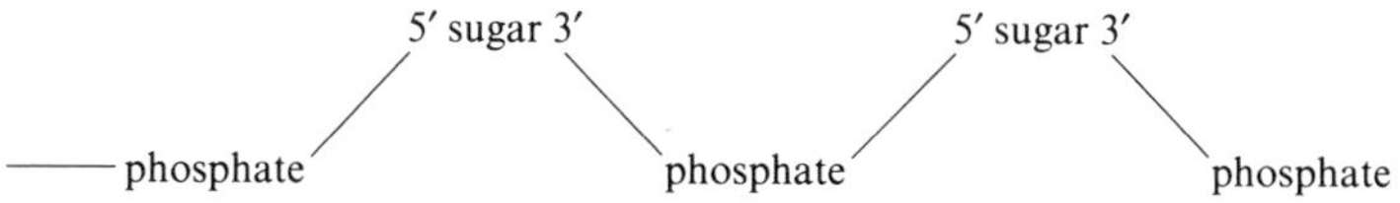

— phosphate — 5′ sugar 3′ — phosphate — 5′ sugar 3′ — phosphate

that is,

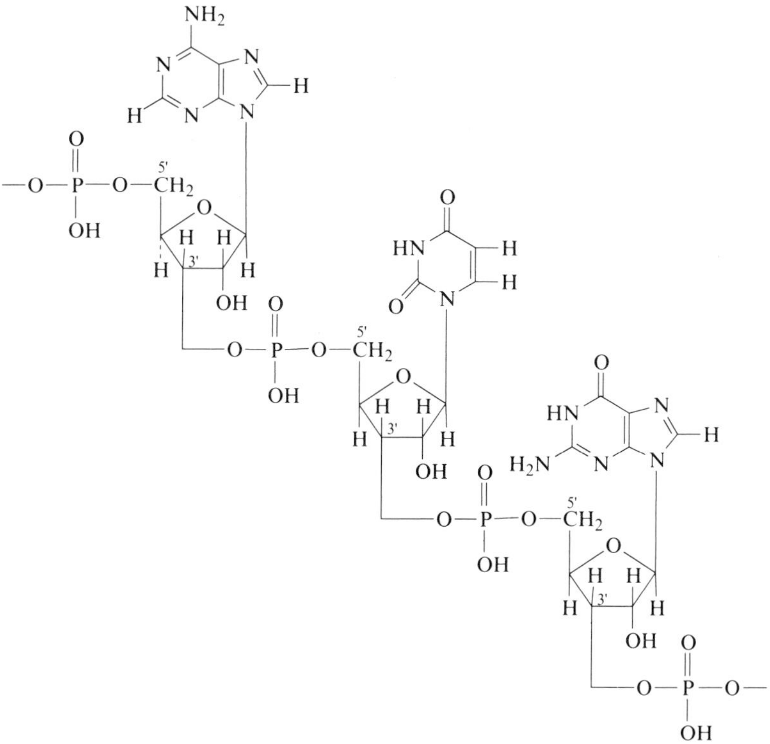

Each DNA molecule consists of two strands twisted by hydrogen bonds between the two base pairs. The base pairing occurs between T and A, and between C and G:

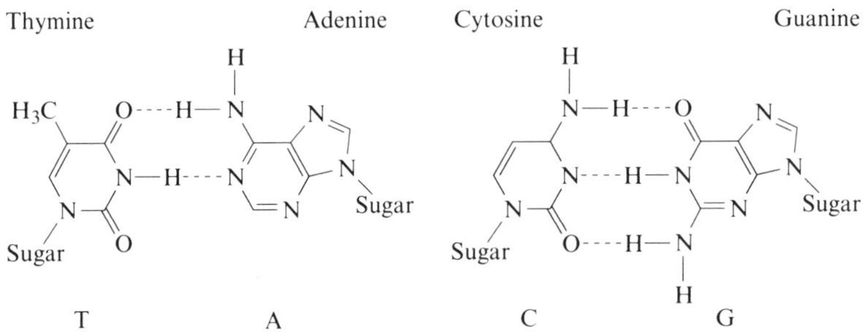

Thymine
Adenine
Cytosine
Guanine
Sugar
Sugar
Sugar
Sugar
T
A
C
G

The overall structure of DNA is believed to follow the Watson–Crick model (Figure 1.2).

A—T
G—C
T—A
C—G

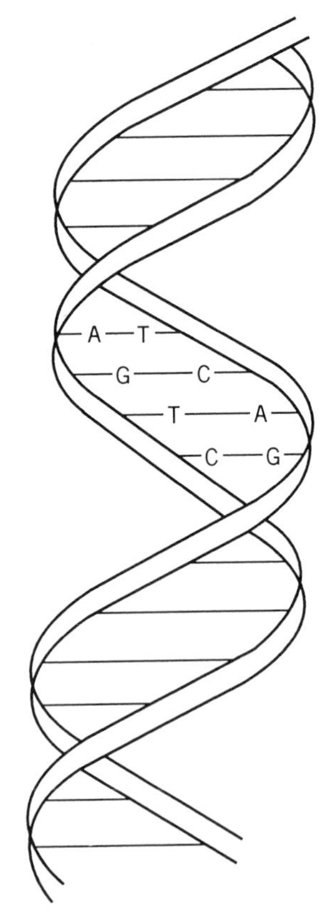

FIGURE 1.2 The Watson–Crick model of DNA.

RNA RNA is a single-stranded nucleic acid. It contains the pentose ribose, in contrast to the 2-deoxyribose of DNA. It has the base uracil instead of thymine. The purine–pyrimidine ratio in RNA is not 1:1 as in the case of DNA. There are three types of RNA, based on their biochemical function:

1. Messenger RNA—very little intramolecular hydrogen bonding and the molecule is in a fairly random coil
2. Transfer RNA—low molecular weight, carrying genetic information (that is, genetic code), highly coiled, and with base pairing in certain regions
3. Ribosomal RNA—spherical particles, site for biosyntheses

Polymers of Sugars Polymers of sugars are often called polysaccharides. They are high molecular weight (25 000–15 000 000) polymers of monosaccharides. The synthesis of polysaccharides involves the synthesis of hemiacetal and acetal. When an aldehyde reacts with an alcohol, the resulting product is hemiacetal. Upon further reaction with an alcohol, a hemiacetal is converted to an acetal. The general mechanism of acetal formation is shown in the following equation:

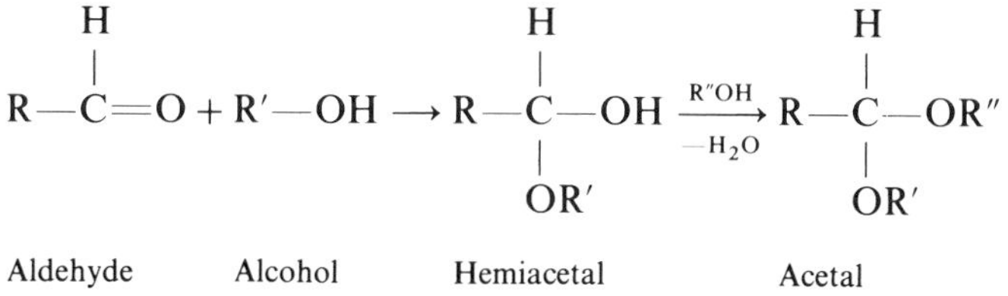

$$\underset{\text{Aldehyde}}{R{-}\overset{\displaystyle H}{\overset{|}{C}}{=}O} + \underset{\text{Alcohol}}{R'{-}OH} \longrightarrow \underset{\text{Hemiacetal}}{R{-}\underset{\underset{\displaystyle OR'}{|}}{\overset{\overset{\displaystyle H}{|}}{C}}{-}OH} \xrightarrow[-H_2O]{R''OH} \underset{\text{Acetal}}{R{-}\underset{\underset{\displaystyle OR'}{|}}{\overset{\overset{\displaystyle H}{|}}{C}}{-}OR''}$$

The sugar linkage is basically the formation of acetals:

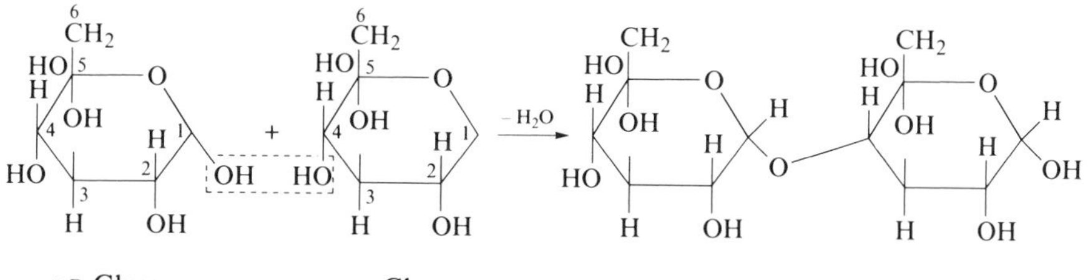

α-D-Glucose acting as hemiacetal + α-D-Glucose acting as alcohol $\xrightarrow{-H_2O}$ α-Maltose-D-glucosyl-(1 → 4)-α-D-glucose

Among the well-known polysaccharides are the three homopolymers of glucose: starch, glycogen, and cellulose. Starch is a mixture of two polymers: amylose (formed by α-1,4-glucosidic linkage) and amylopectin (a branched-chain polysaccharide formed by α-1,4-glucosidic bonds together with some α-1,6-glucosidic linkage). Glycogen is animal starch, similar to amylopectin, but more highly branched. Cellulose is a fibrous carbohydrate composed of chains of D-glucose units joined by β-1,4-glucosidic linkages. The structures of amylose, amylopectin, and cellulose are shown in the following formulas:

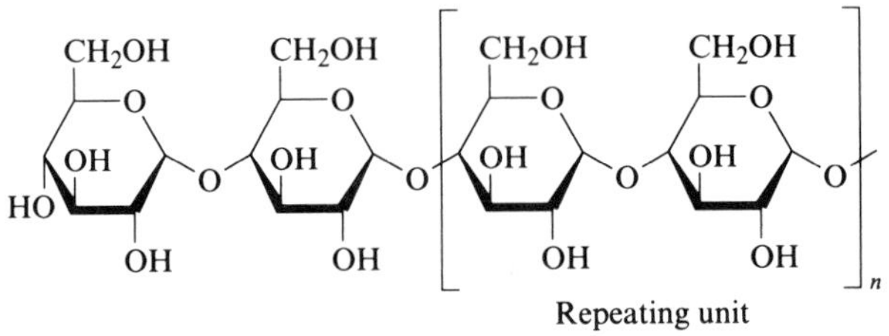
CH2OH
O
OH
HO
OH
n
Repeating unit

Amylose

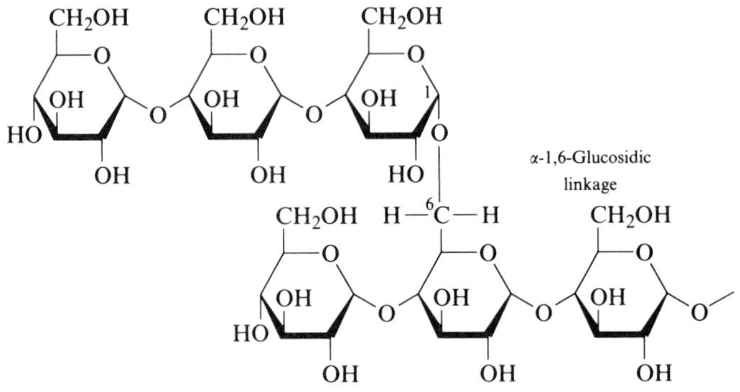
CH2OH
O
OH
HO
1
6
H—C—H
α-1,6-Glucosidic linkage

Amylopectin

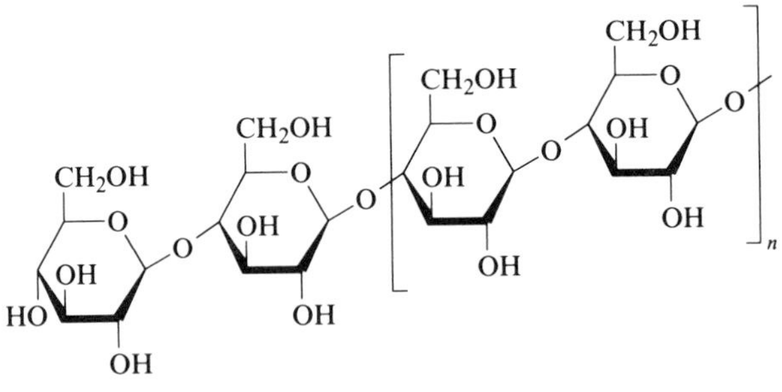
CH2OH
O
OH
HO
n

Cellulose

1.3 MACROMOLECULAR SCIENCE

Three branches of science deal with colloids and macromolecules: colloid science, surface science, and macromolecular science. Colloid science is the study of physical, mechanical, and chemical properties of colloidal systems. Surface science deals with phenomena involving macroscopic surfaces. Macromolecular science investigates the methods of syntheses in the case of synthetic polymers (or isolation and purification in the case of natural products such as proteins, nucleic acids, and carbohydrates) and the characterization of macromolecules. It includes, for example, polymer chemistry, polymer physics, biophysical chemistry, and molecular biology. These three branches of science overlap. What one learns from one branch can often be applied to the others.

The subject matter covered in this book belongs basically to macromolecular science. Emphasis is placed on the characterization of macromolecules (synthetic and biological polymers). Hence, the material also belongs to the realm of physical chemistry.

REFERENCES

Adamson, A. W., *Physical Chemistry of Surfaces*, 2d ed. New York: Wiley – Interscience, 1967.

Billmeyer, F. W., Jr., *Textbook of Polymer Chemistry*, 2d ed. New York: Wiley, 1985.

Carothers, W. H., *J. Am. Chem. Soc.* **51**, 2548 (1929).

Helenius, A., and K. Simons, *Biochim. Biophys. Acta* **415**, 29 (1975).

Shaw, D. J., *Introduction to Colloids and Surface Chemistry*, 2d ed. Stoneham, MA: Butterworth, 1978.

Tanford, C., *The Hydrophobic Effect*. New York: Wiley, 1981.

2

SYNTHESES OF MACROMOLECULAR COMPOUNDS

The first three sections of this chapter deal with addition polymerization, the fourth with condensation polymerization, the fifth with kinetics of the syntheses of polymers, the sixth with polypeptide synthesis, and the seventh with nucleic acid synthesis. Readers should be familiar with these subjects before going on to the major topics of this book. The chapter itself could be considered as a book in miniature on synthetic chemistry. Important synthetic methods and well-known chemical compounds are covered.

2.1 RADICAL POLYMERIZATION

The general reaction scheme for free radical polymerization can be expressed as follows:

$$\text{Initiator} \longrightarrow \text{R}\cdot \qquad \text{Initiation}$$

$$\left.\begin{array}{l}\text{R}\cdot + \text{M} \longrightarrow \text{MR}\cdot \\ \text{MR}\cdot + \text{M} \longrightarrow \text{M}_2\text{R}\cdot\end{array}\right\} \qquad \text{Chain propagation}$$

$$\cdots$$

$$\text{M}_n\text{R}\cdot + \text{M}_m\text{R}\cdot \longrightarrow \text{M}_{n+m} \qquad \text{Chain termination}$$

where M represents a monomer molecule and R· a free radical produced in the initial step. An example of free radical polymerization is the synthesis of polyethylene:

Initiation

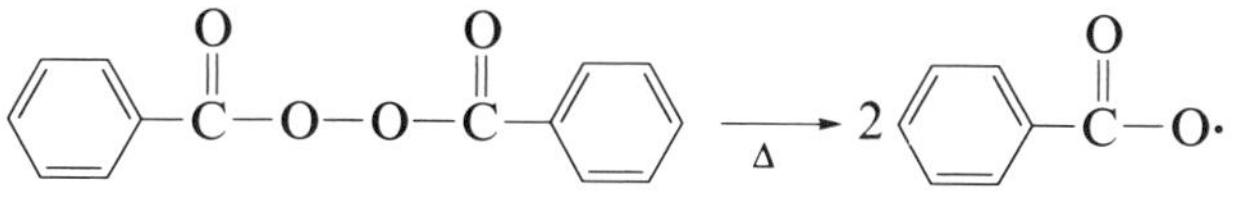

$$C_6H_5-\overset{O}{\overset{\|}{C}}-O-O-\overset{O}{\overset{\|}{C}}-C_6H_5 \xrightarrow[\Delta]{} 2\,C_6H_5-\overset{O}{\overset{\|}{C}}-O\cdot$$

Benzoyl peroxide

$$C_6H_5-\overset{O}{\overset{\|}{C}}-O\cdot + CH_2{=}CH_2 \longrightarrow C_6H_5-\overset{O}{\overset{\|}{C}}-O-CH_2-CH_2^{\cdot}$$

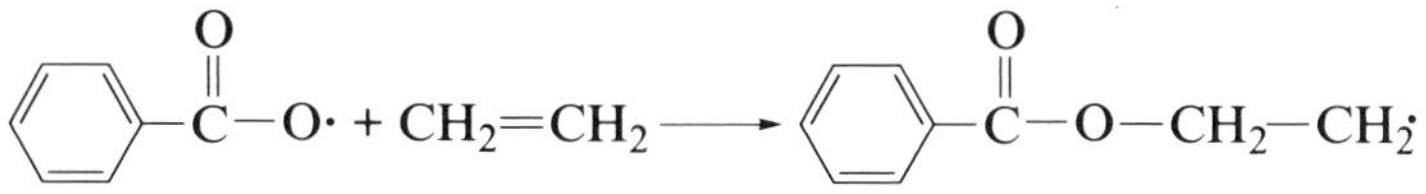

Ethylene (monomer)

Propagation

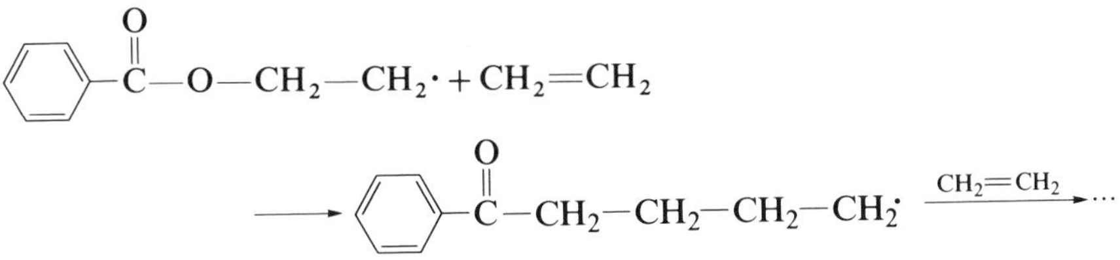

$$C_6H_5-\overset{O}{\overset{\|}{C}}-O-CH_2-CH_2\cdot + CH_2{=}CH_2$$

$$\longrightarrow C_6H_5-\overset{O}{\overset{\|}{C}}-CH_2-CH_2-CH_2-CH_2^{\cdot} \xrightarrow{CH_2{=}CH_2} \cdots$$

Termination

$$2R \sim CH_2-CH_2\cdot \longrightarrow R \sim CH_2-CH_2-CH_2-CH_2 \sim R-$$

Most of the initiators are peroxides and aliphatic azo compounds, such as the following:

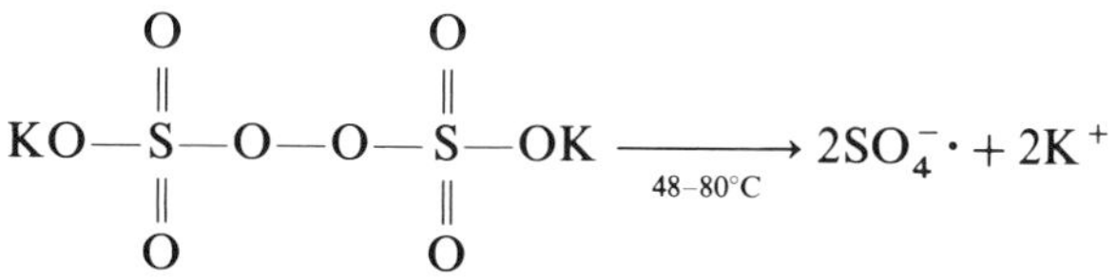

$$KO-\underset{\underset{O}{\|}}{\overset{O}{\overset{\|}{S}}}-O-O-\underset{\underset{O}{\|}}{\overset{O}{\overset{\|}{S}}}-OK \xrightarrow[48\text{–}80^\circ C]{} 2SO_4^{-}\cdot + 2K^+$$

$$(2SO_4^{-}\cdot + 2H_2O \longrightarrow 2HSO_4^{-} + 2HO\cdot)$$

Potassium persulfate

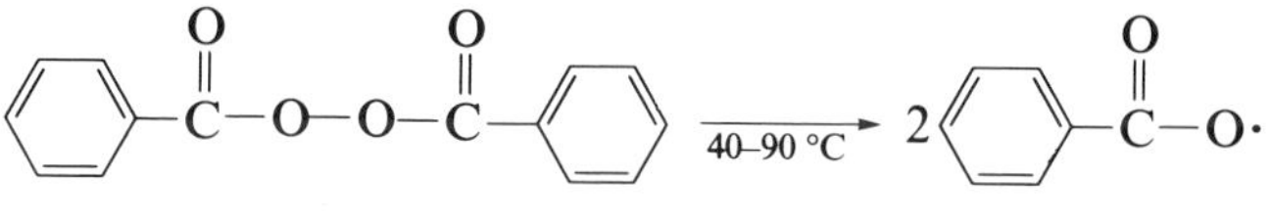

$$C_6H_5-\overset{O}{\overset{\|}{C}}-O-O-\overset{O}{\overset{\|}{C}}-C_6H_5 \xrightarrow[40\text{–}90\ ^\circ C]{} 2\,C_6H_5-\overset{O}{\overset{\|}{C}}-O\cdot$$

Benzoyl peroxide (lucidol)

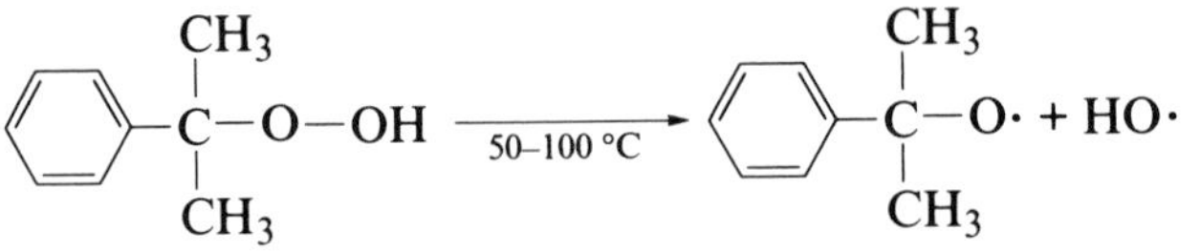

$$C_6H_5-C(CH_3)_2-O-OH \xrightarrow[50\text{–}100\ ^\circ C]{} C_6H_5-C(CH_3)_2-O\cdot + HO\cdot$$

Cumene hydroxyperoxide

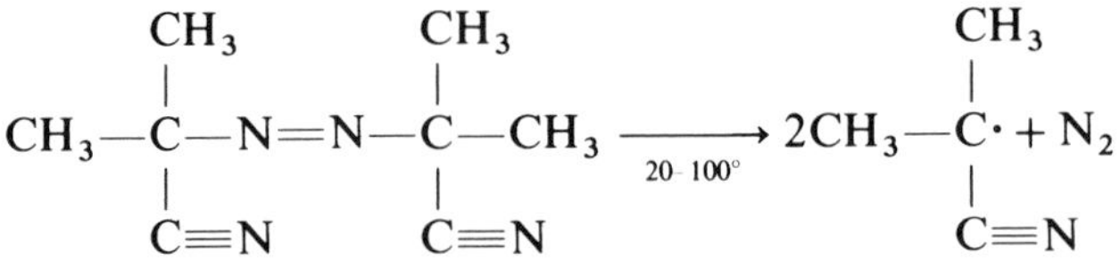

$$CH_3-C(CH_3)(C{\equiv}N)-N{=}N-C(CH_3)(C{\equiv}N)-CH_3 \xrightarrow[20\ 100^\circ]{} 2CH_3-C\cdot(CH_3)(C{\equiv}N) + N_2$$

Azobisisobutyronitrile (AIBN)

2.1.1 Complications

Free radical polymerization often involves complications. Complications may occur during propagation, chain transfer, and chain termination.

Complications in Propagation When there is more than one unsaturated bond in the monomers, propagation can occur in a different mechanism, thereby affecting the chain structure. For example, in the synthesis of polybutadiene, polymerization can lead to three different products:

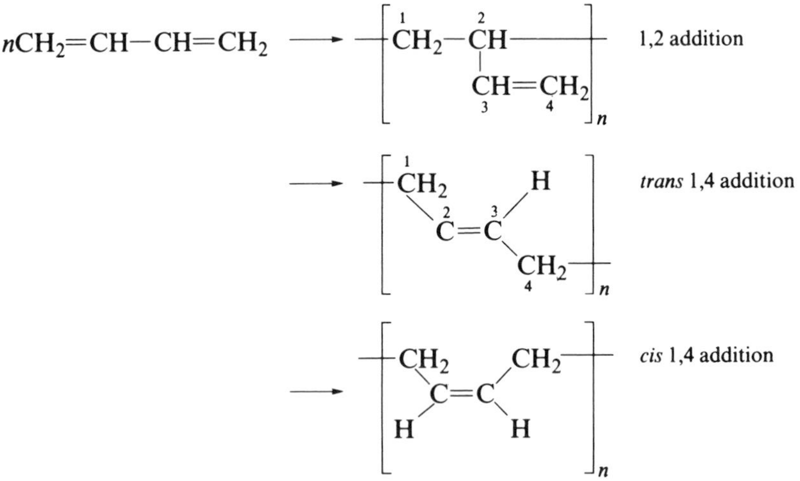

$$nCH_2{=}CH-CH{=}CH_2 \longrightarrow \left[\overset{1}{C}H_2-\overset{2}{C}H(\overset{3}{C}H{=}\overset{4}{C}H_2)\right]_n \quad \text{1,2 addition}$$

$$\longrightarrow \left[\overset{1}{C}H_2-\overset{2}{C}H{=}\overset{3}{C}H-\overset{4}{C}H_2\right]_n \quad \textit{trans}\ \text{1,4 addition}$$

$$\longrightarrow \left[CH_2-CH{=}CH-CH_2\right]_n \quad \textit{cis}\ \text{1,4 addition}$$

The three polymers have different properties. 1,2-Polybutadiene is a hard and rough crystalline compound; 1,4-Polybutadiene is not. The crystalline and glass transition temperatures for *cis*- and *trans*-1,4-polybutadiene are markedly different: T_g is -108°C for *cis* and -18°C for *trans*; T_m 1°C for *cis* and 141°C for *trans*. T_g is the glass transition temperature below which an amorphous polymer can be considered to be a hard glass and above which the material is soft or rubbery. T_m is the crystalline melting point where the crystallinity completely disappears.

The mechanism that the reaction follows depends, among other factors, on the solvent and the temperature. Phenyllithium in tetrahydrofuran favors 1,2 polymers, whereas lithium dispersion or phenyllithium in paraffinic hydrocarbons such as heptane as a solvent favors 1,4 polymers. A higher temperature favors 1,2 polymers; at low temperature the products are predominantly 1,4 repeating units.

Complications in Chain Transfer The reactivity of a radical can be transferred to the monomer, polymer, or solvent, or even to the initiator, as the following examples show:

Transfer to initiator

$$M_n\cdot + C_6H_5-\overset{O}{\overset{\|}{C}}-O-O-\overset{O}{\overset{\|}{C}}-C_6H_5 \longrightarrow M_n-O-\overset{O}{\overset{\|}{C}}-C_6H_5 + C_6H_5-\overset{O}{\overset{\|}{C}}-O\cdot$$

Transfer to monomer

$$M_n\cdot + CH_2=\overset{H}{\overset{|}{C}}-X \longrightarrow M_nH + CH_2=\underset{X}{\underset{|}{C}}\cdot$$

Transfer to solvent

$$M_n\cdot + CCl_4 \longrightarrow M_nCl + Cl_3C\cdot$$

Transfer to polymer

$$M_n\cdot + M_m \longrightarrow M_n + M_m\cdot$$

Chain transfer is the termination of a polymer chain without the destruction of the kinetic chain. Chain transfer does not affect the overall rate of polymerization, but does affect the molecular weight distribution of polymer products. It is related to the efficiency of synthesizing the polymer within a designated range.

Complication in Chain Termination Termination may occur by recombination or by disproportionation:

Recombination

$$\cdot CH_2-CH_2 \sim + \cdot CH_2CH_2 \sim \longrightarrow \sim CH_2CH_2CH_2CH_2 \sim$$

Disproportionation

$$\cdot CH_2{-}CH_2 \sim + \cdot CH_2CH_2 \sim \longrightarrow \sim CH{=}CH_2 + \sim CH_2CH_3$$

Recombination strengthens the chain length, whereas disproportionation gives short chains. Termination could also be carried out with an inhibitor, such as one of the following:

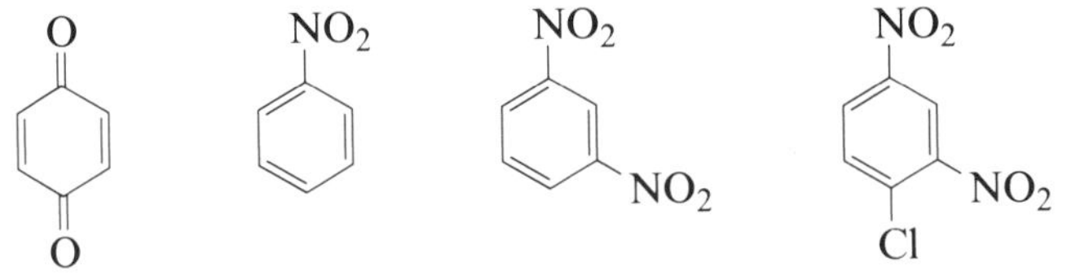

Quinone Nitrobenzene Dinitrobenzene Dinitrochlorobenzene

phenyl-β-naphthalamine, O_2, NO, nitroso compounds, sulfur compounds, amines, and phenols.

2.1.2 Methods of Free Radical Polymerization

There are various ways to carry out free radical polymerization. Here we mention a few of them:

1. Bulk polymerization—synthesis without solvent
2. Solution polymerization—synthesis with (inert) solvent
3. Precipitation polymerization—using solvent (such as methanol) to precipitate out the polymer
4. Suspension polymerization—adding an initiator to the suspension in aqueous solution
5. Emulsion polymerization—adding an initiator (such as potassium persulfate) to the emulsion of water insoluble monomers (such as styrene) in aqueous soap solution

2.1.3 Some Well-known Overall Reactions of Addition Polymers

The following are the overall reactions for the synthesis of typical (also well-known in our daily life) polymers. All of them undergo the mechanism of addition polymerization.

$$nCH_2{=}CH_2 \xrightarrow{O_2,\ \text{heat, pressure}} \sim CH_2CH_2{-}CH_2H_2{-}CH_2CH_2{-}\sim$$

Ethylene

$$-\!\!(CH_2{-}CH_2)\!\!-_n$$

Polyethylene

(Plastic materials—films, housewares)

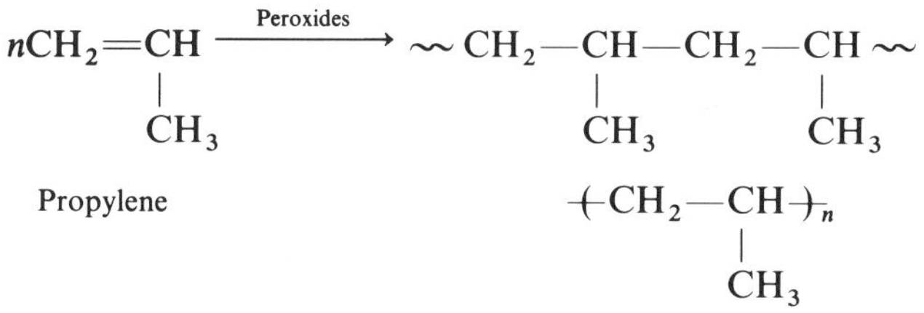

$$nCH_2{=}CH(CH_3) \xrightarrow{\text{Peroxides}} \sim CH_2{-}CH(CH_3){-}CH_2{-}CH(CH_3) \sim$$

Propylene

$$\text{+}CH_2{-}CH(CH_3)\text{+}_n$$

Polypropylene
(Ropes, appliance parts)

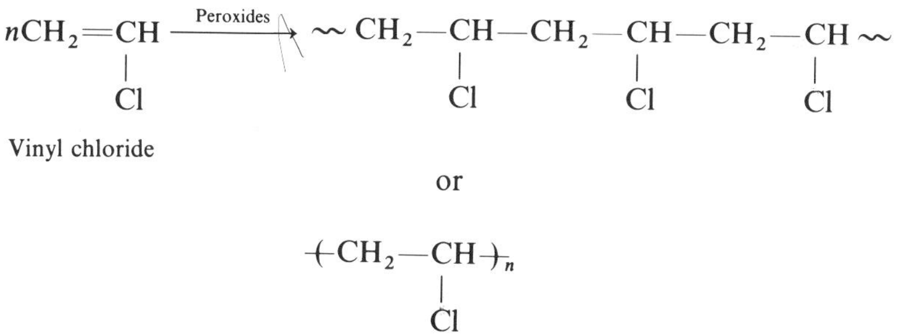

$$nCH_2{=}CH(Cl) \xrightarrow{\text{Peroxides}} \sim CH_2{-}CH(Cl){-}CH_2{-}CH(Cl){-}CH_2{-}CH(Cl) \sim$$

Vinyl chloride

or

$$\text{+}CH_2{-}CH(Cl)\text{+}_n$$

Poly(vinyl chloride)
[Floor coverings, phonograph records, plastic pipes (when plasticized with high-boiling esters), raincoats, upholstery fabrics]

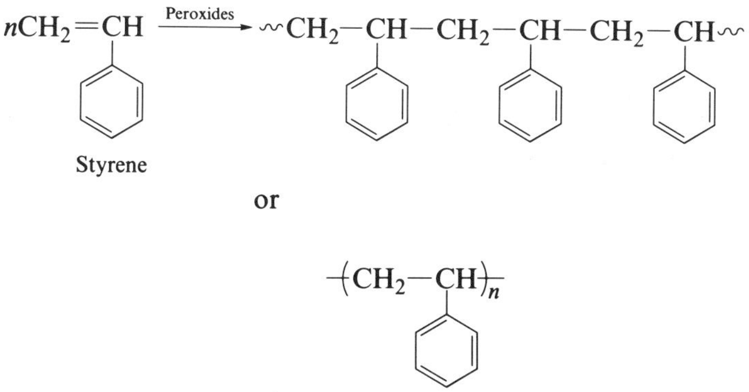

$$nCH_2{=}CH(C_6H_5) \xrightarrow{\text{Peroxides}} \sim CH_2{-}CH(C_6H_5){-}CH_2{-}CH(C_6H_5){-}CH_2{-}CH(C_6H_5) \sim$$

Styrene

or

$$\text{+}CH_2{-}CH(C_6H_5)\text{+}_n$$

Polystyrene
(Coffee cups, packages, insulation)

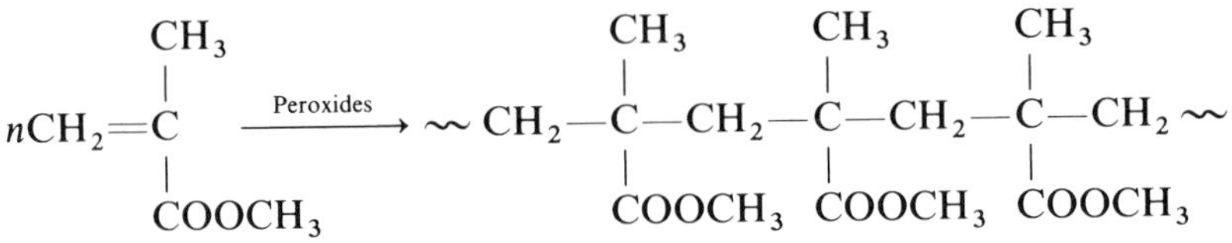

$$nCH_2{=}C(CH_3)(COOCH_3) \xrightarrow{\text{Peroxides}} \sim CH_2{-}C(CH_3)(COOCH_3){-}CH_2{-}C(CH_3)(COOCH_3){-}CH_2{-}C(CH_3)(COOCH_3){-}CH_2 \sim$$

Methyl methacrylate

or

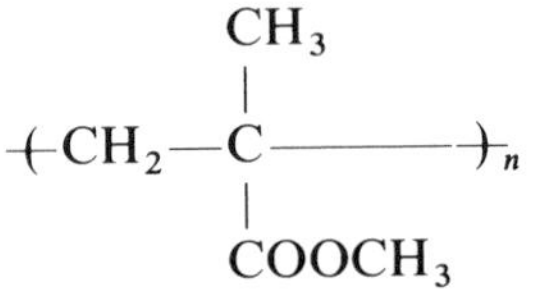

$$\left(CH_2{-}C(CH_3)(COOCH_3)\right)_n$$

Poly(methyl methacrylate)
(Plexiglass (Lucite))

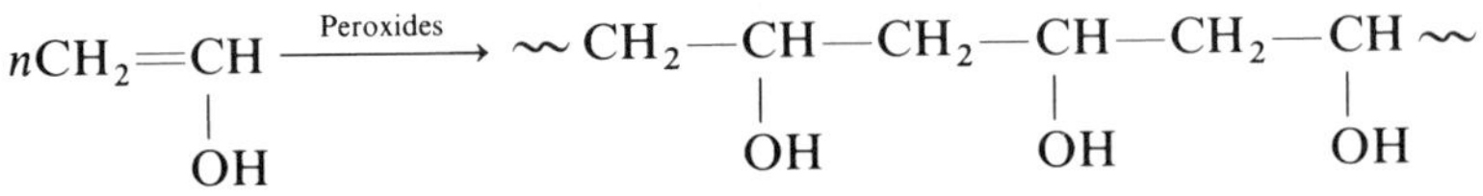

$$nCH_2{=}CH(OH) \xrightarrow{\text{Peroxides}} \sim CH_2{-}CH(OH){-}CH_2{-}CH(OH){-}CH_2{-}CH(OH) \sim$$

Vinyl alcohol

or

$$\left(CH_2{-}CH(OH)\right)_n$$

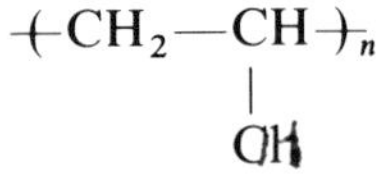

Poly(vinyl alcohol)
(Water-soluble thickening agent)

2.2 IONIC POLYMERIZATION

The two types of ionic polymerizations are anionic and cationic. The former involves carbonions $C^{\ominus}$ and the latter involves carbonium $C^{\oplus}$ ions. Catalysts and cocatalysts are needed in ionic polymerization.

2.2.1 Anionic Polymerization

The catalysts for anionic polymerization are alkali metals, alkali metal amides, alkoxides, and cyanides. The cocatalysts are organic solvents, such as heptane. An example of anionic polymerization is the synthesis of polystyrene:

Initiation

$$NaNH_2 \xrightarrow{NH_3} Na^+NH_2^-$$

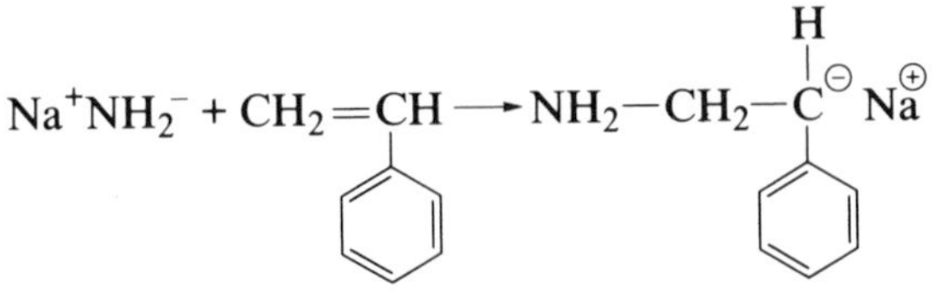

$$Na^+NH_2^- + CH_2{=}CH(C_6H_5) \longrightarrow NH_2{-}CH_2{-}CH(C_6H_5)^{\ominus}\ Na^{\oplus}$$

Propagation

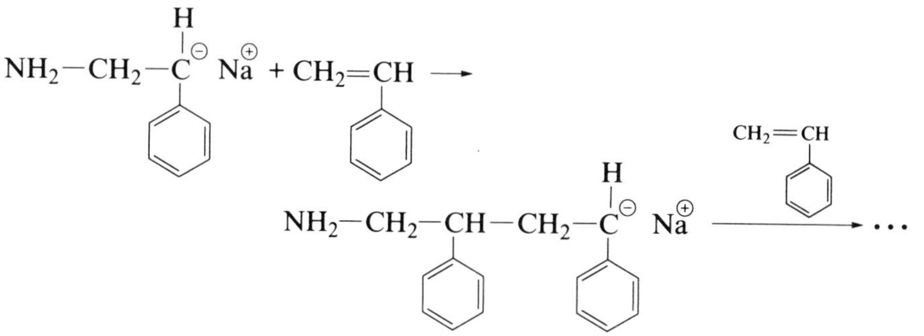

$$NH_2-CH_2-\underset{\displaystyle C_6H_5}{\overset{\displaystyle H}{C^{\ominus}}}\ Na^{\oplus} + \underset{\displaystyle C_6H_5}{CH_2{=}CH} \longrightarrow$$

$$NH_2-CH_2-\underset{\displaystyle C_6H_5}{CH}-CH_2-\underset{\displaystyle C_6H_5}{\overset{\displaystyle H}{C^{\ominus}}}\ Na^{\oplus} \xrightarrow{CH_2=CH(C_6H_5)} \cdots$$

Termination

$$\sim\!CH_2-\underset{\displaystyle C_6H_5}{\overset{\displaystyle H}{C^{\ominus}}}\ Na^{\oplus} + NH_3 \longrightarrow \sim\!CH_2-\underset{\displaystyle C_6H_5}{CH_2} + NH_2^-\,Na^+$$

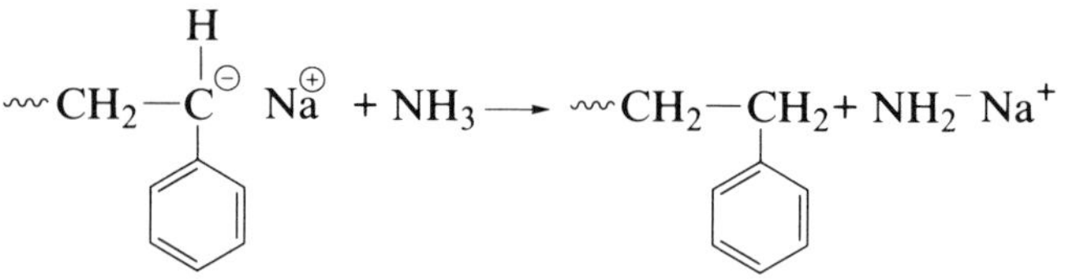

The chain growth in anionic polymerization does not necessarily have to go in one direction as shown in the above example. It can go through two, three, four, or more directions, depending on the catalysts:

Two-way growth—bivalent ions are used as initiator

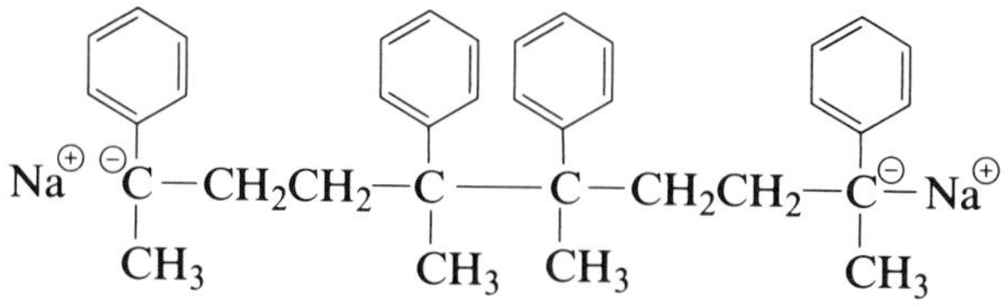

$$Na^{\oplus}\ {}^{\ominus}\underset{\displaystyle CH_3}{\overset{\displaystyle C_6H_5}{C}}-CH_2CH_2-\underset{\displaystyle CH_3}{\overset{\displaystyle C_6H_5}{C}}-\underset{\displaystyle CH_3}{\overset{\displaystyle C_6H_5}{C}}-CH_2CH_2-\underset{\displaystyle CH_3}{\overset{\displaystyle C_6H_5}{C}}{}^{\ominus}-Na^{\oplus}$$

(sodium salt of α-methyl styrene tetramer)

Three-way growth—polyfunctional initiator

$$Na^{\oplus}\ {}^{\ominus}O-CH_2-CH_2-N\begin{cases}CH_2-CH_2O^{\ominus}Na^{\oplus}\\ CH_2-CH_2O^{\ominus}Na^{\oplus}\end{cases}$$

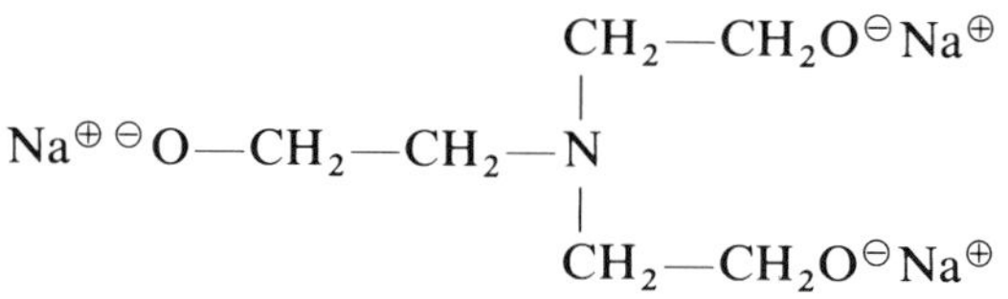

Four-way growth—polyfunctional initiator

$$C(C_6H_4Li)_4$$

(Li at the para position of each of the four phenyl rings)

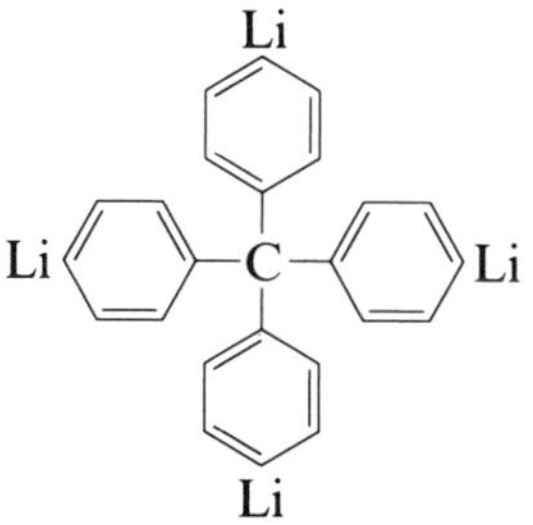

2.2.2 Cationic Polymerization

The catalysts for cationic polymerization are Lewis acids and Friedal-Crafts catalysts such as BF_3, $AlCl_3$, and $SnCl_4$, and strong acids, such as H_2SO_4. The co-catalysts are, for example, water and isobutene. An example of cationic polymerization is the synthesis of isobutene:

Initiation

$$H_2O + BF_3 \longrightarrow H^+(BF_3OH)^-$$

$$H^+(BF_3-OH)^- + CH_2{=}C(CH_3)_2 \longrightarrow CH_3{-}\overset{\displaystyle CH_3}{\underset{\displaystyle CH_3}{C^{\oplus}}}(BF_3OH)^{\ominus}$$

Propagation

$$CH_3{-}\overset{\displaystyle CH_3}{\underset{\displaystyle CH_3}{C^{\oplus}}}(BF_3OH)^{\ominus} + CH_2{=}\overset{\displaystyle CH_3}{\underset{\displaystyle CH_3}{C}} \longrightarrow CH_3{-}\overset{\displaystyle CH_3}{\underset{\displaystyle CH_3}{C}}{-}CH_2{-}\overset{\displaystyle CH_3}{\underset{\displaystyle CH_3}{C^{\oplus}}}(BF_3OH)^{\ominus}$$

$$\xrightarrow{CH_2=C(CH_3)_2} \cdots$$

Termination

$$\sim CH_2{-}\overset{\displaystyle CH_3}{\underset{\displaystyle CH_3}{C^{\oplus}}}(BF_3OH)^{\ominus} \longrightarrow \sim CH_2{-}\overset{\displaystyle CH_3}{C}{=}CH_2 + H^+(BF_3OH)^-$$

2.2.3 Living Polymers

Anionic polymerization is terminated not by the reaction of two growing species, but by chain transfer to the solvent or to impurities present in the system. If an inactive solvent (such as tetrahydrofuran or dioxane) is chosen and the impurities are removed, the polymerization will not stop until all the monomers are consumed. Such a polymer, if kept in an appropriate condition, would always have a reactive end. If additional monomers are introduced to the system, the polymer keeps growing. The polymer is therefore called a living polymer. To meet these requirements, synthesis is usually performed under high vacuum and the product is also kept in a container sealed under high vacuum.

An example is the polymerization of styrene with sodium naphthalene at a pressure of 1×10^{-6} mm Hg:

Initiation via electron transfer

$CH_2{=}CH$

\+ Na $\xrightarrow{THF}$ $\cdot^{-}$ Na^+ $\longrightarrow$

Naphthalene

Electron transfer (not a radical)

$CH_2{=}CH$

Radical anion

Na^+ +

$$2\left[C_6H_5\!-\!\dot{C}H\!-\!\ddot{C}H_2 \longleftrightarrow C_6H_5\!-\!\ddot{C}H\!-\!\dot{C}H_2\right]Na^+$$

Dianion (red color)

$$\longrightarrow Na^+\left[{}^{-}\!:\!\underset{C_6H_5}{\overset{H}{C}}\!-\!CH_2\!-\!CH_2\!-\!\underset{C_6H_5}{\overset{H}{C}}\!:^{-}\right]Na^+$$

Propagation at both ends

$$Na^+\left[{}^{-}\!:\!\underset{C_6H_5}{\overset{H}{C}}\!-\!CH_2CH_2\!-\!\underset{C_6H_5}{\overset{H}{C}}\!:^{-}\right]Na^+ + (n+m)\,CH_2{=}\underset{C_6H_5}{C} \longrightarrow$$

$$Na^+\left[{}^{-}\!:\!\underset{C_6H_5}{\overset{H}{C}}\!-\!CH_2\!\left(\underset{C_6H_5}{CH}\!-\!CH_2\right)_n\!\left(CH_2\!-\!\underset{C_6H_5}{CH}\right)_m\!CH_2\!-\!\underset{C_6H_5}{\overset{H}{C}}\!:^{-}\right]Na^+$$

Since there is no chain transfer involved, preparation through living polymer results in a narrow distribution of molecular weight. It is the best way to synthesize di- and triblock copolymers, star-shaped and comb-shaped polymers, and homopolymers with high molecular weight. The following is an example of the synthesis of di- and triblock copolymers:

$$C_6H_5-Li + CH_2=C(CH_3)(C_6H_5) \longrightarrow C_6H_5-CH_2-C^{\ominus}(CH_3)(C_6H_5) + Li^{\oplus}$$

Phenyl lithium

α-Methyl styrene

$$\xrightarrow{n\text{-}\alpha\text{-Methyl styrene}} C_6H_5\left[CH_2-C(CH_3)(C_6H_5)\right]_n CH_2-C^{\ominus}(CH_3)(C_6H_5) + Li^{\oplus}$$

$$\xrightarrow{m\text{-Styrene}} C_6H_5\left[CH_2-C(CH_3)(C_6H_5)\right]_{n+1}\left[CH_2-C(H)(C_6H_5)\right]_{m-1} CH_2-C^{\ominus}(H)(C_6H_5) + Li^{\oplus}$$

$$\longrightarrow C_6H_5\left[CH_2-C(CH_3)(C_6H_5)\right]_{n+1}\left[CH_2-C(H)(C_6H_5)\right]_m H$$

A diblock copolymer

$$\xrightarrow{S\text{-Isoprene}}$$

$$C_6H_5\left[CH_2-C(CH_3)(C_6H_5)\right]_{n+1}\left[CH_2-C(H)(C_6H_5)\right]_m\left[CH_2-C(CH_3)=CH-CH\right]_s H$$

A triblock copolymer

2.3 COORDINATION POLYMERIZATION

Coordination polymerization is also called the stereospecific or stereoregular polymerization. The essential feature is a directing force to a growing chain end. Coordination polymerization is carried out using a catalyst called the Ziegler-Natta catalyst. Typical Ziegler-Natta catalysts are transition metal halides, such as TiX_4, TiX_3, VX_4, VX_3, VOX_3, Co, and Ni complexes, and organometallic compounds, such as AlR_3, AlR_2X, ZnR_2, and LiR.

It is generally believed that the following oxidation–reduction reaction is responsible for chain growth:

$$TiCl_3 + (CH_3CH_2)_3Al \longrightarrow TiCl_2(CH_2CH_3)\cdot(CH_3CH_2)_2ClAl$$

Titaniumtrichloride (a transition metal salt) reacts with triethylaluminum (a metal alkyl) to form an active catalyst, a titanium complex holding an ethyl group. If an alkene, such as ethylene, is introduced, the alkene attaches itself to titanium by a π bond. With the alkene and ethyl both held by the metal, the alkene unit inserts itself between the metal and the ethyl group. There is now an n-butyl group (in the case of ethylene) attached to the titanium. The bonding site where ethylene was held is vacant. The catalyst is ready to work again and the process continues until the insertion of hydrogen. The long chain then separates from the metal and a molecule of polyethylene has been formed. The mechanism may be better described in equations:

Initiation

$$TiCl_3 + (CH_3CH_2)_3Al \longrightarrow >Ti\!\begin{matrix} CH_3 \\ | \\ CH_2 \\ \cdot\cdot\quad\cdot\cdot \\ \quad\quad \\ \cdot\cdot\quad\cdot\cdot \\ CH_2 \\ | \\ CH_3 \end{matrix}\!Al<$$

An active complex

$$\xrightarrow{CH_2=CH_2} >Ti\!\begin{matrix} CH_2=CH_2 \downarrow \quad CH_3 \\ | \\ CH_2 \\ \cdot\cdot\quad\cdot\cdot \\ \quad\quad \\ \cdot\cdot\quad\cdot\cdot \\ CH_2 \\ | \\ CH_3 \end{matrix}\!Al< \longrightarrow >Ti\!\begin{matrix} (CH_2CH_2)CH_2CH_3 \\ \cdot\cdot\quad\cdot\cdot \\ \quad\quad \\ \cdot\cdot\quad\cdot\cdot \\ CH_2 \\ | \\ CH_3 \end{matrix}\!Al<$$

Propagation

$$>Ti\!\begin{matrix} (CH_2CH_2)\,CH_2CH_3 \\ \cdot\cdot\quad\cdot\cdot \\ \quad\quad \\ \cdot\cdot\quad\cdot\cdot \\ CH_2 \\ | \\ CH_3 \end{matrix}\!Al< \xrightarrow{nCH_2=CH_2} >Ti\!\begin{matrix} CH_2CH_2\!\left(CH_2CH_2\right)_{n}\!CH_2CH_3 \\ \cdot\cdot\quad\cdot\cdot \\ \quad\quad \\ \cdot\cdot\quad\cdot\cdot \\ CH_2 \\ | \\ CH_3 \end{matrix}\!Al<$$

Termination

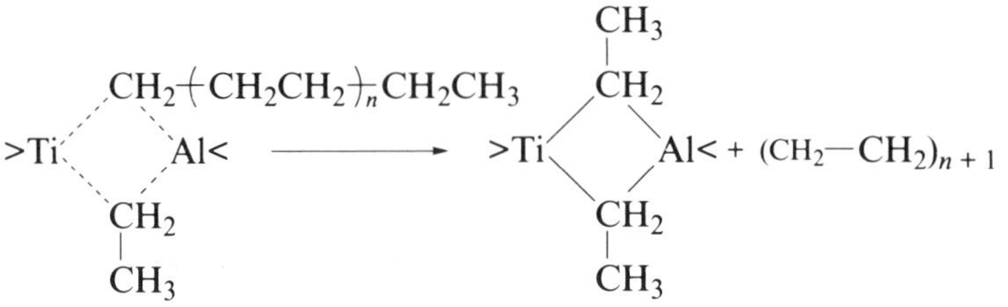

$$>Ti\overset{CH_2\!-\!(CH_2CH_2)_n\!-\!CH_2CH_3}{\underset{\underset{CH_3}{|}}{\underset{CH_2}{\cdots}}}Al< \longrightarrow >Ti\overset{\overset{CH_3}{|}}{\underset{\underset{CH_3}{|}}{\underset{CH_2}{\overset{CH_2}{\diamond}}}}Al< + (CH_2-CH_2)_{n+1}$$

$$(CH_2-CH_2)_{n+1} \longrightarrow CH_2=(CH_2-CH_2)_{n-1}=CH_2$$

There are two important features of coordination polymerization that are different from free radical polymerization: (1) The product is a linear polymer molecule and (2) there is a stereochemical control. For example, branching is almost unavoidable when synthesizing polyethylene with free radical polymerization, because free radicals generate several centers from which branches can grow. As a result, polyethylene synthesized by free radical polymerization has low crystallinity and a low melting point and is mechanically weak. Polyethylene synthesized by the coordination polymerization method, on the other hand, is unbranched and the product has a high degree of crystallinity, a high melting point, and mechanical strength.

Free radical polymerization arranges functional groups, such as alkyls, in a random manner, whereas coordination polymerization can exercise stereochemical control over functional groups. With the proper choice of experimental conditions, such as temperature, solvent, and catalyst, monomers can polymerize to any of three arrangements: isotactic, syndiotactic, and atactic. (For polyethylene, there are no such stereoisomers, since the monomeric units are identical, $-CH_2-$.) Isotactic and syndiotactic polypropylenes are highly crystalline; atactic polypropylene is a soft, elastic, and rubbery material. Following are the three stereoisomers of polypropylene:

Isotactic polypropylene

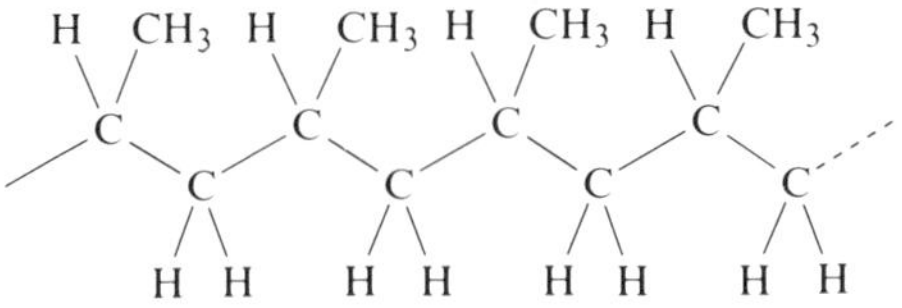

```
 H  CH3  H  CH3  H  CH3  H  CH3
  \/      \/      \/      \/
   C       C       C       C
  / \     / \     / \     / \
     C       C       C       C
    /\      /\      /\      /\
   H  H    H  H    H  H    H  H
```

Syndiotactic polypropylene

```
 H  CH3 CH3  H   H  CH3  H  CH3 CH3  H   H  CH3  H  CH3
  \/      \/      \/      \/      \/      \/      \/
   C       C       C       C       C       C       C
  / \     / \     / \     / \     / \     / \     / \
     C       C       C       C       C       C       C
    /\      /\      /\      /\      /\      /\      /\
   H  H    H  H    H  H    H  H    H  H    H  H    H  H
```

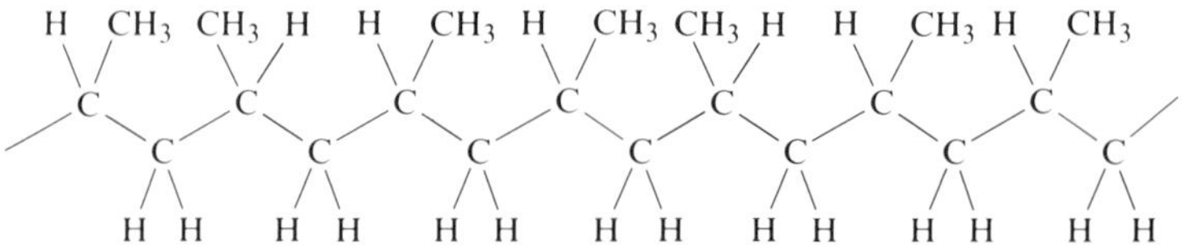

Atactic polypropylene

$$-\overset{CH_3\ \ H}{\underset{}{C}}-\underset{H\ \ H}{C}-\overset{CH_3\ \ H}{C}-\underset{H\ \ H}{C}-\overset{H\ \ CH_3}{C}-\underset{H\ \ H}{C}-\overset{CH_3\ \ H}{C}-\underset{H\ \ H}{C}-\overset{H\ \ CH_3}{C}-\underset{H\ \ H}{C}-$$

2.4 STEPWISE POLYMERIZATION

In stepwise polymerization, there is no initiation, propagation, or termination as is the case in chain reaction polymerization. The polymerization depends entirely on the individual reactions of the functional groups of monomers. The four types of stepwise polymerization are the synthesis of polyester, polyamide, polyurethane, and polycarbonate.

1. Polyester is synthesized by the direct reaction of a diacid and a diol at high temperatures. An example is the synthesis of Dacron:

$$nCH_3CH_2O-\overset{O}{\overset{\|}{C}}-C_6H_4-\overset{O}{\overset{\|}{C}}-OCH_2CH_3 + n\underset{OH}{CH_2}-\underset{OH}{CH_2} \xrightleftharpoons{H^+ \text{ or } OH^-}$$

Diethyl terephthalate — Ethylene glycol

$$\sim CH_2CH_2-O\left[\overset{O}{\overset{\|}{C}}-C_6H_4-\overset{O}{\overset{\|}{C}}-OCH_2CH_2O\right]_n\overset{O}{\overset{\|}{C}}-C_6H_4-\overset{O}{\overset{\|}{C}}-OCH_2CH_2\sim$$

Polyester (Dacron)

$$+\ 2\,nCH_3CH_2OH$$

Polyester is used for clothing and tire cord.

2. Polyamide is synthesized using two difunctional monomers. An example is the synthesis of 66 nylon:

$$nHO-\overset{O}{\overset{\|}{C}}-(CH_2)_4-\overset{O}{\overset{\|}{C}}-OH + nH_2N(CH_2)_6NH_2 \longrightarrow$$

Adipic acid — Hexamethyl diamine

$$\left[\overset{H}{\overset{|}{N}}-\overset{O}{\overset{\|}{C}}-(CH_2)_4-\overset{O}{\overset{\|}{C}}-N(CH_2)_6\right]_n + 2nH_2O$$

Polyhexamethylene adipamide

The number n is usually between 50 and 65, corresponding to a molecular weight of 12 000–15 000. Like polyester, nylon is used for clothing and tire cord.

3. Polyurethanes, also called polycarbamates, are synthesized by the reaction of a diisocyanate with a diol. Experiments are usually carried out in solutions. The following is an example of polyurethane synthesis:

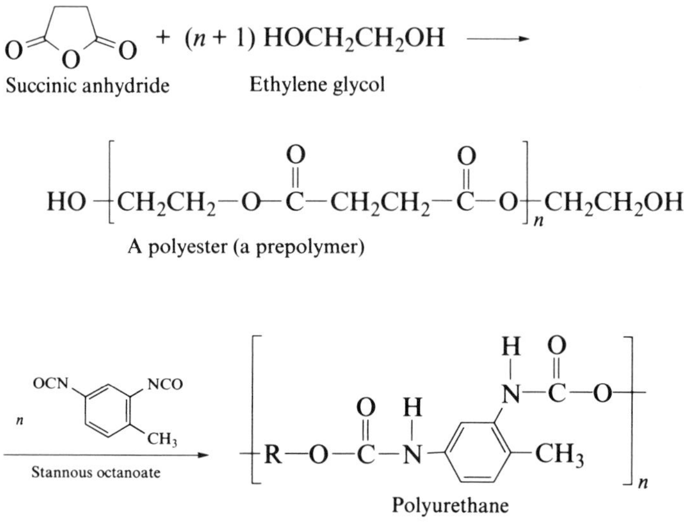

$$\text{Succinic anhydride} + (n+1)\,HOCH_2CH_2OH\ (\text{Ethylene glycol}) \longrightarrow$$

$$HO\!-\!\!\left[CH_2CH_2-O-\overset{\overset{\displaystyle O}{\|}}{C}-CH_2CH_2-\overset{\overset{\displaystyle O}{\|}}{C}-O\right]_n\!\!-CH_2CH_2OH$$

A polyester (a prepolymer)

$$\xrightarrow[\text{Stannous octanoate}]{n\ OCN-C_6H_3(CH_3)-NCO} \left[R-O-\overset{\overset{\displaystyle O}{\|}}{C}-\overset{\overset{\displaystyle H}{|}}{N}-C_6H_3(CH_3)-\overset{\overset{\displaystyle H}{|}}{N}-\overset{\overset{\displaystyle O}{\|}}{C}-O\right]_n$$

Polyurethane

Polyurethane is used for rubber (vulcollanes), elastic fibers (Lycra), hard or elastic forms (Moltoprene), and stain, flooring, and wood and fabric coating.

4. Polycarbonate is synthesized by the reaction of the simplest diacidchloride, phosgene, with bisphenol A in the presence of a base:

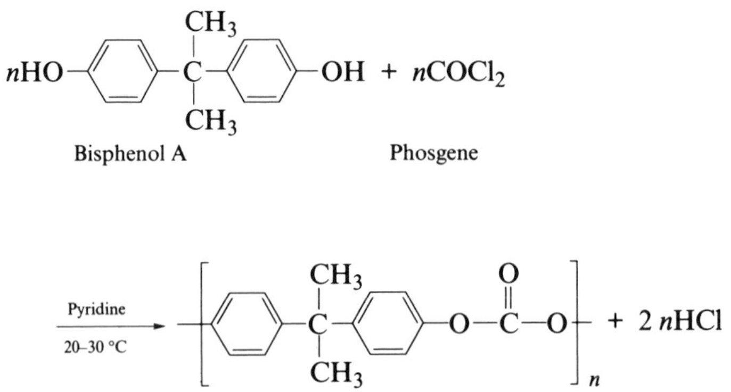

$$nHO-C_6H_4-C(CH_3)_2-C_6H_4-OH + nCOCl_2$$

Bisphenol A Phosgene

$$\xrightarrow[20\text{–}30\ ^\circ C]{\text{Pyridine}} \left[C_6H_4-C(CH_3)_2-C_6H_4-O-\overset{\overset{\displaystyle O}{\|}}{C}-O\right]_n + 2nHCl$$

Property: electric resistance

2.5 KINETICS OF THE SYNTHESES OF POLYMERS

The kinetics of a condensation reaction is similar to those of small molecular reactions. It is basically a simple order reaction (first order, second order, or third order). On the other hand, free radical polymerization, ionic polymerization, and coordination polymerization are all chain reactions. Their mechanism is very complicated.

2.5.1 Condensation Reactions

A typical condensation polymerization may run as follows:

$$\underset{\text{Glycol}}{\text{HO—R—O}}\,\boxed{\text{H + HO}}\,\underset{\text{Diacid}}{\text{OC—R}'\text{—COOH}} \longrightarrow \text{HO—R—O—COR}'\text{—CO(OH)} + H_2O \longrightarrow$$

$$\text{HOOC—R}'\text{—CO(OH)} \quad \text{H/O—R—OH} \longrightarrow$$

$$\text{HOOC—R}'\text{—CO—OR—CO—R}'\text{—CO—OR—OH} \longrightarrow \cdots$$

Let A represent HO–R–OH and B represent COOH–R′–COOH.Then the rate law of condensation polymerization is a simple third-order reaction.

$$\frac{-\,\mathrm{d}[\mathrm{B}]}{\mathrm{d}t} = k[\mathrm{B}]^2[\mathrm{A}]$$

If $[\mathrm{B}] = [\mathrm{A}] = c$, where c is the concentration in moles/liter, then

$$\frac{-\,\mathrm{d}c}{\mathrm{d}t} = kc^3$$

Upon integration we obtain

$$2kt = \frac{1}{c^2} + \text{constant}$$

Let P be the extent of reaction, or probability that the functional groups will react, that is, the fraction of the functional groups that has reacted at time t; then

$$c = c_0(1 - P)$$

where c is the concentration of monomers at any time t and c_0 is the initial concentration. The quantity $1 - P$ is the probability that the groups will not react. Substituting c into the previous equation, we obtain

$$2c_0^2kt = \frac{1}{(1 - P)^2} + \text{constant}$$

The degree of polymerization DP for condensation reaction can be defined in terms of $1 - P$ by the equation $\text{DP} = 1/(1 - P)$.

If we plot $1/(1 - P)^2$ versus t (time), a linear graph is obtained:

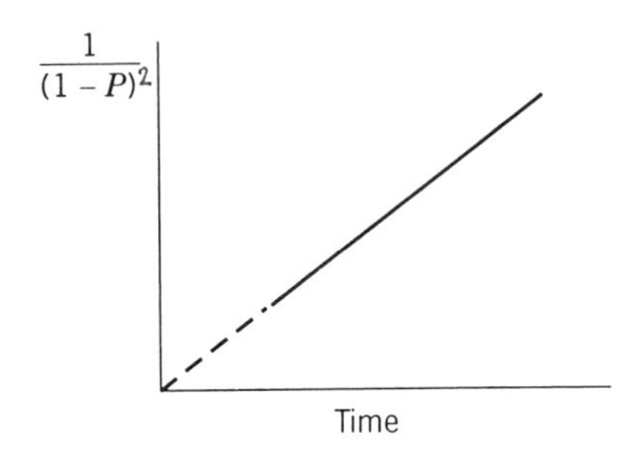

The rate constant k can be obtained from the slope. The third-order reaction can be reduced to the second-order reaction, if a strong acid catalyst is added to the reaction system. The rate equation then becomes

$$-\frac{dc}{dt} = kc^2$$

and

$$c_0k't = \frac{1}{1 - P} + \text{constant}$$

The plot of $1/(1 - P)$ versus t yields a straight line:

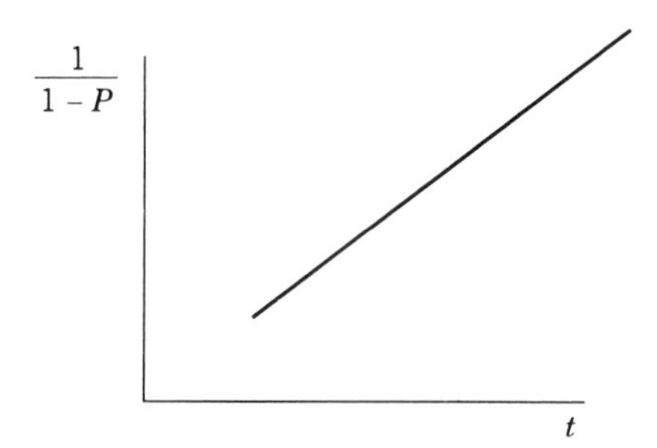

2.5.2 Chain Reactions

The mechanism of the basic feature of chain reaction may be illustrated by that of free radical reaction, particularly the polymerization of styrene, which has been extensively investigated for years by many investigators. Here we describe the mechanism proposed by Mayo and coworkers (1959).

A free radical reaction may go through all or some of the following steps:

Initiation

$$\mathrm{I} \xrightarrow{k_\mathrm{d}} 2\mathrm{R}\cdot \tag{2.1}$$

for example,

$$\mathrm{BZ_2O_2} \xrightarrow{k_\mathrm{d}} 2\mathrm{BZO}\cdot$$

$$2\mathrm{M} \xrightarrow{k_\mathrm{i}} 2\mathrm{M}\cdot \qquad \text{(Biradical from thermal initiation)} \tag{2.2}$$

$$2\mathrm{R}\cdot \xrightarrow{k_\mathrm{r}} \qquad \text{Nonradical products} \tag{2.3}$$

$$\mathrm{M}\cdot + 2\mathrm{M} \xrightarrow{k_\mathrm{a}} 2\mathrm{M}\cdot \qquad \text{(Monoradical)} \tag{2.4a}$$

$$\mathrm{R}\cdot + \mathrm{M} \xrightarrow{k_\mathrm{a}} \mathrm{M}\cdot \qquad \text{(Chain radical)} \tag{2.4b}$$

Monoradicals and chain radicals are the same. The chain has not propagated as yet; hence, both equations are labeled (2.4). BZ_2O_2 is benzoyl peroxide. The term k_d is the rate constant of the decomposition of peroxide, k_i is the rate constant of the thermal initiation of biradical, k_r is the rate constant of the first-order recombination of radicals from the peroxide, and k_a is the rate constant of the reaction of these radicals with monomer.

Propagation

$$\mathrm{M}\cdot + \mathrm{M} \xrightarrow{k_\mathrm{p}} \mathrm{M}_2 \cdots \tag{2.5}$$

Chain transfer to monomer

$$\mathrm{M}_n\cdot + \mathrm{M} \xrightarrow{k_{\mathrm{tr,m}}} \text{Polymer} + \mathrm{M}\cdot \tag{2.6}$$

Chain transfer to solvent

$$\mathrm{M}_n\cdot + \mathrm{S} \xrightarrow{k_{\mathrm{tr,s}}} \text{Polymer} + \mathrm{S}\cdot$$

$$\mathrm{S}\cdot + \mathrm{M} \xrightarrow{k_p} \mathrm{SM}\cdot \xrightarrow{\mathrm{M}, k_{p'}} \cdots$$

Chain transfer to initiator

$$M_m\cdot + I \xrightarrow{k_I} \text{Polymer} + R\cdot \tag{2.7}$$

where k_p is the rate constant of propagation, k_{tr} is the rate constant of transfer, and k_I is the rate constant of the chain transfer to initiator.

Termination

$$M_m\cdot + M_n\cdot \xrightarrow{k_{td}} 2\,\text{polymers (that is, } M_m + M_n) \tag{2.8}$$

where k_{td} is the rate constant of termination by disproportionation,

$$M_m\cdot + M_n\cdot \xrightarrow{k_{tc}} 2\,\text{polymers (that is, } M_{m+n}) \tag{2.9}$$

where k_{tc} is the rate contant of termination by coupling.

The rate of initiation R_i is

$$R_i = \frac{d[M\cdot]}{dt} = 2fk_d[I]$$

where f is the efficiency of the initiation and is given in the expression

$$f = \frac{k_a[2R\cdot][M]}{k_r[2R\cdot] + k_a[2R\cdot][M]} = \frac{1}{(k_r/k_a[M]) + 1} \tag{2.10}$$

The value of f ranges between 0.5 and 1, and is usually 0.70. The rate of termination is

$$R_t = -\frac{d[M\cdot]}{dt} = 2k_t[M\cdot]^2$$

where $k_t = k_{td} + k_{tc}$. At a steady state, $R_i = R_t$, that is,

$$[M\cdot] = \left(\frac{k_i[M]^2 + k_d f[I]}{k_t}\right)^{1/2}$$

The overall rate is

$$-\frac{d[M]}{dt} = k_p[M\cdot][M] = k_p\left(\frac{k_i[M]^2 + k_d f[I]}{k_t}\right)^{1/2}[M] \tag{2.11}$$

Since k_i is usually small, we may assume that $k_i = 0$. Equation (2.11) then becomes

$$-\frac{d[M]}{dt} = k_p \left(\frac{f k_d}{k_t}\right)^{1/2} [M][I]^{1/2} \qquad (2.12)$$

This is the rate law of free radical polymerization. In general, the polymerization is in $\frac{1}{2}$ order with respect to [I] and first order with respect to [M]. However, it depends on f. If f is low, k_i is no longer negligible, and the polymerization could be in $\frac{3}{2}$ order with respect to [M], while still in $\frac{1}{2}$ order with respect to [I].

Of particular interest is the study of chain transfer during the process of chain propagation. Let us define the degree of addition polymerization DP as

$$\text{DP} = \frac{\text{Rate of polymerization } (\text{R}_p)}{\text{Half the rate of formation of chain ends}}$$

$$= \frac{k_p[M\cdot][M]}{f k_d[I] + k_{tr,I}[M\cdot][I] + k_{tr,m}[M\cdot][M] + k_{tr,s}[M\cdot][S] + k_{td}[M\cdot]^2}$$

where $k_{tr,I}$ is the rate constant of chain transfer to the initiator, and $k_{tr,I}[M\cdot][I] = k_I[M\cdot][I]$. Similarly, $k_{tr,m}$ is the rate constant of chain transfer to monomer, and $k_{tr,m}[M\cdot][M] = k_{tr,m}[M_n\cdot][M]$; $k_{tr,s}$ is the rate constant of chain transfer to the solvent, and $k_{tr,s}[M\cdot][S] = k_{tr,s}[M_n\cdot][S]$. Taking the reciprocal of DP and neglecting the solvent effect, we have

$$\frac{1}{\text{DP}} = \frac{k_{tr,I}[M\cdot][I] + k_{tr,m}[M\cdot][M] + k_d f[I] + k_{td}[M\cdot]^2}{k_p[M\cdot][M]} \qquad (2.13)$$

which leads to

$$\frac{1}{\text{DP}} = c_M + \frac{c_I[I]}{[M]} + \frac{k_d f[I]}{-d[M]/dt} + \frac{k_{td}(k_i[M]^2 + k_d f[I])}{(-d[M]/dt)(k_t)}$$

$$= c_M + \frac{c_I[I]}{[M]} + \frac{k_d f[I]}{-d[M]/dt} + \frac{k_{td}(-d[M]/dt)}{k_p^2[M]^2} \qquad (2.14)$$

where $c_M = k_{tr}/k_p$ and $c_I = k_i/k_p$. If the solvent effect is stressed, the reciprocal of DP is in the form (from the equation for [M·] and Eq. (2.12))

$$\frac{1}{\text{DP}} = \frac{R_p k_t}{k_p^2[M]^2} + \frac{k_{tr,m}}{k_p} + \frac{k_{tr,s}[S]}{k_p[M]} \qquad (2.15)$$

Equations (2.12), (2.14), and (2.15) can be experimentally tested. From Eq. (2.12) we plot

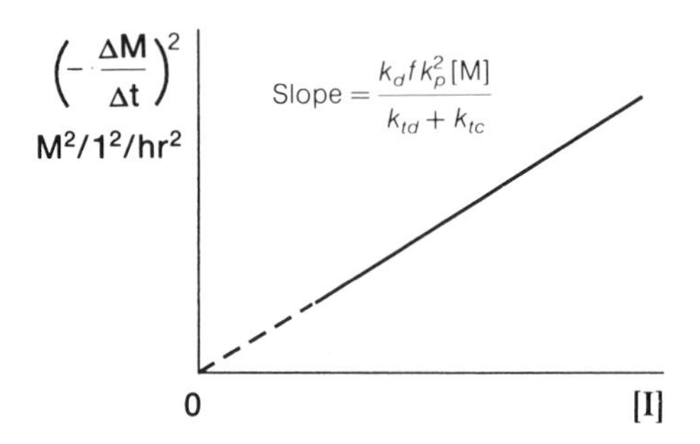

at the same monomer concentration, that is, [M] is constant. From Eq. (2.14) we plot (assuming $k_{td} = 0$).

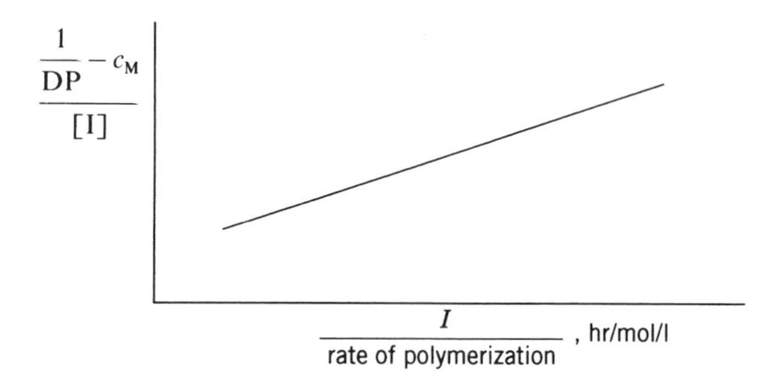

If the value of c_M is known, for example, $c_M = 6 \times 10^{-5}$ for polymerization of styrene, we obtain

$$\text{Slope} = k_d f$$

$$\text{Intercept} = \frac{c_I}{[M]}$$

In obtaining this plot, we assume, of course, that all chains are terminated by coupling, that is, $k_{td} = 0$.

In the use of Eq. (2.15) we assume that $k_{tr,m}$ is negligible. This can be done by running the experiments of uncatalyzed polymerization at different initial monomer concentrations. Assuming that $R_p \propto [M]^2$, Eq. (2.15) becomes

$$\frac{1}{DP} = \left(\frac{k_t}{k_p^2} + \frac{k_{tr,m}}{k_p}\right) + \frac{k_{tr,s}}{k_p}\frac{[S]}{[M]}$$

The plot of 1/DP versus [S]/[M] will yield

$$\text{Slope} = \frac{k_{tr,s}}{k_p}$$

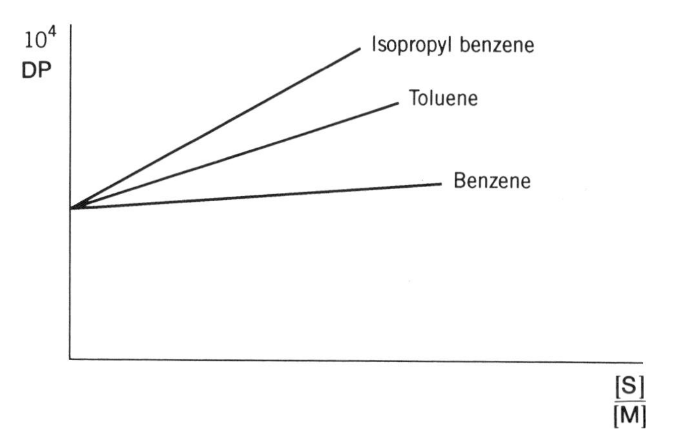

FIGURE 2.1 The effect of solvent on the styrene polymerization at 100°C.

which is the solvent transfer constant. The effect of solvent is shown in Figure 2.1.

It is customary to define the chain length v as the number of monomer units consumed per active center:

$$v = \frac{R_p}{R_i} = \frac{R_p}{R_t} \qquad \text{(steady state)}$$

$$v = \frac{k_p}{2k_t} \frac{[M]}{[M\cdot]}$$

$$v = \frac{k_p^2}{2k_t} \frac{[M]^2}{R_p}$$

$$v = \frac{k_p}{2(f k_d k_t)^{1/2}} \frac{[M]}{[I]^{1/2}}$$

In general, the chain length is related to DP, that is, $v \sim$ DP. The proportionality constant depends on the mode of termination: For termination by coupling, DP $= 2v$, and for termination by disproportionation, DP $= v$.

On the basis of the above mechanism, we can write a simple equation to describe the probability p that a polymer radical $M_n\cdot$ may undergo transfer rather than addition of another monomer:

$$1 - p = \frac{k_{tr}[A]}{k_p[M] + k_{tr}[A]}$$

where A can be an initiator, solvent, radical, or even a monomer. The degree of polymerization $\bar{x}_n$ may be written as before:

$$\bar{x}_n = \frac{1}{1-p}$$

The chain termination by coupling can be described by

$$1 - p = \frac{2k_t[M\cdot]}{k_p[M] + 2k_t[M\cdot]}$$

$$\bar{x}_n = \frac{2}{1-p} = 2 + \frac{k_p}{k_t}\frac{[M]}{[M\cdot]}$$

2.6 POLYPEPTIDE SYNTHESIS

Polypeptide synthesis is basically the work of coupling of two amino acids or peptides in sequence. The most frequently used reagent for coupling is acid chloride. An example is the coupling of gly and ala:

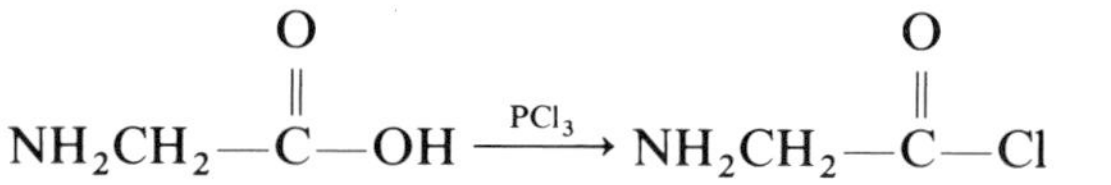

$$NH_2CH_2{-}\overset{\overset{\displaystyle O}{\|}}{C}{-}OH \xrightarrow{PCl_3} NH_2CH_2{-}\overset{\overset{\displaystyle O}{\|}}{C}{-}Cl$$

Glycine — Acid chloride of glycine

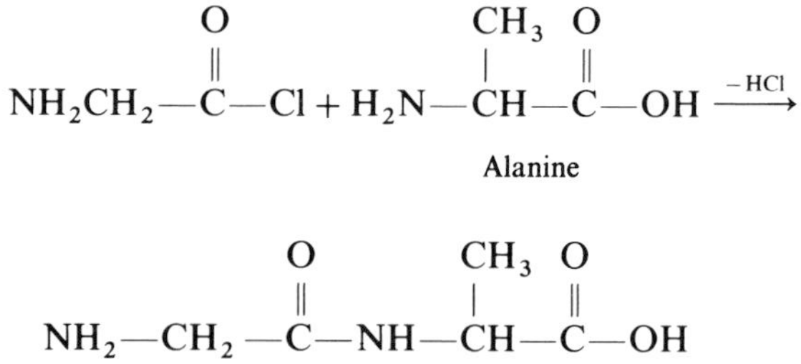

$$NH_2CH_2{-}\overset{\overset{\displaystyle O}{\|}}{C}{-}Cl + H_2N{-}\overset{\overset{\displaystyle CH_3}{|}}{CH}{-}\overset{\overset{\displaystyle O}{\|}}{C}{-}OH \xrightarrow{-HCl}$$

Alanine

$$NH_2{-}CH_2{-}\overset{\overset{\displaystyle O}{\|}}{C}{-}NH{-}\overset{\overset{\displaystyle CH_3}{|}}{CH}{-}\overset{\overset{\displaystyle O}{\|}}{C}{-}OH$$

Gly–ala

However, the side reaction of coupling glycine often occurs also:

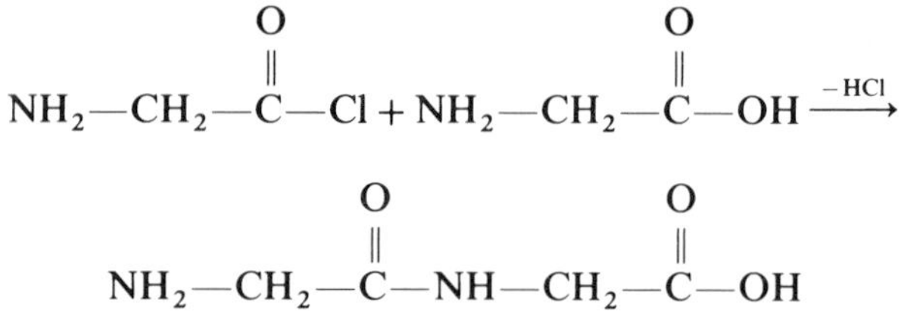

$$NH_2{-}CH_2{-}\overset{\overset{\displaystyle O}{\|}}{C}{-}Cl + NH_2{-}CH_2{-}\overset{\overset{\displaystyle O}{\|}}{C}{-}OH \xrightarrow{-HCl}$$

$$NH_2{-}CH_2{-}\overset{\overset{\displaystyle O}{\|}}{C}{-}NH{-}CH_2{-}\overset{\overset{\displaystyle O}{\|}}{C}{-}OH$$

Gly–gly

To avoid the side reaction, a protecting or blocking group must be substituted on the amino function of the group, for example, $-NH_2-$:

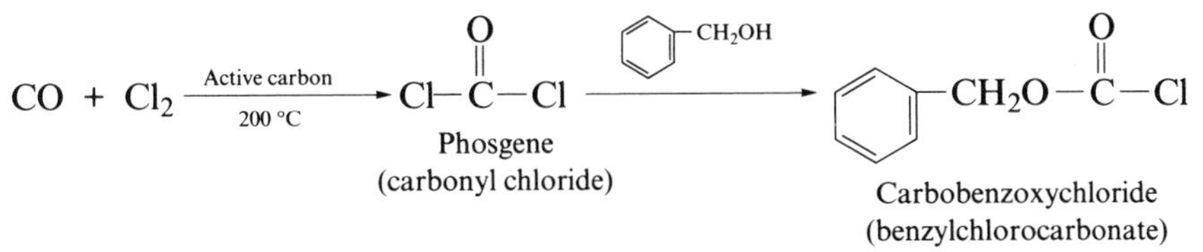

$$CO + Cl_2 \xrightarrow[200\ ^\circ C]{\text{Active carbon}} Cl-\overset{O}{\overset{\|}{C}}-Cl \xrightarrow{C_6H_5-CH_2OH} C_6H_5-CH_2O-\overset{O}{\overset{\|}{C}}-Cl$$

Phosgene (carbonyl chloride)

Carbobenzoxychloride (benzylchlorocarbonate)

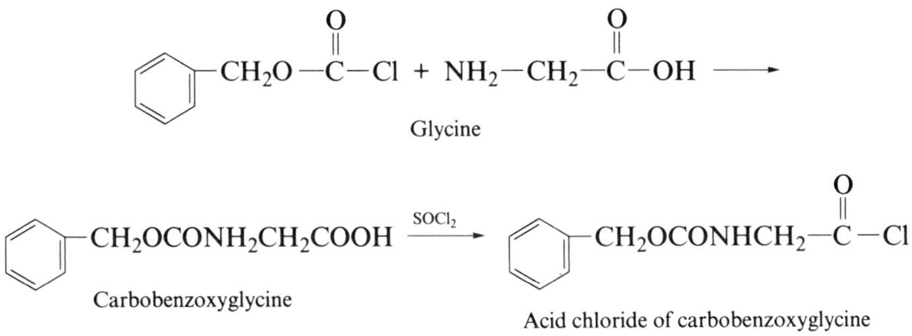

$$C_6H_5-CH_2O-\overset{O}{\overset{\|}{C}}-Cl + NH_2-CH_2-\overset{O}{\overset{\|}{C}}-OH \longrightarrow$$

Glycine

$$C_6H_5-CH_2OCONH_2CH_2COOH \xrightarrow{SOCl_2} C_6H_5-CH_2OCONHCH_2-\overset{O}{\overset{\|}{C}}-Cl$$

Carbobenzoxyglycine

Acid chloride of carbobenzoxyglycine

Once the amino group is protected, we can couple gly and ala without a side reaction:

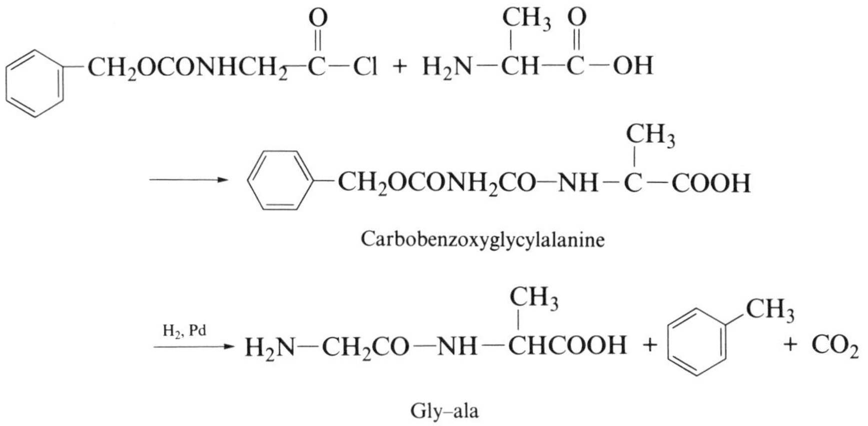

$$C_6H_5-CH_2OCONHCH_2-\overset{O}{\overset{\|}{C}}-Cl + H_2N-\overset{CH_3}{\overset{|}{CH}}-\overset{O}{\overset{\|}{C}}-OH$$

$$\longrightarrow C_6H_5-CH_2OCONH_2CO-NH-\overset{CH_3}{\overset{|}{C}}-COOH$$

Carbobenzoxyglycylalanine

$$\xrightarrow{H_2,\ Pd} H_2N-CH_2CO-NH-\overset{CH_3}{\overset{|}{C}}HCOOH + C_6H_5-CH_3 + CO_2$$

Gly–ala

Since amino acids and peptides often possess a variety of chemically reactive substituents (in addition to amino groups $-NH_2$), such as carboxyl groups ($-COOH$), thiol groups ($-SH$), and hydroxy groups ($-OH$), a number of chemical blocking reagents other than carbobenzoxy chloride must be used. The following are some examples:

To block an N terminal

$$(CH_3)_3{-}CO{-}\overset{\overset{\large O}{\|}}{C}{-}Cl$$

tert-Butoxycarbonyl chloride

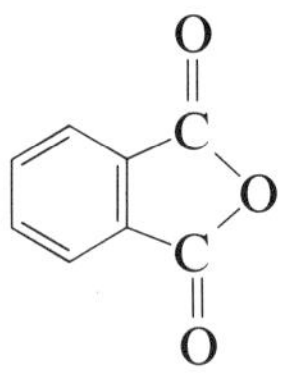

Phthalic anhydride

To block a C terminal

$$CH_3{-}OH$$
$$CH_3{-}CH_2{-}OH$$
$$C_6H_5{-}CH_2{-}OH$$

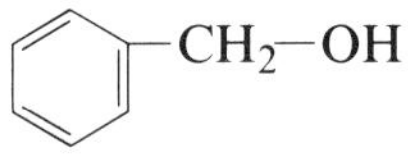

To block a side chain amino group

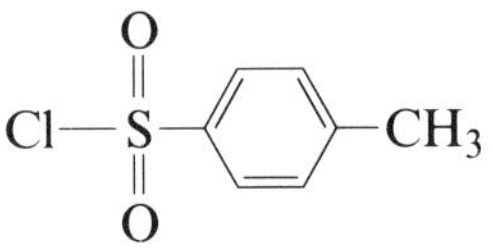

$$Cl{-}\underset{\underset{\large O}{\|}}{\overset{\overset{\large O}{\|}}{S}}{-}C_6H_4{-}CH_3$$

p-Toluenesulfonyl chloride

To block side chains–COOH, –SH, –OH, and (imidazole)

$$C_6H_5{-}CH_2Cl$$

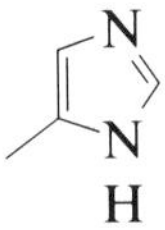

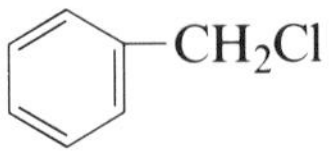

Benzyl chloride

Suppose we have a polypeptide available:

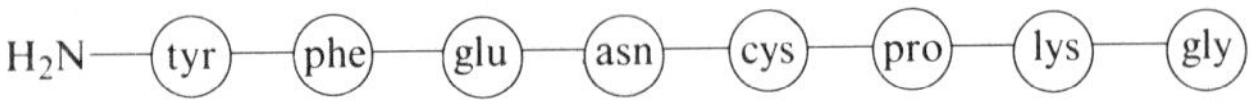

H_2N—(tyr)—(phe)—(glu)—(asn)—(cys)—(pro)—(lys)—(gly)

and we want to add one group, cys, to tyr (N terminal). We first protect the cys and lys groups in the given polypeptide:

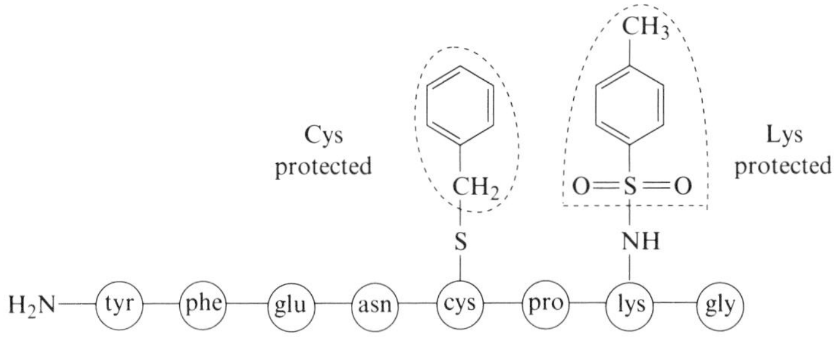

Cys protected: cys—S—CH_2—C_6H_5

Lys protected: lys—NH—SO_2—C_6H_4—CH_3

H_2N—tyr—phe—glu—asn—cys—pro—lys—gly

Then we activate the C terminal end of an independent protected cys:

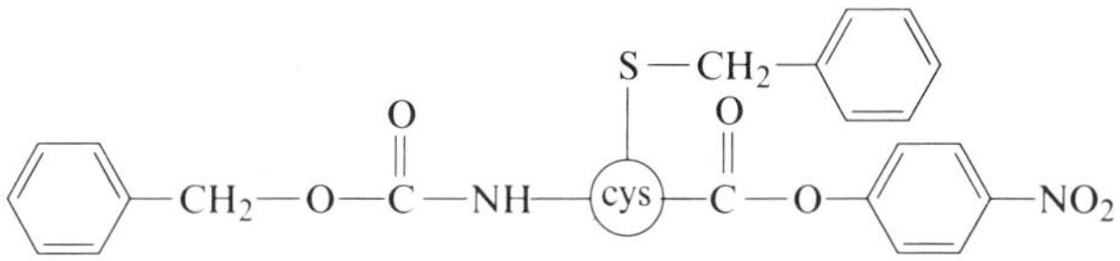

C_6H_5—CH_2—O—C(=O)—NH—cys(S—CH_2—C_6H_5)—C(=O)—O—C_6H_4—NO_2

which will undergo coupling as shown in Fig. 2.2.

Polypeptide synthesis is one of the exciting and challenging areas in modern chemistry. Here we describe the two most well-known polypeptide syntheses: the synthesis of insulin and the synthesis of ribonuclease.

2.6.1 Synthesis of Insulin

The synthesis of insulin was reported by three groups: Zahn (1963), Katsoyannis and Dixon (1964), and Niu et al. (1965). Insulin consists of an A chain and a B chain:

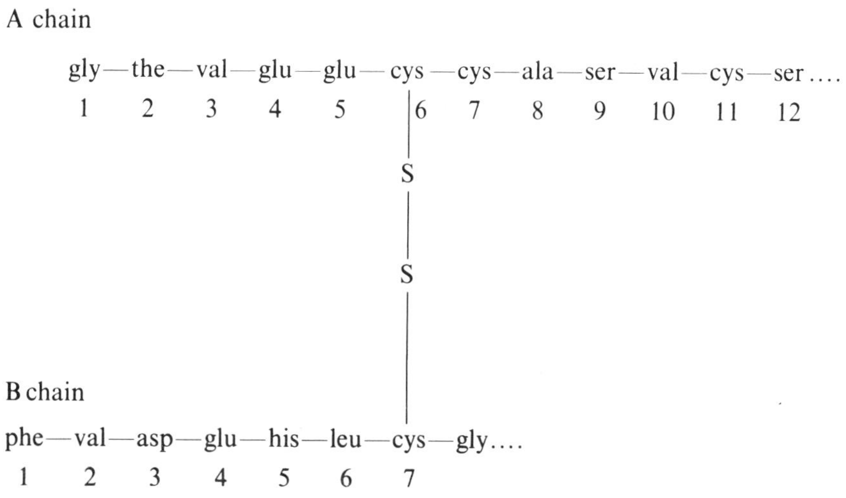

A chain

gly—the—val—glu—glu—cys—cys—ala—ser—val—cys—ser....
1 2 3 4 5 6 7 8 9 10 11 12

S
|
S

B chain

phe—val—asp—glu—his—leu—cys—gly....
1 2 3 4 5 6 7

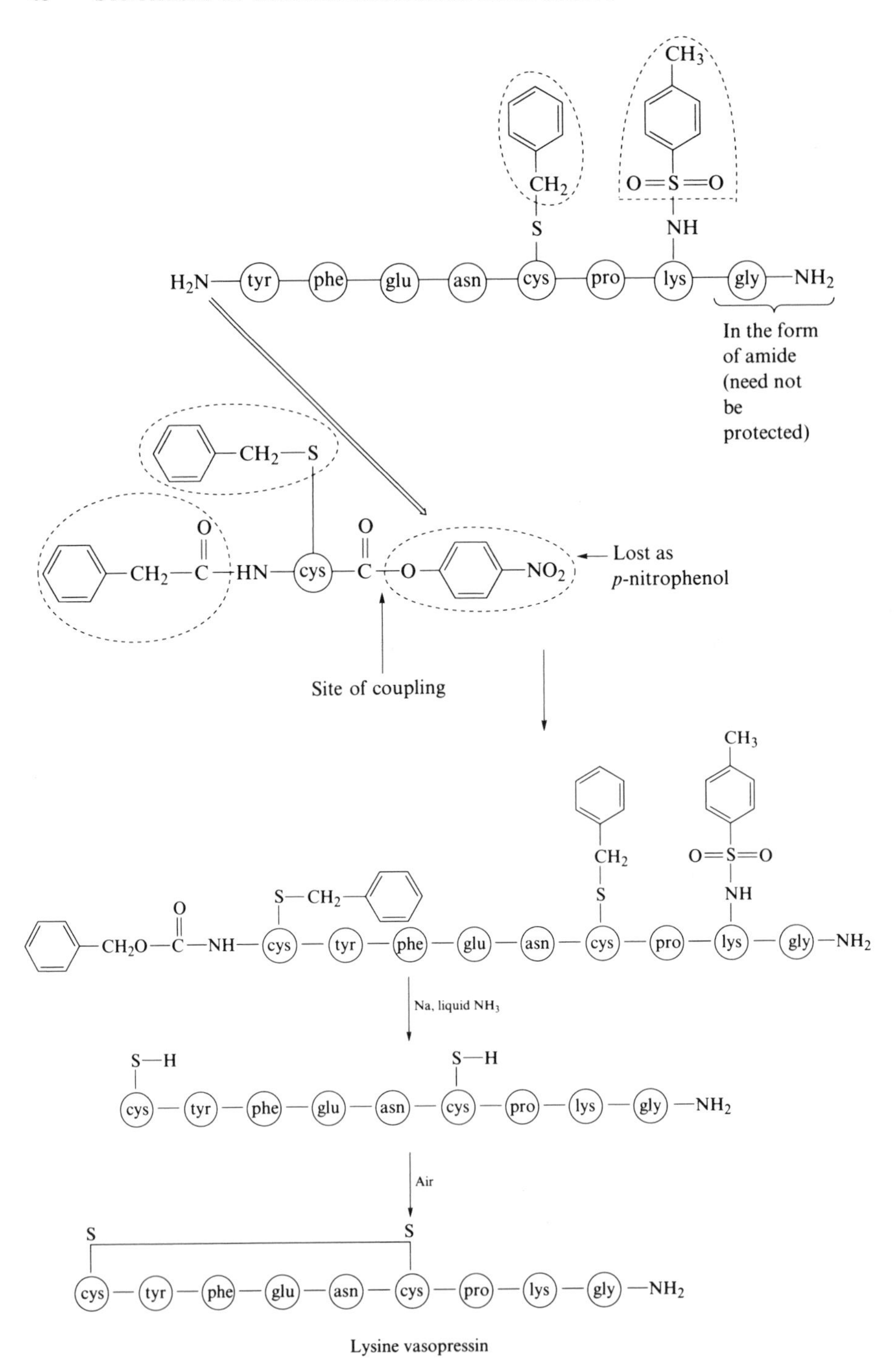

FIGURE 2.2 Coupling of polypeptides. (*Source*: Wingrove and Caret (1981).)

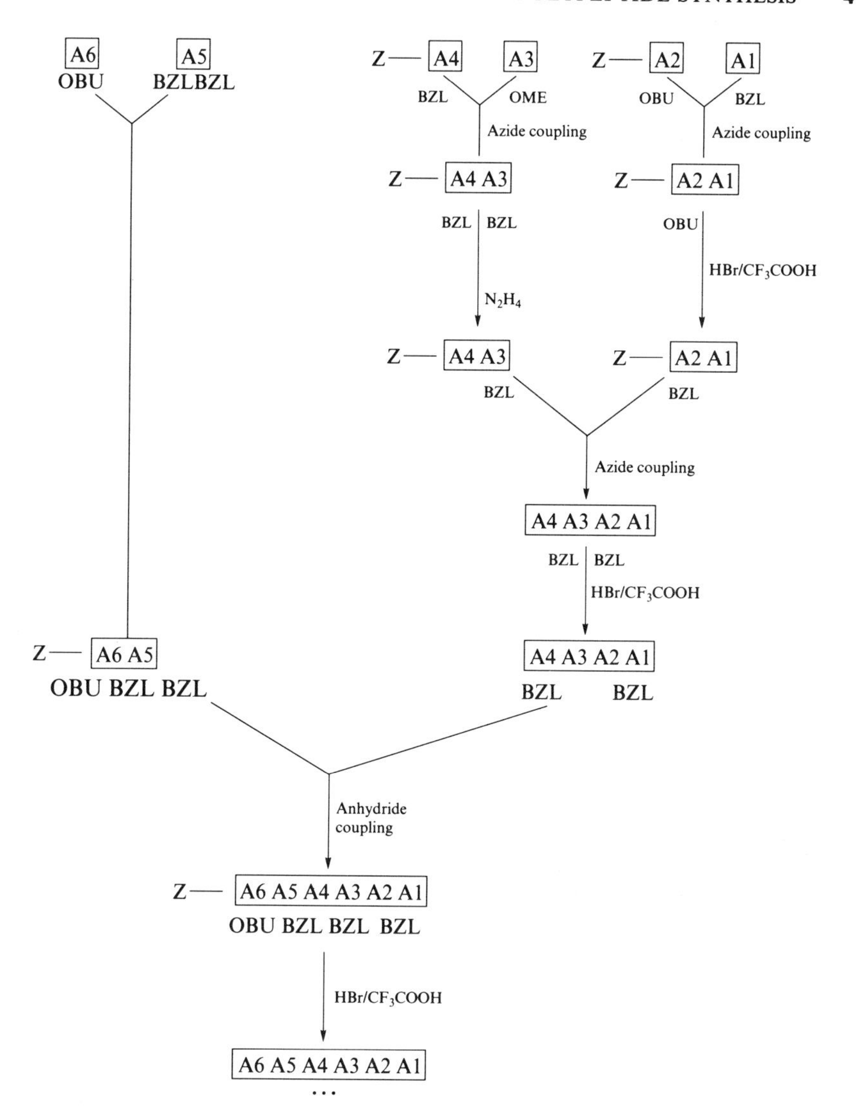

Abbreviations: BZL, benzyl; OBU, butyl ester; OME, methyl ester; A, amino acid; Z, benzyloxy-carbonyl.

FIGURE 2.3 Synthesis of insulin. (*Source*: Vollmert (1973).)

The scheme of synthesis is illustrated in Figure 2.3.

2.6.2 Synthesis of Ribonucleus

The successful synthesis of ribonucleus is attributed to Merrifield (1963), who designed the solid-phase method in polypeptide chemistry. The method starts

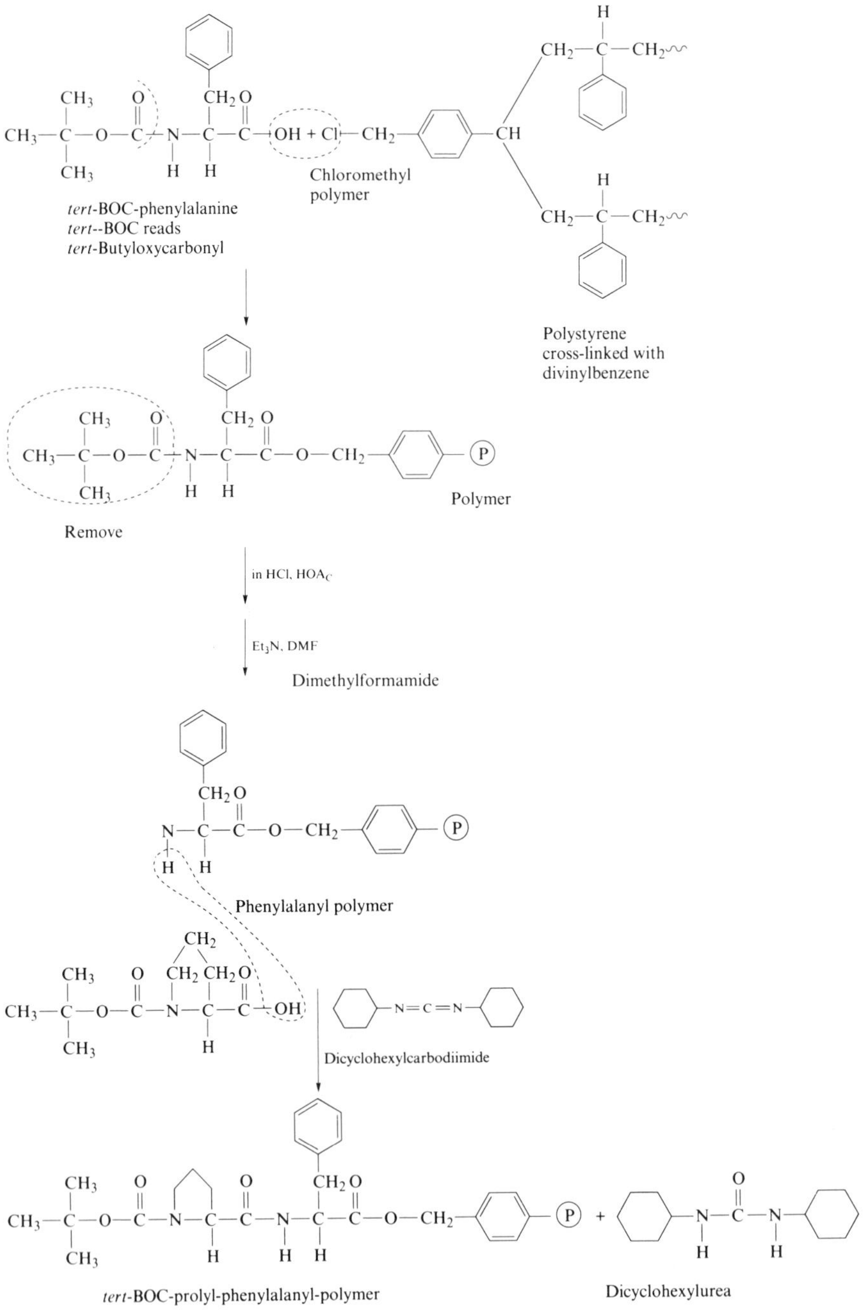

tert-BOC-phenylalanine
tert--BOC reads
tert-Butyloxycarbonyl

Chloromethyl polymer

Polystyrene cross-linked with divinylbenzene

Remove

Polymer

in HCl, HOA_C

Et_3N, DMF

Dimethylformamide

Phenylalanyl polymer

Dicyclohexylcarbodiimide

tert-BOC-prolyl-phenylalanyl-polymer

Dicyclohexylurea

(*A*)

FIGURE 2.4 (*A*) The first steps of ribonuclear synthesis; (*B*) The general diagram. (*Source*: Merrifield (1963).)

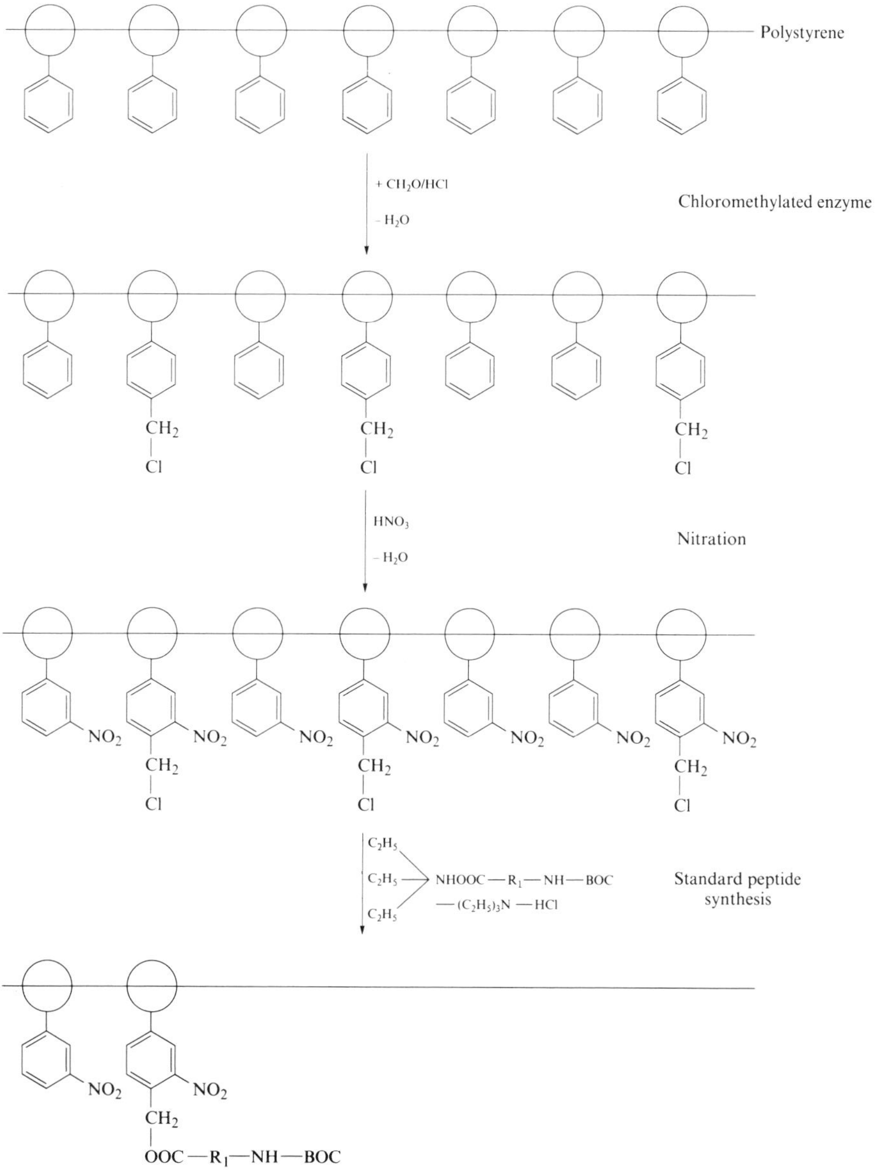

Polystyrene

$+ CH_2O/HCl$

$- H_2O$

Chloromethylated enzyme

CH_2 — Cl

HNO_3

$- H_2O$

Nitration

NO_2

CH_2 — Cl

$(C_2H_5)_3N$ + $NHOOC—R_1—NH—BOC$ — $(C_2H_5)_3N—HCl$

Standard peptide synthesis

NO_2

CH_2

$OOC—R_1—NH—BOC$

(B)

FIGURE 2.4 (*Continued*)

with the synthesis of a polymeric material (polystyrene). The polymer is chloromethylated and nitrated. The first amino acid is attached to the CH_2 group of the chloromethyl polystyrene by esterification. Then the other amino acids are added step by step (Figure. 2.4).

2.7 DNA SYNTHESIS

There are three methods for synthesizing DNA: phosphate diester, phosphate triester, and phosphite triester. To illustrate, we describe the phosphite triester method. This method is parallel to Merrifield's synthesis of ribonucleus in the sense that the strategy involves adding mononuclestides in sequence to a deoxynucleoside, which is covalently attached to a polymeric material. The polymeric material is usually silica-based, such as Vydak, Fractosil, and poreglass.

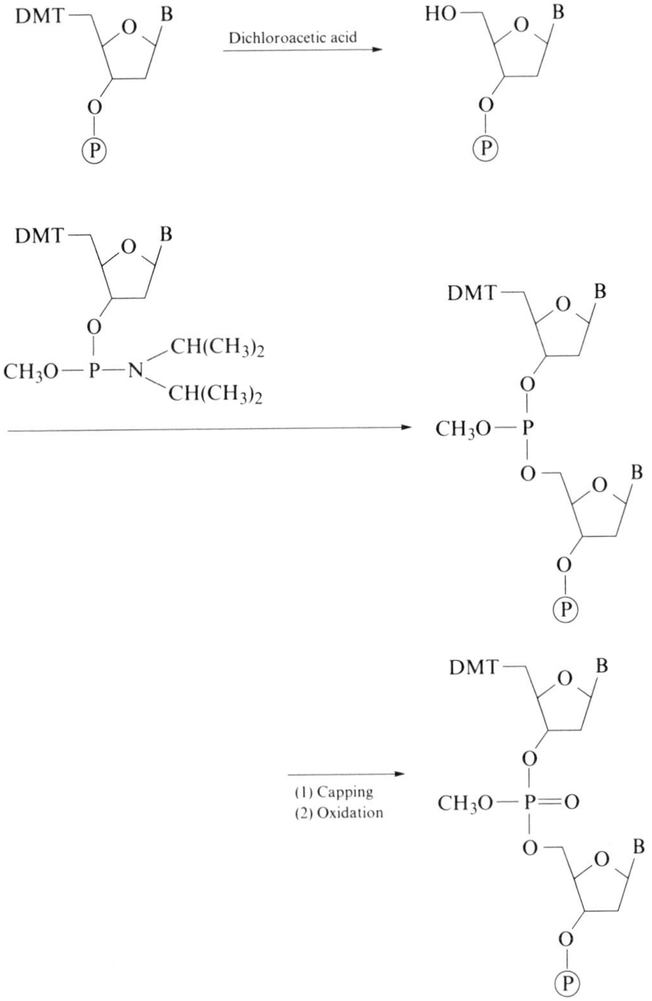

DMT—[deoxyribose]—B, O—(P) —Dichloroacetic acid→ HO—[deoxyribose]—B, O—(P)

DMT—[deoxyribose]—B, O—P(OCH$_3$)—N[CH(CH$_3$)$_2$]$_2$ → DMT—[deoxyribose]—B, O—P(OCH$_3$)—O—[deoxyribose]—B, O—(P)

—(1) Capping (2) Oxidation→ DMT—[deoxyribose]—B, O—P(=O)(OCH$_3$)—O—[deoxyribose]—B, O—(P)

FIGURE 2.5 Capping of unreactive deoxynucleoside followed by oxidation of the phosphite triester to the phosphate triester. (*Source*: Caruthers (1985).)

It consists of coupling a deoxynucleoside that contains a 3′-*p*-nitrophenylsuccinate ester with an amino group attached to the polymer support.

The starting material is a compound that contains a deoxynucleoside covalently joined to silica gel Ⓢ through an amide bond:

where B is the thymine or appropriately protected adanine, cytosine, or guanine, and DMT is the dimethoxytrityl. The site for synthesis is DMT. A certain amount of dichloracetic acid is added to remove the DMT, and then the condensation reaction is allowed to occur with another DMT protected deoxynucleoside 3′-phosphoramidite. Finally, capping of the unreactive deoxynucleoside, or acylation, is followed by oxidation of the phosphite triester to the phosphate triester, thereby adding one nucleotide in the cycle in a 3′ to 5′ direction. The overall reaction is given in Fig. 2.5.

REFERENCES

Caruthers, M. H., *Science* **230**, 281 (1985).

Dostal, H., and R. Raff, *Monatsh. Chem.* **68**, 188 (1936a); *Z. Phys. Chem. B* **32**, 11 (1936b).

Flory, P. J., *Principles of Polymer Chemistry*. Ithaca, NY: Cornell University Press, 1953.

Laidler, K. J., *Chemical Kinetics*. New York: McGraw-Hill, 1965, pp. 425–426.

Lenz, R. W., *Organic Chemistry of Synthetic Polymers*. New York: Wiley, 1967.

Mayo, F. R., *J. Chem. Educ.* **36**, 157 (1959).

Mayo, F. R., R. A. Gregg, and M. S. Matheson, *J. Am. Chem. Soc.* **73**, 1691 (1951).

Merrifield, R. B., *J. Am. Chem. Soc.* **85**, 2149 (1963).

Odian, G., *Principles of Polymerization*, 2d ed. New York: Wiley, 1981.

'State of the Art Symposium: Polymer Chemistry', *J. Chem. Educ.* **58** (1981).

Szwarc, M., pp. 303–325 in H. F. Mark (Ed.), *Encyclopedia of Polymer Science and Technology*, Vol. 8. New York: Wiley, 1968.

Vollmert, B., *Polymer Chemistry*. New York: Springer, 1973.

Wingrove, A. S., and R. L. Caret, *Organic Chemistry*. New York: Harper Row, 1981.

PROBLEMS

2.1 Plot the extent of conversion P (x coordinate), versus the degree of polymerization DP (y coordinate) and interpret the resulting graph.

2.2 In condensation polymerization, if 99% of one of the functional groups has reacted (and, therefore, 99% of the other groups), what is the number average degree of polymerization?

(a) No monofunctional impurity is involved.
(b) 3% impurity is involved.

2.3 The mechanism of condensation polymerization may be expressed as

$$M_m + M_n \xrightarrow{k} M_{m+n}$$

where M_m and M_n represent a chain containing m and n monomers, respectively, and k is the rate constant. Show that the fraction of functional groups f that are reacted at time t can be given by

$$f = \frac{[M_1]_0 kt}{2 + [M_1]_0 kt}$$

where $[M_1]_0$ is the initial concentration of monomers. (*Source*: Dostal and Raff (1936a,b); the articles were cited in Laidler (1965).)

2.4 Following are the data of polymerization of styrene by benzoyl peroxide at 60°:

$[Bz_2O]$ (mL)	$-\Delta[M]/\Delta t$ (m/L h^{-1})	Molecular weight
0.000 126 5	0.0181	800 774
0.000 252 5	0.0248	714 737
0.000 500	0.0328	641 264
0.001 010	0.0466	485 738
0.020 0	0.203	121 750
0.080 0	0.404	53 052

(a) Calculate the degree of polymerization of each polymer product.
(b) Assuming the value of c_M to be 6×10^{-5}, determine the constant of chain transfer with benzoyl peroxide $c_I = k_I/k_{p'}$ and the constant of the decomposition of radical k_d. The parameter c_M is the chain transfer with styrene monomer, $c_M = k_{tr}/k_p$. The fraction of the radicals f that is successful in initiating chains is 0.70 (70%). (*Source*: Mayo et al. (1951).)

3

DISTRIBUTION OF MOLECULAR WEIGHT

For small molecules, such as ethane, there is no problem of molecular weight distribution. All ethane molecules have the same molecular weight. The problem exists for macromolecules, especially synthetic polymers. Not all polymer molecules will have the same molecular weight, even if they are prepared the same way. In step polymerization (condensation reaction), any two molecular species (monomer and monomer, monomer and growing polymer, growing polymer and growing polymer) can react and stop at any moment. As a result, different species of polymer molecules, ranging from those with very long chains to those with very short chains, exist simultaneously. Similarly, in a chain reaction (addition polymerization), a radical, an anionic, or a cationic reactive center adds a monomer unit to a growing polymer. The termination could occur at any stage: chain radical combination, chain radical disproportionation, chain transfer to monomer, to solvent or to any impurity. All of these reactions produce polymer molecules with different chain lengths.

While the problem of molecular weight distribution is serious with synthetic polymers, it is not so with proteins and nucleic acids. However, biological polymers in aqueous solutions under certain conditions often form dimers and trimers; thus, the solution may not be homogeneous either; for example, most bovine serum albumin (BSA) samples contain 10% of dimers. Knowledge about the distribution of molecular weight may apply equally to biological polymers. Since molecular weight is directly related to the size of the chain, the approach to the distribution problem is statistical in nature.

3.1 REVIEW OF MATHEMATICAL STATISTICS

To explain the rationale behind the proposals for describing molecular weight distribution, a simple review of mathematics and statistics is given in this section. The following equations are useful in formulating statistical distribution.

Factorial

$$N! = (N)(N-1)(N-2)\cdots(1)$$

Combination

$$C_n^N = \binom{N}{n} = \frac{N!}{(N-n)!n!}$$

Binomial theorem

$$(a+b)^N = \sum_{n=0}^{\infty} \binom{N}{n} a^{N-n} b^n = \sum_{n=0}^{\infty} \frac{N!}{(N-n)!n!} a^{N-n} b^n$$

Three types of distributions are frequently used in statistics: binomial, Poisson, and Gaussian. They are all relevant to the study of molecular weight distribution.

3.1.1 Binomial Distribution

Let N be the total number of events, independent of each other, p be the probability of success, and x be the number of successful events out of N. Then $1-p$ is the probability of failure and $N-x$ is the number of events that fail. The probability that exactly x events will succeed from the group of N is

$$f(x) = \binom{N}{x} p^x (1-p)^{N-x} \qquad 0 \leqslant x \leqslant n \tag{3.1}$$

where $f(x)$ refers to the binomial distribution because it is closely related to the binomial theorem.

The binomial distribution function is not continuous; hence to calculate the average (mean value of $\bar{x}$) we have to use operator summation, instead of integration:

$$\bar{x} = \sum_{x=0}^{N} xf(x) = \sum x \binom{N}{x} p^x (1-p)^{N-x} = Np = m$$

Note:

$$\bar{x} = \sum_{x=0}^{N} xf(x)$$

$$\overline{x^2} = \sum_{x=0}^{N} x^2 f(x)$$

$$\overline{x^3} = \sum_{x=0}^{N} x^3 f(x)$$

...

3.1.2 Poisson Distribution

The terms in Eq. (3.1) may be rearranged in the following form

$$f(x) = \binom{N}{x} p^x (1-p)^{N-x} = \frac{N!}{(N-x)!x!} p^x \frac{(1-p)^N}{(1-p)^x}$$

Consider three conditions:

1. N is large and x is small. For the first approximation

$$\frac{N!}{(N-x)!} = (N)(N-1)(N-2)\cdots(N-x+1) \simeq N^x$$

Then

$$\frac{N!}{(N-x)!x!} p^x = \frac{(Np)^x}{x!}$$

2. p is small. Hence, the factor $(1-p)^x$ is nearly equal to unity, $(1-p)^x = 1$
3. Take the average value of m (the number of successful events) from the binomial distribution:

$$Np = m$$

from which we obtain

$$N = \frac{m}{p}$$

When we apply all three of these conditions, Eq. (3.1) becomes

$$f(x) = \frac{m^x}{x!}(1-p)^{m/p}$$

Note the definition of e:

$$\lim_{p \to 0} (1+p)^{1/p} = e \Rightarrow \{[1+(-p)]^{m/-p}\}^{-1} = e^{-m}$$

Hence we have

$$f(x) = \frac{m^x e^{-m}}{x!} \qquad x = 0, 1, 2, 3, \ldots \tag{3.2}$$

Equation (3.2) represents the Poisson distribution and $f(x)$ is the Poisson distribution function. Like binomial distribution, Poisson distribution is not continuous. To calculate the average (mean) value of $\bar{x}$ we must use the operator summation. The value of m that is the product of Np plays a characteristic role in Poisson distribution. An important assumption is that the distribution $f(x)$ is through the area with uniform probability. In comparison, binomial distribution has a much wider spread from the lowest value to the highest value than Poisson distribution.

3.1.3 Gaussian Distribution

Gaussian distribution is also called the normal distribution or normal error distribution. It is associated with a limiting form of binomial distribution. The conditions for Gaussian distribution are N very large and $p = \frac{1}{2}$, that is, the probability of success $= \frac{1}{2}$ and the probability of failure $= \frac{1}{2}$; the chances for success and failure are absolutely at random, no bias. Gaussian distribution is a continuous function:

$$f(x) = \frac{1}{\sqrt{2\pi\sigma}} e^{-(x-m)^2/2\sigma^2} \tag{3.3}$$

where x is a continuous variable rather than integer as with the binomial and Poisson distributions and σ is the standard deviation. Gaussian distribution is based on two parameters: m (mean) and σ^2 (variance). As shown in Figure 3.1, if m changes while σ is constant, the curve shifts to the right or left without changing its shape. If the standard deviation σ changes while m is constant, the

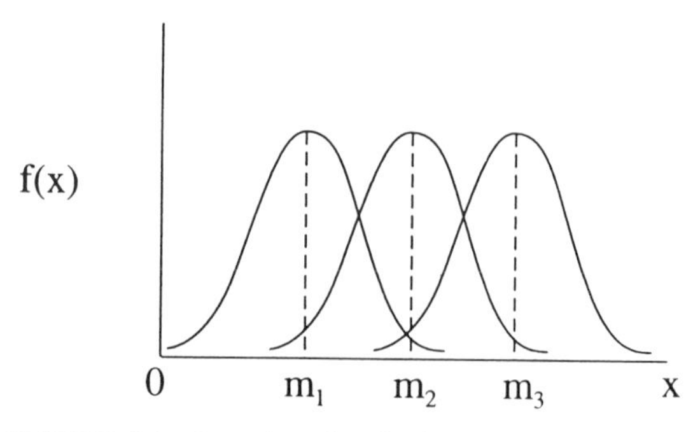

FIGURE 3.1 Gaussian distribution curve: σ constant.

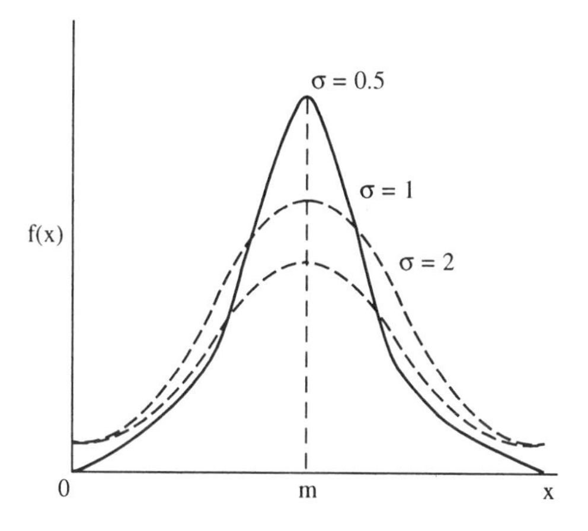

FIGURE 3.2 Gaussian distribution curve: *m* constant.

shape of the curve changes (Figure 3.2). Note that in Gaussian distribution $f(x)$, is a density function and $\int_{-\infty}^{\infty} f(x)\,dx = 1$. The area A under the curve is represented by

$$A = \frac{1}{\sqrt{2\pi\sigma}} \int_{-\infty}^{\infty} e^{-(x-m)^2/2\sigma^2}\,dx$$

If $y = (x - m)/\sigma$, then

$$A = \frac{1}{\sqrt{2\pi}} \int_{-\infty}^{\infty} e^{y^2}\,dy$$

Transforming this expression into polar coordinates, we obtain

$$A^2 = \frac{1}{2\pi} \int_0^{2\pi} \int_0^{\infty} re^{-r^2/2}\,dr\,d\theta = 1$$

So the area is the same whether $\sigma = 0.5$, 1, or 2. To calculate the average (mean) value of $\bar{x}$, we use the operator integration, not summation as in the case of binomial and Poisson distribution.

3.2 ONE-PARAMETER EQUATION

To illustrate our discussion so far, we describe molecular weight distribution with a one-parameter equation, first suggested by Flory (1936, 1940).

3.2.1 Condensation Polymers

Condensation polymers are those produced by a reaction of the type

$$M_x + M_y = M_{x+y}$$

where M is monomer and x and y are integers from 0 to ∞. The growing polymer M_x can be terminated by adding a monomer ($y = 1$), a dimer ($y = 2$), or another growing polymer ($y = n$). That is, the product of condensation polymer is a mixture of various sizes:

$$\begin{array}{ll} M_{x1} & M_{x3} \\ M_{x2} & M_{x4} \\ M_{x1} + M_{x2} & M_{x3} + M_{x4} \\ M_{x1} + M_{x1} + M_{x2} & \cdots \end{array}$$

Consider one single-polymer molecule:

	M—M—M—M—M—···	—M
Monomeric unit	1 2 3 4 5	x
Polymer segment	1 2 3 4 5	$x - 1$

The number of monomeric units is always one unit larger than the number of segments. If x is large, the difference is negligible.

The probability that the functional group of the first unit has reacted is equal to p, for the second unit it is equal to p^2, for the third unit it is equal to p^3, and so on. For the $x - 1$ unit the probability is p^{x-1} that the molecule contains at least $x - 1$ reactive groups, or at least x units. The probability that the xth unit has not reacted is $1 - p$. That is, the probability for the last (xth) unit to be terminated is $1 - p$. Then the probability that among the x monomeric units, $x - 1$ units have reacted and one (the end unit) has not reacted is

$$n_x = p^{(x-1)}(1 - p) \tag{3.4}$$

where n_x is called the number average distribution or most probable distribution of polymer molecules. It is very close to the binomial distribution, but it is not the binomial distribution because there is no combination term $\binom{N}{x}$. The term n_x may also be considered the probability that the molecule consists of exactly x units.

If N_0 is the total number of monomers (for example, OH–R–COOH units) and N is the total number of macromolecules, then the total number of x-mers is

$$N_x = Np^{(x-1)}(1 - p) = Nn_x$$

and the total number of macromolecules is

$$N = N_0(1 - p)$$

Hence

$$N_x = N_0(1-p)^2 p^{(x-1)}$$

If w is the mass of a monomer unit, then the total mass of N_0 monomers is $N_0 w$. The mass of one x-mer is xw; the mass of N_x x-mers is $N_x xw$. We now have the ratio

$$w_x = \frac{N_x xw}{N_0 w} = x(1-p)^2 p^{(x-1)} \tag{3.5}$$

where w_x is the weight fraction of x-mer. Equation (3.4) describes the number average distribution n_x, and Eq. (3.5) describes the weight average distribution function w_x.

3.2.2 Addition Polymers

Addition polymers are obtained from a reaction process of the type

$$\mathrm{M}_1 \xrightarrow{\mathrm{M}} \mathrm{M}_2 \xrightarrow{\mathrm{M}} M_3 \xrightarrow{M} \cdots \xrightarrow{\mathrm{M}} \mathrm{M}_x$$

The monomers M are added to the growing chain one at a time. According to Poisson distribution, the probability of finding x events within a fixed interval of specified length is statistically independent of the number of other events. This clearly fits the description of addition polymerization. Hence, we have

$$n_x = \frac{\mathrm{e}^{-m} m^{x-1}}{(x-1)!} \tag{3.6}$$

$$w_x = \frac{[m/(m+1)]x\mathrm{e}^{-m} m^{x-2}}{(x-1)!} \tag{3.7}$$

The number average degree of polymerization $\bar{x}_n$ can be derived:

$$\bar{x}_n = \sum_{x=1}^{\infty} x n_x = \sum_{x=1}^{\infty} \frac{x\mathrm{e}^{-m} m^{x-1}}{(x-1)!}$$

$$= \mathrm{e}^{-m} \sum_{x=1}^{\infty} \frac{x m^{x-1}}{(x-1)!}$$

Note:

$$\sum_{x=1}^{\infty} \frac{x A^{x-1}}{(x-1)!} = (1+A)\exp A$$

Hence

$$\bar{x}_n = e^{-m}(1 + m)e^m = m + 1$$

Similarly, we can derive the weight average degree of polymerization:

$$\bar{x}_w = \sum_{x=1}^{\infty} x w_x = \sum_{x=1}^{\infty} \frac{x[m/(m+1)]xe^{-m}m^{x-2}}{(x-1)!}$$

$$= \frac{m}{m+1} e^{-m} \sum_{x=1}^{\infty} \frac{x^2 m^{x-2}}{(x-1)!}$$

$$= \frac{1}{m+1} e^{-m} \sum_{x=1}^{\infty} \frac{x^2 m^{x-1}}{(x-1)!}$$

Note:

$$\sum_{x=1}^{\infty} \frac{x^2 A^{x-1}}{(x-1)!} = (1 + 3A + A^2)\exp A$$

$$\bar{x}_w = \frac{1}{m+1} e^{-m}(1 + 3m + m^2)e^m$$

$$= \frac{1}{m+1}(1 + 3m + m^2)$$

The ratio x_w/x_n for the Poisson molecular weight distribution is

$$\frac{\bar{x}_w}{\bar{x}_n} = \frac{[1/(m+1)](1 + 3m + m^2)}{m+1} = \frac{1 + 3m + m^2}{1 + 2m + m^2} = 1 + \frac{m}{(m+1)^2}$$

The second term is always less than 0.5. This indicates a very narrow molecular weight distribution, much narrower than that of a condensation polymer.

3.3 TWO-PARAMETER EQUATIONS

Since there are numerous ways for synthesizing polymers, molecular weight distribution does not always follow a one-parameter equation. Even the condensation polymers do not necessarily follow the most probable distribution, nor do addition polymers follow the Poisson distribution. In many cases we naturally consider classical statistics, the normal distribution.

3.3.1 Normal Distribution

Normal (or Gaussian) distribution is given by

$$n_x = \frac{1}{\sqrt{2\pi}\sigma} e^{-(x-m)^2/2\sigma^2} \tag{3.8}$$

where the two parameters are m and σ. The standard deviation σ is the half-width of the normal curve:

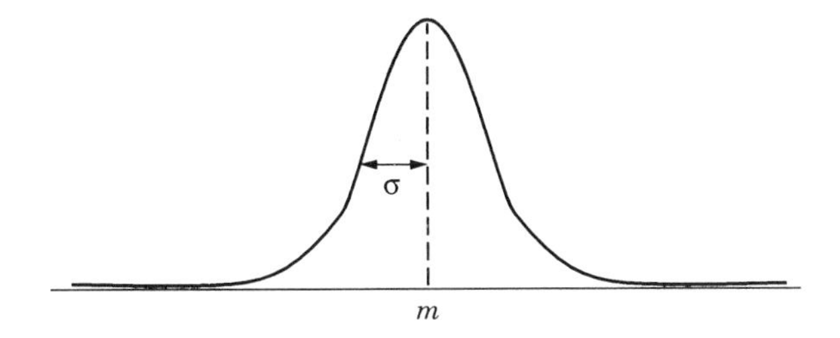

Here, the weight average distribution is given by

$$w_x = \frac{x}{m} n_x$$

The number average degree of polymerization is identical with the mean:

$$\bar{x}_{\mathrm{n}} = \int_{-\infty}^{\infty} x n_x \, \mathrm{d}x = m$$

and the weight average degree of polymerization is

$$\bar{x}_{\mathrm{w}} = \frac{\sigma^2}{m} + m$$

where

$$\sigma^2 = \int_{-\infty}^{\infty} (x - m)^2 n_x \, \mathrm{d}x$$

3.3.2 Logarithm Normal Distribution

The normal distribution function extends to both the positive and negative sides. To avoid the negative molecular weights (which do not exist), an assumption is made that the logarithm of molecular weight is normally distributed. Thus, we replace x by $\ln x$, and m by $\ln m$. Then the weight distribution becomes

$$w_{\ln x} = \frac{1}{\sqrt{2\pi\sigma}} \exp\left(\frac{-(\ln x - \ln m)^2}{2\sigma^2} \right) \tag{3.9}$$

and

$$\int_{1}^{\infty} w_{\ln x} \, \mathrm{d} \ln x = 1$$

Now, the average degrees of polymerization are

$$\bar{x}_n = m \exp\left(\frac{-\sigma^2}{2}\right)$$

$$\bar{x}_w = m \exp\left(\frac{\sigma^2}{2}\right)$$

from which we can show that

$$m = (\bar{x}_n \bar{x}_w)^{1/2}$$

$$\frac{\bar{x}_w}{\bar{x}_n} = \exp \sigma^2$$

where

$$\sigma^2 = \int_0^\infty (\ln x - \ln m)^2 w_x \, dx$$

and

$$w_x = \frac{1}{x\sqrt{2\pi\sigma}} \exp\left(\frac{-(\ln x - \ln m)^2}{2\sigma^2}\right)$$

$$\int_0^\infty w_x \, dx = 1$$

3.4 TYPES OF MOLECULAR WEIGHT

There are at least four types of molecular weight: number average, weight average, z average, and intrinsic viscosity. The number average molecular weight is given by

$$\bar{M}_n = M_0 \bar{x}_n$$

where M_0 is the molecular weight of the structural unit. For example,

$$\bar{x}_n = \sum x p^{x-1}(1-p) = \frac{1}{1-p}$$

$$\bar{M}_n = \frac{M_0}{1-p}$$

The weight average molecular weight is given by

$$\bar{M}_w = M_0 \bar{x}_w$$

For example,

$$\bar{x}_w = \sum x w_x = \sum x^2 p^{x-1}(1-p)^2 = \frac{1+p}{1-p}$$

$$\bar{M}_w = M_0\left(\frac{1+p}{1-p}\right)$$

The z average molecular weight is given by

$$M_z = \frac{\int_{x_0}^{x_n} M_{zx} z \, \mathrm{d}x}{\int_{x_0}^{x_n} z \, \mathrm{d}x}$$

or

$$M_z = \frac{\int_{x_0}^{x_n} M_{zx} z f(x)(\mathrm{d}x/x)}{\int_{x_0}^{x_n} z f(x)(\mathrm{d}x/x)}$$

The z value and the $z+1$ value of molecular weight are obtained by successively taking all the differences between the reading for a certain line and a reference line under the experimental curve. They are often related to the evaluation of an area that corresponds to the molecular weight. In the ultracentrifuge sedimentation experiment, the z value and the $z+1$ value of molecular weight are often the results of such evaluation.

The intrinsic viscosity $[\eta]$ (see Chapter 6) may be expressed as

$$[\eta] = K\bar{x}_v^a$$

where K and a are constants, and $\bar{x}_v$ is the viscosity degree of polymerization, which can be calculated as follows:

$$\bar{x}_v = \left[\frac{\sum_{x=1}^{\infty} x^{1+a} F(x)}{\sum_{x=1}^{\infty} x F(x)}\right]^{1/a}$$

$$\bar{x}_v = \left(\frac{\sum_{x=1}^{\infty} x^a w_x}{\sum_{x=1}^{\infty} w_x}\right)^{1/a}$$

where $F(x)$ is the fraction of molecules of size x. If $f(x)$ is not continuous, we have

$$\sum F(x_i) = 1$$

If $f(x)$ is continuous, we have

$$\int_0^{\infty} F(x) \, \mathrm{d}x = 1$$

Note that

$$M_v = M_0 \bar{x}_v$$

As a comparison of the types of molecular weight, we let N_i be the number of molecules of kind i (i being the degree of polymerization) present in mixture, M_i be their molecular weight, and c_i be the concentration or weight. Then,

$$\bar{M}_n = \frac{\sum_i N_i M_i}{\sum_i N_i} = \frac{\sum_i c_i}{\sum_i c_i / M_i}$$

$$\bar{M}_w = \frac{\sum_i N_i M_i^2}{\sum_i N_i M_i} = \frac{\sum c_i M_i}{\sum c_i}$$

$$\bar{M}_z = \frac{\sum_i N_i M_i^3}{\sum_i N_i M_i^2} = \frac{\sum c_i M_i^2}{\sum c_i M_i}$$

$$\bar{M}_{z+1} = \frac{\sum_i N_i M_i^4}{\sum_i N_i M_i^3} = \frac{\sum c_i M_i^3}{\sum c_i M_i^2}$$

$$\bar{M}_v = \left(\frac{\sum_i N_i M_i^{1+a}}{\sum_i N_i M_i} \right)^{1/a}$$

The average degree of polymerization $\bar{i}$ is related to the average molecular weight by

$$\bar{M} = M_0 \bar{i}$$

where M_0 is the molecular weight of the repeating unit. Therefore, we may compare various types of average degree of polymerization corresponding to average molecular weight as follows:

$$\bar{i}_n = \frac{\sum w_i}{\sum (w_i / i)}$$

$$\bar{i}_w = \frac{\sum w_i i}{\sum w_i}$$

$$i_{zx} \cong \bar{i}_z = \frac{\sum w_i i^2}{\sum w_i i}$$

$$\bar{i}_{(z+1)x} \cong \mathbf{i}_{z+1} = \frac{\sum w_i i^3}{\sum w_i i^2}$$

where w_i denotes the average mass in grams of molecules of size i. In general,

$$M_n < M_w < M_z < M_{z+1} < M_v$$

If there is no molecular weight distribution, namely, if the molecular weight of all species in a sample is the same, then

$$\bar{M}_n = \bar{M}_w = \bar{M}_z = \bar{M}_{z+1} = \bar{M}_v$$

The ratio $\bar{M}_w/\bar{M}_n$, called the polydispersity, is a useful measure of the spread of a polymer distribution. Larger values of $\bar{M}_w/\bar{M}_n$ indicate a very wide spread, with substantial amounts of materials at both extremes. When $M_w/M_n = 1$, all polymeric molecules have the same molecular weight and there is no spreading.

3.5 EXPERIMENTAL METHODS FOR DETERMINING MOLECULAR WEIGHT AND MOLECULAR WEIGHT DISTRIBUTION

Here we list some of the well-known experimental methods for determining molecular weight and molecular weight distribution. Details are discussed in later chapters.

1. Number average molecular weight and molecular weight distribution
   a. Osmotic pressure
   b. Intrinsic viscosity
2. Weight average molecular weight and molecular weight distribution
   a. Ultracentrifuge sedimentation
   b. Diffusion
   c. Light scattering
3. z and $z + 1$ average molecular weight and molecular weight distribution
   a. Ultracentrifuge sedimentation

Methods 1–3 are all primary methods.

4. Secondary methods
   a. Fractional precipitation
   b. Gel permeation chromatography
   c. HPLC
   d. Electrophoresis

The following are some useful sum terms:

$$\sum_{x=1}^{\infty} p^{x-1} = \frac{1}{1-p} \qquad p < 1$$

$$\sum_{x=1}^{\infty} x p^{x-1} = \frac{1}{(1-p)^2} \qquad p < 1$$

$$\sum_{x=1}^{\infty} x^2 p^{x-1} = \frac{1+p}{(1-p)^3} \qquad p < 1$$

$$\sum_{x=1}^{\infty} \frac{A^{x-1}}{(x-1)!} = \exp A$$

$$\sum_{x=1}^{\infty} \frac{xA^{x-1}}{(x-1)!} = (1+A)\exp A$$

$$\sum_{x=1}^{\infty} \frac{x^2A^{x-1}}{(x-1)!} = (1+3A+A^2)\exp A$$

REFERENCES

Feller, W., *An Introduction to Probability Theory and its Applications*, Vol. 1. New York: Wiley, 1950.

Flory, P. J., *J. Am. Chem. Soc.* **58**, 1877 (1936).

Flory, P. J., *J. Am. Chem. Soc.* **62**, 1561 (1940).

Flory, P. J., *Principles of Polymer Chemistry*. Ithaca, NY: Cornell University Press, 1953.

Mood, A. M., *Introduction to the Theory of Statistics*. New York: McGraw-Hill, 1950.

Peebles, L. H., Jr., *Molecular Weight Distribution in Polymers*. New York: Wiley-Interscience, 1971.

PROBLEMS

3.1 Show that (a) The equation $\bar{r}_n = \sum_1^{\infty} rp^{r-1}(1-p)$ leads to the equation $\bar{r}_n = 1/(1-p)$, and (b) The equation $\bar{r}_w = \sum_1^{\infty} r^2p^{r-1}(1-p)^2$ leads to $r_w = (1+p)/(1-p)$, where $\bar{r}_n$ is the number average degree of polymerization and $\bar{r}_w$ is the weight average degree of polymerization.

3.2 Show that $\bar{x} = m$ for binomial, Poission, and normal distribution.

3.3 The most probable distribution function of the molecular weight of condensation polymers is given by

$$w_r = rp^{r-1}(1-p)^2$$

(a) Plot w_r (in the range between 0 and 0.20) versus r (in the range between 0 and 50) for $p = .5$, .8, and .9, respectively.

(b) Plot w_r $(0 \leqslant w_r \leqslant .04)$ versus $r(0 \leqslant r \leqslant 250)$ for $p = .9$. (*Source*: Flory (1936).)

3.4 The Poisson distribution of the molecular weight of addition polymers (which do not have termination) is given by

$$w_r = \frac{\gamma}{\gamma+1} e^{-\gamma} \frac{r\gamma^{r-2}}{(r-1)}!$$

Plot w_r in percentage ($0 \leqslant w_r \leqslant 6$) versus r ($0 \leqslant r \leqslant 140$) for $\gamma = 50$, 100, and 500, respectively. (*Source*: Flory (1940).)

3.5 Show that the mole fraction distribution X_x is the same as number average distribution n_x, that is, $X_x = n_x$.

3.6 Show that the number average molecular weight M_n is given by

$$M_n = \frac{\sum M_i N_i}{\sum N_i} = M_0 \sum_1^{\infty} X n_x = \frac{M_0}{1-p}$$

and that the weight average molecular weight M_w is given by

$$M_w = \frac{\sum M_i^2 N_i}{\sum M_i N_i} = M_0 \sum X w_x = M_0 \frac{1+p}{1-p}$$

where M_0 is the molecular weight of a monomer unit.

3.7 Consider a solution containing equal numbers of molecules of molecular weights 50×10^3, 100×10^3, 200×10^3, and 400×10^3. Calculate $\bar{M}_n$, $\bar{M}_w$, and $\bar{M}_z$. Assume that the solution contains equal weight concentrations of the four species, and calculate $\bar{M}_n$, $\bar{M}_w$, and $\bar{M}_z$.

3.8 A protein sample consists of 90% by weight of 100 000 molecular weight material and 10% by weight of dimer of 200 000 molecular weight. Calculate $\bar{M}_w$ and $\bar{M}_n$.

4

MACROMOLECULAR THERMODYNAMICS

In describing the thermodynamic behavior of a system (here, macromolecules), the three most important state functions to be specified are ΔS, ΔH, and ΔG. When applied to solutions, the partial molar quantities of these three functions, particularly $\Delta\bar{G}$ (the chemical potential), must be added as the bases for the interpretation of polymer phenomena. In this chapter we begin with a brief review of the principles of thermodynamics in general. This is basically a review of thermodynamic terms. Since Flory, among others, laid the foundation of physical polymer chemistry, we describe in detail his lattice theory of ΔS of mixing, his concept of contact energy ΔH of mixing, and his utilization of ΔG and $\Delta\bar{G}$ in the treatment of dilute polymer solutions. His contribution to creating and explaining the two parameters χ_1, and θ has left a deep imprint in polymer language for many generations to come. We believe that it is beneficial for a reader or research worker to become familiar with Flory's work before embarking on any advanced subject. In conjunction with Flory's concept of contact energy, we also describe Hildebrand's concept of solubility parameter δ, not only for comparison, but also for application.

Chapters 4 and 5 are closely related in providing fundamental principles of modern polymer physical chemistry. Chapter 4 duly emphasizes Flory's work and Chapter 5 emphasizes de Gennes' work. de Gennes' work, in a sense, refutes some of Flory's theory and orients polymer physical chemistry in a new direction.

4.1 A REVIEW OF THERMODYNAMICS

Let us first define two important terms in thermodynamics: system and surroundings. Any material body (for example, one mole of gas, 500 mL of a protein solution) under study is called a system or a thermodynamic system. With respect to a given system, the rest of the world is called the surroundings. In a diagram, a system and its surroundings may be expressed as

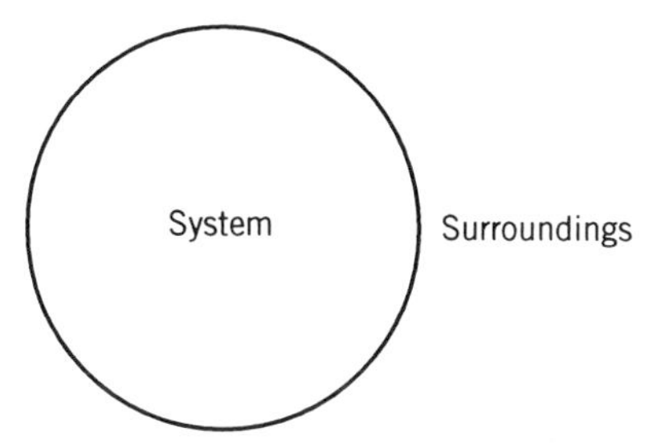

A system plus its surroundings is called the universe. Thermodynamics is the science of the change of energy in a system with regard to its surroundings.

We choose three parameters to characterize a system: P (pressure), V (volume), and T (temperature). For convenience, a fourth parameter is also chosen, C (the heat capacity), which is closely related to T. With P, V, and T as three independent variables and C as an auxiliary variable given, we can now define two thermodynamic quantities, w (work) and q (heat) in a differential equation form:

$$\mathrm{d}w = P\,\mathrm{d}V$$

$$\mathrm{d}q = C\,\mathrm{d}T$$

The letter w refers to work done on or by the system, and q refers to the heat (energy) entering or released from the system. Whether "done on or by" or "entering or released from," w or q is always related to the surroundings. The quantities w and q have positive or negative signs. The convention we follow is shown Figure 4.1. If heat enters the system from the surroundings, q is positive;

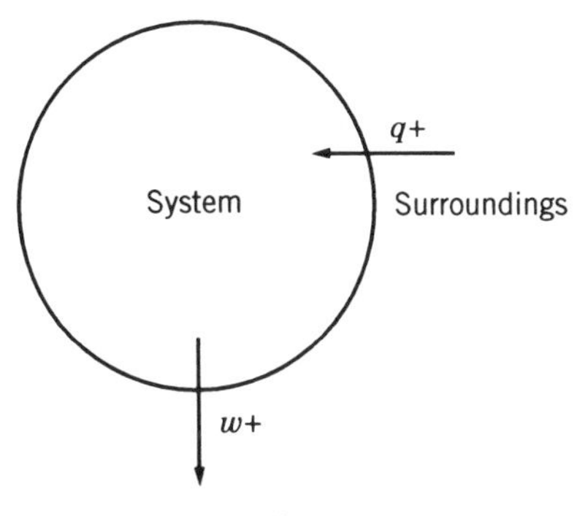

FIGURE 4.1 Convention of the sign of q and w.

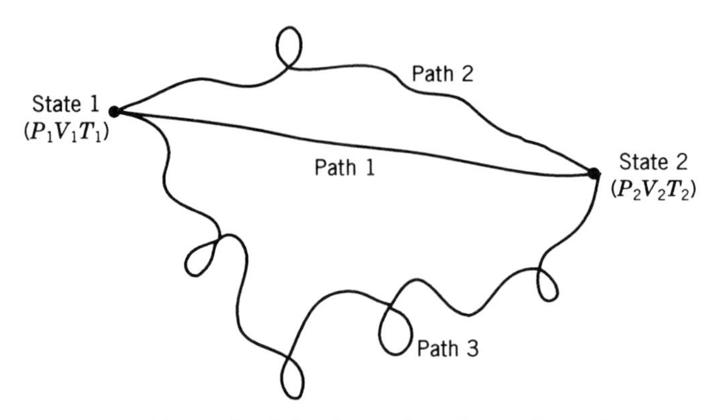

FIGURE 4.2 Path of the change in a thermodynamic state.

if heat is released from the surroundings, q is negative. Similarly, if work is done on the surroundings by the system, w is positive; if work is done on the system by the surroundings, w is negative.

We say that the state of a system is known if the three independent variables are specified. For example, one mole of gas is in state 1, when we know the exact values of P_1, V_1, and T_1. Similarly, the same one mole of gas is in state 2 if we know the exact values of P_2, V_2, and T_2. The change in a system occurs when the values change from P_1, V_1, and T_1 to P_2, V_2, and T_2. We say that the system changes from state 1 to state 2. In calculus, the term *path* is used to mean the route of change from state 1 to state 2. This term is important to our discussion because the two quantities $\mathrm{d}w$ and $\mathrm{d}q$ both depend on the path, as shown in Figure 4.2.

A system can follow many paths to change from state 1 to state 2. In Figure 4.2, path 1 is shorter than path 2, which is shorter than path 3. Consequently, for a system to arrive at state 2 from state 1, path 1 requires much less work ($\mathrm{d}w$) and energy ($\mathrm{d}q$) than path 2 or path 3, if other conditions are equal. In chemistry, examples of paths are constant volume, constant temperature, and constant pressure.

An interesting fact is that while $\mathrm{d}w$ and $\mathrm{d}q$ both depend on the path, the combination of the two quantities in the form of $\mathrm{d}q - \mathrm{d}w$ is independent of the path. The combination $\mathrm{d}q - \mathrm{d}w$ depends only on the states involved. If the change occurs from state 1 through a certain path back to state 1 as shown in Figure 4.3, then $\mathrm{d}q$ and $\mathrm{d}w$ are not zero, but $\mathrm{d}q - \mathrm{d}w$ is zero, because there is no change in state. Because of its importance, we introduce a special term, $\mathrm{d}E$, to represent $\mathrm{d}q - \mathrm{d}w$:

$$\mathrm{d}E = \mathrm{d}q - \mathrm{d}w \tag{4.1}$$

This is the first law of thermodynamics. The term E represents internal energy.

All processes by which changes occur in a system from one state to another

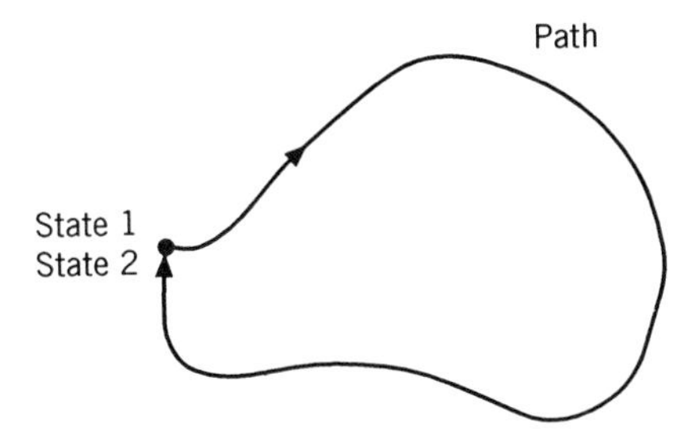

FIGURE 4.3 Path of a state from one position back to the original position.

may be classified into two types: reversible and irreversible. A revesible process from state 1 to state 2 consists of an infinite number of equilibrium states; an irreversible process does not. That is, a system can change back to any previous equilibrium state in the reversible process but not in the irreversible process if the change of energy is reversed in direction. Now, while dq_{rev} (the infinitesimal change in heat in a reversible process) depends on the path, the quantity dq_{rev}/T (that is, dq_{rev} multiplied by a factor $1/T$, which in mathematics is called the integrating factor) is independent of the path. The quantity dq_{rev}/T is the change in entropy dS:

$$dS = \frac{dq_{rev}}{T} \tag{4.2}$$

This is the second law of thermodynamics. The term S represents entropy. Note that dq is not necessarily equal to dq_{rev}. It may be equal to dq_{irrev}. The calculation of dq_{irrev} is more complicated than that of dq_{rev}.

These two laws form the basis of equilibrium thermodynamics. In this chapter, the term thermodynamics is understood to mean equilibrium thermodynamics; we do not consider nonequilibrium thermodynamics here. The quantities dE and dS are state functions, because their values depend only on the change in states; they are independent of path.

It is very difficult to measure E. For practical purposes three more measurable thermodynamic quantities are introduced, all of which are based on the two laws:

$$H = E + PV$$

$$G = H - TS$$

$$A = E - TS$$

where H is the enthalpy or heat content, G is the Gibb's free energy, and A is the Helmholtz free energy. When these three quantities are written in differential form we have to specify conditions for simplicity:

At constant pressure

$$dH = dE + P\,dV \tag{4.3}$$

At constant temperature

$$dG = dH - T\,dS \tag{4.4}$$

$$dA = dE - T\,dS \tag{4.5}$$

All five quantities (dE, dS, dH, dG, and dA, or in integral forms ΔE, ΔS, ΔH, ΔG, and ΔA) are state functions; they do not depend on the path. Among the five, Eq. (4.4) is the most important in the study of macromolecules. It is usually put in integrated form:

$$\Delta G = \Delta H - T\Delta S \tag{4.6}$$

where ΔG is an indicator as to whether the change of state is favorable or not. If ΔG is negative, the change is favored and will occur spontaneously; if ΔG is positive, the change is not favored and large amounts of energy are needed to force the state to change; if $\Delta G = 0$, the system is at equilibrium. ΔH is the energy term. If ΔH is negative, energy will be released when a change occurs, for example, when a bond is broken or formed. If ΔH is positive, energy must be supplied from the surroundings for a change to occur. $T\Delta S$ is related to the molecular configuration. If $\Delta S = 0$, there is no change in configuration. If ΔS is positive, a spontaneous change in the configuration of the molecules occurs.

In the above description, the state is specified by P, V, and T because historically thermodynamics starts with the observation of change in the properties of gases. A gas is described or specified by P, V, and T only. If we deal with solutions (a system of two or more components), we have to introduce one more independent variable, n, the number of moles of a component, which we discuss in later sections.

We begin our study of macromolecular thermodynamics by discussing these three basic quantities: ΔS, ΔH, and ΔG. Our interpretation of the change in a polymeric system is based on the change in the values of ΔS, ΔH, and ΔG. A change in the properties of the polymer always occurs whenever there is a change in the surroundings (environment). In latter secions we discuss two more thermodynamic quantities: $\bar{v}$, partial specific volume, and μ, the chemical potential. Both are related to the behavior of polymer solutions.

4.2 ΔS OF MIXING: THE FLORY THEORY

The Flory model* for ΔS of mixing a polymer chain with solvents has been influential in polymer chemistry for several decades. The model assumes the

*Huggins (1942) proposed a similar model to calculate ΔS_m. The Flory model is, therefore, also named Flory-Huggins model. The same idea was also expressed by Miller (1943).

validity of the lattice theory to describe the change in molecular configuration of the polymer in the presence of a solvent, just as it describes the patterns of the crystal structure of molecules. The central point is the filling of lattice sites in a three-dimensional space by polymer segments and solvent molecules; that is, how many ways can we fill up the lattice sites?

Consider a binary solution consisting of two types of molecules, the polymer chain and the solvent, and neglect the interaction potentials existing between polymer segments. Here the meaning of segment is slightly different from that described in previous chapters. The emphasis here is on the space that a polymer segment occupies. Thus, a segment is not a structural or monomeric unit. A segment is defined as "that portion of a polymer molecule requiring the same space as a molecule of solvent" (Flory, 1953). The difference may be shown as follows:

Monomeric units	0——0——0 1 2 3
Polmyer segment in the general sense	0——0——0 (1: 0——0) (2: 0——0)
Polymer segment in lattice theory	0——0——0··· 1 2 3

Assume that the polymer segment in a chain and the solvent molecule are of equal volume. Let us fill lattice sites, one segment or one solvent molecule per site. While the solvent molecules can be placed in any unoccupied site available, the possibilities for placing the segment are limited, not only by the availability of sites, but by linkage to adjacent segments. Figure 4.4 shows the random filling of the sites.

Let y be the number of segments in a chain, that is, the ratio of molar volumes of solute to solvent, n_1 the number of solvent molecules, n_2 the number of solute chain molecules, and n_0 the total number of lattice sites. Then

$$n_0 = n_1 + yn_2$$

Suppose i polymer molecules (chains) have been inserted previously at random. Now consider the insertion of the $(i+1)$th polymer. There are n_0-yi vacant cells for the first segment of molecule $i+1$. Let z be the lattice coordination number or number of cells that are first neighbors to a given cell in a three-dimensional space. The range of z is from 6 to 12: $z = 6$ for a regular cubic lattice, $z = 12$ for a hexagonal lattice. Let f_i be the probability that a given site adjacent to the first segment is occupied (not available). Then $(1 - f_i)$ is the probability that a given site is available for a segment. The number of ways that the $(i+1)$th polymer molecule (chain) would fill the cell sites, v_{i+1}, may be expressed as

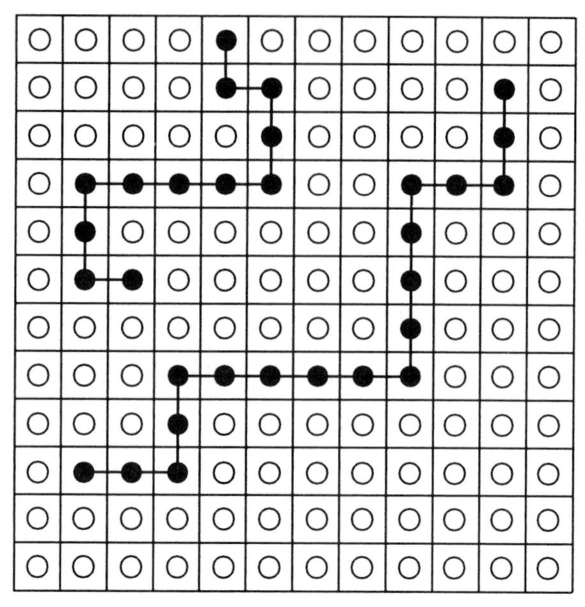

FIGURE 4.4 Random filling of the sites: ●, polymer segment; ○, solvent molecule.

follows:

$$\nu_{i+1} = \underbrace{(n_0 - yi)}_{\substack{\text{Vacant}\\ \text{for the first}\\ \text{segment}}}\underbrace{z(1-f_i)}_{\text{Second}}\underbrace{(z-1)(1-f_i)}_{\text{Third}}\underbrace{(z-1)(1-f_i)}_{\text{Fourth}}\cdots\underbrace{(z-1)(1-f_i)}_{(y-2)\text{th}}$$

$$= (n_0 - yi)z(z-1)^{y-2}(1-f_i)^{y-1}$$

where, for example, z represents 6 sites if the coordination number is 6.

The segments of the $(i+1)$th polymer molecule have been so far treated as if they were distinguishable. But they are indistinguishable, in that we cannot distinguish configurations that involve the interchange of polymer segment positions. For that reason, we have to introduce the term $1/n_2!$ The number of ways in which all the polymer molecules n_2 could be arranged in the lattice is

$$\Omega = \frac{1}{n_2!}\prod_{i=1}^{n_2} \nu_i = \frac{1}{n_2!}\prod_{i=0}^{n_2-1} \nu_{i+1}$$

Let $\bar{f}_i$ be the average probability of a site not available. We make a simple approximation,

$$1 - f_i \cong 1 - \bar{f}_i = \frac{\text{Number of sites available}}{\text{Total number of sites}} = \frac{n_0 - yi}{n_0}$$

Hence,

$$v_{i+1} = (n_0 - yi)z(z-1)^{y-2}\left(\frac{n_0 - yi}{n_0}\right)^{y-1}$$

Replacing the lone factor z by $z - 1$, we obtain

$$v_{i+1} = (n_0 - yi)^y\left(\frac{z-1}{n_0}\right)^{y-1}$$

With a second approximation,

$$(n_0 - yi)^y = (n_0 - yi)(n_0 - yi)\cdots(n_0 - yi) \cong \frac{(n_0 - yi)!}{[n_0 - y(i+1)]!}$$

we have

$$v_{i+1} = \frac{(n_0 - yi)!}{[n_0 - y(i+1)]!}\left(\frac{z-1}{n_0}\right)^{y-1}$$

And

$$\begin{aligned}
\Omega &= \frac{1}{n_2!}\prod_{i=0}^{n_2-1} v_{i+1} \\
&= \frac{1}{n_2!}\left\{\underset{i=0}{\frac{n_0!}{(n_0-y)!}}\;\underset{i=1}{\frac{(n_0-y)!}{(n_0-2y)!}}\;\underset{i=2}{\frac{(n_0-2y)!}{(n_0-3y)!}}\cdots\underset{i=n_2-1}{\frac{[n_0-(n_2-1)y]!}{(n_0-n_2y)!}}\right\}\left(\frac{z-1}{n_0}\right)^{n_2(y-1)} \\
&= \frac{1}{n_2!}\frac{n_0!}{(n_0-n_2y)!}\left(\frac{z-1}{n_0}\right)^{n_2(y-1)}
\end{aligned}$$

Using the Boltzmann-Planck law, the entropy of configuration of polymer molecules can be expressed as

$$S_{\text{conf}} = k \ln \Omega$$

Introducing the Stirling's formula

$$\ln n! = n \ln n - n$$

and replacing n_0 with $n_1 + yn_2$, we obtain

$$S_{\text{conf}} = \underbrace{-kn_1 \ln\left(\frac{n_1}{n_1 + yn_2}\right) + kn_2 \ln\left(\frac{yn_2}{n_1 + yn_2}\right)}_{\Delta S_{\text{mix}}}$$

$$\underbrace{+ kn_2 \ln y - kn_2(y-1)\ln\left(\frac{z-1}{e}\right) - kn_2 \ln y}_{\Delta S_{\text{diso}}}$$

where ΔS_{mix} is the entropy of mixing (of polymer and solvent) and ΔS_{diso} is the entropy of disorientation of polymer molecules. What is important to us is the expression of the entropy of mixing, $(\Delta S_{\text{mix}} \equiv \Delta S_{\text{m}})$:

$$\Delta S_{\text{m}} = -k(n_1 \ln \phi_1 + n_2 \ln \phi_2) \tag{4.7}$$

where ϕ_1 and ϕ_2 are volume fractions:

$$\phi_1 = \frac{n_1}{n_1 + yn_2}$$

$$\phi_2 = \frac{yn_2}{n_1 + yn_2}$$

Equation (4.7) gives the final result of the Flory model of ΔS of mixing. The importance of this model is to explain the very large deviation in the behavior of polymer solutions experimentally observed from that of an ideal solution. An ideal solution is defined as a solution that obeys the Rauolt's law:

$$p_1 = x_1 p_1^0$$

where p is the vapor pressure, x is the mole fraction, and the superscript (°) refers to the substance in its pure phase. Van't Hoff describes the ΔS of mixing for an ideal solution as

$$\Delta S_{\text{m}} = -k(n_1 \ln x_1 + n_2 \ln x_2)$$

When we compare Eq. (4.7) with Van't Hoff's equation, we notice that for polymer solutions, we cannot use mole fraction x as an independent variable to describe the ΔS of mixing. Instead, we have to use volume fraction ϕ as an independent variable. The mole fraction x_i does not give room for molecular interactions, whereas the volume fraction ϕ_i depends very much on molecular interactions.

4.3 ΔH OF MIXING

Again, in comparison with an ideal solution for which $\Delta H_{\mathrm{mix}} = 0$, polymer solutions cannot have $\Delta H_{\mathrm{mix}} = 0$. In polymer solutions there is an interaction not only between the macromolecule (chain) and the solvent, but also between the polymer segment themselves within the chain. The former is called intermolecular interaction and the latter is called intramolecular interaction.

All the interactions, whether intermolecular or intramolecular, are essentially electrostatic in origin. If the molecules carry charge (electrolytes), the interaction is Coulombic. Even if the molecule is neutral, electrons are in motion all the time around the centers of atoms and their bonds. Under certain environments the centers of positive and negative charges do not coincide, thus turning the molecule into a dipole:

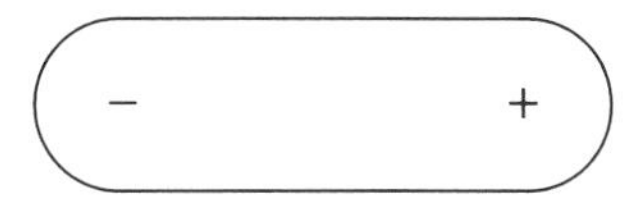

Hence, there are always dipole–dipole interactions between molecules. As a result, there is always a weak force, called van der Waals force, existing between molecules. Before we describe ΔH_{mix}, we first discuss van der Waals force as a background.

van der Waals force involves three factors: permanent dipole–permanent dipole interaction, induced dipole interaction, and dispersion effect. Permanent dipole–permanent dipole interaction exerts an orientation effect. It may be expressed as

$$E = -\frac{\mu_1 \mu_2}{r^3}[2\cos\theta_1 \cos\theta_2 - \sin\theta_1 \sin\theta_2 \cos(\phi_1 - \phi_2)]$$

where E is the interaction energy, μ_1 and μ_2 refer to the dipole moments of two permanent dipoles, and r, θ, and ϕ (here) are spherical polar coordinates. A permanent dipole in one molecule can induce a dipole in another molecule. The interaction energy E may be expressed in the form

$$E = -\frac{1}{r^6}(\alpha_1 \mu_2^2 + \alpha_2 \mu_1^2)$$

where α_1 and α_2 refer to the polarizability of molecules 1 and 2. The interaction energy of dispersion effect may be expressed as

$$E = -\frac{3\alpha_1 \alpha_2}{2r^6} \frac{h\nu_{0,1} \times h\nu_{0,2}}{h\nu_{0,1} + h\nu_{0,2}}$$

where ν_0 is the frequency of a molecule in its unperturbed state and h is Planck's constant.

We now introduce two approaches of ΔH of mixing for macromolecules: cohesive energy density and contact energy. Both are related to van der Waals force.

4.3.1 Cohesive Energy Density

The van der Waals equation of real gas is

$$\left(P + \frac{n^2 a}{V^2}\right)(V - nb) = nR'T$$

where R' is the gas constant, a is the van der Waals constant, n is the number of moles, b is the excluded volume, and a/V^2 represents the attractive force between the two molecules. In 1906 van Laar derived a parallel equation to describe the heat involved in the vapor pressure of binary liquid mixture:

$$\Delta H_{\text{m}} = \frac{n_1 V_1 n_2 V_2}{n_1 V_1 + n_2 V_2}\left(\frac{a_1^{1/2}}{V_1} - \frac{a_2^{1/2}}{V_2}\right)$$

where V is the volume and subscripts 1 and 2 refer to solvent and solute molecules, respectively.

Scatchard (1931) and Hildebrand (1933) independently suggested that this equation, with some modification, can equally be valid to describe the energy ΔE involved in the mixing of solute and solvent. They derived a similar equation:

$$\Delta E = (n_1 V_1 + n_2 V_2)\left[\left(\frac{\Delta E_1^{\text{v}}}{V_1}\right)^{1/2} - \left(\frac{\Delta E_2^{\text{v}}}{V_2}\right)^{1/2}\right]^2 \phi_1 \phi_2 \tag{4.8}$$

where ΔE^{v} is the energy of vaporization, and all the other symbols in the equation were defined previously. Hildebrand called the quantity $\Delta E^{\text{v}}/V$ (the energy of vaporization per mL) a measure of "internal pressure" and used the symbol $\delta = (\Delta E^{\text{v}}/V)^{1/2}$, which is now called the solubility parameter, a measurable quantity. Thus, we have

$$\left(\frac{\Delta E^{\text{v}}}{V}\right)^{1/2} = \delta = \frac{a^{1/2}}{V}$$

The term E_{m}/V (hence $\Delta E^{\text{v}}/V$) is also called the cohesive energy density (E being cohesive energy).

Hildebrand further derived an equation to relate the heat of mixing ΔH_{mix} with the solubility parameter δ, which is now called the Hildebrand equation:

$$\Delta H_{\text{m}} = V_{\text{m}}(\delta_1 - \delta_2)^2 \phi_1 \phi_2 \tag{4.9}$$

where V_m is the volume of solution. Both δ_1 and δ_2 are measurable quantities. They can be used to estimate qualitatively whether a compound can be dissolved in a solvent. As a rough rule of thumb, a polymer is miscible with the solvent if the following condition is satisfied:

$$(\delta_2 - 1.1) < \delta_1 < (\delta_2 + 1.1) \tag{4.10}$$

The solubility parameters of some solvents δ_1 are given in Table 4.1; the solubility parameters of some polymers are given in Table 4.2.

TABLE 4.1 Values of the Solubility Parameter δ_1 of Some Common Solvents

Solvent	δ_1(25°C)		V_1 (25°C) (mL)
N-Pentane	7.05		116
N-Hexane	7.3	(7.24)	132
N-Heptane	7.45		147
Ethyl ether	7.5		105
Cyclohexane	8.2		109
Carbon tetrachloride	8.65	(8.58)	97
Toluene	8.9		107
Benzene	9.15		89
Chloroform	9.3	(9.24)	81
Acetone	9.71		
Acitic acid	10.5		
Acetonitrile	11.9		
Ethanol	14.3		
Ethylene glycol	16.3		
Water	23.3		

Source: Hildebrand and Scott (1950).

TABLE 4.2 Values of the Solubility Parameter δ_2 of Some Common Polymers

Polymer	δ_2
Polyisobutylene	8.1
Polystyrene	9.2
Poly(vinyl chloride)	9.5
Polyethylene	7.9
Poly(methyl methacrylate)	9.1
Poly(ethylene terephthalate)	10.7
66 Nylon	13.6
Polyacrylonitrile	15.4

Source: Hildebrand and Scott (1950).

Several experimental methods can be used to determine δ_1 and δ_2. Since the values of δ_1 are tabulated in the literature for many common solvents and the values of δ_2 for a newly prepared polymer often need to be determined, we confine our discussion to two methods for determining δ_2. One measures the density ρ and the other measures the refractive index n. We can determine δ_2 for a polymer of known structure by measuring the density ρ, using the equation

$$\delta_2 = \rho \frac{\sum k'}{M} \tag{4.11}$$

where M is the molecular weight of the polymer and k' is the molar attraction constant of the structural configuration of the repeating unit in the polymer chain. The typical values of the molar attraction constants k' are given in Table 4.3. We can determine δ_2 by measuring the refractive index n of the polymer solution using the equation

$$\delta_2 = 30.3 \frac{n^2 - 1}{n^2 + 2}$$

For solubility of solids in liquid, Hildebrand and Scatchard independently derived another equation, now called the Hildebrand-Scatchard equation:

$$\log \frac{1}{x_2} = \frac{\Delta H_m^F}{4.575}\left(\frac{T_m - T}{T_m T}\right) - \frac{\Delta C_p}{4.575}\frac{T_m - T}{T} + \frac{\Delta C_p}{1.987}\log\frac{T_m}{T} + \frac{V_2}{4.575T}(\delta_1 - \delta_2)^2 \phi_1^2 \tag{4.12}$$

where x_2 is the mole fraction solubility, ΔH_m^F is the heat of fusion at the melting point T_m in Kelvins, ΔC_p is the difference between the heat capacities of solid and liquid, T is the temperature of solution in Kelvins, and V_2 is the molar

TABLE 4.3 Molar Attraction Constants k' $(cal/cm^3)^{1/2}/mol$

Group	k'
$-CH_3$	148
$-CH_2-$	131.5
$>CH-$	86
$>C<$	32
$CH_2=$	126.5
$-CH_2=$	121.5
$>C=$	84.5
$>C=O$	263

Source: Hoy (1976) and Billmeyer (1984).

volume of the solute as a hypothetical supercooled liquid. This equation is often used in pharmacology among other fields.

4.3.2 Contact Energy (First-Neighbor Interaction or Energy Due to Contact)

Flory advanced the idea of contact energy as the cause of the heat of mixing for polymer solutions. Consider a structural unit in a chain molecule and label it with a number 2:

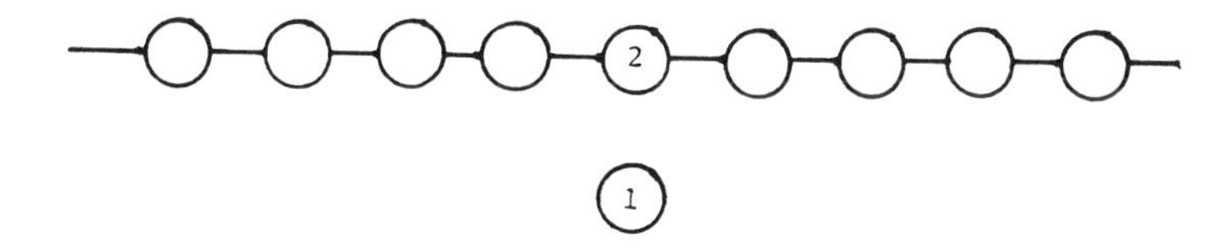

Consider also another chain with the same labeling:

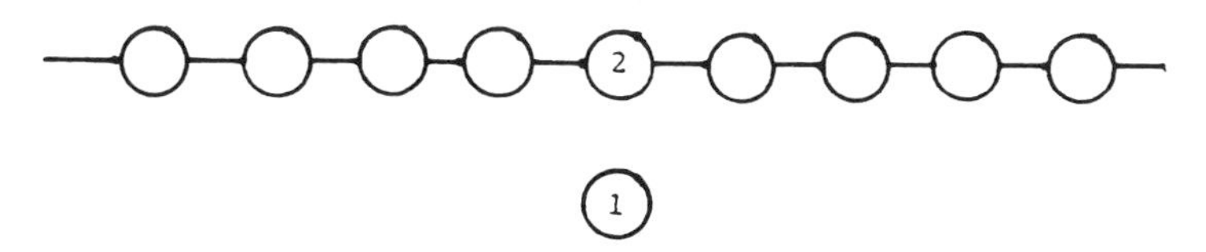

Let number 1 refer to the solvent molecule, and number 2 refer to any structural unit with 3 as its next neighbor, 4 next to 3, and so forth. We focus on structural unit 2. Within the same chain, 2 and 3 are the next neighbors, but 2 and 4 and 2 and 5 are not. Structural unit 2 of one chain can have next-neighbor contact with structural unit 2 of other chains, as well as with solvent molecule n_1:

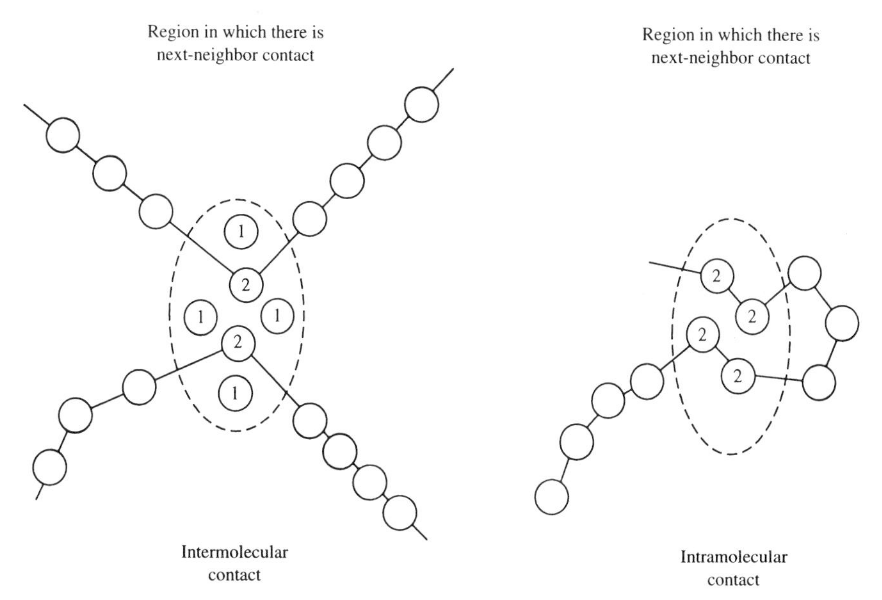

First-neighbor contact is called short-range interaction and non-first-neighbor contact is called long-range interaction. For short-range interaction there are three types of first-neighbor contact, 1–1, 1–2, 2–2, and three corresponding kinds of energy: w_{11}, w_{12}, and w_{22}. The energy associated with mixing solute (polymer segments) and solvent is Δw_{12}, defined as

$$\Delta w_{12} = w_{12} - \tfrac{1}{2}(w_{11} + w_{22})$$

The total 1–2 pairs contact is given by

$$zyn_1\phi_2 \equiv zn_1\phi_1$$

where n_1 is the number of the solvent molecules, z is the coordination number, y is the number of segments per polymer chain ($y = V_2/V_1$), and V_2 and V_1 are the molar volumes of polymer and solvent, respectively. The product of the total number of contact pairs and Δw_{12} is the result that we are looking for, namely, the heat of mixing ΔH_{m}. The equation is

$$\Delta H_{\text{m}} = z\Delta w_{12} y_1 n_1 \phi_2$$

where y_1 is the number of segments that have contacted with solvent molecules. This is van Laar expression.*

We may group the terms into a new form:

$$\Delta H_{\text{m}} = kT\chi_1 n_1 \phi_2 \tag{4.13}$$

where

$$\chi_1 = \frac{z\Delta w_{12} y_1}{kT}$$

It should be pointed out that the equation of ΔH_{m} in terms of Δw_{12} is not important, because it is not feasible to calculate Δw_{12} as such. Equation (4.13) in terms of χ_1 is important in polymer chemistry: ΔH_{m} here is called the contact energy. χ_1 is a dimensionless quantity, which characterizes the interaction energy per solvent molecule divided by kT (k being the Boltzmann constant). ΔH_{m} is also called the Flory interaction energy. The physical meaning of χ_1 is

$$kT\chi_1 = (\text{Energy of the solvent in solution}) - (\text{Energy of the solvent in its pure state})$$

In the literature of polymer chemistry, the two parameters δ_2 (the solublity parameter of polymer) and χ_1 (the interaction energy of the solvent) are

*It is interesting to note that both Hildebrand and Flory's theories were originated from van Laar's work.

frequently investigated. Both are measurable quantities. The quantity ΔH_m, on the other hand, is only of secondary importance.

4.4 ΔG OF MIXING

Once ΔS_m and ΔH_m are known, ΔG_m can easily be calculated. Thus, according to Hildebrand,

$$\Delta S_m = -R'(x_1 \ln x_1 + x_2 \ln x_2)$$
$$\Delta H_m = V_m(\delta_1 - \delta_2)^2 \phi_1 \phi_2$$
$$\Delta G_m = R'T(x_1 \ln x_1 + x_2 \ln x_2) + V_m(\delta_1 - \delta_2)^2 \phi_1 \phi_2$$

According to Flory,

$$\Delta S_m = -k(n_1 \ln \phi_1 + n_2 \ln \phi_2)$$
$$\Delta H_m = kT\chi_1 n_1 \phi_2$$
$$\Delta G_m = kT[\underbrace{(n_1 \ln \phi_1 + n_2 \ln \phi_2)}_{\text{Combinatory term}} + \underbrace{\chi_1 n_1 \phi_2}_{\text{Contact term}}] \tag{4.14}$$

Equation (4.14) is further explored in Section 4.6.

4.5 PARTIAL MOLAR QUANTITIES

When we deal with solutions, we need another independent variable to specify the composition of the system, in addition to P, V, and T. This additional independent variable is n, the number of moles of a component in the system. A thermodynamic property in which n is chosen as an independent variable is called the partial molar quantity. Consider an extensive property g:

$$g = g(T, P, n_1 n_2, \ldots)$$

The partial molar quantity of the ith component is defined as

$$g_i = \left(\frac{\partial g}{\partial n_i}\right)_{T,P,n_i \neq n_j}$$

Applying Euler's theorem, we have

$$g = n_1 g_1 + n_2 g_2 + \cdots = \sum n_i g_i$$

In macromolecular chemistry, two partial molar quantities are especially important: partial molar volume (particularly partial specific volume) and partial molar free energy, also called chemical potential. The partial specific volume $\bar{v}_i$ is defined as

$$\bar{v}_i = \left(\frac{\partial v}{\partial w_i} \right)_{T,P,w_i \neq w_j}$$

(where w_i is the weight in grams). The partial molar Gibbs free energy, also called the chemical potential μ, is defined as

$$\bar{G}_i = \mu_i = \left(\frac{\partial G}{\partial n_i} \right)_{T,P,n_i \neq n_j}$$

Partial molar volume, $\bar{V}_i = (\partial V / \partial n_i)_{T,P,n_i \neq n_j}$ and partial specific volume, $\bar{v}_i$, are closely related; one is expressed in moles and the other is expressed in grams.

4.5.1 Partial Specific Volume

An exact description of partial specific volume in terms of molecular structure is not known. The parameter $\bar{v}_i$ alone does not seem to be clear enough to characterize the behavior of the component i in the solution in relation to its size, shape, and chemical reactivity. However, it is an indispensable quantity which permits interpretation of other experimental parameters, such as light scattering, diffusion, and sedimentation, as we will see later.

The experimental methods for measuring partial specific volume include density gradient columns, pyconometry, magnetic floatation method, and the vibration method. Regardless of which method we use, the measurement is often tedious and difficult. In protein chemistry, instead of measurement, a reasonable

TABLE 4.4 Partial Specific Volume of Amino Acid Residue, $\bar{v}_i$

Amino Acid	$\bar{v}_i$ (mL/g)	Amino Acid	$\bar{v}_i$ (mL/g)
Glycine	0.64	Cysteine	0.61
Alanine	0.74	Methionine	0.75
Valine	0.86	Aspartic acid	0.60
Leucine	0.90	Glutamic acid	0.66
Isoleucine	0.90	Lysine	0.82
Serine	0.63	Arginine	0.70
Threonine	0.70	Histidine	0.67
Phenylalanine	0.77	Proline	0.76
Tyrosine	0.71	Hydroxyproline	0.68
Tryptophan	0.74	Glutamine	0.67

estimation of $\bar{v}$ is sometimes obtained by using the equation

$$\bar{v}_{\mathrm{p}} = \frac{\sum \bar{v}_i w_i}{\sum w_i} \tag{4.15}$$

where $\bar{v}_{\mathrm{p}}$ is the partial specific volume of protein, $\bar{v}_i$ is the partial specific volume of amino acid i, and w_i is the weight percent of ith amino acid residue in the protein. The values of $\bar{v}_i$ are presented in Table 4.4.

4.5.2 The Chemical Potential

The chemical potential μ is related to a measurable quantity, the activity a, which is also called the effective concentration:

$$\begin{aligned} \mu_i &= \mu_i^\circ + R'T \ln a_i \\ &= \mu_i^\circ + R'T \ln \gamma_i x_i \end{aligned}$$

where i refers to the ith component, μ_i° is the chemical potential in the reference state ($a_i = 1$) and is a function of temperature and pressure only, γ_i is the activity coefficient, and x_i is the mole fraction. For a very dilute solution, $\gamma_i \to 1$, we have

$$\mu_i = \mu_i^\circ + R'T \ln x_i$$

Consider a two-component system in which $x_1 \gg x_2$. We can write

$$\ln x_1 = \ln(1 - x_2) = -x_2 - \tfrac{1}{2}x_2^2 - \cdots$$

Note:

$$\ln(1 - x) = -x - \frac{x^2}{2} - \frac{x^3}{3} - \frac{x^4}{4} - \cdots$$

Substituting the expanded $\ln x_1$ term into the equation of chemical potential, we have

$$\mu_1 - \mu_1^\circ = -R'T(x_2 + \tfrac{1}{2}x_2^2 + \cdots)$$

Now the mole fraction of component 2 can be converted to the concentration c_2 (in g/mL), by the following approximation:

$$x_2 = \frac{n_2}{n_1 + n_2} = \frac{(g/M_2)/\mathrm{mL}}{(n_1 + n_2)/\mathrm{mL}}$$

where M_2 is the molecular weight of the solute (that is, polymer). Since

$n_1 + n_2 \simeq n_1$, we have

$$x_2 = \frac{g/\text{mL}}{M_2}\frac{\text{mL}}{n_1} = \frac{c_2}{M_2} V_1^\circ$$

where V_1° is the molar volume in milliliters of solvent.

The equation of $\mu_1 - \mu_1^\circ$ then becomes

$$\mu_1 - \mu_1^\circ = -R'TV_1^\circ\left[\left(\frac{1}{M_2}\right)c_2 + \left(\frac{V_1^\circ}{2M_2^2}\right)c_2^2 + \cdots\right]$$

or

$$\mu_1 - \mu_1^\circ = -R'TV_1^\circ\left(\frac{1}{M_2} + A_2c_2 + A_3c_2^2 + \cdots\right) \tag{4.16}$$

Comparing Eq. (4.16) with the virial equation for one mole of gas,

$$PV = R'T(1 + A_2P + A_3P^2 + \cdots)$$

we realize that A_2 is the second virial coefficient, A_3 is the third virial coefficient, and so forth. Equation (4.16) will be used in the interpretation of osmotic pressure (Chapter 7) and light scattering (Chapter 8).

4.6 THERMODYNAMICS OF DILUTE POLYMER SOLUTIONS

We now differentiate Flory's equation of ΔG_m

$$\Delta G_m = kT(n_1 \ln \phi_1 + n_2 \ln \phi_2 + \chi_1 n_1 \phi_2)$$

with respect to n_1 of solvent molecules, keeping in mind that ϕ_1 and ϕ_2 are both of functions of n_1,

$$\phi_1 = \frac{n_1}{n_1 + n_2 y}$$

$$\phi_2 = \frac{n_2 y}{n_1 + n_2 y}$$

and multiply by Avogadro's number N_A to obtain the chemical potential per mole. We find

$$\mu_1 - \mu_1^\circ = R'T\left[\ln(1 - \phi_2) + \left(1 - \frac{1}{y}\right)\phi_2\right] + R'T\chi_1\phi_2^2 \tag{4.17}$$

Expanding the term $\ln(1-\phi_2)$ in series leads to

$$\mu_1 - \mu_1^\circ = -R'T\left[(\tfrac{1}{2}-\chi_1)\phi_2^2 + \frac{\phi_2^3}{3} + \cdots\right] \tag{4.18}$$

But we know that

$$\mu_1 - \mu_1^\circ = \Delta\bar{H}_1 - T\Delta\bar{S}_1 \tag{4.19}$$

(from $\Delta\bar{G}_1 = \Delta\bar{H}_1 - T\Delta\bar{S}_1$), where $\Delta\bar{H}_1$ is the partial molar enthalpy and $\Delta\bar{S}_1$ is the partial molar entropy. For that reason, Flory defined these terms as

$$\Delta\bar{H}_1 = R'T\kappa_1\phi_2^2$$
$$\Delta\bar{S}_1 = R'\psi_1\phi_2^2$$

where κ_1 is the heat of dilution parameter and ψ_1 is the entropy of dilution parameter. Then Eq. (4.19) becomes

$$\begin{aligned}\mu_1 - \mu_1^\circ &= R'T\kappa_1\phi_2^2 - R'T\psi_1\phi_2^2 \\ &= -R'T(\psi_1 - \kappa_1)\phi_2^2\end{aligned} \tag{4.20}$$

Comparison of Eq. (4.18) with Eq. (4.20) leads to

$$\kappa_1 - \psi_1 = \chi_1 - \tfrac{1}{2} \tag{4.21}$$

This is equivalent to assuming that χ_1 consists of two parts, the entropy and the enthalpy. Flory further defined the ideal temperature θ by

$$\theta = \frac{\kappa_1 T}{\psi_1} \tag{4.22}$$

We then have

$$\psi_1 - \kappa_1 = \psi_1\left(1 - \frac{\theta}{T}\right) \tag{4.23}$$

Equations (4.21), (4.22), and (4.23) have formd an important background in polymer chemistry for several decades. We shall further discuss the physical meanings of χ_1 and θ in later chapters. Here we briefly describe how they are related to the quality of solvent in polymer solutions. The value of χ_1 is useful for indicating whether a solvent is good or poor for a particular polymer: A good solvent has a low value of χ_1, while a poor solvent has a high value of χ_1. The borderline is $\chi_1 = \frac{1}{2}$.

In good solvents molecular interaction between polymer segment and solvent molecule is favored. Solvent molecules can pass through the holes or cavities formed by a polymer chain, as shown in the following diagram (○ is a solvent molecule):

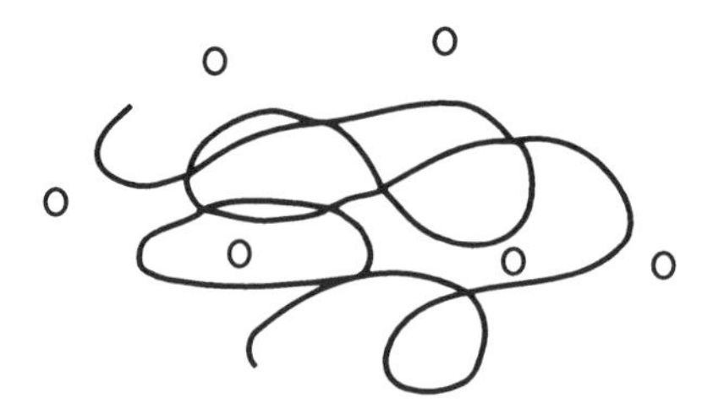

At the same time two polymer segments compete for space occupation. No segment allows the other segment of the same chain to be too close, because of a repulsive force of polymer segments against each other. This repulsive force results in the excluded volume u, which is defined as

$$u = \int_0^\infty \frac{U(r)}{kT} r^2 \, dr$$

where $U(r)$ is the potential existing between the two segments, r is the distance, and k is the Boltzmann constant. The overall polymer molecule is in an extended conformation. Its end-to-end distance (its size), which we describe in Chapter 5, is large; that is, the polymer chain occupies a large space in the solution.

In poor solvents, molecular interaction between the polymer segment and solvent molecule is not favored. The polymer segment does not attract the solvent molecule to pass through the holes or cavities created by the polymer chain. To resist the approaching solvent molecules, the polymer segments reduce their resistance against each other and an attractive force develops between them. Consequently, the excluded volume gets smaller. The overall polymer shrinks and the end-to-end distance becomes smaller. If a particular temperature θ is set (Eq. (4.22)), then the polymer molecule further shrinks to a state in which the attractive force cancels the repulsive force. The polymer molecule now exists in a compact form:

Such a state is called the θ state (theta state). At the θ state, the excluded volume effect vanishes and the dimensions of the polymer molecule are called unperturbed dimensions. Thus, the θ state is also called the unperturbed state or the (pseudo) ideal state.

It should be pointed out that for experimental reasons we group the following three terms together: good solvent, poor solvent, and θ temperature. We could have θ temperature for good solvents as well as for poor solvents. But the θ temperature for good solvents is often in a range where the experiment is difficult or inconvenient to carry out, for example, 100 K. On the other hand, the θ temperature for poor solvents is often in a range that is convenient for experiments, for example, 20–80°C. For this reason the θ temperature is often associated with poor solvents and poor solvents are often called theta solvents. As mentioned before, it is only with respect to a given polymer that the solvent is good or poor. For example, cyclohexane is poor solvent for polystyrene, but a good solvent for poly(methyl methacrylate). Table 4.5 lists good solvents, poor solvents, and θ temperature for a few common polymers.

The experimental methods used for the study of thermodynamic parameters such as χ_1 and θ of a polymer in dilute solutions are numerous. Among them are intrinsic viscosity, light scattering, diffusion, sedimentation, vapor pressure, and phase equilibrium. Here we discuss vapor pressure and phase equilibrium, leaving the other methods for later chapters.

TABLE 4.5 Solvents and θ Temperature for Some Common Polymers

Polymer	Good Solvent	Poor Solvent	θ Temperature (°C) (related to poor solvent)
Polyisobutylene	Cyclohexane	Anisole	105
		Benzene	24
Polypropylene	Benzene	1-Chloronaphthalene	74
	Cyclohexane		
Polystyrene	Benzene	Cyclohexane	34
	Toluene	Toluene/methanol	
		76.9/23.1	25
		75.2/24.8	34
		72.8/27.2	45
		trans-Decalin	20.5
Poly(vinyl chloride)	Cyclohexane	Benzyl alcohol	155.4
Poly(vinyl acetate)	Dioxane	Ethanol	56.9
Poly(methyl methacrylate)	Benzene	*n*-Butyl chloride	35.4
	Cyclohexane	3-Heptanone	33.7
		3-Octanone	71
		n-Propanol	84.4

4.6.1 Vapor Pressure

The classical way to describe thermodynamic properties of a solution, such as vapor pressure and osmotic pressure, is to describe the behavior of solvent activity a_1 over the whole concentration range. By definition

$$\ln a_1 = \frac{\mu_1 - \mu_1^\circ}{R'T}$$

which, using Flory's equation, can be expressed as

$$\ln a_1 = \ln(1 - \phi_2) + \left(1 - \frac{1}{y}\right)\phi_2 + \chi_1 \phi_2^2 \tag{4.24}$$

The parameter $\ln a_1$ can be determined by measuring the vapor pressure of the solvent in the polymer solution, p_1, and in its pure phase, p_1°:

$$\ln a_1 = \frac{p_1}{p_1^\circ}$$

Since ϕ_1 and ϕ_2 are known from the preparation of the solution and y can be calculated from V_1 and V_2, χ_1 can be calculated from Eq. (4.24) once the value of a_1 is determined by vapor pressure measurement. The two other thermodynamic parameters, $\Delta \bar{H}_1$ and $\Delta \bar{S}_1$, can be calculated using the following equations:

$$\Delta \bar{H}_1 = R'T\chi_1\phi_2^2 \tag{4.25}$$

$$\Delta \bar{S}_1 = -R'\left[\ln(1 - \phi_2) + \left(1 - \frac{1}{y}\right)\phi_2\right] \tag{4.26}$$

The quantities $\Delta \bar{H}_1$ and $\Delta \bar{S}_1$ can also be calculated from the temperature coefficient of the activity a_1:

$$\Delta \bar{H}_1 = -R'T\left(\frac{\partial \ln a_1}{\partial T}\right)_{P,\phi_2} \tag{4.27}$$

$$\Delta \bar{S}_1 = -R'\left(\frac{\partial (T \ln a_1)}{\partial T}\right)_{P,\phi_2} \tag{4.28}$$

Thus, we can check whether χ_1 gives a reasonable value to characterize the interaction between the solvent and solute in dilute polymer solutions.

4.6.2 Phase Equilibrium

Equation (4.18) can be put in a slightly different form:

$$\frac{\mu_1 - \mu_1^\circ}{R'T} = -\phi_2 - \frac{\phi_2^2}{2} - \frac{\phi_2^3}{3} + \cdots + \phi_2 - \frac{1}{y}\phi_2 + \chi_1\phi_2^2 \tag{4.29}$$

If we plot $-(\mu_1 - \mu_1^\circ)/(R'T)$ versus ϕ_2, we obtain a curve like the one shown in Figure 4.5. The curve is a phase separation curve, in which a maximum, a minimum, or an inflection will be shown. The conditions for incipient phase separation are

$$\left(-\frac{\partial \mu_1}{\partial \phi_2}\right)_{T,P} = 0$$

$$\left(\frac{\partial^2 \mu_1}{\partial \phi_2^2}\right)_{T,P} = 0$$

Differentiating Eq. (4.29) to obtain $(\partial \mu_1/\partial \phi_2)_{T,P}$ and equating to zero, we obtain

$$\frac{1}{1-\phi_2} - \frac{1}{1-y} - 2\chi_1\phi_2 = 0 \tag{4.30}$$

Further differentiating to obtain $(\partial^2 \mu_1/\partial \phi_2^2)_{T,P}$ and equating to zero will lead to

$$\frac{1}{(1-\phi_2)^2} - 2\chi_1 = 0 \tag{4.31}$$

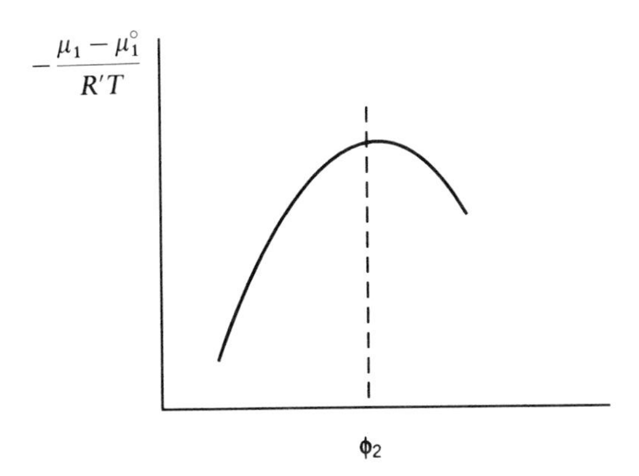

FIGURE 4.5 Phase separation curve.

Eliminating χ_1 from Eqs. (4.30) and (4.31), we have

$$\phi_2 = \frac{1}{1 + y^{1/2}}$$

For large y, the equation of ϕ_2 can further be reduced to

$$\phi_2 = \frac{1}{y^{1/2}} \tag{4.32}$$

We now substitute Eq. (4.32) into Eq. (4.31):

$$\chi_1 = \frac{(1 + y^{1/2})^2}{2y} = \frac{1}{2y} + y^{-1/2} + \tfrac{1}{2}$$

or

$$\chi_1 - \tfrac{1}{2} = \frac{1}{2y} + y^{-1/2}$$

This leads to

$$-\psi_1\left(1 - \frac{\theta}{T}\right) = \frac{1}{2y} + \frac{1}{y^{1/2}}$$

At the temperature where phase separation occurs, we have $T = T_c$, where T_c

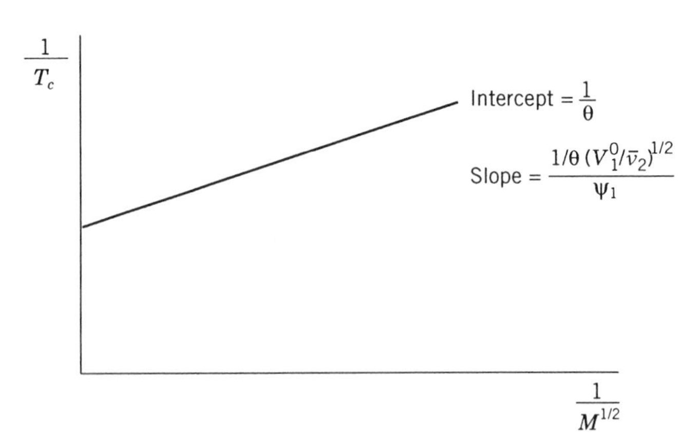

FIGURE 4.6 Phase equilibrium plot for determining θ and ψ_1.

is the critical temperature. We thus have

$$\frac{1}{T_c} = \frac{1}{\theta}\left[1 + \frac{1}{\psi_1}\left(\frac{1}{y^{1/2}} + \frac{1}{2y}\right)\right] \tag{4.33}$$

For large $y(1/2y \simeq 0)$, Eq. (4.33) becomes

$$\begin{aligned} \frac{1}{T_c} &= \frac{1}{\theta}\left(1 + \frac{1}{\psi_1 y^{1/2}}\right) \\ &= \frac{1}{\theta}\left(1 + \frac{b}{M^{1/2}}\right) \end{aligned} \tag{4.34}$$

where

$$\begin{aligned} b &= \frac{(V_1^\circ/\bar{v}_2)^{1/2}}{\psi_1} \\ y &= \frac{M\bar{v}_2}{V_1^\circ} \end{aligned} \tag{4.35}$$

The terms V_1° (the molar volume of the solvent), $\bar{v}_2$ (the partial specific voume of the polymer molecule), and M (the molecular weight of the polymer) have their usual meanings. Thus, if we determine the consolute temperatures for a series of fractions covering an extended range in molecular weight, we can determine the θ temperature by extrapolating these critical temperatures to infinite molecular weight (Figure 4.6).

APPENDIX: THERMODYNAMICS AND CRITICAL PHENOMENA

The direct application of thermodynamics to the study of polymer solutions has been demonstrated in Flory's theories of ΔS_{mix} and ΔH_{mix}, as described in this chapter. Here, as a prelude to Chapter 5, we describe the connection of thermodynamics and critical phenomena in phase transition. The study of phase transition is one of the most challenging problems in statistical mechanics. Phase transition such as liquid–gas transition and magnetism transition involves an order parameter. For liquid–gas transition the order parameter is the density ρ, and for magnetism transition the parameter is the magnetism M. Above the critical temperature, both of the order parameters vanish:

$$\begin{aligned} \rho_{(\text{liquid})} - \rho_{(\text{gas})} &\neq 0, \qquad \text{for } T < T_c \\ &= 0, \qquad \text{for } T > T_c \\ M &\neq 0, \qquad \text{for } T < T_c \\ &= 0, \qquad \text{for } T > T_c \end{aligned}$$

Even though the phase transition (liquid–gas) and the magnetism transition are different phenomena, there is a one-to-one correspondence between the two parameters:

$$\rho \leftrightarrow M$$

Near the critical point (critical temperature) almost all physical quantities (such as ρ and M) obey some sort of power laws:

$$\rho \sim (T_c - T)^{\alpha}$$
$$M \sim (T_c - T)^{\beta}$$

The exponents α and β are called critical exponents. We discuss power laws and critical exponents on polymer configurations in the next chapter. Here we point out that all these laws have their roots in the concept of thermodynamic equilibrium. Consider the equation

$$A = E - TS \qquad (\text{or } G = H - TS)$$

where A is the free energy, E is the internal energy, T is the temperature, and S is the entrophy. The equation implies that to reach equilibrium, A must be minimized ($A \to 0$). At high temperature, the term TS dominates and the minimum value of A is related to the maximum value of S, which, in turn, leads the system into a disordered state. The order parameter is zero. At low temperatures the internal energy parameter E is the dominating factor. The state with the minimum internal energy is the ordered state. Hence, the order parameter is not zero.

REFERENCES

Aminabhavi, T. M., and P. Munk, *Macromolecules* **12**, 1186 (1979).

Bawn, C. E. H., R. F. J. Freemand, and A. R. Kamaliddin, *Trans. Faraday Soc.* **46**, 677 (1950).

Billmeyer, F. W., Jr., *Textbook of Polymer Chemistry*, New York: Wiley, 1984.

Brandrup, J., and E. N. Immergut, *Polymer Handbook*, 2d ed. New York: Wiley-Interscience, 1975.

Cohn, J., and J. T. Edsall, *Proteins, Amino Acids and Peptides*. New York: Reinhold, 1943.

Dayhoff, M. O., G. E. Perlmann, and D. A. MacInnes, *J. Am. Chem. Soc.* **74**, 2515 (1952).

Flory, P. J., *J. Chem. Phys.* **9**, 660 (1941); **10**, 51 (1942).

Flory, P. J. *Principles of Polymer Chemistry*. Ithaca, NY: Cornell University Press, 1953.

Flory, P. J., and T. G. Fox, Jr. *J. Am. Chem. Soc.* **73**, 1915 (1951).

Hildebrand, J. H., and R. L. Scott, *The Solubility of Non-Electrolytes*. New York: Reinhold, 1950.

Hildebrand, J. H., and S. E. Wood, *J. Chem. Phys.* **1**, 817 (1933).

Hoy, K. L., *J. Paint Technol.* **42**, 76 (1976).

Huggins, M. L., *J. Chem. Phys.* **9**, 440 (1941); *J. Phys. Chem.* **46**, 151 (1942); *J. Am. Chem. Soc.* **64**, 1712 (1942).

Krause, S., and E. Cohn-Ginsberg, *J. Phys. Chem.* **67**, 1479 (1963).

Lawson, D. D. and J. D. Ingham, *Nature* **223**, 614 (1969).

Martin, A., and M. J. Miller, *J. Pharm. Sci.* **71**, 439 (1982).

Miller, A. R. *Proc. Camb. Philos. Soc.* **38**, 109 (1942); **39**, 54, 131 (1943).

Newing, M. J., *Trans. Faraday Soc.* **46**, 613 (1950).

Scatchard, G., *Chem. Rev.* **8**, 321 (1931).

Schultz, A. R., and P. J. Flory, *J. Am. Chem. Soc.* **74**, 4767 (1952).

van Laar, J. J., *Sechs Vorträge über das Thermodynamische Potential*, Braunschweig, 1906.

Wall, F. T., *Chemical Thermodynamics*, 3d ed. San Francisco: Freeman, 1974.

Yalkowsky, S. H., and S. C. Valvani, *J. Pharm. Sci.* **69**, 912 (1980).

PROBLEMS

4.1 The values of the solubility parameter δ_1 in $(\text{cal/cm}^3)^{1/2}$ are given as follows:

n-Hexane	7.24
Carbon tetrachloride	8.58
Benzene	9.15
Acetone	9.71
Methanol	14.5

The solubility parameters δ_2 in $(\text{cal/cm}^3)^{1/2}$ are 8.6 for polystyrene and 9.1 for poly(methyl methacrylate). predict by calculation whether each of the two polymers will dissolve in the above five solvents.

4.2 Let $\phi_i = N_i V_i / \sum N_i V_i$, where the subscript i refers to the ith component in solution (e.g., for a two component system, 1-solvent, 2-solute), ϕ_i is the volume fraction, N is the number of moles, and V is the partial molar volume. Let the notations be changed:

$$\Delta W_{12} = h_{12}$$

$$\chi_1 = g_{12}$$

Show that Flory's equations of ΔG_m and ΔH_m can be transformed into

$$\Delta G_m = R'T\left[\sum_{i=1}^{2} N_i \ln \phi_i + \left(\sum_{i=1}^{2} N_i V_i\right)\phi_1\phi_2\left(\frac{g_{12}}{V_1}\right)\right]$$

$$\Delta H_m = R'\left(\sum_{i=1}^{2} N_i V_i\right)\phi_1\phi_2\left(\frac{h_{12}}{V_1}\right)$$

(*Source*: Aminabhavi and Munk (1979).)

4.3 Derive the equation

$$\mu_1 - \mu_1^\circ = R'T\left[\ln(1-\phi_2) + \left(1 - \frac{1}{y}\right)\phi_2 + \chi_1\phi_2^2\right]$$

4.4 Derive the equation

$$\psi - \kappa_1 = \psi\left(1 - \frac{\theta}{T}\right)$$

4.5 Silicone, $CH_3[(CH_3)_2SiO]_nSi(CH_3)_3$ and benzene are mixed in the following proportion: $\phi_1 = 0.2, 0.4, 0.6, 0.7, 0.8$. Calculate $\Delta\bar{S}_1$ and plot $\Delta\bar{S}_1/\phi_2^2$ versus ϕ_2 using

$$\Delta\bar{S}_1 = -R'\left[\ln(1-\phi_2) + \left(1 - \frac{1}{y}\right)\phi_2\right] \qquad \text{(Flory)}$$

$$\Delta\bar{S}_1 = -R'\left[\ln(1-\phi_2) - \frac{z}{2}\ln\left[1 - \frac{2}{z}\left(1 - \frac{1}{n}\right)\phi_2\right]\right] \qquad \text{(Miller)}$$

Compare the results of the two equations. Assume that $y = n = 100$, $z = 4$. (*Source*: Newing (1950).)

4.6 Measurement of the lowering of vapor pressure by silicones (consisting of 100 structural units, $y = 100$) in benzene yields the following set of data:

ϕ_1	0.0843	0.171	0.266	0.356	0.427	0.820
a_1	0.400	0.635	0.850	0.940	0.955	0.995

Calculate χ_1 values and plot χ_1 versus ϕ_2.

4.7 Precipitation temperature–concentration measurement on fractions of different molecular weight of isotactic poly(methyl methacrylate) in acetonitrile yields the following set of data:

T_c(°C)	16.3	18.2	21.2	22.0	23.4
Mol wt	0.90×10^5	1.61×10^5	2.62×10^5	4.56×10^5	5.35×10^5

Using these data determine θ (the Flory temperature) and ψ_1 (the entropy parameter). (*Source*: Krause and Cohn-Ginsberg (1963).)

4.8 Critical temperatures T_c for polystyrene (mol.wt 89,000) in cyclohexane solutions with volume fractions ϕ_2 were determined by observing the solutions to be turbid on cooling. Data were obtained as follows:

ϕ_2	0.17	0.20	0.28	0.32
T_c(°C)	22.5	21.8	17.5	15

What is the θ temperature of polystyrene in cyclohexane? (*Source*: Flory and Fox (1951).)

4.9 From the measurements of amino acid analysis, the weight percent of amino acid residues in insulin was found as follows:

Amino Acid	Weight Percent of Amino Acid Residues
Serine	2.96
Threonine	2.26
Leucine and isoleucine	25.88
Cystine	11.56
Tyrosine	11.25
Histidine	9.46
Arginine	2.73
Lysine	1.10
Glutamic acid	13.80
Glutamine	12.43

Estimate the partial specific volume of the protein.

5

CHAIN CONFIGURATIONS

Flory's influence on polymer chemistry is again detailed in this chapter. His interpretation of the polymer chain configurations in terms of the expansion factor α (or swelling ratio) constitutes the central point of general polymer chain dynamics. Before we introduce Flory's theory, we describe the random flight model of chain configurations; after we introduce Flory's theory, we describe the two-parameter theory (α and z). We then focus on the new idea of complex dumping and internal movements of polymer chain molecules. This new challenge came from various related areas, among them the development of renormalization group theory, the modern concept of chaos, and the universality of critical exponents on scaling laws. Along with Flory's work, we discuss in detail de Gennes' work on chain configurations. It was de Gennes' theories that pointed to similarities between motion of long chains and magnetic movements undergoing a phase transition. His theories of semidilute solution and reptation in the entangled chain have stimulated new approaches and new ways of interpreting chain configurations. This chapter also covers many other important topics on chain configurations. As is our normal way of presentation, background materials are given first, namely, random walk statistics and a description of the Markov chain. In the appendices we provide the mathematics of the scaling relation and correlation function.

5.1 PRELIMINARY DESCRIPTIONS OF A POLYMER CHAIN

A polymer chain consists of many structural units (repeating units or monomeric units). The chain may take on many different configurations, for example:

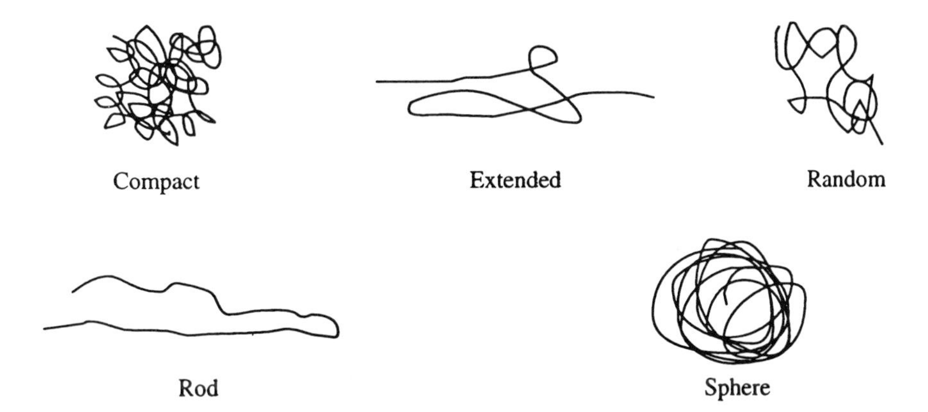

The configurations are restricted by chemical bonds. The double bond C=C has little or no rotational freedom. The single bond C—C, which has rotational freedom, is constrained due to geometric reasons. The interaction among the hydrogen or other atoms leaves certain rotational angles preferred on energetic grounds, as in the case of a small molecule, ethane. The two well-known configurations of ethane are shown in Figure 5.1. The staggered configuration is more stable than the eclipsed, because it offers the least interaction. A molecule may possess a variety of structures, but certain configurations are preferred because of the interference and restrictions imposed by rotational angles under certain conditions.

Figure 5.2 shows a simple model of a chain consisting of four carbons connected with three single C–C bonds. C_1, C_2, and C_3 lie on a plane; that is, they are constrained in a planary position. C_4 may not necessarily be on the same plane. It can take many positions, for example, C_4, C'_4, and C''_4, and the C_3–C_4 bond is free to rotate about C_3 or about the entire C_1–C_2–C_3 plane.

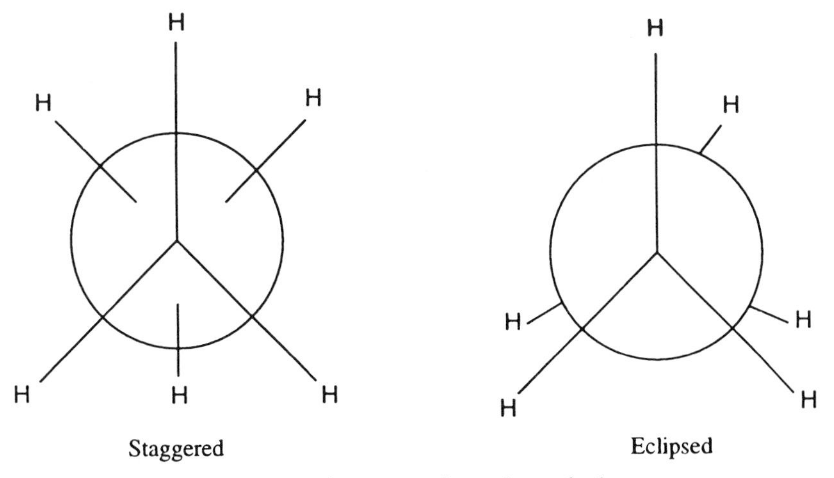

Figure 5.1 The two configurations of ethane.

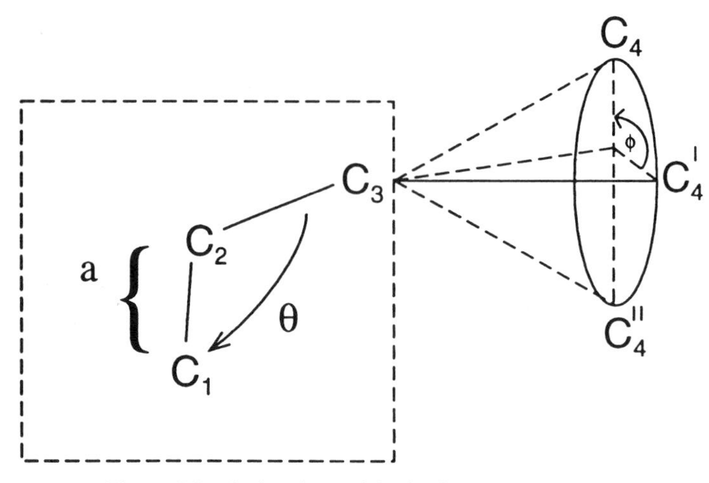

Figure 5.2 A simple model of a four-carbon chain.

Thus, the configuration of the four-carbon chain is determined by three parameters: the bond length a, the rotational angle θ, and the hindrance angle ϕ. This model can be applied to chains containing any number of carbons greater than 3. But, as the number of bonds increases in a chain, the situation becomes more complicated.

The modern approach to chain configuration is to use techniques from probability theory or statistical mechanics. For the first approximation, chain configuration is studied in terms of its distribution of chain length only, ignoring θ and ϕ as independent variables. The second assumption is that chains are supposed to be completely in the form of a random coil, ignoring other molecular shapes such as spheres and rods. Furthermore, since we cannot investigate polymer molecules in the gaseous phase, the study of configuration and its relationship to molecular properties has to be confined to polymers dissolved in solutions. According to Kuhn (1930), chemical bonds of polymers in solution (or in melts) are flexible, and this flexibility gives rise to the variety of properties. In Sections 5.2 and 5.3, we discuss the configuration of an ideal (or rather oversimplified) polymer chain that is based on the above assumptions. From Section 5.4 on, we introduce more advanced theories that are no longer entirely based on the above assumptions. For the most part, these theories are presented in chronological order, which also happens to follow the logical order. Recent advances strongly suggest that many properties of polymer solutions depend on the configuration of the chain, rather than the nature of the chain atoms.

5.2 RANDOM WALK AND THE MARKOV PROCESS

The model of random walk is closely related to the Bernoulli trials. It has been utilized in physics and chemistry as an approximation to one-dimensional diffusion and Brownian motion. When the theory was advanced by Markov, random walk became a special case of what is now called the Markov chain.

5.2.1 Random Walk

For simplicity we describe first the one-dimensional random walk (see Chandrasekhar, 1943). Consider n steps of random walk along an axis, with each step either in the forward or the backward direction. Each step has equal probability 1/2:

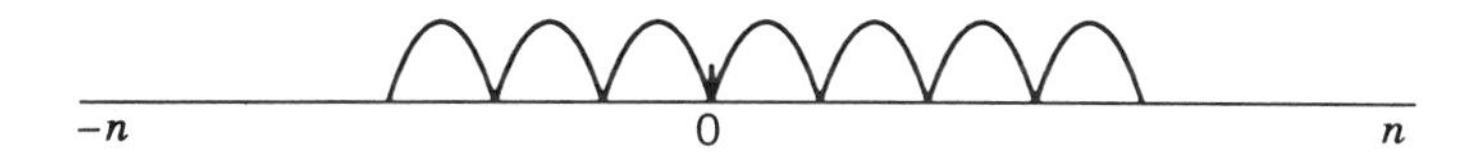

The two extremes are all n steps directed to the right (positive) or all n steps directed to the left (negative) from the origin 0 along one dimension (x axis). Most probably n steps end in between n and $-n$, since the walk is random. Let n_+ be the positive step, n_- be the negative step, and m be the last step. Then m is the distance from the origin and $m = n_+ - n_-$ or $m = n_- - n_+$. If $n_+ = n_-$, $m = 0$. We take

$$m = n_+ - n_- \tag{5.1}$$

where

$$n_+ + n_- = n \tag{5.2}$$

From Eqs. (5.1) and (5.2), we obtain

$$n_+ = \frac{n+m}{2}$$

$$n_- = \frac{n-m}{2}$$

That is, among n steps of random walk,

$\frac{n+m}{2}$ steps are in the positive direction

$\frac{n-m}{2}$ steps are in the negative direction

The probability of leading to the value of m after n steps of random walk can be expressed in terms of $w(n, m)$, which is the Bernoulli's probability (or the

Bernoulli distribution).* Thus, for the random walk distribution (probability distribution) equation, we have

$$w(n, m) = \frac{n!}{((n+m)/2)!\,((n-m)/2)!}\left(\frac{1}{2}\right)^n \tag{5.3}$$

By using Stirling's formula,

$$\log n! = (n + \tfrac{1}{2})\log n - n + \tfrac{1}{2}\log 2\pi$$

Eq. (5.3) is converted to

$$\begin{aligned}\log w(n, m) = {} & \left(n + \frac{1}{2}\right)\log n - \frac{1}{2}(n + m + 1)\log\left[\frac{n}{2}\left(1 + \frac{m}{n}\right)\right] \\ & - \frac{1}{2}(n - m + 1)\log\left[\frac{n}{2}\left(1 - \frac{m}{n}\right)\right] - \frac{1}{2}\log 2\pi \\ & - n\log 2\end{aligned}$$

Since $m \ll n$, we can use the series expansion

$$\log(1 + x) = x - \frac{x^2}{2} + \frac{x^3}{3} - \cdots$$

to convert Eq. (5.3) further:

$$\begin{aligned}\log w(n, m) = {} & \left(n + \frac{1}{2}\right)\log n - \frac{1}{2}\log 2\pi - n\log 2 - \frac{1}{2}(n + m + 1) \\ & \times\left(\log n - \log 2 + \frac{m}{n} - \frac{m^2}{2n^2}\right) - \frac{1}{2}(n - m + 1) \\ & \times\left(\log n - \log 2 - \frac{m}{n} - \frac{m^2}{2n^2}\right)\end{aligned}$$

* *Mathematical note*: Suppose that there are only two possible outcomes to a trial, success or failure. Let p be the probability of success; then 1-p will be the probability of failure. The probability that among n trials there are k successes is

$$w(n, k) = \binom{n}{k} p^k (1 - p)^{n-k}$$

where

$$\binom{n}{k} = \frac{n!}{k!\,(n-k)!}$$

and $w(n, k)$ is called the Bernoulli probability. It is identical to the binomial function.

Simplifying, we obtain

$$\log w(n,m) = \frac{1}{2}\log n + \log 2 - \frac{1}{2}\log 2\pi - \frac{m^2}{2n}$$

Taking exponential functions on both sides, we get

$$w(n,m) = \left(\frac{2}{\pi n}\right)^{1/2} \exp\left(\frac{-m^2}{2n}\right) \tag{5.4}$$

This is exactly the Gaussian distribution function. Instead of m, we now introduce the net displacement x from the starting point as the variable; that is, we change the variable:

$$x = ma$$

where a is the length of step. We realize that $w(n,x)\,\Delta x = w(n,m)(\Delta x/2a)$; hence, we obtain

$$w(n,x) = \frac{1}{(2\pi n a^2)^{1/2}} \exp\left(\frac{-x^2}{2na^2}\right)$$

Here Δx represents the intervals along the straight line and $w(n,x)\,\Delta x$ is the probability that the random walk ends in the interval between x and $x + \Delta x$ after n steps. Thus, we conclude that the random walk is governed by a distribution function $w(n,x)$ which is Gaussian in nature.

5.2.2 The Markov Chain

An important feature of the random walk theory is that in a sequence of trials, each trial is independent of the others. Markov advanced the theory by generalizing the condition that the outcome of any trial may depend on the outcome of the preceding trial; that is, the probability of an event is conditioned by the previous event. This idea clearly fits the description of the configuration of a polymer chain.

Let us first change the term "random walk" into "random flight," because the latter term is used more often in polymer chemistry, though both terms mean the same thing. Consider the configurations of a given chain characterized by the position vector **R**, as shown in Figure 5.3. The general problem of random flight of a Markov chain may be expressed as follows. After N displacement, the position **R** of the particle is given by

$$\mathbf{R} = \sum_{i=1}^{N} \mathbf{r}_i \qquad i = 1, \ldots, N$$

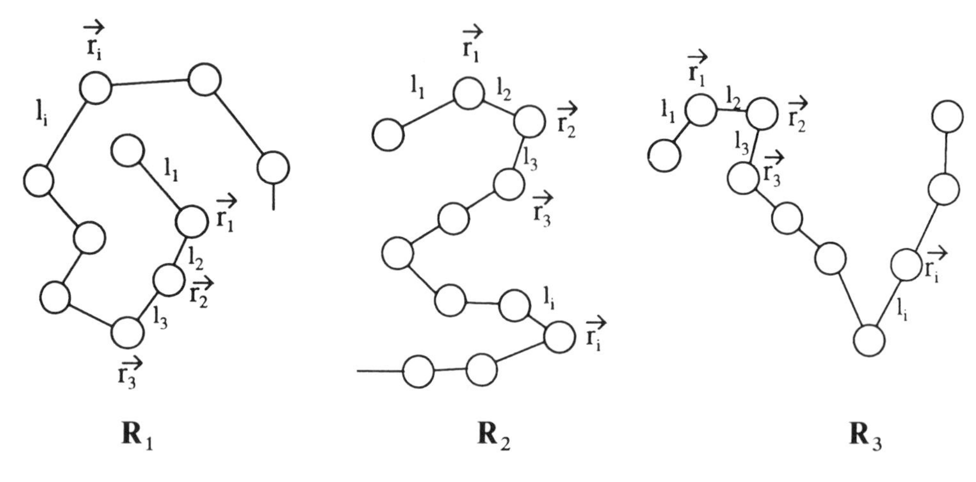

Figure 5.3 Configurations of a chain.

where $\mathbf{r}_i$ refers to the individual displacement. The probability $\tau_i(x_i, y_i, z_i)$ that the ith displacement is found between $\mathbf{r}_i$ and $\mathbf{r}_i + \mathrm{d}\mathbf{r}_i$ is

$$\tau_i(x_i, y_i, z_i)\mathrm{d}x_i\,\mathrm{d}y_i\,\mathrm{d}z_i = \tau_i\,\mathrm{d}\mathbf{r}_i$$

According to the fundamental limit theorem (an extension of the Bernoulli theorem by Laplace) if an event A occurs m times in a series of n independent trials with constant probability p, and if $n \to \infty$, then the distribution function tends to be

$$\frac{1}{\sqrt{2\pi}}\int_{-\infty}^{t} \mathrm{e}^{-(1/2)u^2}\,\mathrm{d}u$$

as a limit. That is, the distribution function tends to be a normal (Gaussian) function. The term u here is a dummy variable. It has no physical meaning. Markov further extends it to the sums of dependent variable, in our case, $\mathbf{R}$ and $\Sigma\mathbf{r}_i$. We take a special case of τ_i as the Gaussian distribution of displacement $\mathbf{r}_i$, that is, $\tau_i(\mathbf{r}_i)$:

$$\tau_i = \left(\frac{3}{2\pi l_i^2}\right)^{3/2} \exp\left(-\frac{3r_i^2}{2l_i^2}\right) \tag{5.5}$$

where l is the length of displacement. Applying the displacement of $\mathbf{r}_i$ to $\mathbf{R}$, we have

$$w_N(R) = \left(\frac{3}{2\pi N\langle l^2\rangle}\right)^{3/2} \exp\left(-\frac{3R^2}{2N\langle l^2\rangle}\right)$$

where $w_N(\mathbf{R})d\mathbf{R}$ is the probability that after N displacements the particle lies in the interval between $\mathbf{R}$ and $\mathbf{R} + d\mathbf{R}$, and $\langle l^2\rangle$ is the mean square displacement. If all the individual displacements have the same length, then the distribution function is

$$\tau_i = \frac{1}{4\pi l^3}\delta(|\mathbf{r}_i|^2 - l_i^2)$$

where δ is the Dirac function,

$$\begin{aligned} r_{ij} &= 0 \qquad \text{if } i \neq j \\ &= 1 \qquad \text{if } i = j \end{aligned}$$

and the probability function is

$$w(R) = \left(\frac{3}{2\pi N l^2}\right)^{3/2} \exp\left(\frac{-3R^2}{2Nl^2}\right) \tag{5.6}$$

5.3 RANDOM FLIGHT CHAINS

We assume that each bond between the two monomeric units of a polymer has a constant length a and that τ_i is the bond probability. This enables the configuration of a polymer molecule in the form of a flexible random coil to be described in terms of a Markov chain; namely, the probability $w(R)dR$ that after N displacements the position of the particle lies between R and $R + dR$ can be described by Eqs. (5.5) and (5.6). All what we have to do is to change r_i to r in Eq. (5.5) and l_i to a in Eq. (5.6) and to rewrite the two equations as follows:

$$\tau(\mathbf{r}) = \left(\frac{3}{2\pi a^2}\right)^{3/2} \exp\left(\frac{-3r^2}{2a^2}\right) \tag{5.7}$$

$$w(R) = \left(\frac{3}{2\pi N a^2}\right)^{3/2} \exp\left(\frac{-3R^2}{2Na^2}\right) \tag{5.8}$$

With the distribution function $w(R)$ available, we can now calculate the mean square end-to-end distance $\langle R^2 \rangle$ of a polymer chain in the following way:

$$\langle R^2 \rangle = \int_0^\infty R^2 w(R) 4\pi R^2 \, dR = Na^2 \tag{5.9}$$

the function $w(R)$ is being normalized,

$$\int_0^\infty w(R) 4\pi R^2 \, dR = 1$$

The root-mean-square end-to-end distance of a chain is the square root of $\langle R^2 \rangle$:

$$\langle R^2 \rangle^{1/2} = N^{1/2} a$$

The radius of gyration $\langle S^2 \rangle^{1/2}$, which is the root-mean-square distance of an end from the center of gravity, is related to the root-mean-square end-to-end distance by

$$\langle S^2 \rangle^{1/2} = \frac{1}{\sqrt{6}} \langle R^2 \rangle^{1/2}$$

or

$$\frac{\langle S^2 \rangle}{\langle R^2 \rangle} = \frac{1}{6}$$

Equation (5.9) represents an ideal polymer chain, a random flight chain. It is also regarded as a freely jointed chain. To apply Eq. (5.9) to the real chain, several modifications have been suggested by various investigators. Here we list three of them.

1. On the basis of statistical reason, for Eq. (5.9) to be true, the value of a must be large and that of N must be small. One way to solve this problem is to introduce a proportionality constant c, which is called the characteristic ratio. Equation (5.9) is modified to the form

$$\langle R^2 \rangle = cNa^2 \tag{5.10}$$

If $c = 1$, the chain is, needless to say, ideal. If $c \neq 1$, we must correct our estimation of the value of $\langle R^2 \rangle$. Thus, the value of c provides some information on the dimension of a polymer chain in relation to its deviation from ideality: the larger the value of c, the larger the chain extended.

2. In case the bond angles θ of a chain are restricted but rotations about

the bonds are not restricted, it is suggested that Eq. (5.9) be further modified:

$$\langle R^2 \rangle_f = cNa^2 \frac{1 - \cos\theta}{1 + \cos\theta} \tag{5.11}$$

Such a chain is often called freely rotating chain.

3. We can write the random flight chain in a more general way:

$$\langle R^2 \rangle = \sum_{i=1}^{N} \langle \mathbf{r}_i^2 \rangle + 2 \sum\sum_{1 \leqslant i \leqslant j \leqslant N} \langle \mathbf{r}_i \mathbf{r}_j \rangle \tag{5.12}$$

where $\langle \mathbf{r}_i \mathbf{r}_j \rangle$ is the correlation factor (see Appendix B). In obtaining Eq. (5.9), we tacitly assume that

$$\langle \mathbf{r}_i \mathbf{r}_j \rangle = 0 \qquad \text{for } i \neq j$$

and that

$$\langle \mathbf{r}_i^2 \rangle = a^2$$

However, this assumption is not true, because we neglect the excluded volume effect which is expressed in the terms $\langle \mathbf{r}_i \mathbf{r}_j \rangle$. That is, we should not assume that $\langle \mathbf{r}_i \mathbf{r}_j \rangle \neq 0$ for $i \neq j$. Hermans (1950) and Grimley (1952) suggested independently that we include $\langle \mathbf{r}_i \mathbf{r}_j \rangle (i \neq j)$ in Eq. (5.9). Herman's equation is

$$\langle R^2 \rangle = Na^2 \left[1 + 0.78 \left(\frac{\beta_0}{a^3 N^{1/2}} \right) \right] \tag{5.13}$$

where β_0 is the volume excluded to one segment by the another. Grimley's equation is almost identical to Herman's, except for the numerical coefficient:

$$\langle R^2 \rangle = Na^2 \left[1 + 0.143 \left(\frac{\beta_0}{a^3 N^{1/2}} \right) \right] \tag{5.14}$$

In both cases, as $N \to \infty$, $\langle R^2 \rangle \to Na^2$.

5.4 THE WORMLIKE CHAINS

The wormlike chain model is an extension of the random flight chain model. It was proposed by Kratky and Porod (1949) for the purpose of interpreting the experimental data of small-angle x-ray scattering (see Chapter 9) from biological polymers. It is well known that biological polymers such as proteins and DNA have stiff chains. The wormlike chain resembles a freely rotating chain: The chain consists of N bonds of fixed length joined at fixed bond angles

θ. All bond lengths and all bond angles are taken to be equal. The chain can rotate freely about single bonds. The general expression for the mean-square end-to-end distance of such a chain is

$$\langle R^2 \rangle = Na^2 \left[\frac{1-\cos\theta}{1+\cos\theta} + \frac{2\cos\theta}{N} \frac{1-(1-\cos\theta)^N}{(1+\cos\theta)^2} \right] \tag{5.15}$$

The wormlike chain is a limiting case of such a chain, namely, $a \to 0$ and $\theta \to \pi$. The chain is characterized by the contour length L, defined as

$$L = Na$$

and the persistence length $(2\lambda)^{-1}$, defined as

$$(2\lambda)^{-1} = \frac{a}{1+\cos\theta}$$

By substituting the two parameters, L and $(2\lambda)^{-1}$, into Eq. (5.15) and simplifying, we obtain the general expression of $\langle R^2 \rangle$ for the wormlike chain:

$$\langle R^2 \rangle = \frac{L}{\lambda} - \frac{1}{2\lambda^2}(1 - e^{-2\lambda L}) \tag{5.16}$$

Unlike other flexible chains, the dimensions of the wormlike chain are greatly influenced by the shape of the molecules. The parameter λ is a measure of the chain stiffness.

5.5 FLORY'S MEAN FIELD THEORY

Flory pointed out that volume exclusion must cause $\langle R^2 \rangle$ to increase. The amount of increase can be expressed by the expansion factor α^2, defined as

$$\alpha^2 = \frac{\langle R^2 \rangle}{\langle R^2 \rangle_0} \tag{5.17}$$

where $\langle R^2 \rangle$ is the mean-square end-to-end distance of the chain, which has been perturbed by the excluded volume effect, and $\langle R^2 \rangle_0$ is the distance for the unperturbed chain, that is, the ideal chain, ($\langle R^2 \rangle_0 = Na^2$). Thus, the value of $\langle R^2 \rangle_0$ can only be used as a reference. Once the value of $\langle R^2 \rangle_0$ is known, the value of $\langle R^2 \rangle$, which is the real chain dimension, should solely depend on the value of α. To understand the average dimension of a chain over various configurations, we need to understand α.

Consider the segments of a chain, x in number, which pervade a volume V.

Let ρ be the segment density, which is uniform throughout the volume V. Then $\rho = x/V$ within V and $\rho = 0$ outside V. The volume is related to $\langle R^2 \rangle$ by

$$V = A\langle R^2 \rangle^{3/2}$$

where A is the proportionality constant. The probability that an arbitrary distribution of the centers of segments within V is

$$P_{(i)} \cong \prod_{i=1}^{x} \left(1 - \frac{i\beta}{V}\right) \cong \exp\left(\frac{-\beta x^2}{2V}\right) = \exp\left(\frac{-\beta x^2}{2A\langle R^2 \rangle_0^{3/2}\alpha^3}\right)$$

where β is the volume excluded by a segment. In terms of the conventional parameter z, defined as

$$z = \left(\frac{3}{2\pi}\right)^{3/2} \left(\frac{\langle R^2 \rangle_0}{x}\right)^{-3/2} x^{1/2}\beta$$

we have

$$P_{(i)} = \exp\left[-2^{1/2}\left(\frac{\pi}{3}\right)^{3/2} A^{-1} z\alpha^{-3}\right] \tag{5.18}$$

Equation (5.18) describes the average density corresponding to $\langle R^2 \rangle$. (*Note*: This is the major point of mean field theory; see Chapter 7, Appendix E.) The distribution of chain vector $\mathbf{R}$ for the unperturbed chain is assumed to be Gaussian in nature, that is

$$w(\mathbf{R})\,d\mathbf{R} = \text{const.} \times \exp\left(\frac{-3R^2}{2\langle R^2 \rangle_0}\right) d\mathbf{R}$$

The probability $P_{(ii)}$ of a set of configurations corresponding to $\langle R^2 \rangle$ relative to the probability of a set of configurations corresponding to $\langle R^2 \rangle_0$ is

$$P_{(ii)} = \frac{(d\mathbf{R})}{(d\mathbf{R})_0} \exp\left[\frac{-3(\langle R^2 \rangle - \langle R^2 \rangle_0)}{2\langle R^2 \rangle_0}\right] = \alpha^3 \exp\left[-\frac{3}{2}(\alpha^2 - 1)\right]$$

The combined probability leads to

$$P_{(i)}P_{(ii)} = \alpha^3 \exp\left[-2^{1/2}\left(\frac{\pi}{3}\right)^{3/2} A^{-1} z\alpha^{-3} - \frac{3}{2}(\alpha^2 - 1)\right]$$

Maximizing the expression of $P_{(i)}P_{(ii)}$ (that is, setting $\partial P_{(i)}P_{(ii)}/\partial\alpha = 0$), and solving for α gives

$$\alpha^5 - \alpha^3 = 2^{1/2}\left(\frac{\pi}{3}\right)^{3/2} A^{-1} z \tag{5.19a}$$

or

$$\alpha^5 - \alpha^3 = Bx^{1/2}\beta \tag{5.19b}$$

where

$$B = \left(\frac{\langle R^2 \rangle_0}{x}\right)^{-3/2} (2A)^{-1}$$

The quantity B is a constant for a given series of polymer homologs. Equations (5.19a) and (5.19b) are Flory's theory to describe the polymer chain configuration. de Gennes commented that because of the assumption of Gaussian distribution with respect to $\langle R^2 \rangle_0$, Flory's theory is still intrinsically a random flight chain in nature.

5.6 PERTURBATION THEORY

Perturbation theory (Zimm, 1953; Yamakawa, 1971) is an extension of Flory's theory. It is based on mathematical analysis of the transition probability, which indicates that the excluded volume effect on $\langle R^2 \rangle$ becomes asymptotically proportional to a power of N higher than the first. Flory's term α may be expanded in a power series by the excluded volume parameter z. The underlying idea is to describe α as a universal function of z. For this reason, the theory is also called the two-parameter (α and z) theory. So far it has only partially reached this goal, because it is a many-body problem and the mathematics involved are difficult.

5.6.1 First-Order Perturbation Theory

In this theory the coupling parameter method is used to derive an equation for the distribution function. The coupling parameter method considers only those configurations in which a single pair of segments is interacting. It neglects the higher order approximations for multiple contacts. The following equations are the results obtained by Zimm and coworkers (1953):

$$\alpha_R^2 = 1 + \tfrac{4}{3}z + \cdots \tag{5.20}$$

where

$$z = \left(\frac{3}{2\pi a^2}\right)^{3/2} \beta N^{1/2} = \left(\frac{3}{2\pi \langle R^2 \rangle_0}\right)^{3/2} \beta N^{1/2}$$

$$\beta = \int 1 - g(\mathbf{R}_{ij}) \mathrm{d}\mathbf{R}_{ij}$$

The term $g(\mathbf{R}_{ij})$ is defined as

$$g(\mathbf{R}_{ij}) = \exp\left[-\frac{w(R_{ij})}{kT}\right]$$

where $w(R_{ij})$ is the pair potential of mean force between the ith and jth segments as a function of R_{ij}. β is expected to have large positive values in good-solvent systems and to have small positive, zero, or negative values in poor-solvent systems. In terms of Flory's expressions, the first-order perturbation theory can be rephrased:

$$\begin{aligned} \alpha_R^5 - \alpha_R^3 &= \tfrac{4}{3}z \\ \alpha_S^5 - \alpha_S^3 &= \tfrac{134}{105}z \end{aligned} \tag{5.21}$$

5.6.2 Cluster Expansion Method

In this theory the single contact term, the double contact term, and so forth are taken into consideration. The cluster diagrams developed by Ursell (1927) and Mayer (1940) are employed to calculate the mean-square end-to-end distance. Here we give the results obtained by Yamakawa and coworkers (1971):

$$\alpha_R^2 = 1 + 1.333z - 2.075z^2 + 6.459z^3 - \cdots \tag{5.22}$$

$$\alpha_S^2 = 1 + 1.276z + \cdots \tag{5.23}$$

In term's of the Flory theory, the two-parameter theory of Yamakawa may be expressed as

$$\alpha^5 - \alpha^3 = 2.60z \tag{5.24}$$

At small z,

$$\alpha^2 = 1 + 2.60z - \cdots \tag{5.25}$$

5.7 CHAIN CROSSOVER AND CHAIN ENTANGLEMENT

5.7.1 Concentration Effect

So far our description of the random coil chain basically assumes a dilute solution and we have not yet defined the term dilute solution. It has been discovered that when the concentration increases to a certain point, interesting phenomena occur: chain crossover and chain entanglement. Chain crossover refers to the transition in configuration from randomness to some kind of order, and chain entanglement refers to the new statistical discovery of self-similar property of the random coil (for example, supercritical conductance and percolation theory in physics). Such phenomena also occur to the chain near theta temperature. In this section, we describe the concentration effect on chain configurations on the basis of the theories advanced by Edwards (1965) and de Gennes (1979). In the next section, we describe the temperature effect, which is parallel to the concentration effect.

In 1966, S. F. Edwards proposed classifying the concentrations of polymer solutions into three broad types in terms of the total number of micromolecules (monomers) N, the number of polymer chains n, the effective length l of a micromolecule, the excluded volume per micromolecule u', and the total volume V. The length of the single chain L is defined as

$$L = \frac{N}{n} l$$

The three types of concentration can be described as follows:

1. Dense solutions in which $V/N < u'$
2. Intermediate concentrations in which

$$L^{9/5} u'^{3/5} l^{-3/5} > \frac{V}{n} > u'$$

3. Dilute solutions in which

$$\frac{V}{n} < L^{9/5} u'^{3/5} l^{-3/5}$$

Edwards discussed intermediate concentrations in detail, for this is the region in which chain crossover and chain entanglement occur. To describe this region, he introduced two derived quantities:

$$\xi = \left(\frac{12 n u' L}{V l^3} \right)^{-1/2}$$

$$g = \left(\frac{u' L}{l} \right)^{1/3}$$

We thus have four basic quantities; $l_0 = (u')^{1/3}$, $l, r_0 = (V/n)^{1/3}$, and $r_1 = (Ll)^{1/2}$, and two derived quantities, ξ and g. The quantity ξ is now called the screen length and the quantity g^3 is the effective volume of the molecules in one chain. Edwards showed that the intermediate region exists provided that

$$l_0 < l \ll g < r_0 < \xi < r_1$$

Edwards reached a conclusion for the mean-square end-to-end distance of a polymer in intermediate solution by giving

$$\langle r^2 \rangle = c_2 \left(\frac{u'}{l} \right)^{2/5} L^{6/5} \tag{5.26}$$

where c_2 is a numerical constant. This equation is very similar to Flory's equation

$$\langle R^2 \rangle \sim N^{12/10} \tag{5.27}$$

but Edwards and Flory used different approaches. Flory adopted the mean field model and the chain is intrinsically Gaussian, as we mentioned before. Edwards adopted the self-consistent argument and the chain is not Gaussian.

A more detailed and elegant description of the concentration effect on the chain configuration is offered by de Gennes, particularly in the intermediate region. de Gennes pointed out that at a certain concentration, the behavior of a polymer chain is analogous to the magnetic critical and tricritical phenomena. He applied a method to study chain behavior similar to that used by Wilson (1971) to study magnetic critical and tricritical phenomena, namely, the adoption of renormalization group theory to derive equations (Chapter 7, Appendix D). The techniques focus on the length scale L as a basic parameter on which polymer properties are related and the power laws are obtained for asymptotic behavior by taking $N \to \infty$. de Gennes classifies the concentration c into three categories: the dilute solution c', the semidilute solution c^*, and the concentrated solution c''. They are related in the following way:

$$c' < c^* < c''$$

The concentration c^* is equivalent to the critical point where the crossover phenomenon occurs from randomness to order. It is also equivalent to p_c (the critical probability) in the percolation theory, where the crossover phenomenon occurs from the finite cluster (such as a macromolecule containing a finite number of monomers) to the infinite cluster (such as the network of an entangled macromolecule, which extends from one end to the other). The three regions are characterized by three important quantities: the number of statistical elements per chain N, the number of statistical elements per unit volume ρ (density), and the correlation or screen length ξ.

Dilute Solutions In a dilute solution c', a polymer chain behaves like a small hard sphere, that is, a single blob, with a radius r in the form of a power law:

$$r \sim N^{\nu} \quad \text{and} \quad \nu = \frac{3}{d+2} \qquad \text{for } d \leqslant 4 \tag{5.28}$$

where ν is an excluded volume exponent (or the critical exponent for the mean dimension of a polymer chain), and d is the dimensionality ($d = 1, 2, 3, \ldots$). For example, for $d = 3, \nu = \frac{3}{5}$. Each polymer in a dilute solution is an isolated chain. Here, the length scale L is represented by the radius r and is equal to the correlation length ξ, that is, $r = \xi$. The chain concentration is equal to ρ/N, that is, the coils are separated from each other.

Semidilute Solutions As concentration increases, the blobs begin to overlap each other. There are interchain interactions; that is, the chains begin to interpenetrate. If the concentration is increased to a point slightly higher than c^*, the coils become entangled and a network of mesh ξ is formed. The parameter ξ now refers to the distance between two entanglement points or two intersection points. At the entangled point it can no longer be distinguished which chain is tangled with which other chain. Yet, just as in fractal geometry and percolation theory, at such a random state, a strange phenomenon happens: An individual long chain (originally one blob, that is, in dilute solution) can be visualized as a succession of sequences of well-ordered and isolated blobs.

The formation of the long chain in the entanglement may be sketched as in Figure 5.4. In the dilute solution region (1), an individual polymer chain A may be conceived as confined in a blob of radius r or ξ. In the semidilute solution region (2), many chains interpenetrate into each other to form a mesh of size ξ. In (3), still in the semidilute solution region but with a slightly higher than c^*, the same individual chain A is now in the form of a succession of subchains or a sequence of blobs of radius ξ. The parameter ξ (the correlation length) is equivalent to the mean free path between two collisions in the kinetic theory of gas.

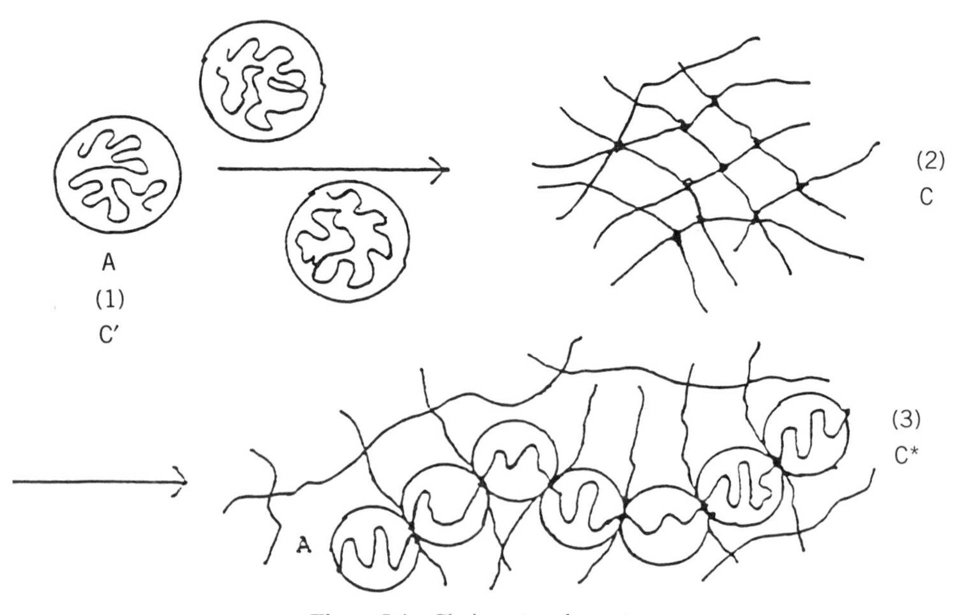

Figure 5.4 Chain entanglement.

The concentration c^* can be defined by (a power law)

$$c^* = \frac{\rho^*}{u'} \sim N^{1-\nu d} \tag{5.29}$$

where ρ^* is the overlap threshold density and u' is the excluded volume parameter, which involves monomer–monomer interactions. The degree of interpenetration as shown in (2) in Figure 5.4 can be measured in terms of ρ/ρ^*.

Assume that we are dealing with polymers in good solvents and in the semidilute solution. If r is a scale to measure, then the chain entanglement shows the following properties. At $r > \xi$, that is, outside the blob, the repulsive interactions between monomers are "screened out" by other chains in the solution so that the whole chain is composed of blobs connected in an ordinary random walk without excluded volume effect. Overall, the chain follows Gaussian statistics. At $r < \xi$, that is, within the blob, the chain does not interact with other chains, but there is a strong excluded volume effect.

In a semidilute solution the system dynamics are controlled by interaction contacts:

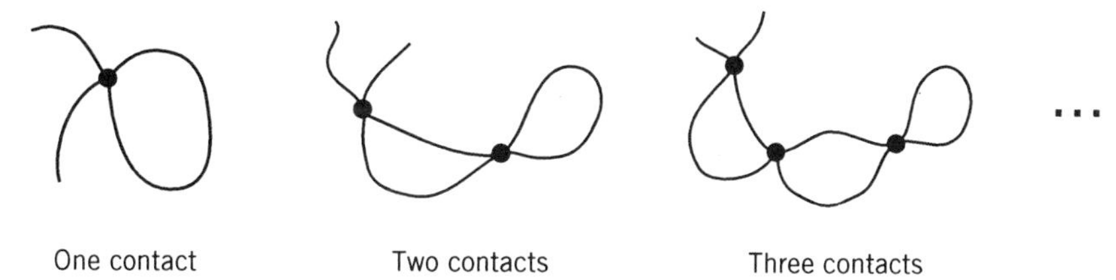

The end-to-end distance or radius of gyration (R or S) is no longer important. It is the screen length ξ (the diameter of blob), which is related to R, that plays an important role. The concentration dependence of ξ is obtained from the scaling relationship,

$$\xi \sim R\left(\frac{c}{c^*}\right)^x \tag{5.30}$$

If $R \sim M^{-3/5}$ and $c^* \sim M^{-4/5}$, then $x = -\frac{3}{4}$ (where M is the molecular weight). We thus have the power law

$$\xi \sim c^{-3/4} \tag{5.31}$$

This power law, Eq. (5.31), is the central theme of the scaling theory for understanding entangled polymer chains in good solvents. Furthermore, at concentration c^*, the screen length ξ is equal to the coil size.

Concentrated Solutions In the concentrated solution c'', the length ξ is smaller than the coil size. When the concentration is very high, ρ equals unity and ξ becomes of order one, which is comparable to the monomer size, as shown in the following diagram:

where ● is a monomeric unit and --- indicates no more stable bonds. Here the chain collapses, or is better said to be in a collapsed state.

Comments on Flory's Theory by de Gennes According to de Gennes, Flory did not realize the existence of the critical point c^* in the polymer solution. His mean field theory of α in the dilute polymer solution leads, indeed, to a correct expression of scaling law,

$$R \sim N^{-6/5}$$

because luckily the errors introduced in ΔS_{mix} and ΔH_{mix} for a polymer chain canceled out. Flory's approach to the polymer solution problem is basically wrong. According to de Gennes, he neglected the changes in the "pair correlation function $g(\mathbf{r})$ that occur when we incorporate the effects of excluded volume." For a correct approach, we must not neglect the correlations between all pairs of monomers. The pair correlation may be defined as

$$g(\mathbf{r}) = \tfrac{1}{2}[\langle c(0)c(\mathbf{r})\rangle - c^2]$$

de Gennes' own theory may be summarized in a slightly different way when correlation function is taken into consideration. In the dilute polymer solutions, coils are separated, while in the more concentrated solutions, coils overlap. At the threshold ($c = c^*$), the coils begin to be densely packed. In terms of correlation function $g(\mathbf{r})$, for $\mathbf{r} < \xi$ (that is, inside the blob), we have

$$g(\mathbf{r}) \cong \frac{1}{r^{4/3}a^{5/3}}$$

where a is the length of a statistical unit (equivalent to bond length). For $r > \xi$ (that is, outside the blob), where the chain begins to feel the effects of surrounding chains,

$$g(\mathbf{r}) = c\frac{\xi}{r}$$

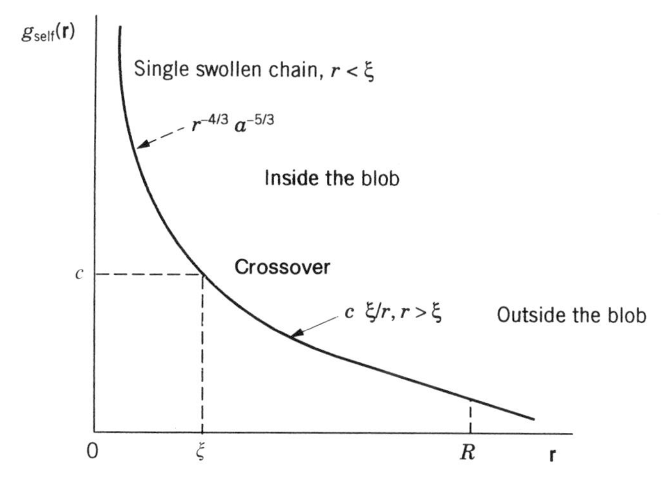

Figure 5.5 A description of crossover phenomena, **r** is scale of measure.

The two coils crossover smoothly at $r = \xi$. For $r > R$ (where R is the overall size of the chain),

$$g(\mathbf{r}) \to 0$$

In Figure 5.5, the crossover is shown as a smooth curve.

5.7.2 Temperature Effect

The conformation* of a single polymer chain in solutions is affected by temperature just as much as by concentration, de Gennes pointed out that the θ point is in fact a tricritical point. Experimentally, while it is difficult to find the sharp point of c^*, it is easy to find the sharp point of θ for a particular polymer solution system. The equation of $\alpha^5 - \alpha^3$ developed by Flory on the basis of mean field approximation is modified by de Gennes with the new approach, renormalization group theory, to describe the effect of temperature on chain configuration which is parallel to concentration effect. de Gennes' equation is given as

*We agree with Flory in the use of the terms configuration and conformation. Flory said, "The term *conformation* connotes form and a symmetrical arrangement of parts. The alternative term, *configuration*, is perhaps the more general of the two in referring to the disposition of the parts of the object in question without regard for shape and symmetry." For practical reasons, he considered these two terms interchangeable (Flory, 1969). In recent literature of polymer science, both terms are used. Authors customarily use the term conformation with respect to temperature effect. We use both terms in this chapter so that readers will feel confortable with both.

$$\underbrace{\alpha^5 - \alpha^3}_{\text{(i)}} - \underbrace{\frac{y}{\alpha^3}}_{\text{(ii)}} = \underbrace{kN^{1/2}\frac{w_1}{a^3}}_{\text{(iii)}} \tag{5.32}$$

where

$$y = k^2 w_2 a^{-6}$$

In these equations y is a parameter ($y = 0, 0.01, 0.038, 0.1$, or 1), w_1 and w_2 are successive virial coefficients, N is the number of monomers along the chain, k is a numerical constant, and a is the monomer length. Equation (5.32) can be used to describe the single chain behavior in infinite dilution. But, instead of using the concentration c as a scaling variable, here we choose T as a scaling variable.

According to Perzynski et al. (1982), Eq. (5.32) can be interpreted as follows: The term $\alpha^5 - \alpha^3$ (term i) describes the elasticity of the chains. The negative term $-\alpha^3$ prohibits large swelling. Term i is important when $T > \theta$. The term y/α^3 (term ii) describes a hard core repulsion and slows down the collapse of the chain. Term ii is important when $T < \theta$. Term iii is proportional to the reduced temperature τ, defined as

$$\tau = \frac{T - \theta}{T} \qquad \text{for } T > \theta$$

$$\tau = \frac{\theta - T}{\theta} \qquad \text{for } T < \theta$$

This term is important both for $T > \theta$ (the swelling of the chain) and $T < \theta$ (the contraction of the chain). It is a description of the crossover phenomena (Figure 5.5)

We can now introduce a reduced variable that involves temperature: $\tau\sqrt{N}$. The expansion factor α is simply a function of this single reduced variable $\tau\sqrt{N}$:

$$\alpha = f(\tau\sqrt{N})$$

At θ temperature, $f(\tau\sqrt{N}) \cong 1$ and $\tau\sqrt{N} \ll 1$. We have

$$R \sim R_\theta \sim N^{1/2} \qquad \text{(Gaussian chain)}$$

At $T < \theta, f(\tau\sqrt{N}) \cong (\tau\sqrt{N})^{-1/3}$ and $\tau\sqrt{N} \gg 1$, and we have

$$R \sim N^{-1/3}\tau^{-1/3} \qquad \text{(Globule chain)}$$

The collapse state at which Eq. (5.32) is aimed can be visualized by plotting the experimental data of $\log R$ versus $\log(-\theta\tau)$ as shown in Figure 5.6. At $T > \theta$,

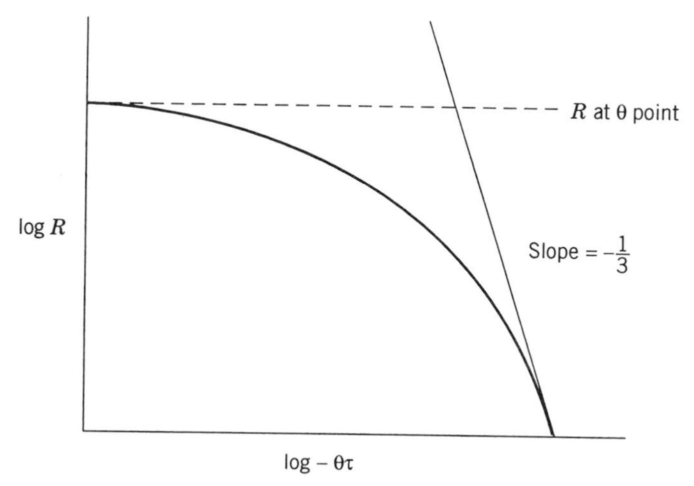

Figure 5.6 The collapse state.

$f(\tau\sqrt{N}) \cong (\tau\sqrt{N})^{3/5}$, and $\tau\sqrt{N} > 1$, we have

$$R \sim N^{3/5}$$

The value of R decreases slowly at first as the temperature decreases. At a very narrow range of the temperature $\theta - T$, the value of R rapidly drops, which is an indication of the collapse state of the polymer chain.

The question is, what about the combined effects of concentration and temperature on the chain conformation? This has been described in terms of the coexistence curve, which is shown in Figure 5.7. The coexistence curve describes the states of temperature and solute concentration at which the attractive parts of segment–segment interactions just begin to dominate. The polymer begins to form aggregates, but no phase separation occurs as yet.

5.7.3 Tube Theory (Reptation Theory)

Tube theory describes the motion of a chain in an entanglement of mesh ξ. It was originally proposed by de Gennes (1971) and was advanced by Doi and Edwards (1978). A chain in the entanglement is viewed as the chain being confined in an environment densely filled with the strands of other chains, which, in a sense, form a tube. It is the center line of the tube that tracks the current chain conformation. The chain can move laterally, but the displacements can take place only by diffusion along its contour.

The short-range interaction of the chain segments causes the molecule to accommodate with the tube; that is, the length of the chain is identical with the mesh size ξ. The long-range interaction, however, makes the chain capable of sliding along the contour like a snake until it finally escapes from the tube.

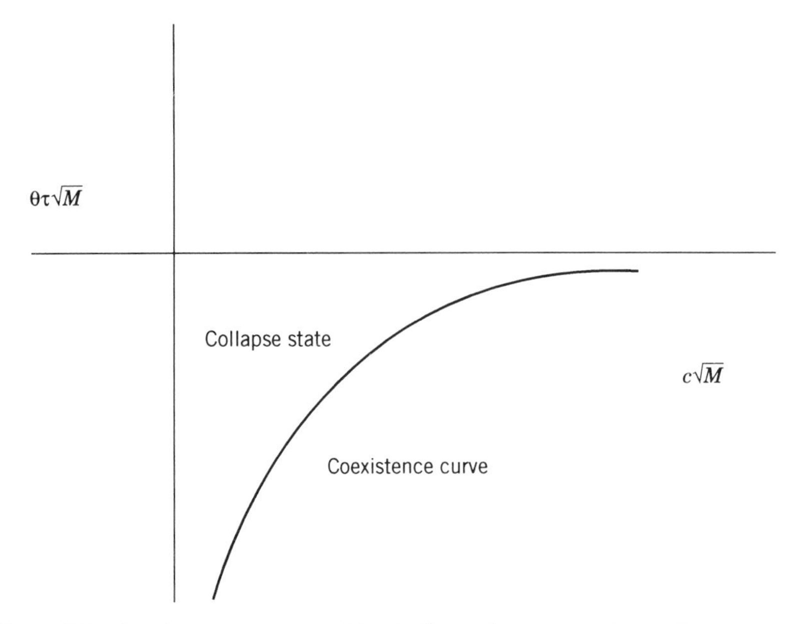

Figure 5.7 Coexistence curve: combined effects of concentration and temperature.

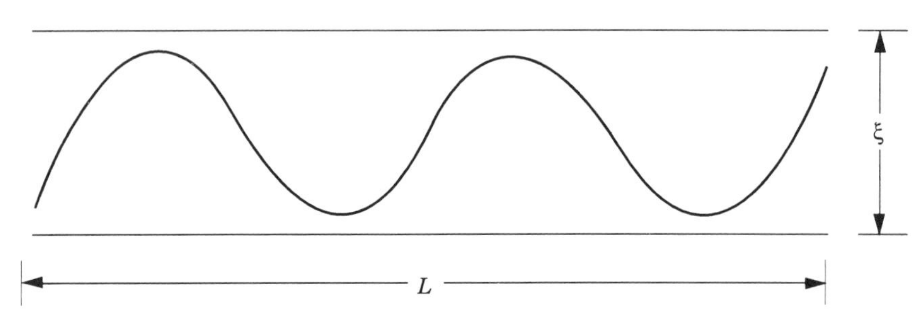

Figure 5.8 The reptation model.

This process is called reptation. The entangled chains rearrange their conformations from time to time by reptation. In technical terms, reptation is a curvilinear diffusion along their own contour. The important parameters in this model are the tube diameter ξ and the tube or curvillinear length L (see Figure 5.8).

The tube diameter and the tube length are related by

$$L = \left(\frac{N}{g}\right)^{1/2} \xi$$

where N is the total number of chains in consideration, and g is the number of monomers in a subunit. Let μ_t be the mobility of the chain along the tube and μ_1 the mobility of one unit. Then

$$\mu_t = \left(\frac{N}{g}\right)\mu_1$$

But

$$\mu_1 = \frac{1}{\eta_s \xi}$$

where η_s is the solvent viscosity. The diffusion coefficient D (Chapter 10) is related to μ_t in terms of the mean-square distance by

$$D = kT\mu_t = \frac{kT}{\eta_s \xi}\frac{g}{N} \tag{5.33}$$

Assume that the primitive path of the chain in the tube is a random walk sequence of N steps with step length ξ. Then we have

$$R = \left(\frac{N}{g}\right)^{1/2}\xi \tag{5.34}$$

Thus, L, the end-to-end vector of the tube, and R, the root-mean-square end-to-end distance, coincide. The reptation time T_R, which is the time for the chain to renew its configuration, is related to L by

$$T_R D \sim L^2 \tag{5.35}$$

The tube theory has been applied successfully to the mechanical properties in concentrated polymer liquids. It also has been proved useful for analyzing the motion of DNA in a gel which includes the effect of molecular fluctuations.

5.8 SCALING AND UNIVERSALITY

To summarize what we have discussed so far, the configuration of a polymer chain depends on two basic quantities: molecular weight M (which, in turn, is related to the number of polymeric links N) and the end-to-end distance R or radius of gyration S (which, in turn, is related to both inter- and intramolecular interactions). At the asymptotic region there is not much difference between R and S.

Although the random motion of a flexible polymer chain is complex, recent theoretical developments predict that the universal behavior should be observed in certain dynamic regimes. Such a behavior is related to the critical exponents as expressed in power laws.

The universal laws of current interest for a polymeric system are in the form

$$p \sim M^{\nu}$$

where p refers to R_g, R_H, A_2, or $[\eta]$, M is the molecular weight, and ν is the critical exponent. These laws govern which scale we choose to measure the macrostructure of a system. Experimentally, R_g is the radius of gyration, which is an average measure of the size of the macromolecule. The instrument used to measure R_g is light scattering. R_H is also the radius of gyration, but it is a measure of the hydrodynamic size of a polymer as it drifts through a fluid. The instruments that measure R_H are ultracentrifuge sedimentation and diffusion (for example, laser light scatterings). A_2 is the second virial coefficient, which is a direct measure of excluded volume effect. The instrument that best measures this quantity is osmotic pressure. The term $[\eta]$ is the intrinsic viscosity due to the rate of dissipated energy in shear flow of the polymer solution, which is another measure of the hydrodynamic volume of a power law.

Since M is related both to N (the number of links) and c (the concentration), the independent variable M can be converted to N or c, and the power laws are in the form

$$p \sim N^{\nu} \quad \text{or} \quad p \sim c^{\nu}$$

The quantities N and c vary from material to material (or from system to system), but the exponent ν stays the same. The exponent ν is the most crucial term describing the state of the polymeric system (for example, randomness, crossover, order). It is the critical exponent that describes the universal behavior of polymer chains, and it is independent of the details of the materials and the models that we use.

Chapters 6 through 11 describe the experimental methods that, in addition to the classical exposition, are closely related to current thinking on universal laws that govern polymer behavior.

APPENDIX A SCALING CONCEPTS

Scaling Relationship

For any two quantities A and B that are related;

$$A = f(B) \quad \text{or} \quad B = g(A)$$

and that characterize a regime (for example, a molecular system), we can write their relationship in the form of a power law:

$$A \sim B^{\nu} \qquad (\text{or } B \sim A^{\nu'})$$

where ν (or ν') is the exponent of the scaling variable (A or B). If a regime is in chaos (in a random state), the power law has a special meaning; namely, ν represents a critical value, which may be called the fractal dimension, the critical correlation length or the measure, and with which we observe an order in randomness. The value ν can be determined experimentally or by computer simulation.

Dimensions of a Polymer Chain

In a very dilute solution c' with a good solvent, the polymer molecule swells and its size can be expressed (in terms of mean-square end-to-end distance) by

$$\langle R^2 \rangle \sim N^{2\nu}$$

(ν being the excluded volume effect exponent). When the solution is very concentrated (c''), the chain, which is not collapsed yet, is in the ideal (or unperturbed) state and its size can be expressed as

$$\langle R^2 \rangle \sim N$$

In between we have a semidilute solution c^* with the threshold density ρ^*. A given chain now has two different sizes:

$$\langle R^2 \rangle \sim N^{-(2\nu-1)/(d\nu-1)} \qquad \text{Inside the blob}$$

$$\langle R^2 \rangle \sim N \qquad \text{Outside the blob}$$

APPENDIX B CORRELATION FUNCTION

Let x and y be two independent variables, and $P_1(x)$ and $P_2(y)$ be probability density distributions. Then the joint probability distribution of x and y, $P(x, y)$, is related to $P_1(x)$ and $P_2(y)$ by

$$P(x, y)\,\mathrm{d}x\,\mathrm{d}y = P_1(x)P_2(y)\,\mathrm{d}x\,\mathrm{d}y$$

If, however, x and y are not independent, then $P_1(x)$ and $P_2(y)$ can still be correlated provided that there exists a function $g(x, y)$, called the correlation function, such that

$$P(x, y)\,\mathrm{d}x\,\mathrm{d}y = P_1(x)P_2(y)g(x, y)\,\mathrm{d}x\,\mathrm{d}y$$

Example 1

Consider the expression

$$\int_0^t \mathrm{e}^{-\zeta(t-t')/m} f(t')\,\mathrm{d}t'$$

where ζ is the frictional coefficient, m is the mass, and t and t' are different times. Neither t nor t' is an independent variable; $(t - t')$ is a variable of interest. Let $\rho = t - t'$. Then

$$\int_0^t e^{-\zeta(t-t')/m} f(t')\,dt' = \int_0^t e^{-\zeta\rho/m} f(t-\rho)\,d\rho$$

We introduce another dummy variable σ such that

$$\sigma = t - t'$$

Although both ρ and σ are expressed in terms of $t - t'$, they are not equal; they are two different variables. Squaring the function, we obtain

$$\int_0^t d\rho \int_0^t d\sigma e^{-\xi\rho/m} e^{-\xi\sigma/m} \langle f(t-\rho) f(t-\sigma)\rangle$$

If we impose a condition that $f(t)$ is random, then the function $\langle f(t-\rho)f(t-\sigma)\rangle$ depends only on the time interval $|\rho - \sigma|$. We now have

$$g(|\rho - \sigma|) \equiv \langle f(t-\rho) f(t-\sigma)\rangle$$

The function in the form $\langle f(t-\rho)f(t-\sigma)\rangle$ is called the autocorrelation function. It involves double integrals.

Example 2

Let the two functions be

$$x_1(t) = \int_0^t f(t')\,dt'$$

$$x_2(t) = \int_0^t f(t'')\,dt''$$

Here t, t', and t'' are not true independent variables, but the time interval $|t'' - t'|$ is. The parameter t'' is equivalent to ρ and t' is equivalent to σ in the previous example. Hence, the functions Δx_1 and Δx_2 are correlated by

$$\langle [\Delta x(t)]^2\rangle = \int_0^t dt' \int_0^t dt'' \langle f(t') f(t'')\rangle$$

Again, if t is completely at random, the expression $\langle f(t')f(t'')\rangle$ is the correlation function. If $\sigma \equiv t'' - t'$, we can write the autocorrelation function (Berry et al.,

1980), as

$$g(f) \equiv \langle f(t')f(t' + \sigma) \rangle = \langle t(0)t(\sigma) \rangle = \langle f(t' - \sigma)f(t') \rangle$$

This topic correlation function, will further be described in the Appendix to Chapter 11, Fourier Series, Fourier Integral, and Fourier Transform.

REFERENCES

Berry, R. S., S. A. Rice, and J. Ross, *Physical Chemistry*. New York: Wiley, 1980.

Chandrasekhar, S., *Rev. Mod. Phys.* **15**, 1 (1943).

de Gennes, P. G. *J. Chem. Phys.* **55**, 572 (1971).

de Gennes, P. G. *Phys. Lett. A* **38**, 399 (1972); *J. Phys. Lett.* **36**, L-55 (1975); *Macromolecules* **9**, 564 (1976).

de Gennes, P. G., *Scaling Concepts in Polymer Physics*. Ithaca, NY: Cornell University Press, 1979.

Doi, M., and S. F. Edwards, *J. Chem. Soc. Faraday Trans. II*, **74**, 1789, 1802, 1818 (1978); *The Theory of Polymer Dynamics*, New York: Oxford University Press, 1988.

Edwards, S. F., *Proc. Phys. Soc. London* **85**, 613 (1965); **88**, 265 (1966).

Feller, W., *An Introduction to Probability Theory and Its Applications*. New York: Wiley, 1950.

Fixman, M., *J. Chem. Phys.* **23**, 1656 (1955).

Flory, P. J., *Principles of Polymer Chemistry*. Ithaca, NY: Cornell University Press, 1953.

Flory, P. J., *Statistical Mechanics of Chain Molecules*. New York: Interscience, 1969; Hanser Publishers, new edition, 1989.

Flory, P. J., *Science* **188**, 1268 (1975).

Freed, K. F., *Renormalization Group Theory of Macromolecules*. New York: Wiley, 1987.

Graessley, W. W., *Adv. Polym. Sci.* **16**, 1 (1974).

Grimley, T. B., *Proc. R. Soc. London Ser. A* **212**, 339 (1952).

Hermans, J. J., *Rec. Trav. Chim.* **69**, 220 (1950).

Kac, M., *Am. Math. Monthly* **54**, 369 (1947).

Kratky, O. and G. Porod, *Rec. Tran. Chim.* **68**, 1106 (1949).

Kuhn, W., *Ber.* **63**, 1503 (1930).

Kurata, M., *J. Polym. Sci. A* **2–6**, 1607 (1968).

Leger, L., H. Hervel, and F. Rondelez, *Macromolecules* **14**, 1732 (1981).

Mandelbrot, B. B., *The Fractal Geometry of Nature*. New York: Freeman, 1982.

Mayer, J. E., and M. G. Mayer, *Statistical Mechanics*, New York: John Wiley, 1940.

Miyaki, Y., Y. Einaga, and H. Fujita, *Macromolecules* **6**, 1180 (1978).

Nierlich, M., J. P. Cotton, and B. Farnoux, *J. Chem. Phys.* **69**, 1379 (1978).

Noolandi, J., G. W. Slater, H. A. Lim, and J. L. Viovy, *Science* **243**, 1456 (1989).

Perzynski, R., M. Adam, and M. Delsanti, *J. Phys.* **43**, 129 (1982).

Stauffer, D., *Introduction to Percolation Theory*. London, Taylor & Francis, 1985.

Sun, S. F., C. C. Chow, and R. A. Nash, *J. Chem. Phys.* **93**, 7508 (1990).

Ursell, H. D., *Proc. Cambridge Phil. Soc.* **23**, 685 (1927).

Uspensky, J. V., *Introduction to Mathematical Probability*. New York: McGraw-Hill, 1937.

Wilson, K. G., *Phys. Rev. B* **4**, 3174, 3184 (1971).

Yamakawa, H., *Modern Theory of Polymer Solutions*. Harper & Row, 1971.

Yamakawa, H., and M. Fujii, *Macromolecules* **7**, 128 (1974).

Zimm, B. H., *J. Chem. Phys.* **14**, 164 (1946).

Zimm, B. H., W. H. Stockmayer, and M. Fixman, *J. Chem. Phys.* **21**, 1716 (1953).

PROBLEMS

5.1 The distribution for a random chain in one dimension (random walk) may be expressed in the form of Bernoulli's equation:

$$w(n, m) = \left(\frac{1}{2}\right)^n \frac{n!}{[(n+m)/2]!\,[(n-m)/2]!}$$

where $w(n, m)$ is the probability of m displacement out of n steps. Using Stirling's formula

$$n! = \sqrt{2\pi}\,\frac{n^{n+1/2}}{\mathrm{e}^n}$$

show that Bernoulli's equation can be converted to

$$w(n, m) = \left(\frac{2}{\pi n}\right)^{1/2} \left[1 - \left(\frac{m}{n}\right)^2\right]^{-(n+1)/2} \left(\frac{1 - m/n}{1 + m/n}\right)^{m/2}$$

which, in turn, can be converted to a Gaussian distribution function (or normal error function):

$$w(n, m) = \left(\frac{2}{\pi n}\right)^{1/2} \mathrm{e}^{-m^2/2n}$$

Assume that $|m/n| \ll 1$.

Hint: $(1 + x)^m = 1 + mx + \cdots \qquad x \ll 1$

$$\lim_{h \to 0} (1 + h)\frac{1}{h} = e$$

5.2 Suppose a person walks from the origin in one dimension one unit step each, forward or backward. The probability for each direction is 1/2. Find the probability function of the person being at each of the possible points between $-n$ and n (where n is the number of total steps).

Hint: $x = \sum_{i=1}^{n} x_i$ where x is the net displacement after n steps and x_i are the individual steps.

5.3 Show that the root-mean-square end-to-end distance is

$$\langle R^2 \rangle^{1/2} = N^{1/2} a$$

if the probability function $w(R)$ for an ideal polymer chain (random flight chain) is given by

$$w(R) = \left(\frac{3}{2\pi N a^2} \right)^{3/2} \exp\left(\frac{-3R^2}{2Na^2} \right)$$

5.4 Show that for wormlike chains,

$$\lim_{\lambda L \to \infty} \langle R^2 \rangle = \frac{L}{\lambda} \qquad \text{for random flights}$$

$$\lim_{\lambda L \to 0} \langle R^2 \rangle = L^2 \qquad \text{for rods}$$

(*Source*: Yamakawa (1971).)

5.5 (a) For the random flight chain model, Kurata (1968) derived an equation to relate α_R with z:

$$\tfrac{1}{5}(\alpha_R^5 - 1) + \tfrac{1}{3}(\alpha_R^3 - 1) = \tfrac{4}{3} z$$

where

$$\alpha_R = \frac{\langle R^2 \rangle^{1/2}}{\langle R^2 \rangle_0^{1/2}}$$

show that at large z,

$$\alpha_R^2 = (\text{constant}) z^{2/5}$$

(b) The parameter z is defined as

$$z = \left(\frac{1}{4\pi} \right)^{3/2} \frac{\beta}{M_0^2 A'^{3/2}} M^{1/2},$$

where A' is a constant and β is the binary cluster integral. Since z is proportional to $M^{1/2}$ for a given polymer–solvent system, the plot of α_R^2 versus $M^{1/5}$ should be linear. Plot the following data of polystyrene in benzene to test the theory:

In benzene at 25°C		In cyclohexane at 34.5°C	
Mol wt $\times 10^{-6}$	$\langle R^2\rangle^{1/2}$ (nm)	Mol wt $\times 10^{-6}$	$\langle R^2\rangle_0^{1/2}$ (nm)
56	50	55	22
39	39	38	18
31	35	32	16
23	29	23	14
15	23	15	11
9	16	9	87

(c) An empirical equation was suggested whereby

$$\alpha_R^2 = (1.64)z^{2/5}$$

Plot $(1.64)z^{2/5}$ versus $z^{2/5}$ for z from 6 to 40 and read the values of z for α_R^2 in part (b). (*Source*: Miyaki (1978).)

5.6 The radius of gyration for a polystyrene (Mol wt 30×10^6) in benzene solution was found to be 1010 Å in a semidilute solution. Estimate the overlapping concentration c^* in g cm^3.

5.7 The theoretical equation of the end-to-end distance $R(\rho)$ for a polymer in a semidilute solution may be expressed in the form

$$R^2(\rho) = \frac{N}{\rho(\xi)}\xi^2 \sim N\rho^{(1-2\nu)/(\nu d-1)}$$

Show that this equation can be reduced to an equation that can be tested experimentally in the form

$$R^2(c) \sim M_w c^{-1/4} \qquad \text{for } d = 3$$

that is,

$$\frac{R^2(c)}{M_w} \sim c^{-.25}$$

Note: $\rho(\xi) \sim \xi^{1/\nu}$.

6

VISCOSITY AND VISCOELASTICITY

Consider the motion of a body in space against an external force. If the vector sum of all force components (translational and rotational motions) in all directions is not zero, then the body will accelerate in motion or change in state, that is, change in force components. If, on the other hand, the vector sum is zero, then body will be stressed. If that body is not rigid, then it will undergo a deformation, it will be strained. We call this body a viscoelastic body. The mechanical properties of a polymer may be described by three quantities: stress σ, strain ε, and rigidity or modulus E. The units of stress and modulus are both dyn/cm^2 (or g/cm s^{-2}). Strain is dimensionless.

In this chapter we devote two-thirds of the space to viscosity and one-third to viscoelasticity of a polymer in solution. Viscosity is related to shear stress σ and shear strain ε, without reference to E. Viscoelasticity is related to all the three parameters, with E being of primary importance.

6.1 VISCOSITY

Imagine a fluid as a stack of parallel plates separated in coordinate x. The plates move in the same direction y but with different velocity, as seen in Figure 6.1. Each plate exerts force on the plate underneath. The shear stress σ is the force F per unit area A,

$$\sigma = F/A \quad \text{dyne/cm}^2 \tag{6.1}$$

and the shear strain ε is the derivative of coordinate y with respect to x,

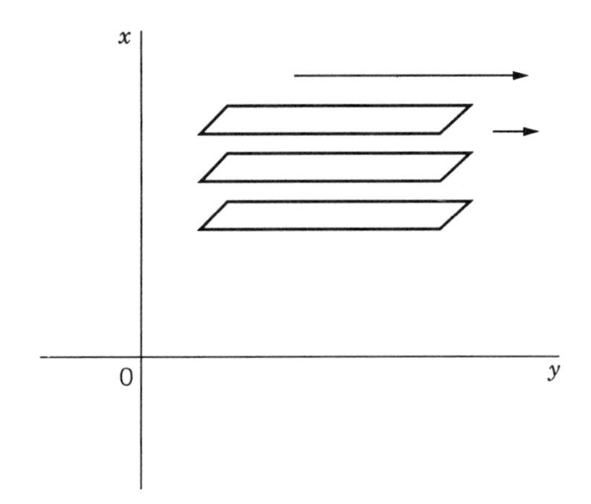

FIGURE 6.1 The fluid as a stack of parallel plates

$$\varepsilon = \frac{dy}{dx} \tag{6.2}$$

The viscosity η is the ratio of shear stress over the rate of shear strain:

$$\eta = \frac{\sigma}{d\varepsilon/dt} \tag{6.3}$$

But notice the identities

$$\frac{d\varepsilon}{dt} = \frac{d}{dt}\left(\frac{dy}{dx}\right) = \frac{d}{dx}\left(\frac{dy}{dt}\right) = \frac{d\mathbf{v}}{dx}$$

where v is velocity of the fluid. We thus have

$$\eta = \frac{\sigma}{d\mathbf{v}/dx} \tag{6.4}$$

A fluid that obeys Eq. (6.4) is called a Newtonian fluid; a fluid that does not obey Eq. (6.4) is called a non-Newtonian fluid. The non-Newtonian behavior of a polymer solution is rarely investigated. It is the Newtonian behavior that is dealt with here.

6.1.1 Capillary Viscometers

The viscometers that are most commonly used for polymers in solution are the Oswald, Fenski, and Ubbelohde types. The devices are all based on the

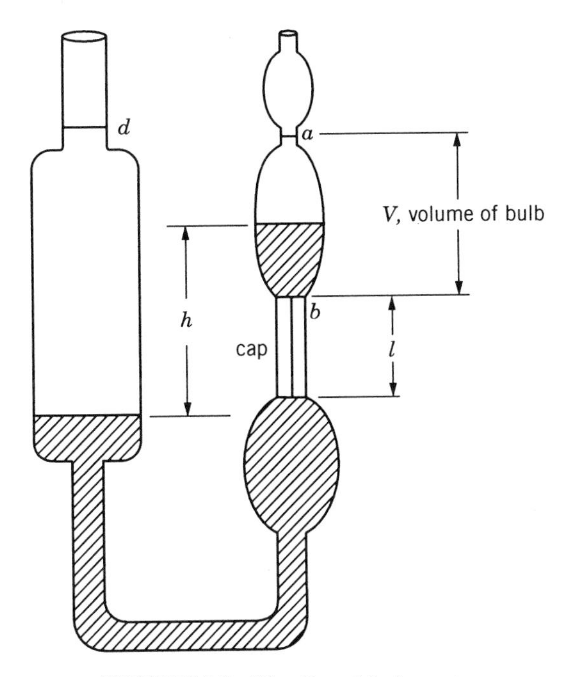

FIGURE 6.2 The Oswald vicometer.

measurement of the flow of liquids through a tube. We describe the Oswald viscometer (Figure 6.2) as an example.

The flow time t of liquid through the capillary of length l is the time taken for the meniscus to fall from point a to point b, that is, the efflux of the bulb. The pressure head p is the difference of pressure at the two ends of the liquid, h. It is calculated from

$$p = h\rho_0 g$$

where ρ_0 is the density of the liquid and g is the acceleration of gravity. Since h is variable as the liquid flows, p is likewise variable.

The viscosity η may be calculated according to Poiseuille's law:

$$\eta = \frac{\pi r^4 tp}{8Vl} \tag{6.5}$$

where r is the radius of the capillary in cm, p is the pressure head in dyn/cm^2, V is the volume of bulb, and V/t is the volume flow rate.

The unit of η is poise P (in honor of Poiseuille) of dyn/s cm^{-2}. For example, the viscosity of water at 20°C is 0.01 P or 1 centipoise (cP). The dimensions of

η are m l^{-1} t^{-1} (m mass; l, length; t, time); hence in the SI system the units are kg/m s and in the c.g.s. system the units are g/cm s.

From Eq. (6.5), two quantities can be defined: the maximum shearing stress σ at the capillary wall

$$\sigma = \frac{h\rho g r}{2l} \tag{6.6}$$

and the rate of shear D (i.e., velocity gradient)

$$D = \frac{4V}{r^3 \pi t} \tag{6.7}$$

Then the viscosity η can be expressed in terms of σ and D:

$$\eta = \frac{\sigma}{D} \tag{6.8}$$

The plot of log D versus log σ will indicate whether the fluid is Newtonian or non-Newtonian (Figure 6.3*a*,*b*). The flow of fluid may depend on the force of shear in different regions, as shown in Fig. 6.3*b*.

If a solution is known to be independent of the force of shear, then the measurement of η based on Poiseuille's law can be made easy by grouping all those terms related to a specific viscometer as a calibration constant A'. Hence, we have

$$\eta = A' h \rho_0 t \tag{6.9}$$

If the polymer solutions is very dilute, ρ_0 may be used as the density of the solution without introducing a significant error. Since h and t are both variables, to use the Oswald viscometer we must fill up the solution (or solvent) to the mark d to keep h constant for each measurement. Only by so doing is h incorporated into the constant A', and we have

$$\eta = A \rho_0 t \tag{6.10}$$

Equation (6.9) can be applied to any solution (not necessarily dilute) in the form

$$\eta = A \rho t \tag{6.11}$$

where ρ is the density of solution. If we divide both sides by ρ, we have

$$\eta_k = \frac{\eta}{\rho} = At \tag{6.12}$$

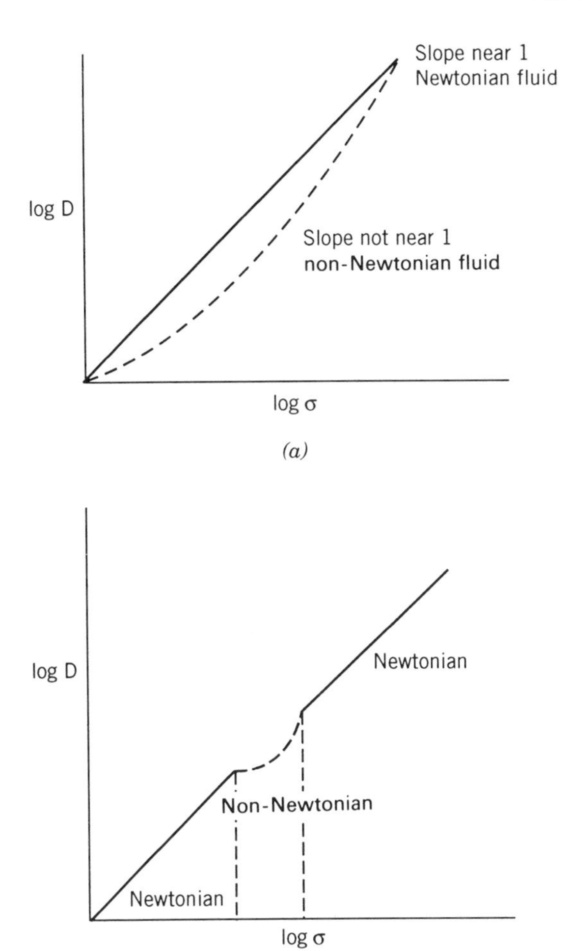

FIGURE 6.3 $\log D$ versus $\log \sigma$.

The quantity η_k is called the kinematic viscosity and the viscosity is proportional to the flow time alone. The units of kinematic viscosity are the stoke (S) and centistoke (cS) or cm^2/s.

To avoid the tedious process of filling the solution to a certain mark d each time, Ubbelohde modified the Oswald viscometer to the form shown in Figure 6.4. The U-tube of the Ubbelohde viscometer eliminates the need to refill the solution each time.

For high precision work, a small correction for kinetic energy may be considered:

$$\eta = A\rho_0 t - \frac{Y}{T} \qquad (6.13)$$

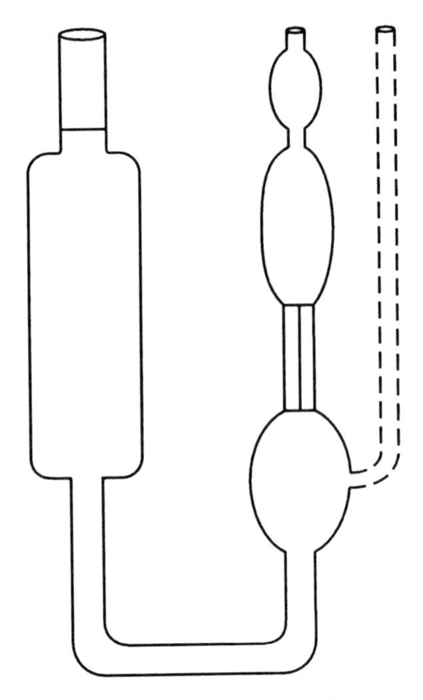

FIGURE 6.4 The Ubbelohde viscometer.

where Y is another calibration constant, and Y/t is the kinetic energy correction term. The values of A and Y can be determined by carrying out two experiments for two liquids (such as water and benzene) of known viscosity at the same temperature or the same liquid (such as water) at two different temperatures.

6.1.2 Intrinsic Viscosity

In macromolecular chemistry, the relative viscosity η_r is often measured. The relative viscosity is the ratio of the viscosity of solution to that of the solvent:

$$\eta_r = \frac{\eta_{\text{solution}}}{\eta_{\text{solvent}}} = \frac{\eta}{\eta_0} = \frac{(\rho t)_{\text{solution}}}{(\rho t)_{\text{solvent}}} \cong \frac{t_{\text{solution}}}{t_{\text{solvent}}} \tag{6.14}$$

The specific viscosity η_{sp} is obtained from the relative viscosity by subtracting one unit from η_r:

$$\eta_{sp} = \eta_r - 1 \tag{6.15}$$

The intrinsic viscosity, denoted by $[\eta]$, is defined as

$$[\eta] = \lim_{c \to 0} \frac{\eta_{sp}}{c} = \lim_{c \to 0} \left(\frac{1}{c} \ln \frac{\eta}{\eta_0} \right) \tag{6.16}$$

where c is the concentration of the polymer in grams per 100 mL or grams per mL of the solution. The quantity η_{sp}/c is called the reduced viscosity. The unit of intrinsic viscosity is deciliters per gram (dL/g) or millimeters per gram (mL/g) depending on the concentration unit of the solution. The intrinsic viscosity is also called the limiting viscosity number.

The plot of η_{sp}/c versus c or $(1/c)$ ln η_r versus c often gives a straight line, the intercept of which is $[\eta]$. Huggins (1942) showed that the slope is

$$\frac{\mathrm{d}}{\mathrm{d}c}\frac{\eta_{sp}}{c} = k[\eta]^2$$

or

$$\frac{1}{[\eta]^2}\frac{\mathrm{d}}{\mathrm{d}c}\frac{\eta_{sp}}{c} = k$$

In integrated form, the equation is

$$\frac{\eta_{sp}}{c} = [\eta] + k[\eta]^2 c \tag{6.17}$$

where k is a dimensionless constant, called the Huggins' constant. The value of k is related to the structures of polymers. It enables an estimate of $[\eta]$ to be made from a single determination of η_{sp}/c. Table 6.1 gives some sample values of k.

Some other equations for the determination of $[\eta]$ are

Schultz and Blaschke (1941)

$$\frac{\eta_{sp}}{c} = [\eta] + k'[\eta]\eta_{sp} \tag{6.18}$$

Kraemer (1938)

$$\frac{1}{c}\log \eta_r = [\eta] + k''[\eta]^2 c \tag{6.19}$$

TABLE 6.1 Some Sample Values of *k*

Polymer	Solvent	Temperature (°C)	Huggins k value
Polystyrene	Toluene	30	0.38
Polystyrene	Methylethyl ketone	40	0.54
Polyisobutene	Benzene	24	0.94
Polyisobutene	Cyclohexane	25	0.405
Polyisobutene	Methylethyl ketone	25	1.598

Martin (1942)

$$\frac{\eta_{sp}}{c} = [\eta]\,e^{k''[\eta]c} \tag{6.20}$$

For molecules of high intrinsic viscosity a correction must be made for the effect of the rate of shear strain. For relatively low intrinsic viscosity, the rate of shear strain does not have any appreciable effect.

6.1.3 Treatment of Intrinsic Viscosity Data

The Mark-Houwink Equation Staudinger (1932) suggested that the molecular weight M of polymers is proportional to the reduced viscosity:

$$\frac{\eta_{sp}}{c} = k'M$$

where k' is a proportionality constant. Mark (1938) and Houwink (1940) independently correlated the intrinsic viscosity with molecular weight:

$$[\eta] = KM^a \tag{6.21}$$

where K and a both are constants. The Mark-Houwink equation is applicable to many polymers and is extensively used to determine molecular weight. The constants K and a both vary with polymers and solvents. Table 6.2 gives values of K and a for a number of polymer–solvent pairs.

TABLE 6.2 Viscosity–Molecular Weight, $[\eta] = KM^a$

Polymer	Solvent	Temperature (°C)	$K \times 10^4$ (dL/g)	a	$M \times 10^{-4}$
Polystyrene Atactic	Benzene	25	2.7	0.71	6–31
	Cyclohexane	25	1.6	0.80	6–31
		34(θ)	7.5	0.50	7
	1-Chloronaphthalene	74(θ)	18	0.50	4–33
	Decalin	135	1.6	0.77	2.39
Isotactic	Decalin	135	1.1	0.80	2.62
Poly(methyl methacrylate)	Benzene	20	0.84	0.73	7–700
Atactic	Chloroform	20	0.96	0.78	1.4–60
	3-Heptanone	34(θ)	6.3	0.50	6.6–171
Polyvinylacetate	Toluene	25	0.71	0.73	4–330
	Benzene	30	2.2	0.65	97–153
Nylon 6	*m*-Cresol	25	3.2	0.62	0.05–0.5

Source: J. Brandrup and E. H. Immergut (Eds.), *Polymer Handbook*. New York: Wiley, 1975.

Equation (6.21) describes the relationship between viscosity and molecular weight. Since molecular weight is related to the size of polymer chain, Eq. (6.21) also describes the relationship between $[\eta]$ and N (the number of the links in a polymer chain) or $[\eta]$ and $\langle R^2 \rangle$ (the mean-square end-to-end distance). This has stimulated a great interest over the last forty years as to how they are related. More specifically, what are the meanings of K and a?

The Flory-Fox Equation As mentioned earlier, at θ temperature there are only short-range steric interactions between neighboring or near-neighboring polymer segments. The polymer molecule shrinks to a compact form. Flory and Fox (1951) suggested that the intrinsic viscosity at θ condition may be regarded as a measure of the ratio of the effective hydrodynamic volume of the polymer to its molecular weight M. They took the radius of hydrodynamic volume as the root-mean-square end-to-end distance $\langle R^2 \rangle^{1/2}$ in random coil:

$$[\eta]_\theta \sim \frac{\langle R^2 \rangle^{1/2}}{M}$$

If the proportionality constant is taken as Φ, the equation becomes

$$[\eta]_\theta = \Phi \frac{\langle R^2 \rangle_0^{3/2}}{M} = \Phi \left(\frac{\langle R^2 \rangle_0}{M} \right)^{3/2} M^{1/2} = K M^{1/2} \tag{6.22}$$

where

$$K = \Phi \left(\frac{\langle R^2 \rangle_0}{M} \right)^{3/2} \tag{6.23}$$

The proportionality constant Φ is a universal constant, called the Flory viscosity constant. Thus, Flory and Fox provided, partially, the meanings of K and a in the Mark-Houwink equation. At θ temperature, K is defined by Eq. (6.23) and $a = \frac{1}{2}$.

The exact value of Φ is very difficult to determine. Rearranging the equation in the form

$$\Phi = \frac{[\eta]_\theta M}{\langle R^2 \rangle_0^{3/2}} \tag{6.24}$$

we see that any uncertainty in experimental values of $[\eta]_\theta$ and $\langle R^2 \rangle_0$ would contribute to the error of Φ. Two different experimental methods are required to determine these two quantities: viscosity measurement for $[\eta]_\theta$ and light scattering measurement for $\langle R^2 \rangle_0$. The sample solution must be identical and the experimental conditions must be the same. It is very difficult to minimize errors to the same extent simultaneously in two different experiments. In using Eq. (6.24) many investigators therefore rely on the theoretical value. But theoreti-

TABLE 6.3 Some Values of Φ

Polymer	Solvents	$M \times 10^{-6}$	$\Phi \times 10^{-21}$[a]
Polystyrene	Cyclohexane	0.20–4.0	1.5–2.9
	Methylethyl ketone	0.20–1.8	2.0–2.5
	Dichlomethane	0.50–1.8	2.0–2.2
Poly(methyl methacrylate)	Chloroform	0.70–1.4	2.0–2.2
	Methylethyl ketone	0.70–1.4	2.0–2.2
Polyisobutylene	Cyclohexane	0.50–.72	2.2
Poly(acrylic acid)	Dioxane	0.80–1.4	2.2

[a] $[\eta]$ in dL/g.

cal values vary with the model assumed (for example, the pearl string model and the hydrodynamic equivalent sphere, which are discussed later). Flory and Fox estimated the value of Φ to be 2.1×10^{21} from a set of experimental data, whereas the Kirkwood and Riseman equation (to be discussed later) would give 3.6×10^{21}. The value of the exponent depends on the unit of $[\eta]_\theta$ and $\langle R^2 \rangle_0$. If $[\eta]_\theta$ is chosen in dL/g and $\langle R^2 \rangle_0$ in cm^2, then Φ is 2.1×10^{21}. If, however, $[\eta]_\theta$ is in cm^3/g and $\langle R^2 \rangle_0$ is (usually) in cm^2, then Φ is 2.1×10^{23}. Table 6.3 gives the values of Φ to show the variations.

In a good solvent (or in a poor solvent at a temperature other than θ), long-range interactions (or excluded volume effect) take place. The polymer molecule expands. The expansion factor α, which describes the excluded volume effect, can now be calculated from the intrinsic viscosity data:

$$\alpha^3 = \left(\frac{\langle R^2 \rangle}{\langle R^2 \rangle_0} \right)^{3/2} = \frac{[\eta]}{[\eta]_0} (= \alpha_\eta^3)$$

Flory and Fox suggested that the Mark-Houwink equation (Eq. (6.21)) can be put in the form

$$[\eta] = K M^{1/2} \alpha^3 \tag{6.25}$$

where

$$K = \Phi \left(\frac{\langle R^2 \rangle}{M \alpha^2} \right)^{3/2} = \Phi \left(\frac{\langle R^2 \rangle_0}{M} \right)^{3/2} \tag{6.26}$$

Equation (6.26) provides new meanings for K and a if $[\eta]$ is not measured at θ temperature. Flory and Fox further suggested that once the value of α^3 is obtained, we can calculate α^5 and use the equation

$$\alpha^5 - \alpha^3 = 2 c_M \psi_1 \left(1 - \frac{\theta}{T} \right) M^{1/2} \tag{6.27}$$

to obtain thermodynamic parameters, ψ_1, κ_1, and θ. The quantity c_M can be calculated from

$$c_M = \left(\frac{27}{2^{5/2}\pi^{3/2}}\right)\left(\frac{\bar{v}^2}{N_A^2 V_0}\right)\left(\frac{\langle R^2 \rangle_0}{M}\right)^{-3/2} \tag{6.28}$$

where V_0 is the molar volume of the solvent, and $\bar{v}$ is the partial specific volume of the polymer and N_A the Avogadro's number. Thus, through the efforts of Mark, Houwink, Flory, and Fox, the measurement of intrinsic viscosities makes it possible to determine the molecular weight M, the unperturbed dimension of the polymer chain $\langle R^2 \rangle_0$, the extent of expansion α, and the thermodynamic parameters ψ_1, κ_1, and θ. Table 6.4 gives the values of thermodynamic parameters for polystyrene and polyisobutylene obtained by intrinsic viscosity measurements.

The Stockmayer and Fixman Equation If several samples of the same polymer with different molecular weight are available, we can determine the structural parameter K and the interaction parameter χ_1 simultaneously by measuring the intrinsic viscosities. This was suggested by Stockmayer and Fixman (1963) in the equation

$$[\eta] = KM^{1/2} + 0.51\Phi BM \tag{6.29}$$

TABLE 6.4 Thermodynamic Parameters Obtained from Intrinsic Viscosities

Solvent	θ(K)	ψ_1	κ_1 at 25°C
POLYSTYRENE			
Cyclohexane	307	0.13	0.134
Benzene	100	0.09	0.03
Toluene	160	0.11	0.06
Dioxane	198	0.10	0.07
Ethyl acetate	222	0.03	0.02
Methylethyl ketone	0	0.006	0
POLYISOBUTYLENE			
Benzene	297	0.15	0.15
Toluene	261	0.14	0.12
Cyclohexane	126	0.14	0.059
n-Hexadecane	175	0.094	0.055
n-Heptane	0	0.035	0

Note: See Problem 6.4. *Source*: Flory, 1953.

where

$$B = \frac{(1 - 2\chi_1)\bar{v}^2}{V_0 N_A}$$

The equation is valid for polymer chains in all solvents (good or poor). It separates the effect of short-range interaction K from that of long-range interaction B. A plot of $[\eta]/M^{1/2}$ versus $M^{1/2}$ would give the intercept K, from which we can determine $\langle R^2 \rangle_0$, and the slope $0.51\Phi B$, from which we can determine χ_1.

Among the three equations we have discussed, only the Mark-Houwink equation is exact; that is, no assumptions have been made. The (Flory-Fox and Stockmayer-Fixman) equations are both based on their models.

6.1.4 Stokes' Law

Let R be the radius of a large particle in the form of a sphere and $\mathbf{v}$ be the velocity of the particle moving through a continuum fluid. Then the frictional force that the particle experiences may be expressed as

$$F = -f\mathbf{v} = -6\pi\eta R\mathbf{v}$$

where f is the frictional coefficient and η is the viscosity of fluid medium. This equation is called Stokes' law. In polymer chemistry the law is more often expressed in the form

$$f = 6\pi\eta_0 R$$

where η_0 is the viscosity of solvent. Note that here the polymer molecule is assumed to be in the shape of a sphere; hence, its size is expressed in terms of radius.

The frictional force increases with velocity, because the moving object experiences more drag as it collides with more molecules. Several physical properties are directly related to the frictional coefficient f, for example, the diffusion coefficient and the sedimentation coefficient. For this reason, the intrinsic viscosity, diffusion coefficient, and sedimentation coefficient are often referred to as frictional properties. Stokes' law will be discussed in this chapter and in the chapters on the diffusion coefficient and sedimentation coefficient (Chapters 10 and 12).

6.1.5 Theories in Relation to the Intrinsic Viscosity of Flexible Chains

Kirkwood-Riseman Theory (1948) This theory is based on a model in which the chain consists of a sequence of monomer units. When a polymer molecule is placed in a fluid of surrounding medium (solvent molecules), the flow is

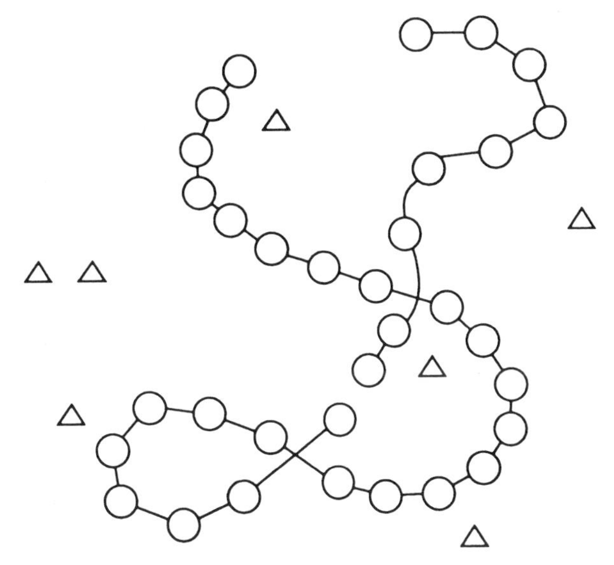

FIGURE 6.5 Pearl string model: Δ, solvent molecule; ○, monomeric unit.

perturbed by the resistance offered by each polymer unit. This model is known as the pearl string (or pearl necklace) model, where each monomer unit is a bead (see Figure 6.5). The emphasis of the Kirkwood-Riseman theory is on the hydrodynamic resistance of the individual beads. When the individual resistance is summed up we obtain the resistance of the whole molecule.

According to Kirkwood and Riseman, the frictional force $\mathbf{F}_l$ exerted on the surrounding medium by the bead l due to the motion of the chain in the medium may be expressed as

$$\mathbf{F}_l = f\,|\mathbf{v}_l - \mathbf{u}_s|$$

where f is the friction coefficient, $\mathbf{v}_l$ is the velocity of the bead, and $\mathbf{u}_s$ is the velocity of the solvent. The perturbation is calculated by a tensor formula advanced by Oseen (1927) which leads to

$$[\eta] = \left(\frac{N_A f b^2}{3600\eta_0 M_0}\right) N F(\lambda_0 N^{1/2}) \tag{6.31}$$

where

$$\lambda_0 = \frac{f}{(6\pi^3)^{1/2}\eta_0 b}$$

$$N = \frac{M}{M_0}$$

In these equations, N_A is the Avogado's number, b is the effective bond length of the chain, N is the number of links in the polymer molecule, M is the molecular weight of the polymer, and M_0 is the molecular weight of the monomer unit. The function $F(\lambda_0 N^{1/2})$ is theoretically tabulated. Its value is between 0 and 1.

Thus, according to this theory, at a given molecular weight, the intrinsic viscosity $[\eta]$ is related to N by the frictional coefficient f of a monomer unit and the effective bond length b of the chain. As N approaches infinity, $[\eta]$ becomes independent of f and the above equation may be simplified to

$$b^3 = \frac{2267 M_0}{(6\pi^3)^{1/2} N_A} \lim_{N \to \infty} \frac{[\eta]}{N^{1/2}} \tag{6.32}$$

By plotting $[\eta]/N^{1/2}$ versus $N^{-1/2}$, the intercept leads to the determination of b. The frictional coefficient of the polymer may also be determined by plotting log–log $[\eta]$ versus N with the assistance of the theoretical table of $F(\lambda_0 N^{1/2})$ given by Kirkwood and Riseman in their original paper.

Debye-Bueche Theory Debye and Bueche (1948) criticize the theory of Kirkwood and Riseman as being unrealistic in assuming that the hydrodynamic resistance of each individual bead is equal. They cannot accept the assumption that there is no mutual interaction among the beads. Instead, they argue that because of the superposition of the disturbances caused by all the other beads, an individual bead deviates from the original undisturbed velocity field during the motion. Following Einstein's method of calculating the effect of a number of immersed impermeable spheres on the overall viscosity (Einstein, 1911), Debye and Bueche considered the polymer molecule as a whole to be an equivalent hydrodynamic sphere of radius R_s, impenetrable to the solvent (Figure 6.6). Inside the sphere the bead density ρ' is constant everywhere; outside the sphere, $\rho' = 0$. The quantity of ρ' may be defined as

$$\rho' = \left(\frac{3}{4\pi}\right)\left(\frac{N}{R_s^3}\right)$$

where N is the number of beads in the string. This model is now called the equivalent sphere model.

The intrinsic viscosity based on this model is given by

$$[\eta] = \frac{(4\pi/3)R_s^3}{M} \phi(\gamma) \tag{6.33}$$

where the argument γ of the function ϕ is the shielding ratio defined as

$$\gamma = \frac{R_s}{L}$$

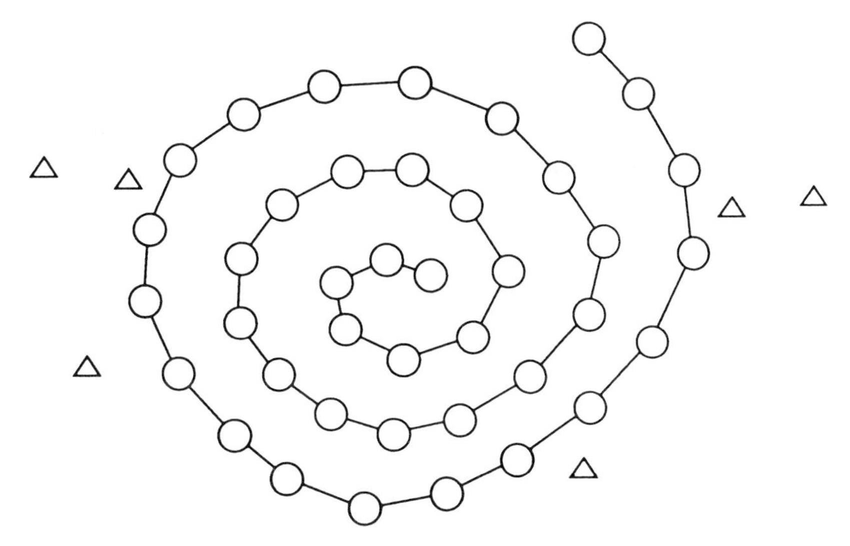

FIGURE 6.6 Equivalent hydrodynamic sphere model. Δ solvent molecule, ○ monomeric unit.

and L is the shielding length related to bead density:

$$\frac{1}{L^2} = \rho'\left(\frac{f}{\eta}\right)$$

The values of γ are in the range between 0 and ∞. For small value of γ, ($\lim \gamma = 0$)

$$\phi(\gamma) = \frac{\gamma^2}{10(1 - \frac{2}{35}\gamma^2 + \cdots)}$$

and we have

$$[\eta] = \left(\frac{1}{10}\right)\left(\frac{f}{\eta_0 m}\right) R_s^2 \tag{6.34}$$

where f is the friction coefficient of one bead, m is the mass of one bead, and η_0 is the viscosity of the solvent. For large values of γ, we have

$$\phi(\gamma) = \frac{5}{2}\left[1 - \left(\frac{3}{\gamma}\right) - \cdots\right]$$

and

$$[\eta] = \frac{5}{2}\left(\frac{4\pi}{3} R_s^3\right)\frac{1}{M} = \frac{5}{2}\left(\frac{4\pi}{3} R_s^3\right)\frac{1}{Nm} \tag{6.35}$$

According to this model, intrinsic viscosity depends not only on the molecular weight, but also on the specific volume of the macromolecule and the shielding ratio. de Gennes considered this model a model of porous sponge with uniform density. The monomers inside the sponge are screened from the flow. Only those on the surface are subject to friction.

6.1.6 Chain Entanglement

Bueche (1952) studied the viscosity behavior of a polymer in concentrated solution, or rather Newtonian melt viscosity. He derived first an equation for unentangled chains:

$$\eta_0 = \left[\frac{N_A}{\sigma} \left(\frac{\langle R^2 \rangle_0}{M\bar{v}_2} \right) z_c \right] f$$

where N_A is Avogado's number, σ is the tensile strength, $\bar{v}_2$ is the specific volume of the liquid, f is the friction factor, z is the chain length (or in a slightly different sense the number of submolecules per polymer chain), and z_c is the critical chain length below which there are no entanglements. This equation predicts that

$$\eta_0 \sim z \qquad \text{(for unentangled chains)}$$

Using the power law we can express more specifically the intrinsic viscosity for unentangled chains as a function of the chain length:

$$\eta_0 \sim z^a$$

where a is between 1 and 2.5 for $z < z_c$.

If there are entanglements involved, as in a polymer melt, the equation of viscosity is modified to

$$\eta_0 = \frac{1}{8} \left(\frac{N_A^2}{6^{1/2}} \right) \left[\left(\frac{\langle R^2 \rangle_0}{M} \right)^{2,5} \left(\frac{z^{3,5} g(z)}{m_0^{1/2} \bar{v}_2^2 z_2^2} \right) \right] f \qquad \text{for } z \geqslant z_c$$

Here m_0 is the mass per chain atom, $g(z)$ is the slippage function. From asymptotic dimensional analysis, this equation may be expressed in the power law as

$$\eta \sim M^{3,4}$$

Experimental results confirm the power law in the form

$$\eta \sim M^{3,4}$$

Theoretical analysis (due to Freed (1980)) in terms of scaling theory describe the complexity of this equation in the following way:

$$\frac{\eta}{c} = \frac{\zeta l^2 N^2 N_A}{\eta_0 M} F\left(\frac{cN_A\alpha^3}{M}, \frac{\omega\zeta^2 l^2 N^2}{kT}\right)$$

where c is the concentration, ζ is the frictional coefficient of the monomer, l is the monomer length, N is the number of monomers, N_A is Avogadro's number, M is the molecular weight of the polymer, ω is the frequency, which corresponds to the elastic modulus of an entanglement network (see Section 6.10), α is the distance in space between entanglement functions, and kT is the absolute temperature in energy unit. The first argument of the scaling function F, $cN_A\alpha^3/M$, is related to probability of entanglements, and the second argument, $\omega\zeta^2 l^2 N^2/kT$, is related to shear modulus. The equation describes the complexity of the dependence of viscosity on the molecular weight in power law form and the justification of the exponent 3.4 in the empirical equation, but it is still impossible to reduce the equation to a simple form with which we can theoretically calculate the viscosity from molecular parameters.

6.1.7 Biological Polymers (Rigid Polymers, Inflexible Chains)

Most of the biological polymers, such as proteins and nucleic acids and some synthetic polymers, have relatively inflexible chains. For rigid particles, the size is no longer of predominant importance, because the polymer chain is no longer in the form of a flexible random coil; instead, shape becomes an important parameter. The shape factor v may be expressed in terms of the axial ratio p, defined as

$$p = \frac{a}{b}$$

where a is the major semiaxis and b is the minor semiaxis. Following are some theoretical proposals for the estimation of the shape factor v from the viscosity measurement.

Einstein (1911) Equations for Spherical Molecules Einstein, in a study of the viscosity of a solution of suspension of particles (colloids), suggested that the specific viscosity η_{sp} is related to a shape factor v in the following way:

$$\eta_{sp} = v\phi \tag{6.36}$$

where ϕ is the volume fraction;

$$\phi = \frac{nv}{V}$$

where n is the number of noninteracting identical particles, v is the volume of each particle, and V is the volume of the solution or suspension. Assume that the molecules are of a spherical shape, rigid and large relative to the size of the solvent molecules, and that the particles are small enough to exhibit Brownian motion, but large enough to obey the laws of macroscopic hydrodynamics. Then

$$v = 2.5$$

The Einstein equation

$$\eta_{sp} = 2.5\phi \tag{6.37}$$

is now used as a reference to estimate the shape of macromolecules. Any deviation can be interpreted as the fact that the molecules are not a sphere.

Peterlin (1938) Equations for Ellipsoids of Revolution in General Under the influence of an applied shearing stress, a gradient of distribution function $\partial F/\partial t$ may be defined to specify the orientation of the particle with respect to two axes (major and minor):

$$\frac{\partial F}{\partial t} = \psi \nabla^2 F - \nabla(F\omega) = 0 \tag{6.38}$$

where F is a function of (ψ, δ, R), the quantity ψ is the rotational diffusion coefficient, δ is the ratio of the rate of shear to the rotational diffusion coefficient, and R is the ellipsoids of rotation. ∇^2 is the Laplacian operator:

$$\nabla^2 = \frac{\partial^2}{\partial x^2} + \frac{\partial^2}{\partial y^2} + \frac{\partial^2}{\partial z^2}$$

and ω is angular velocity of the rotating particles. On the basis of Eq. (6.38), Peterlin derived an equation of the ellipsoids of rotation R from which the shape factor v is determined as

$$v = \frac{p^2 - 1}{p^2 + 1}$$

Simha (1940) Equations for Prolate and Oblate Ellipsoids Simha solved Eq. (6.38) for the viscosities of solutions of ellipsoids of revolution for the limiting case $\delta \to 0$. For very dilute solution at $\delta \to 0$, Simha derived the following two equations of shape factors v:

For prolate ellipsoids

$$v = \frac{p^2}{15(\ln 2p - 1.5)} \tag{6.39a}$$

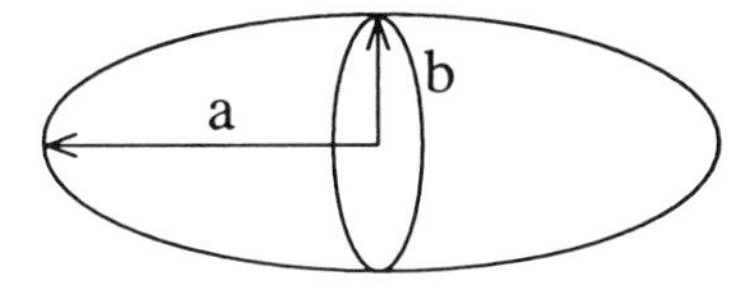

Prolate ellipsoid

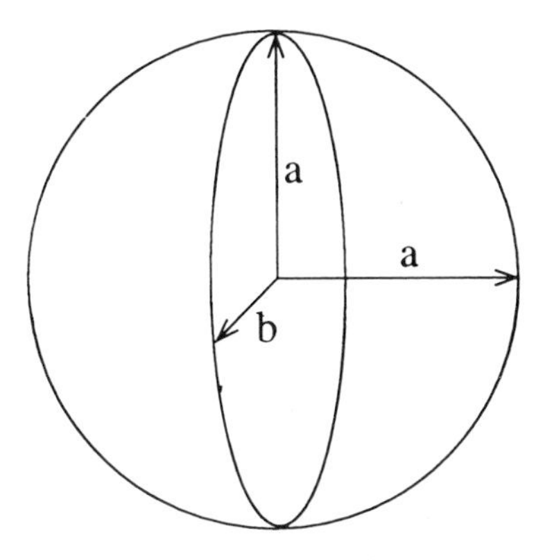

Oblate ellipsoid

FIGURE 6.7 The shapes of macromolecules.

TABLE 6.5 Some Numerical Values of *p*

Axial Ratio	Prolate	Oblate
1	2.50	2.50
10	13.6	8.04
20	38.6	14.8
40	121	28.3
100	593	68.6
300	4278	204

For obtate ellipsoids

$$v = \frac{16p}{15 \arctan p} \tag{6.39b}$$

Two ellipsoids are shown in Figure 6.7. Table 6.5 gives some of the numerical values of p and ellipsoids.

Kuhn and Kuhn (1945) Equations for Rigid Rods and Disks Kuhn and Kuhn derived equations for the shape factors of rigid rods and disks as follows:

$$v = 0.4075(p-1)^{1.508} + 2.5 \qquad 1 < p < 15 \tag{6.40a}$$

$$v = \frac{p^2}{15(\ln 2p - 1.5)} + \frac{p^2}{5(\ln 2p - 0.5)} + 1.6 \qquad p > 15 \tag{6.40b}$$

$$v = \left(\frac{32}{15\pi}\right)(q-1) - \frac{0.628(q-1)}{q-0.075} + 2.5 \qquad q = \frac{1}{p} > 1 \tag{6.40c}$$

Scheraga and Mandelkern (1953) Equations for Effective Hydrodynamic Ellipsoid Factor β Scheraga and Mandelkern suggested that $[\eta]$ is the function of two independent variables: p, the axial ratio, which is a measure of shape, and V_e, the effective volume. To relate $[\eta]$ to p and V_e, Scheraga and Mandelkern introduced f, the frictional coefficient, which is known to be a direct function of p and V_e. Thus, we have for a sphere,

$$\frac{\eta_{sp}}{c} \equiv [\eta] = \left(\frac{N_A}{100}\right)\left(\frac{V_e}{M}\right) v$$

$$f_0 = 6\pi\eta_0 \left(\frac{3V_e}{4\pi}\right)^{1/3}$$

where v is the shape factor (for a sphere, $v = 2.5$). The parameter V_e could be eliminated from the above two equations.

Using the Einstein and Svedberg equations, which we discuss later,

$$D = \frac{kT}{f} \qquad \text{(Einstein's equation of diffusion coefficient } D\text{)}$$

$$S = \frac{M(1-\bar{v}\rho)}{Nf} \qquad \text{(Svedberg's equation of sedimentation coefficient } S\text{)}$$

We obtain

$$\beta \equiv \frac{D[\eta]^{1/3} M^{1/3} \eta_0}{kT} \tag{6.41}$$

or

$$\beta \equiv \frac{NS[\eta]^{1/3}\eta_0}{M^{2/3}(1-\bar{v}\rho)} \tag{6.42}$$

The value of β is a measure of effective hydrodynamic ellipsoid.

6.2 VISCOELASTICITY

The term viscoelasticity is a combination of viscosity and elasticity. In the study of viscosity, we neglect the modulus; in the study of viscoelasticity, we attach

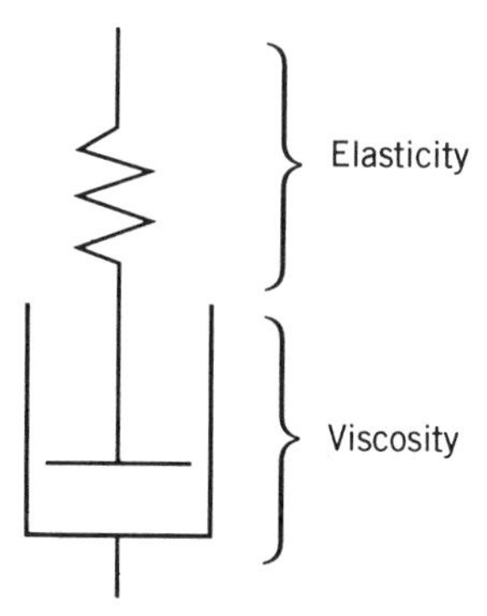

FIGURE 6.8 Maxwell's model of a mechanical body.

great importance to the modulus. A mechanical body contains both elastic springs and viscous damping elements. On this basis several models were developed to describe viscoelastic behavior. Among them are Maxwell's model and the Kelvin-Voigt model. Here we discuss Maxwell's model of a mechanical body under one-dimensional deformation. The diagram of the model is shown in Fig. 6.8.

The elasticity is characterized by Hooke's law:

$$\sigma_0 = \varepsilon_0 E$$

where σ_0 is the instantaneous stress, E is the Young's modulus or the tensile modulus, and ε_0 is the instantaneous strain. The reciprocal of the tensile modulus is called the tensile compliance D:

$$D = \frac{1}{E}$$

In Maxwell's model the two parameters E and η are related by

$$\eta = \tau E \tag{6.43}$$

where τ is the relaxation time. Recalling Eq. (6.3), we can write the equation of motion of the Maxwell model as

$$\frac{d\varepsilon}{dt} = \underbrace{\frac{1}{E}\frac{d\sigma}{dt}}_{\text{Hooke's law}} + \underbrace{\frac{\sigma}{\eta}}_{\text{Newton's law}} \tag{6.44}$$

It is clear that viscoelasticity is basically a relaxation phenomenon. The relaxation time often characterizes the behavior of the material. In comparison to the time scale of measurement, if the relaxation time is long, the material behaves as an elastic solid. If the relaxation time is very short, the material behaves as a viscous liquid. Only when the relaxation time is comparable to the time scale of measurement is the material viscoelastic.

For one-dimensional deformation, the body undergoes uniaxial tension or compression. For a three-dimensional deformation, the body undergoes the shear force. That is, for a three-dimensional deformation, we have shear stress (labeled σ_s) and shear strain (labeled γ). The shear modulus G and the shear compliance J are defined and related by

$$G = \frac{\sigma_s}{\gamma} = \frac{1}{J}$$

The relationship between E and G and between D and J is given by

$$E = 2(1 + \mu)G$$

$$J = 2(1 + \mu)D$$

where μ is Poisson's ratio:

$$\mu = \frac{1}{2}\left[1 - \frac{1}{V}\frac{\mathrm{d}V}{\mathrm{d}\varepsilon}\right]$$

and V is the volume.

If a mechanical body is under a constant stress and if its strain is measured as a function of time, we can introduce two more quantities:

$$J(t) = \frac{\gamma(t)}{\sigma_{s0}}$$

$$D(t) = \frac{\varepsilon(t)}{\sigma_{t0}}$$

where σ_{s0} is the constant shear stress, ε is the tensile strain and σ_{t0} is the constant tensile stress. The quantity $J(t)$ is called the shear creep and the quantity $D(t)$ is called the tensile creep. Both are measured by relaxation experiments and both are known as step functions. In a dynamic relaxation experiment, however, the stress or strain is an oscillating function with an angular frequency ω. Our measurement of dynamic modulus values is often in terms of ω rather than time. The quantities $\sigma(t)$ and $\varepsilon(t)$ for one dimensional are redefined:

$$\sigma(t) = \sigma_0 \mathrm{e}^{\mathrm{i}\omega t} \tag{6.45}$$

$$\varepsilon(t) = \varepsilon_0 \mathrm{e}^{\mathrm{i}(\omega t - \delta)}$$

where δ is the loss angle. Similarly, for three dimensions, we redefine

$$\sigma_s(t) = \sigma_0 \sin \omega t$$

$$\gamma(t) = \frac{\sigma_0}{G} \sin \omega t$$

The viscosity behavior measured by the relaxation experiment is now described by

$$\frac{d\gamma}{dt} = \frac{\sigma_0}{\eta} \sin \omega t$$

which can be integrated to give

$$\gamma(t) = \frac{\sigma_0}{\eta\omega} - \frac{\sigma_0}{\eta\omega} \cos \omega t$$

Substituting Eq. (6.45) into Eq. (6.44), we obtain

$$\frac{d\varepsilon(t)}{dt} = \frac{\sigma_0}{E} i\omega e^{i\omega t} + \frac{\sigma_0}{\eta} e^{i\omega t} \tag{6.46}$$

Solving this equation for a Maxwell body leads to the following equations of modulii, which include the parameter τ:

$$\tan \delta = \frac{E_2}{E_1} = \frac{1}{\tau\omega}$$

$$E_1 = E \frac{\tau^2\omega^2}{1 + \omega^2\tau^2}$$

$$E_2 = E \frac{\tau\omega}{1 + \omega^2\tau^2} \tag{6.47}$$

where δ is the tangent of the phase angle between the stress and strain, that is, the loss angle, E_1 and E_2 are real parts of the complex modulus E^*:

$$E^* = E_1 + iE_2 \tag{6.48}$$

Similar equations can be written for a complex shear modulus G^* (where j is an index of element; $j = 1, 2, \ldots N$):

$$G^* = G_1 + iG_2 = \frac{i\omega\eta_j}{1 + i\omega\tau_j} \tag{6.49}$$

with

$$\tau_j = \frac{\eta_j}{G_j}$$

$$G_1 = G_j \frac{\omega^2 \tau_j^2}{1 + \omega^2 \tau_j^2}$$

$$G_2 = G_j \frac{w \tau_j}{1 + \omega^2 \tau_j^2}$$

$$\tan \delta = \frac{1}{\omega \tau_j} \tag{6.50}$$

6.2.1 The Rouse Theory

In 1953 Rouse published a paper to describe theoretically the flow of polymers in dilute solutions. The polymer molecule is assumed to exist as a statistical coil and is subdivided into N submolecules. Each submolecule is thought of as a solid bead. The beads behave as Gaussian chains and their entropy–elastic recovery can be described by a spring with a spring constant $3kT/a^2$, where a is the average end-to-end distance of a submolecule and k is the Boltzmann constant. The model is shown in Figure 6.9.

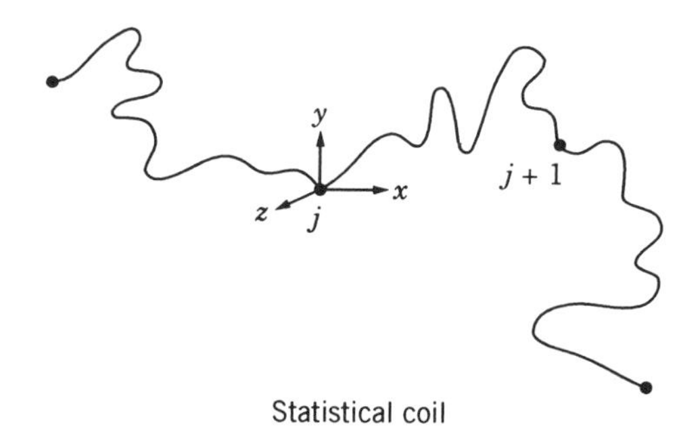

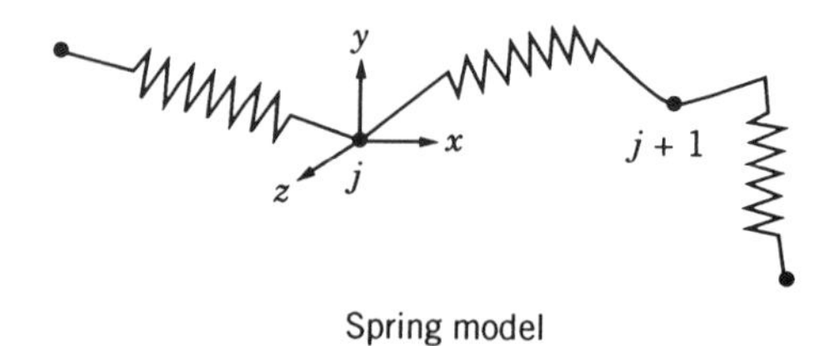

FIGURE 6.9 Spring-bead model.

However, there is no interaction between beads other than the spring force. The restoring force on each of the beads is given by

$$f_{jx} = \frac{-3kT}{a^2}(-x_{j-1} + 2x_j - x_{j+1}) \tag{6.51}$$

When the polymer is emerged in a solvent medium, an additional force, frictional drag, acts on the bead:

$$f_{jx} = \zeta \frac{dx_j}{dt} = \zeta \dot{x}_j \tag{6.52}$$

where ζ is the segmental friction factor. Because the forces arising from the acceleration of the beads are small, the elastic force given in Eq. (6.51) and the viscous force given in Eq. (6.52) must be equal. Thus, we have

$$\zeta \dot{x}_j = \frac{-3kT}{a^2}(-x_{j-1} + 2x_j - x_{j+1}) \tag{6.53}$$

This set of linear first-order differential equations can be summarized in the form of a matrix equation

$$[\dot{X}] = -B[A][X] \tag{6.54}$$

where B represents the mobility of the end of a submolecule. To solve Eq. (6.54), an orthogonal transformation of coordinates was performed (see Yamakawa (1971)). By doing so, the coordination of all the motions of the path of a polymer is resolved into a series of modes. Each mode has a characteristic relaxation time τ_j.

Final results obtained for the viscoelastic properties of dilute solutions of coiling polymers are given in a series of equations. The real and imaginary components of a complex viscosity $\eta^* = \eta_1 - i\eta_2$ are

$$\begin{aligned} \eta_1 &= \eta_s + nkT \sum_{j=1}^{N} \frac{\tau_j}{1 + \omega^2 \tau_j^2} \\ \eta_2 &= nkT \sum_{j=1}^{N} \frac{\omega \tau_j^2}{1 + \omega^2 \tau_j^2} \end{aligned} \tag{6.55}$$

where η_s is the viscosity of the solvent and η is the number of links of a chain. The components for a complex shear modulus

$$G^* = G_1 + iG_2 = i\omega\eta^* \tag{6.56}$$

are

$$G_1 = nkT \sum_{j=1}^{N} \frac{\omega^2 \tau_j^2}{1 + \omega^2 \tau_j^2}$$

$$G_2 = \omega \eta_s + nkT \sum_{j=1}^{N} \frac{\omega \tau_j}{1 + \omega^2 \tau_j^2} \tag{6.57}$$

The relaxation times τ_j in these equations are

$$\tau_j = r^2 \left[24BkT \sin^2 \frac{j\pi}{2(N+1)} \right]^{-1} \qquad j = 1, 2, \ldots, N \tag{6.58}$$

where r^2 is the mean-square end-to-end distance of the submolecule. In a steady flow, $\omega = 0$, we have $G_1 = 0$, $G_2 = 0$, and $\eta_2 = 0$. The real part of η^* is now the steady-flow viscosity η_0:

$$\eta_0 = \eta_s + \frac{nr^2 N(N+2)}{36B} \tag{6.59}$$

Each relaxation time makes a contribution of $\eta kT\tau_j$ to η_0. The subscript runs from 1 to N. τ_1 is the longer relaxation time, which accounts for nearly all of the viscosity. τ_N is the short relaxation time, which accounts for only a small part of the viscosity. The two quantities may be expressed as

$$\tau_1 = nr^2 \left[24B \sin^2 \frac{\pi}{2(N+1)} \right]^{-1}$$

$$\tau_N = \frac{r^2}{24BkT} \qquad \text{if } N \gg 1 \tag{6.60}$$

According to Eq. (6.58), the relaxation times are inversely proportional to temperature and to the mobility of the end of a submolecule. The mobility increases with the decrease in the viscous force that is exerted by the surrounding medium. The relaxation times are directly proportional to r^2 (the mean-square end-to-end distance of the submolecule) and consequently to $\langle R^2 \rangle$ (the mean-square end-to-end distance of the polymer molecule).

If $N \gg 2$, Eq. (6.59) can be rewritten

$$\eta_0 - \eta_s = \frac{\eta N \langle R^2 \rangle}{36B} \tag{6.61}$$

Note that $\langle R^2 \rangle = r^2 N$. This equation is similar to the one obtained by Debye in his hydrodynamic calculation of the viscosity of a solution of free-draining chains. However, N here is the number of submolecules rather than the number of atomic groups, and the frictional coefficient of an atomic group is replaced by B^{-1}, as Rouse pointed out. For experimental study, see Rouse's other two papers (1953, 1954).

6.2.2 The Zimm Theory

Zimm's model (1956) is also a chain of beads connected by ideal springs. The chain consists of N identical segments joining $N+1$ identical beads with complete flexibility at each bead. Each segment, which is similar to a submolecule, is supposed to have a Gaussian probability function. The major difference between the two models lies in the interaction between the individual beads. In the Rouse model, such interaction is ignored; in Zimm's model, such interaction is taken into consideration.

According to Zimm's model, if a chain is suspended in a viscous liquid, each bead j encounters three different forces: mechanical force, Brownian motion, and the motion of a fluid.

Mechanical Force Mechanical force can be expressed by

$$F_{xj} = \zeta(\dot{x}_j - v'_{xj}) \tag{6.62}$$

where ζ is a friction factor, $\dot{x}_j = dx_j/dt$, and v'_{xj} is the velocity component of the fluid if the jth bead were absent. Similar expressions are given for F_{yj} and F_{zj}.

Brownian Motion From the Brownian motion, the beads move, resulting in a force F that involves Hooke's law:

$$F_{xj} = -kT\frac{\partial \ln \psi}{\partial x_j} - \frac{3kT}{a^2}(-x_{j-1} + 2x_j - x_{j+1}) \tag{6.63}$$

where ψ is the probability of finding each lead j with coordinates between x_j and $x_j + dx_j$; the term $3kT/a^2$ is the Hooke's law spring force constant; the x's are coordinates of the jth bead and its next neighbors. Similar equations are given for F_{yj} and F_{zj}.

Motion of a Fluid Zimm adopted the Kirkwood-Riseman approximate form of the Oseen interaction formula to describe the force on the motion of a fluid:

$$v'_{xj} = v_{xj} + \sum_{j \neq k} T_{jk} F_k \quad (j \text{ and } k \text{ both being index numbers}) \tag{6.64}$$

where v_{xj} is the x component of the velocity that the fluid would have in the absence of all forces from a chain and the interaction coefficients T_{jk} are given

$$T_{jk} = \frac{1}{(6\pi^3)^{1/2}\eta a(|j-k|)^{1/2}} \tag{6.65}$$

with η the viscosity of the fluid. Similar equations are given for v'_{yj} and v'_{zj}.

The equation of motion is obtained by combining three equations representing three different forces:

$$\begin{aligned} \dot{x}_j &= v_{xj} - D\frac{\partial \ln \psi}{\partial x_j} - \sigma(-x_{j-1} + 2x_j - x_{j+1}) \\ &\quad - \zeta \sum_{k \neq j} T_{jk}\left[D\frac{\partial \ln \psi}{\partial x_k} + \sigma(-x_{k-1} + 2x_k - x_{k+1})\right] \\ &\quad 0 < j < N \end{aligned} \tag{6.66}$$

where $D = kT/\zeta$ and $\sigma = 3kT/a^2\zeta$. Similar expressions are given for $\dot{y}_j$ and $\dot{z}_j$.

The solutions to the above differential equations are based on a method similar to the one used by Rouse. The equations are first put in the form of vector and matrix notations. Then the matrices are diagonalized by similarity transformation, congruent transformation, and so on, all of which are well-known techniques in matrix algebra. Finally, the matrices are separated by a linear transformation into normal coordinates with which direct integrations become possible.

Zimm's results are

$$\begin{aligned} [\eta] &= \frac{N_A a^2 \zeta}{6M\eta} \sum_{k=1}^{N} \frac{1}{\lambda_k(1 + i\omega\tau_k)} \\ \tan 2\chi &= \frac{\sum_{k=1}^{N} \tau_k/(1 + i\omega\tau_k)}{\kappa \sum_{k=1}^{N} \tau_k^2/(1 - 2\omega^2\tau_k^2 + 3i\omega\tau_k)} \\ \Delta n &= \frac{q'cNa^2[\eta]\kappa\eta}{MR'T}\left(1 + \frac{1}{4}\tan^2 2\chi\right)^{1/2} \end{aligned} \tag{6.67}$$

where χ is the angle between the principal axes of the polarizability tensor and the χ axis, q' is an optical constant, R' is the gas constant, c is the concentration, T is the temperature, and κ is the shear rate. The lost angle δ here is the phase angle between the stress and strain and

$$\tan\sigma = \frac{E''}{E} = \frac{1}{\tau\omega} \qquad (E - \text{tensile modulus})$$

The relaxation times τ_k are given by

$$\tau_k = \frac{1}{2\sigma\lambda_k} = \frac{M\eta[\eta]_0}{R'T\lambda_k(\sum_{k=1}^{N} 1/\lambda_k)} \tag{6.68}$$

where $[\eta]_0$ is the intrinsic viscosity at $\omega = 0$. For the free-draining case,

$$\lambda_k = \frac{\pi^2 k^2}{N^2}$$

and we get

$$[\eta]_0 = \frac{N_A N^2 a^2 \zeta}{36 M \eta}$$

$$\tan 2\chi_0 = \frac{5R'T}{2M[\eta]_0 \kappa \eta}$$

$$\tau_k = \frac{6M\eta[\eta]_0}{\pi^2 R'Tk^2} \tag{6.69}$$

For the non-free-draining case, we have

$$\lambda_k = \left(\frac{4h}{N^2}\right)\lambda'_k$$

where h is a parameter and λ'_k can be tabulated. The intrinsic viscosity is

$$[\eta] = \frac{\pi^{3/2} N_A L^3}{4 \cdot 3^{1/2} M} \sum_{k=1}^{N} \frac{1}{\lambda'_k(1 + i\omega\tau_k)} \tag{6.70}$$

with

$$L = N^{1/2} a \qquad (a = \text{mean segment length})$$

$$\tau_k = \frac{M\eta[\eta]_0}{0.586\, R'T\lambda'_k} \tag{6.71}$$

REFERENCES

Akloins, J. J., W. J. MacKnight, and M. Shen, *Introduction to Polymer Viscoelasticity*. New York: Wiley, 1972.

Baxendale, J. H., S. Bywater, and M. G. Evans, *J. Polyn. Sci.* **1**, 237 (1946).

Brant, D. A., and P. Flory, *J. Am. Chem. Soc.* **87**, 2788 (1965).

Bueche, F., *J. Chem. Phys.* **20**, 1959 (1952); **22**, 603 (1954).

Bueche, F., *Physical Properties of Polymers*. New York: Wiley, 1962.

Debye, P., and A. M. Bueche, *J. Chem. Phys.* **16**, 573 (1948).

DeGennes, P.G., *Introduction to Polymer Physics*, Cambridre, Cambridge University Press, 1990.

Einstein, A., *Ann. Phys.* **19**, 289 (1906); **34**, 591 (1911).

Eisele, U., *Introduction to Polymer Physics*, New York: Springer, 1990.

Ferry, J. D., *Viscoelastic Properties of Polymers*, 3d ed. New York: Wiley, 1980.

Flory, P. J., and T. G. Fox, Jr., *J. Am. Chem. Soc.* **73**, 1904 (1951).

Flory, P. J., *Principles of Polymer Chemistry*, Ithaca: Cornell University Press, 1953.
Fox, T. G., Jr., and P. J. Flory, *J. Am. Chem. Soc.* **73**, 1909, 1915 (1951).
Fox, T. G., Jr., J. C. Fox, and P. J. Flory, *J. Am. Chem. Soc.* **73**, 1901 (1951).
Freed, K. F., *Macromolecules* **13**, 623 (1980).
Houwink, R., *J. Prakt. Chem.* **157**, 15 (1940).
Huggins, M. L., *J. Am. Chem. Soc.* **64**, 2716 (1942).
Kirkwood, J. G., and J. Riseman, *J. Chem. Phys.* **16**, 565 (1948); **17**, 442 (1949).
Kraemer, E. O., *Ind. Eng. Chem.* **30**, 1200 (1938).
Kuhn, W., and H. Kuhn, *Helv. Chim. Acta* **28**, 97 (1945).
Kurata, M., and W. H. Stochmayer, *Fortschr. Hochpolym. Forsch.* **3**, 196 (1963).
Mark, H. Der feste Körper, Leipzig: Hirzel, 1938.
Martin, A. F. *Abstor. 103rd Am. Chem. Soc. Meeting*, p. 1-c (1942).
Oseen, C. W., *Hydrodynamik*, Leipzig: Akademische Verlagsgesellschaft, 1927.
Peterlin, A., *Z. Phys.* **111**, 232 (1938).
Peterlin, A., *J. Polym. Sci.* **5**, 473 (1950).
Rouse, P. E., Jr., *J. Chem. Phys.* **21**, 1272 (1953).
Rouse, P. E., Jr., and K. Sittel, *J. Chem. Phys.* **24**, 690 (1953).
Scheraga, H. A., and L. Mandelkern, *J. Am. Chem. Soc.* **75**, 179 (1953).
Schultz, G. W. and F. Blaschke, *J. Prakt. Chem.* **158**, 130 (1941).
Simha, R., *J. Phys. Chem.* **44**, 25 (1940).
Sittel, K., P. E. Rouse, Jr., and E. D. Bailey, *J. Appl. Physics*, **10**, 1212 (1954).
Staudinger, H. *Die hochmolekularen organischen Verbindungen*, Berlin: Julius Springer, 1932.
Stockmayer, W. H., and M. Fixman, *J. Polymer Sci. C* **1**, 137 (1963).
Sun, S. F., and N. del Rosario, *J. Am. Chem. Soc.* **92**, 1837 (1970).
Sun, S. F., *J. Phys. Chem.* **76**, 128 (1972).
Yamakawa, H., *Modern Theory of Polymer Solutions*. New York: Harper & Row, 1971.
Yang, J. T., *Adv. Protein Chem.* **16**, 323 (1961).
Zimm, B. H., *J. Chem. Phys.* **24**, 269 (1956).

PROBLEMS

6.1 The following data were found using a Cannon-Ubbelohde viscometer N.50 to measure the flow time of bovine serum albumin in 25% dioxane–75% water (at ionic strength 0.03 and temperature 25°C) as a function of the concentration of protein:

c(g/100 mL)	$t(s)$
0.986	337.2
0.657	335.5
0.493	334.6
0.394	334.1
0.329	333.7
0	331.7

The density ρ of a solution may be estimated by

$$\rho = \rho_0 + \left(\frac{1 - \bar{v}\rho_0}{100}\right)c$$

where $\bar{v}$ for BSA is 0.738 mL/g and ρ_0 for 25% dioxane–75% water is 1.0181 g/mL. Determine the intrinsic viscosity. (*Source*: Sun (1972).)

6.2 The following data were obtained in terms of degree of polymerization versus the intrinsic viscosity for poly(methyl methacrylate) in benzene at 25°C:

Degree of Polymerization	$[\eta]$(dL/g)
700	0.334
1450	0.678
2230	0.929
3940	1.153
4080	1.305
9800	2.510

The molecular weight of the monomer is 100. Determine K and a. (*Source*: Baxendale et al. (1946).)

6.3 The intrinsic viscosity of polystyrene in benzene (a good solvent) at 25°C is 5.27 dL/g, and in 27.2% methanol–72.8% toluene (a poor solvent) it is 1.39 dL/g. Calculate the molecular weight of polystyrene. Take $K = 9.71 \times 10^{-5}$ and $a = 0.74$ for benzene, and $K = 8.81 \times 10^{-4}$ and $a = 0.5$ for 27.2% methanol–72.8% toluene.

6.4 The following data were obtained for the intrinsic viscosity of polystyrene in cyclohexane as a function of temperature:

Mol Wt of the Polymer	T (°C)	$[\eta]$ (dL/g)
1 270 000	34	0.89
	45	1.14
	55	1.43
360 000	34	0.47
	45	0.56
	55	0.66
92 000	34	0.23
	45	0.25
	55	0.28

(a) Evaluate K.

(b) Assuming $\phi = 2.1 \times 10^{21}$, calculate $\langle R_0^2 \rangle / M$.

(c) Calculate ψ_1, c_m and θ.
(d) Estimate ψ_1 and κ_1 at 25°C.

For polystyrene at 25°C, $\bar{v}$ is 0.92. Hint: $[\eta]_\theta = [\eta]_{34^\circ C}$. (*Source*: Fox and Flory (1951).)

6.5 For a large value of the degree of polymerization N (where $N = M/M_0$), the Kirkwood-Riseman equation of the intrinsic viscosity is reduced to

$$b^3 = \frac{2267}{(6\pi^3)^{1/2}} \frac{M_0}{N_A} \lim_{N \to \infty} \frac{[\eta]}{\sqrt{N}}$$

where N_A is Avogadro's number. The following experimental data are given for the intrinsic viscosity of polystyrene in benzene as a function of the degree of polymerization:

$[\eta]$(mL/g)	$N \times 10^3$
0.20	0.625
0.60	2.50
1.75	10.0
3.15	22.5

(a) Plot $[\eta]/\sqrt{N}$ versus $1/\sqrt{N}$ and extrapolate the curve to obtain the intercept

$$\left(\frac{[\eta]}{\sqrt{N}}\right)_0 \left(= \lim \frac{[\eta]}{\sqrt{N}}\right)$$

Then calculate the value of b, the effective bond length. $M_0 = 52$ for $-C_6H_5CH-CH_2-$.

(b) Assume that $\lambda_0 = 1.9 \times 10^{-4}$ (see Kirkwood and Riseman (1948)). Calculate ζ, the frictional coefficient of a monomeric unit, from

$$\lambda_0 = \frac{\zeta}{(6\pi^3)^{1/2} \eta_0 b}$$

where η_0 is the viscosity of the solvent ($\eta_0 = 0.602$ cP for benzene).

(c) Estimate the root-mean-square end-to-end distance of a polystyrene with molecular weight 10^6, using

$$R_0 = N^{1/2} b$$

(*Source*: Kirkwood and Riseman (1948).)

6.6 Calculate the root-mean-square end-to-end distance for a polystyrene with molecular weight 2×10^5 in benzene, with an intrinsic viscosity of 107 mL/g. According to Debye and Bueche's analysis, the shielding ratio is in the vicinity of 0.836 with $\sigma = 3.92$. (*Source*: Debye and Bueche (1948).)

6.7 The following data were obtained in the measurement of intrinsic viscosities and sedimentation coefficients of bovine serum albumin in acidic water–dioxane mixtures at 25°C and ionic strength 0.03:

% Dioxane	$[\eta]$ (mL/g)	$S_{25,w}$	$100\eta_0$
0	0.170	2.25	0.893
15	0.198	2.09	1.190
30	0.231	1.98	1.502
40	0.253	2.06	1.705

(a) Calculate the Scheraga-Mandelkern shape factor β.

(b) Discuss the physical significance in the change of β. (*Source*: Sun and del Rosario (1970).)

6.8 The following experimental data were obtained for the study of molecular dimensions of polypeptides poly-β-benzyl-1-aspartate (PBLA) and poly-β-benzyl-1-glutemate (PBLG):

Polymer	M_0	Solvent	Temperature (°C)	$[\eta]$ (dL/g)	$M_n \times 10^{-3}$	$A_2 \times 10^4$ (cm^3 mol/g^2)	α
PBLA	205	*m*-Cresol	100	0.915	187	3.37	1.22
PBLG	219	Dichloro-acetic acid	25	1.84	336	7.5	1.51

where M_0 is the molecular weight of an amino acid residue, $[\eta]$ is the intrinsic viscosity, M_n is the molecular weight obtained from the osmotic pressure measurements, A_2 is second virial coefficient, and α is the linear expansion factor. Assume that M_v (viscosity molecular weight) $= M_n$ and $T = \theta$. Calculate the following:

(a) $\langle R^2 \rangle_0 / M$, the ratio of the mean-square unperturbed distance $\langle R^2 \rangle_0$ between the polymer chain ends and the molecular weight of the chain.

(b) $\langle R^2 \rangle_0$, the unperturbed mean-square end-to-end distance.

(c) nl^2, the theoretical mean-square end-to-end distance where n is the degree of polymerization and l is the fixed distance between the α carbon of the *trans* peptide repeating unit in the chain. Assume that $l = 3.80$ Å.

(d) $\langle R^2 \rangle_0 / nl^2$, the characteristic ratio of the polypeptides. (*Source*: Brant, 1965).

6.9 The shear stress relaxation modulus for one element in the Maxwell model is given by

$$G(t) = Ge^{-t/\tau}$$

For the z element

$$G(t) = \sum_{i=1}^{z} G_i e^{-t/\tau_i}$$

show that the viscosity η, in terms of relaxation time, for this model is

$$\eta = \sum_{i=1}^{z} G_i \tau_i$$

The viscosity is related to $G(t)$ by

$$\eta = \int_0^\infty G(t)\,dt$$

(*Source*: Aklonis et al. (1972).)

7

OSMOTIC PRESSURE

Osmotic pressure is one of four well-known colligative properties of a non-electrolyte solution; the other three are lowering of vapor pressure, elevation of the boiling point, and depression of the freezing point. They are all expressed in terms of the change in the activity of the solvent a_1, when a solute is present. The activity a_1, is defined as

$$\mu_1 - \mu_1^\circ = RT \ln a_1 \tag{7.1}$$

where for a two-component system the subscript 1 refers to the solvent, the subscript 2 refers to the solute, and μ is the chemical potential. The Gibbs-Duhem equation (Gibbs, 1931, Duhem, 1896) relates the activity a_1 to mole fraction x_2:

$$\left(\frac{\delta \ln a_1}{\delta \ln x_1}\right)_{T,P} = \left(\frac{\delta \ln a_2}{\delta \ln x_2}\right)_{T,P} \tag{7.2}$$

If the solution is very dilute, we can write

$$\frac{\ln a_1}{\ln x_1} \cong 1$$

That is, $a_1 \cong x_1$. Since $x_1 + x_2 = 1$, we can now express a_1 (the activity of solvent) in terms of x_2 (the mole fraction of the solute) by

$$a_1 = 1 - x_2$$

By definition the mole fraction is related to the concept of mole, which, in turn, is related to the molecular weight. We thus have the following correlation:

$$a_1 \Rightarrow x_2 \Rightarrow \text{Molecular weight of the solute}$$

The value of a_1 can be determined with any of the four colligative properties. We have already discussed vapor pressure measurement. In this chapter, we discuss the measurement of osmotic pressure, neglecting boiling point elevation and freezing point depression, because they are less important.

The chemical potential, activity, and osmotic pressure are related in the following equation:

$$\mu_1 - \mu_1^\circ = R'T \ln a_1 = -\pi' \bar{v}_1 \tag{7.3}$$

where π' is the osmotic pressure (to be distinguished from $\pi = 3.1416$, an irrational number) and $\bar{v}_1$ is the partial specific volume of the solvent. Equation (7.3) states that because of the difference in chemical potential between the solvent in the solution (μ_1) and that in its pure phase (μ_1°), there is a driving force to push the solvent flow into the solution. This driving force is expressed by the quantity a_1.

7.1 OSMOMETERS

There are various types of osmometers. In Figure 7.1 we show two examples. Some commonly used membranes are cellulose acetate, cellulose hydrate, cellulose nitrate, polyurethanes, gel cellophane membranes, and poly(chlorotrifluoroethylene).

The osmotic pressure is calculated using

$$\pi' = \rho_1 g h \tag{7.4}$$

where ρ_1 is the density of the solvent (g/cm^3 or kg/m^3), g is the gravitational constant, 0.981 m s^2, and h is the height expressed in cm. The unit of osmotic pressure is Pa m^3 kg^{-1} (SI system: 1 Pa = 1 N/m^2).

7.2 DETERMINATION OF MOLECULAR WEIGHT AND THE SECOND VIRIAL COEFFICIENT

For a thermodynamically ideal solution (for example, a very dilute solution), the osmotic pressure obeys Van't Hoff's equation:

$$\frac{\pi'}{c} = \frac{R'T}{M} \tag{7.5}$$

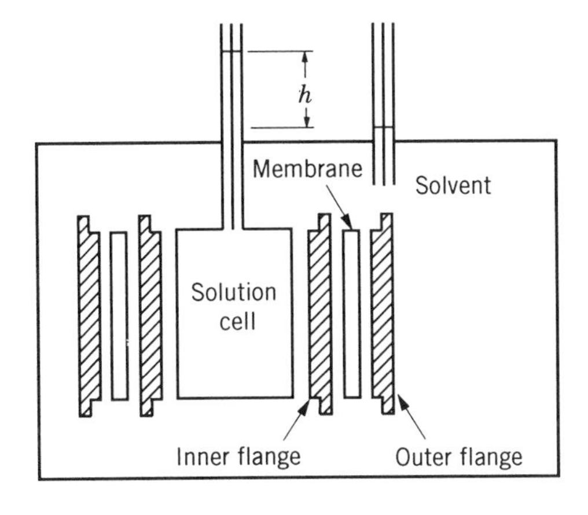

(a)

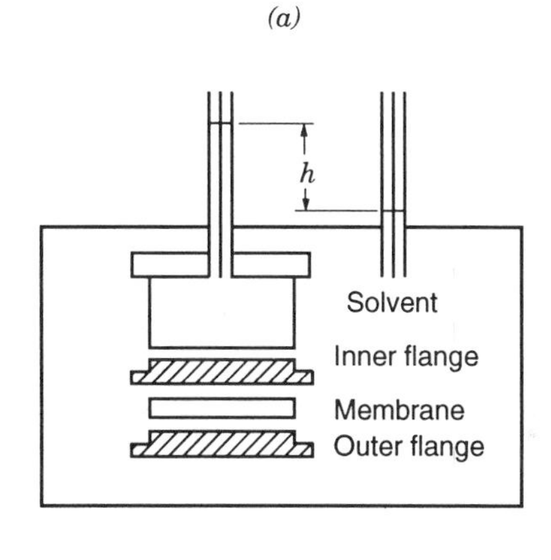

(b)

FIGURE 7.1 Osmometers. (*a*) Helfritz type; (*b*) Schultz type.

where c is the concentration of the solute in g/100 mL, R' is the gas constant (8.314 Pa m^3/(mol K)), T is the temperature in K, and M is the number-average molecular weight. The plot of π' versus c or π'/c versus c gives characteristic straight lines with slopes 1 and 0, respectively. In either case, Van't Hoff's equation is used to determine the molecular weight of the solute, that is, the y intercept of the linear plot. At higher concentrations, deviations from the ideal are usually pronounced and the plot of π'/c versus c, while still a straight line, gives different slopes, depending on the chemical system. Van't Hoff's equation must be modified to

$$\frac{\pi'}{c} = \frac{R'T}{M}(1 + \Gamma_2 c) \tag{7.6}$$

where Γ_2 is the second virial coefficient, which is a measure of the deviation from the ideal. In a complicated system the plot of π'/c versus c may not give a straight line, in which case Van't Hoff's equation must be further modified. Parallel to the equation of state for real gases, the osmotic pressure may be expressed in terms of the virial equation:

$$\pi' = R'T(A_1c + A_2c^2 + A_3c^3 + \cdots) \tag{7.7}$$

where $A_1 = 1/M$, and A_2 and A_3 are the second and third virial coefficients, respectively (cf. Eq. (4.16) in Chapter 4). For the convenience of treating data, the virial equation may be expressed in a slightly different form:

$$\frac{\pi'}{c} = R'T(A_1 + A_2c + A_3c^2 + \cdots) \tag{7.8}$$

$$\frac{\pi'}{c} = A_1(1 + \Gamma_2c + \Gamma_3c^2 + \cdots) \tag{7.9}$$

where

$$\Gamma_2 = A_2M, \quad \Gamma_3 = A_3M, \ldots$$

The coefficients $A_2, A_3, \ldots$ or $\Gamma_2, \Gamma_3, \ldots$ are functions of the molecular weight distribution of solute, the solvent–solute system, and the temperature. Virial coefficients beyond A_3 or Γ_3 are usually of no physical interest and can be neglected.

There are several ways to treat the osmotic pressure data if the plot is not linear:

1. The curve fitting method can be used to determine A_1, A_2 and A_3 directly.

2. In the case where A_1 is known or can be reasonably estimated, Eq. (7.9) can be rearranged in the form

$$\left(\frac{\pi'/c - 1}{A_1}\right)\frac{1}{c} = \Gamma_2 + \Gamma_3c \tag{7.10}$$

and $[(\pi'/c - 1)/A_1](1/c)$ versus c can be plotted. A straight line may be obtained in which the intercept gives the value of Γ_2 and the slope gives the values of Γ_3.

3. Flory (1953) suggested that data be treated in the following way:

$$\frac{\pi'}{c} = \frac{R'T}{M}(1 + \Gamma_2c + j\Gamma_2^2c^2)$$

The value of j depends on the polymer–solvent system. As a trial we may assume that $j \cong 0.25$ and take the square root on both sides. The equation becomes

$$\left(\frac{\pi'}{c}\right)^{1/2} = \left(\frac{\pi'}{c}\right)_0^{1/2} \left(1 + \frac{\Gamma_2}{2} c\right) \tag{7.11}$$

By plotting $(\pi'/c)^{1/2}$ versus c, a straight line may be expected and the values of M and Γ_2 may be determined.

4. The value of j in Eq. (7.11) may not be a constant and may depend on the molecular weight. To avoid complications, Bawn and coworkers suggested another method to treat data:

$$\left(\frac{\pi'}{cR'T}\right)_1 - \left(\frac{\pi'}{cR'T}\right)_2 = A_2(c_1 - c_2) + A_3(c_1^2 - c_2^2)$$

and hence

$$\frac{[(\pi'/cR'T)_1 - (\pi'/cR'T)_2]}{c_1 - c_2} = A_2 + A_3(c_1 + c_2) \tag{7.12}$$

The pairs of points (c_1, c_2), $[(\pi'/c)_1, (\pi'/c)_2]$, are selected on the experimental curve. By plotting the term of the left side of the equation against $c_1 + c_2$, a straight line may be obtained, from which we get intercept A_2 and slope A_3.

Summary The equation of osmotic pressure is usually expressed in one of the following three ways:

$$\frac{\pi'}{c} = \frac{R'T}{M} + Bc + Cc^2 \tag{7.13a}$$

$$\frac{\pi'}{c} = R'T\left(\frac{1}{M} + A_2c + A_3c^2 + \cdots\right) \tag{7.13b}$$

$$\frac{\pi'}{c} = \frac{R'T}{M}(1 + \Gamma_2c + \Gamma_3c^2 + \cdots) \tag{7.13c}$$

The quantity π'/c is called the reduced osmotic pressure. Its unit is J/kg. B, A, and Γ are different symbols for the same quantity, virial coefficient.

7.3 THEORIES OF OSMOTIC PRESSURE AND OSMOTIC SECOND VIRIAL COEFFICIENT

While the measurement of osmotic pressure π' and the calculation of the second virial coefficient A_2 are relatively simple, their theoretical interpretations are rather complicated. Throughout the past half century, many investigators have tried to set up a model and derive equations for π' and A_2. Because of the unsymmetrical nature with respect to the sizes of solute (macromolecule) and solvent (small molecule), polymer solutions involve unusually large intermolecular interactions. Furthermore, since π' is directly related to μ_1, any theoretical knowledge learned from the osmotic pressure and the second virial coefficient contribute to the knowledge of the general thermodynamic behavior of polymer solutions. For this reason, Chapters 4 and 7 are closely related in macromolecular chemistry.

7.3.1 The McMillan-Mayer Theory

In the McMillan-Mayer theory of solutions, the macromolecule is considered as a small gel emerged in a solution. The gel is surrounded by the solvent, but inside the gel there are also solvent molecules (○) as shown in the diagram:

Macromolecule

The gel acts like a semipermeable membrane. Only the solvent molecule can pass through the gel. The temperature remains the same inside and outside. The outside system has a specified value of the chemical potential μ_1, while the inside system has specified values of μ_1 for the solvent and μ_2 for the solute. If the solution is in osmotic equilibrium with pure solvent, the inside (solution) would have pressure $p + \pi'$ to balance the outside (pure solvent) pressure p. The osmotic pressure inside the system is the excess pressure, created to give μ_1 the same value (which tends to be lowered in the presence of the solute) inside as outside.

The grand partition function (see Appendix C) for the inside solution (according to Hill (1960)) is

$$e^{(p+\pi')V/kT} = \sum Q(V, T)\lambda_1^{N_1}\lambda_2^{N_2} \tag{7.14}$$

where Q is the grand canonical ensemble partition function, N is the number of molecules, V is the volume of the solution, and λ is often called the absolute activity, which is defined as $\lambda = e^{kT}$. The expression on the left, $e^{(p+\pi')V/kT}$, is the Boltzmann distribution factor and k is the Boltzman constant.

With a messy mathematical manipulation on the equation that is related to Q and μ_1, one can reach the theoretical expressions for π' and A_2:

$$\pi' = kT\rho_2 + \sum A_2\rho_2^n \tag{7.15}$$

$$A_2 = -\frac{1}{2}\int_0^{\infty}\left[\exp\left(-\frac{w}{kT}\right) - 1\right]4\pi r^2\, dr \tag{7.16}$$

where ρ_2 is the density of the polymer molecule, n is the number of molecules involved in the interaction, and w is the reversible work that is needed to bring two solute molecules together from $r = \infty$ to r in the outside solution.

The value of A_2 depends on w. If the solute molecule is considered as a hard sphere of diameter a, then

$$\begin{aligned} w &= \infty \qquad r < a \\ &= 0 \qquad r \geqslant a \end{aligned}$$

and

$$A_2 = \frac{2\pi a^3}{3} \qquad \text{(here } \pi = 3.1416\text{)} \tag{7.17}$$

7.3.2 The Flory Theory

Flory's theory concerns with the distribution of excluded volume u between n_2 polymer molecules over a volume V of solution. If there is an independent volume exclusion on the part of the individual molecules, the partition function of the system is

$$Q = \text{const.} \times \prod_{i=0}^{n_2-1} (V - iu) \tag{7.18}$$

and

$$\begin{aligned} \Delta G &= -kT \ln Q \\ &= -kT \sum_{i=0}^{n_2-1} \ln(V - iu) + \text{const.} \\ &= -kT\left[n_2 \ln V + \sum_{i=1}^{n_2-1} \ln\left(1 - \frac{iu}{V}\right)\right] + \text{const.} \end{aligned} \tag{7.19}$$

Expanding the logarithmic terms in series and manipulating for the purpose of simplification, we obtain

$$\pi' = -\frac{\mu_1 - \mu_1^\circ}{V_1}$$

$$= -N\frac{(\partial\Delta G/\partial n_1)_{T,P,n_2}}{V_2}$$

That is,

$$\pi' \cong kT\left[\frac{n_2}{V} + \frac{u}{2}\left(\frac{n_2}{V}\right)^2\right] \tag{7.20}$$

Note that $n_2/V = cN_A/M$, where c is the concentration in grams per unit volume and N_A is Avogadro's number. Equation (7.20) can be rewritten as

$$\frac{\pi'}{c} \cong R'T\left[\frac{1}{M} + \left(\frac{N_A u}{2M^2}\right)^2 c\right] \tag{7.21}$$

and is discussed further in a later section.

In comparison of Eq. (7.21) with the virial equation,

$$\frac{\pi'}{c} \cong R'T(A_1 + A_2 c + A_3 c^2 + \cdots)$$

we find

$$A_1 \cong \frac{1}{M}$$

$$A_2 = \frac{N_A u}{2M^2} \tag{7.22}$$

While Eq. (7.22) clearly relates A_2 to u and M, it vaguely indicates that the value of A_2 for a polymer system in solution decreases with the increasing value of M. It is vague, because the theory has not specified yet the significance of u. To interpret u, the model needs to be improved.

7.3.3 The Flory and Krigbaum Theory

In the improved model offered by Flory and Krigbaum (1950) the very dilute polymer solution is considered a dispersion of clouds, or dilute clusters, of segments. The region between swamps contains pure solvent. Consider a pair of polymer molecules l and m, separated by a distance A_{lm}. When they are brought together, the change in entropy $\partial(\Delta S)$ and the change in enthalpy $\partial(\Delta H)$

are given by

$$\partial(\Delta S) = -2k\psi_1\left(\frac{V^2}{V_1}\right)\rho_l\rho_m\partial V$$

$$\partial(\Delta H) = -2kT\kappa_1\left(\frac{V^2}{V_1}\right)\rho_l\rho_m\partial V \tag{7.23}$$

where ψ_1 is the entropy parameter, κ_1 is the enthalpy parameter, V is the volume of the polymer molecule, V_1 is the volume of solvent molecule, ∂V is the volume element in the vicinity of the pair of molecules, and ρ_l and ρ_m are the densities of segments in ∂V. The partition function Q is

$$\ln Q = -2\int_0^{\infty}\int_0^{\pi} JV^2\bar{v}^{-2}\rho_l\rho_m 2\pi r^2 \sin\theta \, \mathrm{d}r \, \mathrm{d}\theta \tag{7.24}$$

where $\pi = 3.1416$, r and θ represent cylindrical coordinates, $\bar{v}$ is the partial specific volume, and J is defined as

$$J = \frac{(\psi_1 - \kappa_1)\bar{v}^2}{V_1} \tag{7.25}$$

The second virial coefficient for homogeneous polymers can now be expressed theoretically:

$$\Gamma_2 = M\left(\frac{\bar{v}^2}{V_1}\right)(\psi_1 - \kappa_1)F(x)$$

or

$$A_2 = \frac{\bar{v}^2}{V_1}(\psi_1 - \kappa_1)F(x) \tag{7.26}$$

where

$$F(x) = 1 - \frac{x}{2!}2^{3/2} + \frac{x^2}{3!}3^{3/2} - \cdots \tag{7.27}$$

The quantity x is related to the expansion factor α by

$$x = 2(\alpha^2 - 1) \tag{7.28}$$

In relation to the dimensions of the molecule over those of its random plight counterpart, we have, as we may recall from Chapter 5,

$$\alpha^5 - \alpha^3 = 2c_M(\psi_1 - \kappa_1)M^{1/2} \tag{7.29}$$

If polymer molecules are not homogeneous in size, we need to modify the second virial coefficient:

$$A_2 = \frac{\bar{v}^2}{V_1}(\psi_1 - \kappa_1)\gamma \tag{7.30}$$

where

$$\gamma = \sum_i \sum_j w_i w_j F(x_{ij}) \tag{7.31}$$

The terms w_i and w_j are weight fractions of polymer species i and j and are related to x_{ij},

$$\frac{x_{ij}}{M_i M_j} = \left[\frac{2(x_{ii}/M_i^2)^{2/3}(x_{jj}/M_j)^{2/3}}{(x_{ii}/M_i^2)^{2/3} + (x_{jj}/M_j)^{2/3}}\right]^{2/3} \tag{7.32}$$

Note: here we have variable x_{ij}, instead of x.

The characteristic features of Flory's theory and Flory and Krigbaum's theory are summarized as follows:

1. These theories introduced the Gaussian approximation for a spatial distribution of segments around the center of the mass.
2. They showed further the dependence of A_2 on M. The relationship between the two parameters may be expressed as

$$A_2 \sim M^{\nu} \tag{7.33}$$

   The value of the exponent ν is between -0.1 and -0.3, which is of great current interest in polymer chemistry.
3. Since $\psi_1 - \kappa_1 = 1 - \theta/T$, Eqs (7.26) and (7.30) can be rewritten respectively as

$$A_2 = \left(\frac{\bar{v}^2}{V_1}\right)\left(1 - \frac{\theta}{T}\right)F(x) \tag{7.34}$$

$$A_2 = \left(\frac{\bar{v}^2}{V_1}\right)\left(1 - \frac{\theta}{T}\right)\gamma \tag{7.35}$$

At θ temperature, $T = \theta$ and $A_2 = 0$. This is now used as a definition of θ temperature. Experimentally, if A_2 of a polymer system vanishes, we say that the temperature at which the measurements (for example, osmotic pressure and light scattering) are taken is the θ temperature.

7.3.4 The Kurata and Yamakawa Theory

Kurata and Yamakawa (1958) criticized McMillan and Mayer's theory for lack of experimental support and Flory and Krigbaum's theory for their assumption of Gaussian distribution of polymer segments around the center of the mass. The introduction of the factorization approximation does not help much. As a result, Flory and Krigbaum's theory underestimates the molecular weight dependence of A_2. For Kurata and Yamakawa, the excluded volume effect has a non-Gaussian character with respect to the chain configuration. Kurata and Yamakawa's approach, however, basically follows the same line as that of McMillan and Mayer.

Consider a polymer molecule that consists of $N + 1$ segments of diameter b. The length of each link between two segments is a constant a. Kurata and Yamakawa assume that the total potential energy of the solution is additive over segment pair, i and j:

$$\begin{aligned} w_{ij} &= w(R_{ij}) \\ &= \infty \qquad \text{if } 0 \leqslant R_{ij} \leqslant b \\ &= -w_0 \exp\left(-\frac{3R_{ij}^2}{2d^2}\right) \qquad \text{if } R_{ij} > b \end{aligned} \tag{7.36}$$

where R_{ij} represents the distance between the segment pair, and w_0 and d are constants. Then a function $\chi(R_{ij})$ can be defined as

$$\chi_{ij} \equiv \chi(R_{ij}) = \exp\left(-\frac{w_{ij}}{kT}\right) - 1 \tag{7.37}$$

Let α and β refer to the configurations of the polymer molecule and let l and m be the specified segments (Note: i and j are for general segments.) Then A_2 can be theoretically expressed as

$$A_2 = -\left(\frac{N_{\mathrm{A}}}{2VM^2}\right)(B_1 + B_2 + \cdots) \tag{7.38}$$

with

$$B_1 = \sum_{l_\alpha l_\beta} \int \chi_{l_\alpha l_\beta} f_\alpha f_\beta \, \mathrm{d}\alpha \, \mathrm{d}\beta \tag{7.39a}$$

$$B_2 = \sum_{l_\alpha l_\beta \; m_\alpha m_\beta} \int \chi_{l_\alpha l_\beta} \chi_{m_\alpha m_\beta} f_\alpha f_\beta \, \mathrm{d}\alpha \, \mathrm{d}\beta \tag{7.39b}$$

The terms f_α and f_β are pairwise distribution functions and are defined as

$$f_\alpha = \text{const.} \times \exp\left(-\sum \frac{w_{i_\alpha, j_\alpha}}{kT}\right) \tag{7.40a}$$

$$f_\beta = \text{const.} \times \exp\left(-\sum \frac{w_{i_\beta, j_\beta}}{kT}\right) \tag{7.40b}$$

B_1 is a single contact term, with one term of interaction function $\chi_{l_\alpha l_\beta}$; B_2 is a double intermolecular contact, with two terms of interaction function, $\chi_{l_\alpha l_\beta}$ and $\chi_{m_\alpha m_\beta}$. The following are diagrams of one-term contact, two-term contact, and three-term contact:

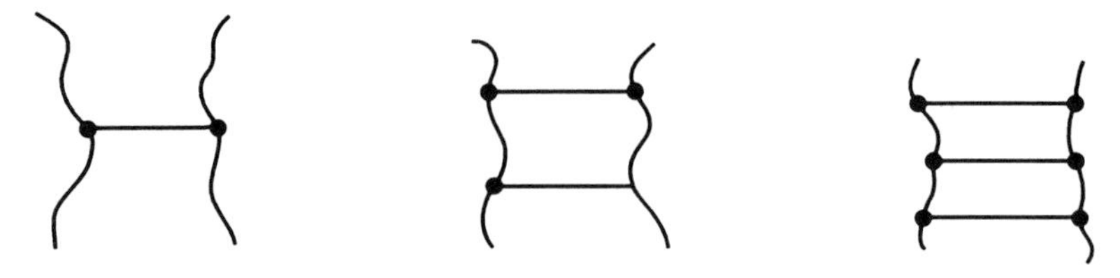

One term contact, B_1　　Two term contact, B_2　　Three term contact, B_3

The final results of these B terms are

$$B_1 = -VN_A^2\left(\frac{4\pi b^3}{3}\right)\left(1-\frac{\theta}{T}\right) \tag{7.41}$$

$$B_2 = B_{20} + B_{21} + B_{30} \tag{7.42}$$

In the expressions B_{20}, B_{21}, and B_{30}, the first subscript refers to the number of intermolecular contacts and the second refers to the number of intramolecular contacts. The numerical estimation of these three terms are

$$B_{20} = -2.865 B_1 z$$

$$B_{21} = 8.78 B_1 z^2$$

$$B_{30} = 9.73 B_1 z^2$$

where z, as defined in Chapter 5, is

$$z = \left(\frac{6}{\pi}\right)^{1/2}\left(\frac{b}{a}\right)^3\left(1-\frac{\theta}{T}\right)N_A^{1/2} \qquad \text{(here } \pi = 3.1416\text{)} \tag{7.43}$$

By substituting these quantities into the equation of A_2, we obtain

$$A_2 = \frac{2\pi N b^3}{3M_0}\left(1-\frac{\theta}{T}\right)h(z) \qquad \text{(here } \pi = 3.1416\text{)} \tag{7.44}$$

$$h(z) = 1 - 2.865z + 18.51z^2 - \cdots \tag{7.45}$$

Again, Eq. (7.44) defines the θ temperature as the temperature at which A_2 for the polymer system is zero.

7.3.5 The des Cloizeaux and de Gennes Scaling Theory

A different approach to the interpretation of osmotic pressure and the second virial coefficient is an attempt to investigate the universal properties: scaling laws and characteristic exponents in polymer solutions. des Cloizeaux (1975) suggested that the properties of a chain with excluded volume effect in a dilute solution can be deduced from Lagrangian theory (Appendix D) for a zero component field in the absence of an external field. If there are n components, the Green's functions (Appendix E) can be used to expand the Lagrangian theory in terms of interactions. In this way, the asymptotic behavior of the correlation function between the extremities of a single chain can be obtained, leading to the scaling law of universal properties.

Consider the polymer solution as a grand canonical ensemble which can be derived by using the Lagrangian theory. The averages of the total number of polymers and the total number of monomers are determined by two chemical potentials, H_0 and s $(= m_0^2 b^2)$, where m_0 is the normalized mass (a quantity used in the renormalization group theory) and b is the monomer length. H_0 is the magnetic field and is chosen because of the similarity between magnetic critical and triclitical phenomena and the behavior of flexible polymer solutions. The magnetic field H_0 is believed to be in correspondence with the chemical potential. The grand partition function $Z(H_0)$ to be introduced to represent polymer solutions is expressed by Des Cloizeaux as

$$Z(H_0) = 1 + \sum_{M=1}^{\infty} \frac{1}{2^M M!} [H_0 b^{(d/2)+1} (2\pi)^{d/2}]^{2M}$$
$$\times \int_0^{\infty} \frac{dL_1}{b} \cdots \int \frac{dL_M}{b} e^{-sLb^{-1}} Z(L_1, \ldots, L_M) \tag{7.46}$$

where M is the index of a chain (not molecular weight), L is the length of the chain, d is the space dimension, and $\pi = 3.1416$. The expression $Z\ (L_1, \ldots, L_M)$ refers to the number of polymer configurations in a box. Using the ordered Green's function, $Z(H_0)$ can be reduced to a simple expression:

$$Z(H_0) = \exp[Vw(H_0)] \tag{7.47}$$

where V is the volume of solution. The function $w(H_0)$ is directly related to the osmotic pressure π':

$$\frac{\pi'}{kT} = w(H_0) \tag{7.48}$$

The concentration of polymers (chains) c_P and the concentration of monomers (links) c_M can also be expressed in terms of the function $w(H_0)$:

$$c_P = \frac{H_0}{2}\frac{\partial w(H_0)}{\partial H_0} \tag{7.49a}$$

$$c_M = -\frac{\partial}{\partial s} w(H_0) \tag{7.49b}$$

If there is no interaction, we have

$$w(H_0) = \frac{b^2 H_0^2}{2s} \tag{7.50}$$

and, therefore, we have

$$\pi' = kTc_P \tag{7.51}$$

Equation (7.51) is the classical van't Hoff equation. To introduce scaling law, we can define the average number of links per chain N as

$$N = \frac{c_M}{c_P} = \frac{1}{2s} \tag{7.52}$$

If there are interactions, we use Green's function, which is included in a term $\Gamma(M_0)$,

$$\Gamma(M_0) = \sum_{M=1}^{\infty} \frac{1}{2^M M!}[M_0(2\pi)^{d/2}]^{2M}\Gamma^{(2m)}(0,\ldots,0) \tag{7.53}$$

The two concentration terms are now expressed as

$$c_P = \frac{M_0}{2}\frac{\partial \Gamma(M_0)}{\partial M_0} \tag{7.54a}$$

$$c_M = \frac{\partial \Gamma(M_0)}{\partial s} \tag{7.54b}$$

The derivation of the equation for osmotic pressure is more complicated. The final result is expressed in a parametric equation:

$$\frac{\pi'}{kTc_P} = F(c_P b^d N^{\nu d}) = F\left(\frac{b^d c_M^{\nu d}}{c_P^{\nu d - 1}}\right) = F(\lambda) \tag{7.55}$$

where F is a scaling function and can be written in a power series

$$F(\lambda) = 1 + F_1\lambda + F_2\lambda^2 + \cdots$$

des Cloizeaux' equation can be compared with Flory's equation (Eq. (7.20)):

Flory's equation

$$\frac{\pi'}{kT} = \frac{n_2}{\mathrm{V}} + \frac{u}{2}\left(\frac{n_2}{\mathrm{V}}\right)^2$$

or

$$\frac{\pi'}{kTc_{\mathrm{P}}} = 1 + \frac{1}{2}uc_{\mathrm{P}}$$

where $c_{\mathrm{P}} = n_2/\mathrm{V}$.

des Cloizeaux' equation

$$\frac{\pi'}{kTc_{\mathrm{P}}} = 1 + F_1 N^{\nu d}(b^d c_{\mathrm{P}}) \tag{7.56}$$

or

$$\frac{\pi'}{kTc_{\mathrm{M}}} = \frac{1}{N} + F_1 N^{\nu d - 2} c_{\mathrm{M}}$$

The interesting part is that des Cloizeaux' equation can be translated into scaling law

$$\frac{\pi'}{kTc_{\mathrm{P}}} \sim N^{\nu d} \tag{7.57}$$

whereas Flory's equation cannot.

It was de Gennes (1972) who first introduced the Lagrangian theory to interpret the behavior of polymer solutions. Although des Cloizeaux showed the similarity between his and Flory's interpretations of the osmotic pressure, de Gennes pointed out that des Cloizeaux' interpretation is superior to Flory's. To compare the two theories, de Gennes first translated the Flory lattice theory into his language. Let Φ represent the fraction of lattice sites occupied by the monomers. Then $\Phi = ca^3$, where c is the concentration in terms of the number of monomers per cm^3, and a^3 is the volume of the unit all in the cubic lattice. Flory's equation of osmotic pressure (Eq. (7.20)) can be expressed in the form

$$\frac{\pi' a^3}{kT} = \frac{\Phi}{N} + \frac{1}{2}(1 - 2\chi)\Phi^2 \cong \frac{1}{2}(1 - 2\chi)\Phi^2 \tag{7.58}$$

$$\left(\text{Note: } \frac{\Phi}{N} \cong 0, \text{ if } \frac{1}{N} \leqslant \Phi \leqslant 1\right)$$

des Cloizeaux' equation (Eq. (7.56)) can be expressed in a similar form:

$$\begin{aligned}\frac{\pi'}{kT} &= \frac{c}{N} + \text{const.}\left(\frac{c}{N}\right)^2 R^3 + \cdots \\ &\cong \frac{c}{N} f_n\left(\frac{cR^3}{N}\right) = \frac{c}{N} f_n\left(\frac{c}{c^*}\right)\end{aligned} \tag{7.59}$$

In a semidilute solution, des Cloizeaux' equation can be further simplified by eliminating coefficients in the form

$$\frac{\pi' a^3}{kT} = \text{const.} \times \Phi^{9/4} \tag{7.60}$$

which de Gennes called the des Cloizeaux law. Now the difference is clear:

For Flory

$$\pi' \sim \Phi^2 \tag{7.61}$$

For des Cloizeaux

$$\pi' \sim \Phi^{9/4} \tag{7.62}$$

Flory neglected certain correlations between adjacent and ever more distant monomers, as criticized by de Gennes.

de Gennes himself is more interested in the relationship between the correlation length ξ and the osmotic pressure, namely, in constructing the scaling form of ξ in the semidilute regime. The scaling relationship between the osmotic pressure and the correlation length, as he suggested, is

$$\pi' = \frac{kT}{\xi^3} \qquad (\Phi^* \leqslant \Phi \leqslant 1) \tag{7.63}$$

7.3.6 Scatchard's Equation for Macro Ions

Scatchard's equation (1946) is derived to explain A_2 for proteins. The change in dimensions (for example, mean-square end-to-end distance) is not of any concern in Scatchard's derivation. No model was assumed and no statistical mechanics* was used. Scatchard successively correlated the osmotic pressure with the distribution of diffusible solutes across a semipermeable membrane by manipulating the terms of activities of the components (such as protein, salt, and water) with changing composition of the solutions. The mathematical detail is simple, but messy. According to Scatchard, the interactions involved in protein solutions are not limited to the exclusion of volume between the segments of macromolecules, but also includes Donnan effect and the binding of small ions to macro ions in a given system. For simplicity, let us consider a three-component system, and let 1 represent the solvent (or buffer), 2 the macromolecule (such as protein), and 3 a salt (for example, NaCl). Scatchard derived an equation of second virial coefficient:

$$A_2 = \frac{1000\bar{v}_1}{M_2^2}\left(\frac{\varepsilon^2}{4m_3} + \frac{\beta_{22}}{2} - \frac{\beta_{23}^2 m_3}{4 + 2\beta_{33}m_3}\right) \tag{7.64}$$

where $\bar{v}_1$ is the partial specific volume of the solvent, ε is the net charge carried by the macro ion (which could be obtained from acid–based titration data), m is the molality of a component in solution, and β's are derivatives of activity coefficients γ:

$$\beta_{22} = \frac{\delta \ln \gamma_2}{\delta m_2} \qquad \text{(Donnan effect)}$$

$$\beta_{23} = \frac{\delta \ln \gamma_2}{\delta m_3} \qquad \text{(Excluded volume effect including charge effect)}$$

$$\beta_{33} = \frac{\delta \ln {}_3}{\delta m_3} \qquad \text{(Binding)}$$

Eq. (7.64) consists of three terms. The first describes the Donnan effect, which states that in osmosis the small ions are not distributed equally on the two sides of the membrane. The Donnan effect always causes an increase in osmotic pressure. To minimize this effect, the protein solution should have high ionic strength or should be near the isoelectric pH. The second term represents the excluded volume effect and the interaction between charges on different macro ions. It is basically similar to the parameter χ_1 in the properties of synthetic polymer solutions, but is a little more complicated because of the charges

*Thermodynamics and statistical mechanics are closely related. They differ only in their approach to a system. Given a mole of gas, thermodynamics describes its behavior in terms of *PVT* and, later, in terms of *E*, *H*, *S*, *G*, and *A*. Given the same mole of gas, statistical mechanics describes its behavior in terms of energy states. The statistical behavior is basically the fluctuations of a random variable which characterizes the system. Eventually, statistical mechanics also leads to the study of change in *E*, *H*, *S*, *G*, and *A*.

involved. The third term involves the interaction between macro ions and the actual binding of small ions. This term is particularly important to the study of the behavior of proteins in dilute salt solutions.

Experimental values for all of the three parameters, $(\beta_{22}, \beta_{23}, \beta_{33})$ and the molecular weight M_2 can be determined from osmotic pressure experments as follows. The plot of π'/c_2 versus c_2 is expected to be a straight line from which the intercept A_1 and the slope A_2 are obtained. From the intercept we obtain M_2 using the relation

$$A_1 = \frac{R'T}{V_m^{\circ} M_2} \tag{7.65}$$

where V_m° is the volume of the solution which contains grams of macro ion in 1000 g of solvent, that is, $V_{\mathrm{m}}^{\circ} = 1$; R' is the gas constant (0.082 atmL/mol K) if π' is in atm. The parameter β_{33}, itself not important, is calculated from the following empirical equation:

$$\frac{\beta_{33}}{2} = \ln \gamma_3 = \frac{-1.172\sqrt{m_3}}{1 + 1.55\sqrt{m_3}} + 0.0215 m_3 + 0.0212 m_3^2 \tag{7.66}$$

This is obtained from a separate and independent experiment (not included in the osmotic pressure measurement). The parameter β_{23} is calculated using

$$\beta_{23} = M_2 b_{23}(2 + \beta_{33} m_3^{\circ}) \tag{7.67}$$

where

$$b_{23} = \frac{1}{c_2} \ln \frac{m'_3}{m_3}$$

In the above equations m_3° is the original concentration of the salt in the protein solution, m_3 is the concentration of the salt in the solution after diffusion, and m'_3 is the concentration of salt diffused out through the membrane. The quantities m_3°, m_3, and m'_3 are all measured in the same osmotic pressure experiment. After all of the above parameters are calculated, the parameter β_{22} can be determined directly from the slope A_2, using Eq. 7.64).

APPENDIX A ENSEMBLES

An ensemble is a collection of systems, each of which contains N particles, occupies a volume of V, and possesses energy E. Each system represents one of the possible microscopic states and each is represented by a distribution of points in phase space. A phase space is defined by $3n$ coordinates and $3n$ momenta for a dynamic system consisting of n particles. There are three types of ensembles:

1. *The microcanonical ensemble* The microcanonical ensemble is the assembly of all states in which the total energy E, the number of molecules N, and the total volume V are all fixed. It is a closed and isolated system. In a microcanonical ensemble; there is no fluctuation of any of the three variables, N, E, and V.
2. *The canonical ensemble* In the canonical ensemble, all states have fixed V (volume) and N (the number of molecules), but the energy E fluctuates. The ensemble could be considered as a closed system in contact with a heat bath that has infinite heat capacity.
3. *The grand canonical ensemble* In a grand canonical ensemble, not only E fluctuates, but also N. The grand canonical ensemble is an open isothermal system. Both heat (energy) and mass (particles) can be transported across the walls of the system.

APPENDIX B PARTITION FUNCTIONS

The partition function Q is defined as

$$Q = \sum_i g_i \mathrm{e}^{-\mu \varepsilon_i}$$

where ε_i is the energy level and μ is identified as

$$\mu = \frac{1}{kT}$$

where k is the Boltzmann constant and T is the temperature. The term g_i refers to the degeneracy of energy levels. The partition function is the sum of the internal energy states of the entire system. The term $\sum_i \mathrm{e}^{-\mu \varepsilon_i}$ is the Boltzmann-weighed sum over all possible fluctuations, that is, all microscopic states permitted by the constraints with which we control the system. Partition function can be used to describe the configuration of a polymer chain, for example,

$$Q = \sum \exp\left[\frac{-U(\{R_n\})}{kT}\right]$$

where U is the potential and $\{R_n\}$ is a set of elements R_n,

$$\{R_n\} = (R_1, R_2, \ldots, R_n) \quad (R = \text{coordinate})$$

The grand ensemble partition function is usually expressed as

$$\begin{aligned} \Xi &= \Xi(V, T, \mu) \\ &= \sum_{j,N} \mathrm{e}^{-E_j/kT} \mathrm{e}^{N\mu/kT} \end{aligned}$$

Partition functions are used to calculate thermodynamic functions. They are related as follows:

$$A(N, V, T) = -kT \ln Q$$

$$p = kT\left(\frac{\partial \ln Q}{\partial V}\right)_{T,N}$$

$$E = kT^2\left(\frac{\partial \ln Q}{\partial T}\right)_{V,N}$$

$$C_V = \left(\frac{\partial E}{\partial T}\right)_{V,N}$$

$$S = k\frac{\partial}{\partial T}(T \ln Q)_N$$

APPENDIX C MEAN FIELD THEORY AND RENORMALIZATION GROUP THEORY

Both of these theories are about the distribution of molecules, which involves interaction among molecules themselves. Both are proposals for solving the equation of the Ising (1925) model about electron spins in two or three dimensions. Both are highly mathematical in nature.

The idea of the mean field theory is to focus on one particular particle in the system. The method singles out only those fluctuations that occur within the cell, and neglects the effects of fluctuations beyond the cell. It reduces the many-body statistical mechanics problem into a one-body problem.

The renormalization group theory, proposed by Wilson in 1971, is about large length scale fluctuations. The method consists of two steps: (1) removing from the partition function a finite fraction of the degrees of freedom by averaging over them and (2) rescaling the resultant partition function by Kandanoff transformation. The transformations contain the group property and provide a renormalization scheme.

APPENDIX D LAGRANGIAN THEORY

Lagrangian theory is basically the variational principle. There are several forms to express this principle. The simplest one, also the earliest one, is the principle of action.

According to Hamilton's principle in mechanics, a dynamical system is characterized by a definite function L and an integral

$$\int_{t_1}^{t} L\,\mathrm{d}t$$

which describes the motion of the system between two positions and which takes the least possible value or least action for its natural path. The product $L\,dt$ is called the action, and the variation of the action leading to its own vanishing is the least action:

$$\delta \int_{t_1}^{t} L\,dt = 0$$

The function L is called the Lagrangian and is defined as

$$L = T - \mathsf{U}$$

where T is the kinetic energy term and U is the potential energy term. U in polymer chemistry refers to interactions and intra-actions between molecules.

APPENDIX E GREEN'S FUNCTION

Green's function is a powerful tool for the solution of second-order partial differential equations, satisfying boundary conditions. For example, given the differential equation

$$\frac{\partial^2 z}{\partial x \partial y} + \frac{2}{x+y}\left(\frac{\partial z}{\partial x} + \frac{\partial z}{\partial y}\right) = 0$$

we wish to find the solution that satisfies the boundary conditions

$$z = 0$$

$$\frac{\partial z}{\partial x} = 3x^2$$

on $y = x$. Since the values of z and $\partial z/\partial x$ (or $\partial z/\partial y$) are prescribed along the line $y = x$, we first find the solution of the differential equation at a point p (ξ, η). Then Green's function is found to be

$$G(x, y; \xi, \eta) = \frac{(x+y)[2xy + (\xi - \eta)(x - y) + 2\xi\eta]}{(\xi + \eta)^3}$$

The term z can now be evaluated with the aid of Green's function, though the process is rather complicated.

REFERENCES

Albrecht, A. C., *J. Chem. Phys.* **27**, 1002 (1957).

Bawn, C. E. H., C. Freeman, and A. Kamaliddin, *Trans. Faraday Soc.* **46**, 862 (1950).

Chandler, D., *Introduction to Modern Statistical Mechanics*, Oxford, UK: Oxford University 1987.

de Gennes, P. G., *Phys. Lett.* **38A**, 339 (1972).

de Gennes, P. G., *Scaling Concepts in Polymer Physics*. Ithaca, NY: Cornell University Press, 1979.

des Cloizeaux, J., *J. Phys. (Paris)* **36**, 281 (1975).

Duhem, P., *Compt. Rend.* **102**, 1449 (1896).

Flory, P. J., and W. R. Krigbaum, *J. Chem. Phys.* **18**, 1086 (1950).

Flory, P. J., *Principles of Polymer Chemistry*. Ithaca, NY: Cornell University Press, 1953.

Freed, K. F., *Renormalization Group Theory of Macromolecules*. New York: Wiley, 1987.

Gibbs, J. W., *Collected Works*, New York: Longmans, Green & Co., 1931.

Hill, T. L., *Introduction to Statistical Thermodynamics*. Reading, MA: Addison-Wesley, 1960.

Ising, E., *Z. Physik* **31**, 253 (1925).

Krigbaum, W. R., and P. J. Flory, *J. Am. Chem. Soc.* **75**, 1775 (1953a).

Krigbaum, W. R., and P. J. Flory, *J. Polym. Sci.* **11**, 37 (1953b).

Kurata, M., and H. Yamakawa, *J. Chem. Phys.* **29**, 311 (1958).

Ma, S. K., *Statistical Mechanics*. Philadelphia, PA: World Scientific, 1985.

McMillan, W. G., and J. E. Mayer, *J. Chem. Phys.* **13**, 276 (1945).

Orofino, T. A., and P. J. Flory, *J. Chem. Phys.* **26**, 1067 (1957); *J. Phys. Chem.* **63**, (1959).

Scatchard, G., *J. Am. Chem. Soc.* **68**, 2315 (1946).

Scatchard, G., A. C. Batchelder, and A. Brown, *J. Am. Chem. Soc.* **68**, 2320 (1946).

Sneddon, I. N., *Elements of Partial Differential Equations*. New York: McGraw-Hill, 1957.

Wilson, K. G., *Phys. Rev.* **B4**, 3174, 3184 (1971).

Zimm, B. H., *J. Chem. Phys.* **14**, 164 (1946).

PROBLEMS

7.1 The osmotic pressures at a series of concentrations from 2 to 10 g of polymer per kilogram of solution of a fractionated vinyl chloride polymer in methyl amyl ketone at 27°C have been determined. The experimental data are as follows:

π' (cm)	0.4	1.1	2	3	4
c (g solute/kg soln)	2	4	6	8	10

(a) Plot π' versus c and π'/c versus c.

(b) Determine the molecular weight of the polymer (using either plot). Suggestion: Consider the empirical equation

$$\frac{\pi'}{c} = \left(\frac{1000R'T}{M}\right) + Ac$$

where R' is 0.0821 lit. atm per mole-degree, T is the absolute temperature, M is the molecular weight, and A is a constant. Which plot is good for the determination of molecular weight?

7.2 The osmotic pressures at a series of concentrations and temperatures of a polymer in water have been determined as follows:

$c\,(\text{kg/m}^3)$	$t\,(^\circ\text{C})$	$\pi'\,(\text{Pa})$
10.00	25	855
	30	827
	35	755
	40	722
50.00	25	7 335
	30	6 360
	35	5 085
	40	3 965
100.0	25	22 340
	30	18 790
	35	14 300
	40	9 670
150.00	25	46 260
	30	38 595
	35	28 845
	40	18 675

Determine (a) the molecular weight and (b) the second and third virial coefficients.

7.3 According to the Flory-Krigbaum theory, the excluded volume for a pair of polymer molecules may be expressed in the form

$$u = \left(\frac{\bar{v}^2}{NV_1}\right)(\psi_1 - \kappa_1)M^2F(x)$$

The function $F(x)$ is defined as

$$F(x) = \left(\frac{4}{\pi^{1/2}}\right)x^{-1}\int_0^\infty \{1 - \exp[-x\exp(-y^2)]\}y^2\,dy$$

Show that

$$F(x) = \sum_{i=1}^{\infty} \frac{(-x)^{i-1}}{i^{3/2}i!}$$

Hence,

$$F(x) = 1 - \frac{x}{2!}2^{3/2} + \frac{x^3}{3!}2^{3/2} - \cdots$$

Hint: The definite integrals

$$I_p = \int_0^\infty y^p\{1 - \exp[-x\exp(-y^2)]\}dy$$

have the series solutions

$$I_p = \frac{1}{2}\Gamma\left(\frac{P+1}{2}\right)\sum_{n=1}^{\infty}\frac{x^n(-1)^{n+1}}{n!n^{(p+2)/2}}$$

where, $p = 0, 1, 2, \ldots$ and Γ is the gamma function.

7.4 The following data were obtained by Flory and Krigbaum for polystyrene in toluene at 30°C:

$\bar{M}_n$	Γ_2	α^3
30 900	0.195	2.56
41 700	0.251	2.80
61 500	0.309	3.00
120 000	0.576	3.56
328 000	1.21	4.20
612 000	2.02	5.56

(a) Show that the second virial coefficient decreases with increasing molecular weight by plotting Γ_2 versus $\bar{M}_n$.

(b) Calculate the theoretical values of A_2 as a function of molecular weight, using

$$A_2 = \frac{\bar{v}^2}{V_1}(\psi_1 - \kappa_1)F(x)$$

Hint: $x = 2(\alpha^2 - 1)$.

(c) Convert the experimental values of Γ_2 to A_2 and compare with the theoretical values of A_2 by plotting A_2 (observed) versus $\bar{M}_n$ and A_2 (calculated) versus $\bar{M}_2$ on the same graph.

Useful information: $\bar{v} = 0.93$ mL/g at 30 °C and $\psi_1 - \kappa_1 = 0.058$. (*Source*: Krigbaum and Flory (1953a, b).)

7.5 A theoretic equation was derived by Kurata and Yamakawa for the osmotic second virial coefficient of flexible polymer chain which takes into account not only the intermolecular interaction, but also the intramolecular interaction of segments (such as the excluded volume effect). The equation is given by

$$A_2 = \frac{2}{3}\pi N b_0^3\left(1 - \frac{\theta}{T}\right)(1 - 2.865z + 18.51z^2 - \cdots)$$

Plot the theoretical curve of A_2 versus the temperature for polystyrene in cyclohexane. The conditions are specified as follows: Temperatures (K): 308, 310, 315, 320, $\theta = 308$; molecular weight of the polymer: 3.20×10^6, of the segment 104, $b_0 = 0.1178$, $z = 0.15$. (*Source*: Kurata and Yamakawa (1958).)

7.6 Osmotic pressure measurements were carried out by Scatchard and coworkers (1946) for bovine serum albumin in various NaCl solutions for the study of molecular interaction between the protein and the salt and for the determination of the molecular weight of BSA. The quantities that were measured are c_2, the concentration of BSA in grams of isoelectric protein, inside the membrane; c_2', the concentration of BSA outside the membrane; m_3, the average concentration of NaCl inside the membrane; m_3, the average concentration of NaCl outside the membrane; and π', the osmotic pressure in mm Hg. Also measured were the pH of the solution inside the membrane and z_2, the net charge of the protein (that is, the valence number). The experimental data are as follows:

c_2	c_2'	pH	z_2	m_3	m_3'	π'
57.42	0.06	4.98	5.3	0.1523	0.1488	18.93
27.28	0.00	5.28	1.2	0.1506	0.1489	8.35
56.20	0.04	5.29	1.0	0.1506	0.1477	19.33
8.95	0.00	5.30	1.0	0.1500	0.1488	2.51
75.03	0.16	5.35	0.3	0.1521	0.1482	28.52
17.69	0.15	5.42	−1.1	0.1498	0.1488	5.07

Calculate

(a) The molecular weight of BSA, M_2.

(b) The distribution of the salt, b_{23}, in each solution.

(c) The molecular interaction between the protein and the salt, β_{23}, in each solution.

(d) The molecular interaction between proteins, β_{22} in each solution.

8

LIGHT SCATTERING

In 1869 Tyndall studied the phenomenon of the scattering by particles in colloidal solution and he demonstrated that if the incident light is polarized, scattering is visible in only one plane. In 1871 and 1881 Rayleigh derived an equation that showed the scattering to be inversely proportional to the fourth power of the wavelength of the incident light. Then in 1944 and 1947 Debye suggested that the measurement of light scattering intensity could be used to determine the molecular weight of a macromolecule in dilute solutions and possibly to determine its size and shape. Since then light scattering has become an important tool in the study of polymer behavior in solutions.

8.1 RAYLEIGH SCATTERING

Light is a form of electromagnetic radiation. The properties of an electromagnetic field may be expressed by two vector quantities: the electric field strength **E** and the magnetic field strength **H**. The two field strengths are related by four differential equations according to Maxwell:

$$\nabla \times \mathbf{E} = \frac{1}{c'}\frac{\partial \mathbf{H}}{\partial t}$$

$$\nabla \cdot \mathbf{H} = 0$$

$$\nabla \times \mathbf{H} = \frac{1}{c'}\frac{\partial E}{\partial t} + 4\pi \mathbf{g}$$

$$\nabla \cdot \mathbf{E} = 4\pi\rho$$

where **g** is the current density, ρ is the charge density, c' is the velocity of light, and ∇ is the mathematical operator:

$$\nabla = \frac{\partial}{\partial x}\mathbf{i} + \frac{\partial}{\partial y}\mathbf{j} + \frac{\partial}{\partial z}\mathbf{k}$$

Rayleigh scattering is best described in terms of three factors: the incident light, the particle (for example, a macromolecule), which serves as an oscillating dipole, and the scattered light. The model is shown in Figure 8.1. The incident light may be expressed by the following well-known equation:

$$E = E_0 \cos 2\pi\left(vt - \frac{x}{\lambda}\right)$$

where E is the electric field or electric intensity, E_0 is the amplitude of the incident wave, v is the frequency of the light in the solution, and t is the propagation time. The frequency v is related to the wavelength λ by

$$v = \frac{c'}{\lambda}$$

The wavelength of the light in the solution is

$$\lambda = \frac{\lambda_0}{n}$$

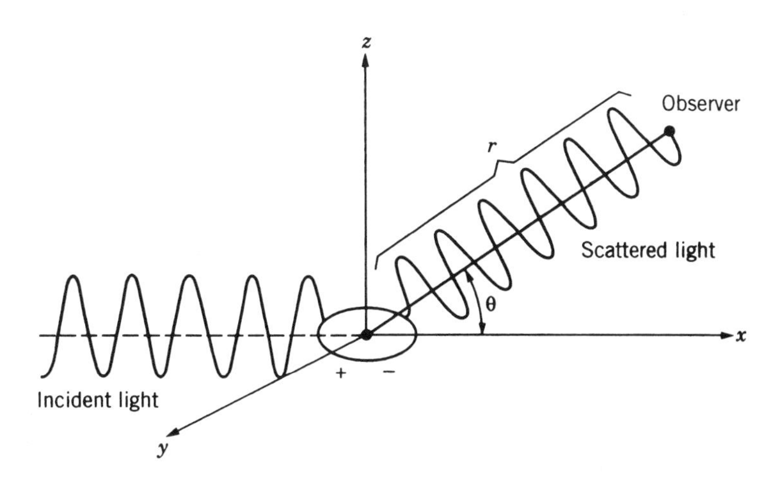

FIGURE 8.1 The Rayleigh scattering model.

where λ_0 is the wavelength of the light in vacuum and n is the refractive index of the medium.

As the incident light hits the molecule, the distribution of electrons in the molecule is distorted, resulting in the polarization of the molecule, which now acts as an oscillating dipole p. The dipole is related to the electric field E_0 by

$$p = \alpha E = \alpha E_0 \cos 2\pi\left(\nu t - \frac{x}{\lambda}\right)$$

Here α is the polarizability of the molecule. The second derivative of the oscillating dipole with respect to time $\mathrm{d}^2p/\mathrm{d}t^2$ describes the electric strength of the scattered light:

$$\frac{\mathrm{d}^2 p}{\mathrm{d}t^2} = \alpha E_0 \cos 2\pi\left(\nu t - \frac{x}{\lambda}\right)(4\pi^2\nu^2)$$

The negative sign in the above equation is dropped because we are interested only in its absolute value.

If the incident light is plane polarized, the scattered light can then be expressed:

$$E_s = \frac{(\mathrm{d}^2p/\mathrm{d}t^2)(\sin\theta/r)}{c'^2}$$

where r is the distance of the dipole from the observer and θ is the angle between the dipole axis and the line r. The division by the square of the velocity of light, c'^2, is a dimensional correction.

Substituting $\mathrm{d}^2p/\mathrm{d}t^2$ in the above equation, we obtain

$$E_s = \frac{4\pi^2\nu^2\alpha E_0 \sin\theta}{c'^2 r} \cos 2\pi\left(\nu t - \frac{x}{\lambda}\right)$$

Since the measurable quantity in a light wave is the intensity I, we convert E to I. The intensity of the light is the amount of energy that falls on a unit area per unit time:

$$I = \varepsilon_0 c' \langle E^2 \rangle$$

In the SI system E is in newtons per coulomb, $\varepsilon_0 = 8.854\,18 \times 10^{-12}\,\mathrm{C^2\,N^{-1}\,m^{-2}}$. The intensity I is also expressed in joules per square meter per second ($\mathrm{J\,m^2\,s^{-1}}$). $\langle E \rangle$ is the field averaged over a period of vibration from $t = 0$ to $t = 1/\nu$.

In measurement we are concerned with the ratio of the intensity of scattered light I over the intensity of the incident light I_0:

$$\frac{I}{I_0} = \frac{E_s^2}{E^2} = \frac{\{[(4\pi^2 v^2 \alpha E_0 \sin\theta)/c'^2 r] \cos 2\pi(vt - x/\lambda)\}^2}{[E_0 \cos 2\pi(vt - x/\lambda)]^2}$$

$$= \frac{16\pi^4 \alpha^2 \sin^2\theta}{\lambda^4 r^2}$$

The equation

$$\frac{I}{I_0} = \frac{16\pi^4 \alpha^2 \sin^2\theta}{\lambda^4 r^2} \tag{8.1}$$

is called the Rayleigh equation for plane polarized light. If the incident beam is unpolarized, the equation is slightly modified:

$$\frac{I}{I_0} = \frac{16\pi^4 \alpha^2}{\lambda^4 r^2}(1 + \cos^2\theta)$$

The correction term is based on the consideration of the total intensity:

$$I\,(\text{total}) = I\ (y\ \text{component}) + I\ (z\ \text{component})$$

$$\rightarrow \sin^2\phi_y + \sin^2\phi_z = 1 + \cos^2\theta$$

The ratio $(Ir^2/I_0)_\theta$ is called the Rayleigh ratio and is often designated as R_θ:

$$R_\theta = \left(\frac{Ir^2}{I_0}\right)_\theta$$

Its unit is cm^{-1} (path length, 1 cm). Thus, in light-scattering experiments, the three quantities R_θ, λ, and θ are the major parameters that we measure.

The physical meaning of the Rayleigh ratio is the attenuation of the incident beam by the loss of intensity after passage through a medium. Though intensity lost here is due to scattering, not absorption, we can still use Lambert's law to describe the phenomenon. Lambert's law states that

$$\frac{I}{I_0} = e^{-\tau l}$$

where τ is the turbidity, which is a measure of the loss of intensity of the incident beam, and l is the path length of the cell. For $l = 1$ (cm), $\tau = -\ln(I/I_0)$. The relationship between the Rayleigh ratio and the turbidity is given by

$$\tau = 2\pi \int_0^\pi \left(\frac{Ir^2}{I_0}\right)_\theta \sin\theta \, d\theta$$

For special scattering angles 90° and 0°, we have

$$\tau = \frac{16\pi}{3} R_{90^\circ} = \frac{8\pi}{3} R_{0^\circ}$$

8.2 FLUCTUATION THEORY (DEBYE)

In 1947 Debye suggested that the amount of light scattered by a solution of high polymers is related to the mass of the solute molecules. Hence, the measurement of the intensity of scattered light enables us to determine the molecular weight of polymers. The difficulty is in how to utilize the Rayleigh equation for that purpose. Debye proposed the following theory.

The Rayleigh scattering equation is applicable to gases where molecules move at random and in near chaos. In a liquid solution the molecules are far from being independent of one another, but unlike crystals, liquids are not well-ordered either. Instead, there is a fluctuation in the concentration of a given volume element. The fluctuation of the concentration $\langle \Delta c \rangle$ results in the fluctuation of polarizability α. Therefore, to apply the Rayleigh scattering equation to the liquid state, we need only to modify the parameter α. For gases, we have $p = \alpha E$; for solutions, we should have $p = \langle \Delta\alpha^2 \rangle E$. The quantity $\langle \Delta\alpha^2 \rangle$ is the fluctuation of the polarizability.

Since $\alpha = \alpha(c, \rho)$, the differential $d\alpha$ is given by

$$d\alpha = \frac{\partial \alpha}{\partial c} dc + \frac{\partial \alpha}{\partial \rho} d\rho$$

where c is the concentration of the solution and ρ is the density. For the first approximation the density term may be ignored, because we are more interested in the composition of solutions not the physical state (liquid) of the solution. Furthermore, we should change notation from $d\alpha$ and dc to $\Delta\alpha$ and Δc:

$$\Delta\alpha = \frac{\partial \alpha}{\partial c} \Delta c$$

Polarizability α is usually measured in terms of refractive index n or dielectric constant ε and the light-scattering apparatus can be designed in relation to the measurement of refractive index. (Note: $\varepsilon^2 = n^2$ according to Maxwell's theory.) For that reason, we utilize the relation

$$\alpha = \frac{n^2 - 1}{4\pi}$$

and convert $\partial\alpha$ to ∂n by

$$\frac{\partial\alpha}{\partial c} = \frac{n}{2\pi}\frac{\partial n}{\partial c}$$

and hence

$$\langle \Delta\alpha^2 \rangle = \frac{n^2}{4\pi^2}\left(\frac{\partial n}{\partial c}\right)^2 \langle \Delta c^2 \rangle$$

We may change ∂ into d, since there is only one independent variable involved now:

$$\langle \Delta\alpha^2 \rangle = \frac{n^2}{4\pi^2}\left(\frac{\mathrm{d}n}{\mathrm{d}c}\right)^2 \langle \Delta c^2 \rangle$$

where $\mathrm{d}n/\mathrm{d}c$ is called the differential refractive index and it is an experimentally measurable quantity. Our attention now is focused on $\langle \Delta c^2 \rangle$.

The fluctuation of concentrations is always accompanied by a change in free energy, ΔG. We now expand ΔG in terms of Δc around the equilibrium concentration $\langle c \rangle$, using the Taylor's series:

$$\Delta G = \frac{\partial G}{\partial c}\Delta c + \frac{1}{2!}\left(\frac{\partial^2 G}{\partial c^2}\right)(\Delta c)^2 + \cdots$$

The first term is zero for a closed term at constant temperature, whereas the higher terms, including $(\Delta c)^3$, may be neglected since the fluctuations are rather small. Thus, only the second term is physically meaningful:

$$\Delta G = \frac{1}{2!}\left(\frac{\partial^2 G}{\partial c^2}\right)(\Delta c)^2$$

Using the Boltzmann expression we obtain a distribution function of concentration:

$$\exp\left(-\frac{\Delta G}{kT}\right) = \exp\left[-\frac{1}{kT}\frac{1}{2!}\left(\frac{\partial^2 G}{\partial c^2}\right)(\Delta c)^2\right]$$

The fluctuation of concentration, $\langle \Delta c^2 \rangle$ can then be evaluated as follows:

$$\langle \Delta c^2 \rangle = \frac{\int_0^\infty (\Delta c)^2 \exp[-(\partial^2 G/\partial c^2)(\Delta c)^2/2kt]dc}{\int_0^\infty \exp[-(\partial^2 G/\partial c^2)(\Delta c)^2/2kT]dc} = \frac{kT}{(\partial^2 G/\partial c^2)_{T,P}}$$

Note:

$$\int_0^\infty \mathrm{e}^{-ax^2}\,\mathrm{d}x = \frac{1}{2}\left(\frac{\pi}{a}\right)^{1/2}$$

$$\int_0^\infty x^2\mathrm{e}^{-ax^2}\,\mathrm{d}x = \frac{1}{4a}\left(\frac{\pi}{a}\right)^{1/2}$$

The two integrals in the equation of $\langle \Delta c^2 \rangle$ can be evaluated by using the two formulas and we obtain

$$\int_0^\infty (\Delta c)^2 \exp\left[-\left(\frac{\partial^2 G/\partial c^2}{2kT}\right)(\Delta c)^2\right] \mathrm{d}c = \frac{1}{4a}\left(\frac{\pi}{a}\right)^{1/2}$$

$$\int_0^\infty \exp\left[-\left(\frac{\partial^2 G/\partial c^2}{2kT}\right)(\Delta c)^2\right] \mathrm{d}c = \frac{1}{2}\left(\frac{\pi}{a}\right)^{1/2}$$

Hence

$$\langle \Delta c^2 \rangle = \frac{(1/4a)(\pi/a)^{1/2}}{(1/2)(\pi/a)^{1/2}} = \frac{1}{2a} = \frac{1}{(\partial^2 G/\partial c^2)/kT}$$

It remains to calculate $(\partial^2 G/\partial c^2)_{T,P}$. We utilize the following three equations at constant T and P:

i. $\mathrm{d}G = \mu_1\,\mathrm{d}n_1 + \mu_2\,\mathrm{d}n_2$

Note: From $G = G(T, p, n_1, n_2)$

$$\mathrm{d}G = \left(\frac{\partial G}{\partial T}\right)_{P,n_i} \mathrm{d}T + \left(\frac{\partial G}{\partial P}\right)_{T,n_i} \mathrm{d}P + \left(\frac{\partial G}{\partial n_1}\right)_{T,P,n_2} \mathrm{d}n_1$$

$$+ \left(\frac{\partial G}{\partial n_2}\right)_{T,P,n_1} \mathrm{d}n_2, \quad n_i = n_1, n_2 \cdots$$

ii. $\mathrm{d}n_1 = \dfrac{\bar{V}_2}{\bar{V}_1}\mathrm{d}n_2$

Note: From $V = n_1\bar{V}_1 + n_2\bar{V}_2$
where μ_1, μ_2 are chemical potentials, n_1, n_2 are the number of moles, and $\bar{V}_1, \bar{V}_2$ are partial molar volumes of 1 and 2, respectively.

iii. $\mathrm{d}n_2 = (V/M)\mathrm{d}c$

Note: From $\dfrac{n_2}{V} = \dfrac{c}{M}$

V is the volume of solution in mL, c is in g/mL, and M is in g/mol.

With a simple manipulation, we obtain

$$\mathrm{d}G = \left(\frac{-\bar{V}_2}{\bar{V}_1}\mu_1 + \mu_2\right)\left(\frac{V}{M}\right)\mathrm{d}c$$

and

$$\frac{\mathrm{d}G}{\mathrm{d}c} = \left(-\frac{\bar{V}_2}{\bar{V}_1}\mu_1 + \mu_2\right)\left(\frac{V}{M}\right)$$

Differentiating this equation with respect to c then gives

$$\left(\frac{\partial^2 G}{\partial c^2}\right)_{T,P} = \frac{V}{M}\left(\frac{\partial \mu_2}{\partial c} - \frac{\bar{V}_2}{\bar{V}_1}\frac{\partial \mu_1}{\partial c}\right)$$

According to the Gibbs-Duhem equation, μ_1 and μ_2 are related in the form

$$n_1\,d\mu_1 + n_2\,d\mu_2 = 0$$

from which we obtain

$$d\mu_2 = -\frac{n_1\,d\mu_1}{n_2}$$

and

$$\frac{\partial \mu_2}{\partial c} = -\frac{n_1}{n_2}\frac{\partial \mu_1}{\partial c}$$

Substituting into $(\partial^2 G/\partial c^2)_{T,P}$ we now have

$$\left(\frac{\partial^2 G}{\partial c^2}\right)_{T,P} = \frac{V}{M}\left(-\frac{n_1}{n_2} - \frac{\bar{V}_2}{\bar{V}_1}\right)\left(\frac{\partial \mu_1}{\partial c}\right)$$

$$= \frac{-V}{M}\left(\frac{n_1\bar{V}_1 + n_2\bar{V}_2}{n_2\bar{V}_1}\right)\left(\frac{\partial \mu_1}{\partial c}\right)$$

Since

$$\frac{n_2 M}{n_1\bar{V}_1 + n_2\bar{V}_2} = c$$

we have

$$\left(\frac{\partial^2 G}{\partial c^2}\right)_{T,P} = -\frac{V}{c\bar{V}_1}\left(\frac{\partial \mu_1}{\partial c}\right)_{T,P}$$

Recall van't Hoff's equation:

$$\mu_1 - \mu_1^\circ = -\bar{V}_1\frac{R'T}{M}c$$

This gives us

$$-\frac{\partial \mu_1}{\partial c} = \frac{\bar{V}_1 R'T}{M}$$

Our equation of $(\partial^2 G/\partial c^2)_{T,P}$ then becomes

$$\left(\frac{\partial^2 G}{\partial c^2}\right)_{T,P} = \frac{V}{c\bar{V}_1}\frac{\bar{V}_1 R'T}{M} = \frac{VR'T}{cM}$$

and

$$\langle \Delta c^2 \rangle = \frac{kT}{(\partial^2 G/\partial c^2)_{T,P}} = \frac{kT}{VR'T/cM} = \frac{N_A kTcM}{N_A VR'T} = \frac{cM}{N_A V}$$

where N_A is Avogadro's number. In comparison, we now have

$$\frac{I}{I_0} = \frac{16\pi^4\alpha^2 \sin^2\theta}{\lambda^4 r^2} \qquad \text{for gases}$$

and

$$\frac{I}{I_0} = \frac{16\pi^4 \langle\Delta\alpha^2\rangle \sin^2\theta V}{\lambda^4 r^2} \qquad \text{for solutions} \tag{8.2}$$

Here again V is the volume of solution in which n scatterers are contained. Since we have already derived the equation for $\langle\Delta\alpha^2\rangle$, Eq. (8.2) can now be put in the form

$$\frac{I}{I_0} = \frac{16\pi^4 \sin^2\theta}{\lambda^4 r^2} \frac{n^2}{4\pi^2}\left(\frac{\mathrm{d}n}{\mathrm{d}c}\right)^2 \frac{cM}{N_A V} V$$

Simplifying and rearranging the terms, we get

$$\frac{Ir^2}{I_0} = \frac{4\pi^2 n^2 (\mathrm{d}n/\mathrm{d}c)^2 \sin^2\theta Mc}{\lambda^4 N_A} \tag{8.3}$$

This equation is for plane polarized incident light. If the incident light is not polarized, we have

$$\frac{Ir^2}{I_0} = \frac{2\pi^2 n^2 (\mathrm{d}n/\mathrm{d}c)^2 \sin^2\theta}{\lambda^4 N_A} Mc(1 + \cos^2\theta) \tag{8.4}$$

If the incident light is not polarized and if $\theta = 90°$, we have

$$\frac{Ir^2}{I_0} = \frac{2\pi^2 n^2 (\mathrm{d}n/\mathrm{d}c)^2 MC}{\lambda^4 N_A} \tag{8.5}$$

Let

$$K = \frac{2\pi^2 n^2 (\mathrm{d}n/\mathrm{d}c)^2}{\lambda^4 N_A}$$

Then the light-scattering equation becomes

$$\frac{Kc}{R_\theta} = \frac{1}{M} \tag{8.6}$$

(cf. the equation for osmotic pressure: $\pi'/R'Tc = 1/M$).

The Rayleigh ratio R_θ may be converted to the turbidity τ using the conversion factors

$$H = \frac{16\pi K}{3} \tag{8.7}$$

$$\tau = \frac{16\pi}{3} R_\theta \tag{8.8}$$

Then we have another form of the light-scattering equation:

$$\frac{HC}{\tau} = \frac{1}{M} \tag{8.9}$$

Both Eqs. (8.6) and (8.9) are very similar to the osmotic pressure equation of van't Hoff.

8.3 DETERMINATION OF MOLECULAR WEIGHT AND MOLECULAR INTERACTION

8.3.1 Two-Component Systems

If the behavior of the solution is not ideal, we have the light-scattering equation in the following forms:

$$\frac{Kc}{R_\theta} = \frac{1}{M} + 2A_2c + 3A_3c^2 + \cdots \tag{8.10}$$

$$\frac{Hc}{\tau} = \frac{1}{M} + 2A_2c + 3A_3c^2 + \cdots \tag{8.11}$$

In comparison with the osmotic pressure equation,

$$\frac{\pi'}{c} = \frac{1}{M} + A_2c + A_3c^2 + \cdots$$

We notice that the slope of the plot in the case of straight lines gives the second virial coefficient directly to the osmotic pressure data, whereas it gives a half value of the second virial coefficient to the light-scattering data. That is,

Second virial coefficient

$= A_2$

$=$ the slope of osmotic pressure linear plot

$=$ twice the slope of the light scattered linear plot

8.3.2 Multicomponent Systems

In multicomponent systems (that is, more than two components in the solution) there is in addition preferential binding (or preferential adsorption) of the solvent components on the polymer chain segment. To obtain accurate molecular weight a correction term of preferential binding must be included in the light-scattering equation. The correction term basically involves dn/dc. If we let component 1 be the solvent, component 2 be the polymer, and component 3 be another solvent (or salt in the case of biological polymers such as protein), we have two slightly different ways to express the light-scattering equation:

$$\left[\frac{K'c}{R_{90}}\left(\frac{dn}{dc}\right)^2\right]_{c=0} = \left(\frac{1}{M_2}\right)_{\text{two}} \qquad \text{for a two-component system}$$

$$\left[\frac{K'c}{R_{90}}\left(\frac{dn}{dc} + D'\frac{dn}{d\phi_1}\right)^2\right]_{c=0} = \left(\frac{1}{M_2}\right)_{\text{three}} \qquad \text{for a three-component system}$$

where $K' = 2\pi^2 n^2/\lambda^4 N_A$, ϕ_1 is the volume fraction of the solvent, and

$$D' = -\frac{d\phi_1}{dc}$$

If we take the ratio of the two equations, we obtain

$$D' = \frac{(dn/dc)[(M_2)_{\text{three}}/(M_2)_{\text{two}} - 1]}{dn/d\phi_1} \tag{8.12}$$

Thus, D', which is the preferential adsorption or preferential binding, can be obtained from the two light-scattering measurements of the same macromolecule, one in a single solvent and the other in a solvent mixture.

In biochemistry there is another method to determine the true molecular weight M_2 and the preferential binding D'. This can be done by carrying out two light-scattering experiments both for the same polymer (for example, a protein) in a solvent mixture. One of them is dialyzed against the solvent system (solvent mixture) prior to the measurement and the other is not. The dialysis is to establish the equilibrium between the solute and solvents. We now have

$$\left(\frac{Kc}{R_{90}}\right)_{c=0} = \frac{1}{M_{\text{app},2}} \qquad \text{Undialyzed solution}$$

$$\left(\frac{Kc}{R_{90}}\right)_{c=0} = \frac{1}{M_2} \qquad \text{Dialyzed solution}$$

and

$$M_{\text{app},2} = M_2(1 + D)^2$$

where

$$D = \frac{(\partial n/\partial g_3)_{T,P,g_2}}{(\partial n/\partial g_2)_{T,P,g_3}} \left(\frac{\partial g_3}{\partial g_2}\right)_{T,\mu_1,\mu_3} \tag{8.13}$$

The symbol g refers to the gram and the term $\partial g_3/\partial g_2$ refers to the preferential binding of component 3 (a solvent) to component 2 (polymer segment). $M_{\text{app},2}$ is the apparent molecular weight of the macromolecule and M_2 is its true molecular weight. The parameter D may also be expressed in the equation

$$D = \frac{(1 - c_2\bar{v}_3)(\partial n/\partial c_3)_{m_2}}{(1 - c_2\bar{v}_2)(\partial n/\partial c_3)_{m_3}} \frac{M_3}{M_2} \left(\frac{\partial m_3}{\partial m_2}\right)_{T,\mu_1,\mu_3} \tag{8.14}$$

where

$$\frac{M_3}{M_2}\left(\frac{\partial m_3}{\partial m_2}\right)_{T,\mu_1,\mu_3} = \left(\frac{\partial g_3}{\partial g_2}\right)_{T,\mu_1,\mu_3}$$

m being the molal concentration (moles of solute per kilogram of the solvent) and $\bar{v}$ the partial specific volume. D and D' are related in the following way:

$$D = D' \frac{\partial n/\partial c_2}{\partial n/\partial \phi_1}$$

and

$$D' = \frac{(\partial n/\partial c_2)_\mu - \partial n/\partial c_2}{\partial n/\partial \phi_1} \tag{8.15}$$

where $(\partial n/\partial c_2)_\mu$ is the refractive index of the polymer solution dialyzed against the solvent mixture and $\partial n/\partial c_2$ is not dialyzed.

8.3.3 Copolymers

The problem with copolymers is also caused by the complexity of the dn/dc values. There are three factors to be considered:

1. The interference in the differential refractive index of one homopolymer by that of the other
2. The low dn/dc value in certain solvents, which is difficult to measure
3. The fluctuation in chain composition. The value of molecular weight of the same copolymer often varies with the dn/dc of the solvent.

Assuming that the refractive index increments of a copolymer chain are proportional to their composition, Stockmayer et al. (1955) suggested the following equation:

$$\left(\frac{I}{K'c}\right)_{\substack{c\to 0\\ \theta\to 0}} = \left(\frac{\mathrm{d}n}{\mathrm{d}c}\right)_0^2 \bar{M}_w + 2b\left(\frac{\mathrm{d}n}{\mathrm{d}c}\right)_0 \langle M\,\Delta x\rangle + b^2\langle M(\Delta x)^2\rangle \tag{8.16}$$

where

$$b = \left(\frac{\mathrm{d}n}{\mathrm{d}c}\right)_\mathrm{A} - \left(\frac{\mathrm{d}n}{\mathrm{d}c}\right)_\mathrm{B}$$

A, B being monomers of two different types. The other two terms are defined as

$$\langle M\,\Delta x\rangle = \sum w_i M_i(\Delta x)_i$$

$$\langle M(\Delta x)^2\rangle = \sum w_i M_i(\Delta x)_i^2$$

where x_i is the composition of the copolymer sample (that is, a volume fraction of monomer i), w is the weight fraction, and M is the molecular weight. In the above equation there are three unknowns: $\bar{M}_w$, $\langle M\,\Delta x\rangle$, and $\langle M(\Delta x)^2\rangle$. If the light-scattering experiment is carried out for the copolymer in three different solvents, we can solve for three unknowns in three linear equations, and thereby get an accurate value of $\bar{M}_w$.

Another proposal was made by Bushuk and Benoit (1958):

$$\bar{M}_\mathrm{ap} = \bar{M}_w + 2P\frac{\nu_\mathrm{A} - \nu_\mathrm{B}}{\nu_0} + Q\left(\frac{\nu_\mathrm{A} - \nu_\mathrm{B}}{\nu_0}\right)^2 \tag{8.17}$$

where

$$P = \tfrac{1}{2}(1 - x_0)(\bar{M}_w - \bar{M}_B) - x_0(\bar{M}_w - \bar{M}_\mathrm{A})$$

$$Q = x_0(1 - x_0)(\bar{M}_\mathrm{A} + \bar{M}_\mathrm{B} - \bar{M}_w)$$

$$\nu_\mathrm{A} = \left(\frac{\mathrm{d}n}{\mathrm{d}c}\right)_\mathrm{A}$$

$$\nu_\mathrm{B} = \left(\frac{\mathrm{d}n}{\mathrm{d}c}\right)_\mathrm{B}$$

$$\nu_0 = \left(\frac{\mathrm{d}n}{\mathrm{d}c}\right)_{\text{copolymer of uniform composition}}$$

In the above equations, x_0 is the average composition of the copolymer, and A and B are two different homopolymers. The equation also contains three parameters to be determined: M_w, P, and Q. Likewise, if we measure the light-scattering intensities from the polymeric solutions in three different solvents, the three parameters could be determined by solving the three equations simultaneously.

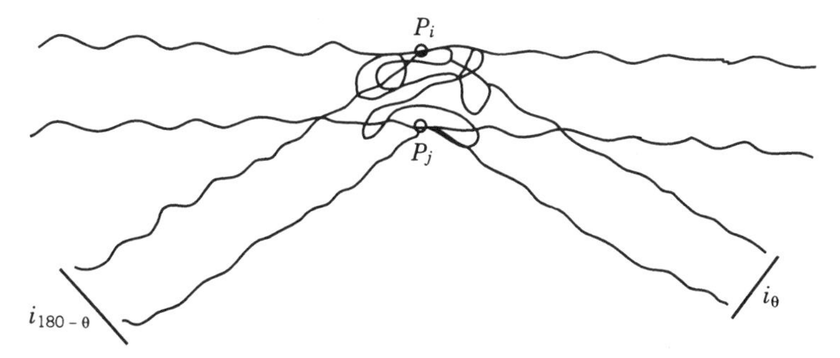

FIGURE 8.2 The model of internal interference.

8.3.4 Correction of Anisotropy and Deporalization of Scattered Light

If the scattering molecules are not isotropic, we must make corrections. The correction factor for anisotropy involves the deporalization ratio P_u, which is defined as the ratio of the horizontal to the vertical components of the scattered light at a 90° angle when the incident light is unpolarized. The correction factor is $(6 - 7P_u)/(6 + 6P_u)$ for $R_{90°}$ and $(3 - 7P_u)/(6 + 7P_u)$ for $\tau_{90°}$.

8.4 INTERNAL INTERFERENCE

For a macromolecule with molecular weight >300 000, the molecule does not act as a single dipole, that is, there is no longer one scattering point. Instead, there may be more than one scattering point. Consider the two-scattering-point system shown in Figure 8.2. When the phase differences between rays that are scattered at two points, P_i and P_j, in a molecule are more than $\lambda/20$ apart, nonspherical scattering envelopes form (Figure 8.3). Mathematical analysis of the

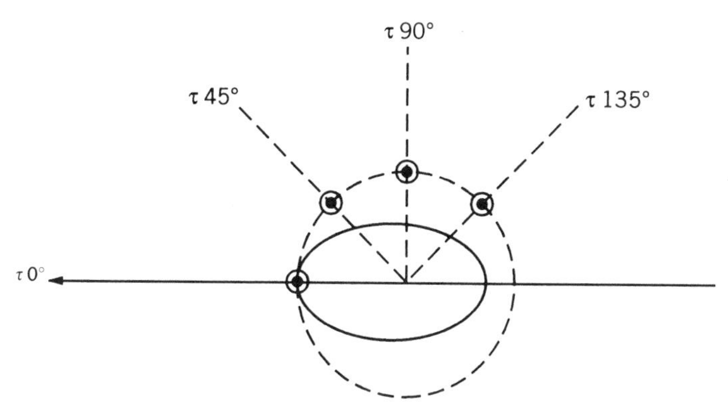

FIGURE 8.3 Scattering envelopes: ---, symmetrical, no destructive interference; — scattering envelop, nonsymmetrical, with destructive interference.

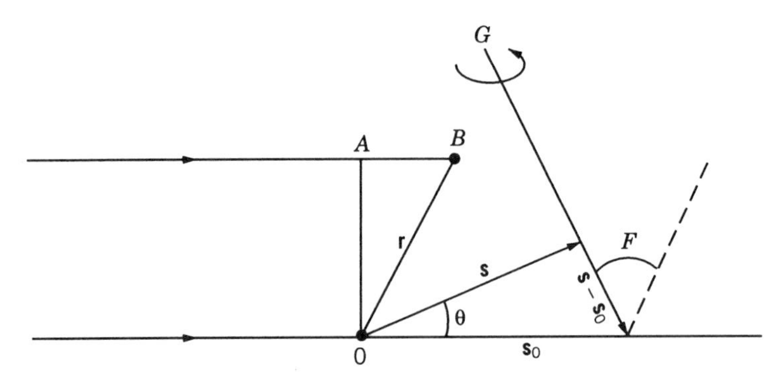

FIGURE 8.4 Coordinates of internal interference.

situation may be described as follows. Consider the two scattering points O and B in Figure 8.4. The vectors $\mathbf{s}_0$ (incident) and $\mathbf{s}$ (scattered) are unit vectors ($|s_0| = |s| = 1$) that define the propagation directions of the incident and scattered rays. The angles F and G specify the orientation of the vector $\mathbf{r}$ relative to the vector $\mathbf{s} - \mathbf{s}_0$. The probability that $\mathbf{r}$ simultaneously points between the angles F and $F + \mathrm{d}F$, and G and $G + \mathrm{d}G$ is

$$\frac{\sin F \, \mathrm{d}F \, \mathrm{d}G}{4\pi}$$

The term 4π is the sum of all possible orientations that can be obtained from

$$\int_0^{2\pi} \int_0^{\pi} \sin F \, \mathrm{d}F \, \mathrm{d}G = 4\pi$$

The amplitude A (that is, OA) is related to the phase shift ϕ', our major interest here, between scattered points (O and B). Their average values are related in the form

$$\langle A \rangle \sim \langle \cos \phi' \rangle$$

On a geometric basis, the phase shift ϕ' can be expressed as

$$\phi' = gr \cos F$$

where

$$g = \frac{4\pi}{\lambda} \sin \frac{\theta}{2}$$

The average phase shift can thus be calculated:

$$\langle \cos \phi' \rangle = \int_0^{2\pi} \int_0^{\pi} \cos(gr \cos F) \frac{\sin F \, \mathrm{d}F \, \mathrm{d}G}{4\pi}$$

Note: $(\sin F \, \mathrm{d}F \, \mathrm{d}G)/4\pi$ is a distribution or probability function.

To carry out the integration we change the variable to $x = gr\cos F$; hence,

$$dx = -gr \sin F \, dF$$

and the integration limits change from $F = 0$ to $x = -gr$ and from $F = \pi$ to $x = gr$. Then

$$\langle \cos \phi' \rangle = \frac{1}{4\pi} \frac{2\pi}{gr} \int_{-gr}^{gr} \cos x \, dx = \frac{\sin gr}{gr}$$

This result shows that the average amplitude scattered by a rotating molecule exhibits g- (or θ) and r-dependent minima and maxima which could possibly be used to determine the internuclear distance r.

Debye suggested the use of $\langle \cos \phi' \rangle$ as a correction factor for the (internal) interference. It is expressed in the form of $1/P(\theta)$ or $1/P(g)$:

$$P(\theta) = P(g) = \sum_i \sum_j \frac{\sin gr_{ij}}{gr_{ij}} \tag{8.18}$$

where r_{ij} is the distance between two elements i and j, and the double summation is performed over all pairs of scattering elements.

The light-scattering equation is now written as

$$\frac{Kc}{R_\theta} = \frac{1}{MP(\theta)} + 2A_2c \qquad \text{(with interference)} \tag{8.19}$$

$$\frac{Kc}{R_\theta} = \frac{1}{M} + 2A_2c \qquad \text{(without interference)}$$

The interference factor $P(\theta)$ depends on the shape of the molecules:

For spheres

$$P(\theta) = \left[\frac{3}{x^3} (\sin x - x \cos x) \right]^2$$

where

$$x = 2\pi \frac{d}{\lambda} \sin \frac{\theta}{2}$$

If the values of x are small, we have

$$\frac{1}{P(\theta)} = 1 + \frac{4\pi^2}{5} \left(\frac{d}{\lambda} \right)^2 \sin^2 \frac{\theta}{2} + \cdots \tag{8.20}$$

For coils

$$P(\theta) = \frac{2}{x^2}[e^{-x} - (1 - x)]$$

$$x = \frac{8}{3}\pi^2 \sin^2\frac{\theta}{2}\frac{\langle R^2 \rangle}{\lambda^2}$$

If the values of x are small, we have

$$\frac{1}{P(\theta)} = 1 + \frac{8\pi^2}{9}\frac{\langle R^2 \rangle}{\lambda^2}\sin^2\frac{\theta}{2} + \cdots \tag{8.21}$$

For rods

$$P(\theta) = \frac{1}{x}\int_0^{2x}\frac{\sin x}{x}\,dx - \left(\frac{\sin x}{x}\right)^2$$

$$x = 2\pi\frac{l}{\lambda}\sin\frac{\theta}{2}$$

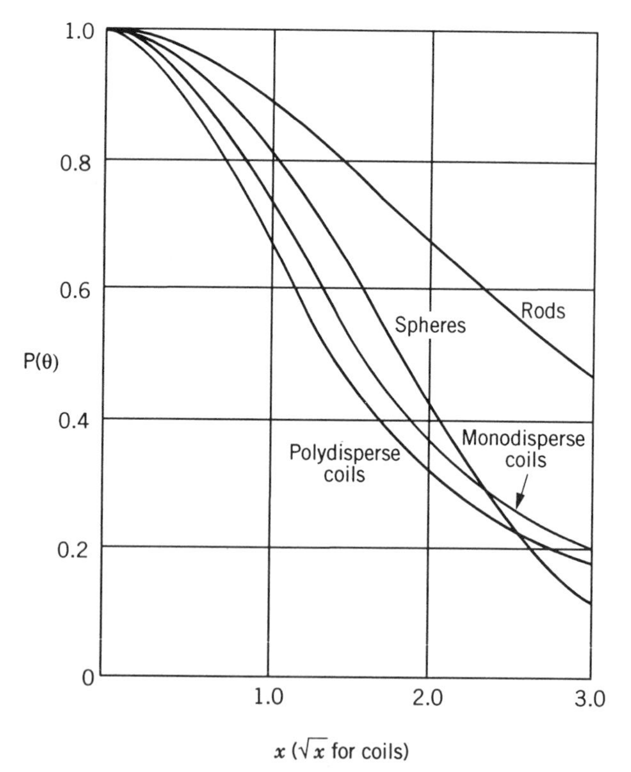

FIGURE 8.5 Theoretical curves for spheres, rods, and coils. (From Doty and Steiner (1950), with permission of Dr. Doty and American Institute of Physics)

If the values of x are small, we have

$$\frac{1}{P(\theta)} = 1 + \frac{4\pi^2}{9}\left(\frac{1}{\lambda}\right)^2 \sin^2\frac{\theta}{2} + \cdots \tag{8.22}$$

Thus, from the interference factor $P(\theta)$ we can obtain the values for d, the diameter if the molecule is in the shape of a sphere; $\langle R^2 \rangle^{1/2}$, the root-mean-square end-to-end distance if the molecule is in the shape of coils; and l, the length of the molecule if the molecule is in the shape of a rod. Figure 8.5 shows the theoretical curves for spheres, rods, and coils.

We may also take the ratio $P_{45}/P_{135} = z$, that is, through substitution of $\theta = 45°$ and $\theta = 135°$ in the above equations, and plot $1/P_{90}$ versus z or P_{90} versus z. The quantity z is called the dissymmetry ratio. If $z = 1$, the particles are small compared to the wavelength of light. If $z > 1$, we can use it to evaluate d, $\langle R^2 \rangle^{1/2}$, or l. In practice, one fits the data by trial and error to the theoretical curves (coils, spheres, or rods) and then determines the dimensions of the macromolecules under study (d, $\langle R^2 \rangle^{1/2}$, l).

8.5 DETERMINATION OF MOLECULAR WEIGHT AND THE RADIUS OF GYRATION BY THE ZIMM PLOT

The Zimm plot is used to determine the molecular weight and radius of gyration of a macromolecule simultaneously regardless of its shape. There is no trial and error method, nor is there any information about the shape. Expanding each term under the summation sign in terms of a power series,

$$\sin x = x - \frac{x^3}{3!} + \frac{x^5}{5!} - \cdots$$

we have

$$P(\theta) = P(g) = \sum_i \sum_j \frac{\sin g r_{ij}}{g r_{ij}}$$

$$= \sum_i \sum_j \left(1 - \frac{g^2 r_{ij}^2}{3!} + \frac{g^4 r_{ij}^4}{5!} - \cdots\right)$$

where

$$g = \frac{4\pi}{\lambda}\sin\frac{\theta}{2}$$

Neglecting the higher terms, we have

$$P(\theta) = 1 - \frac{g^2}{3!}\sum_i \sum_j r_{ij}^2$$

But

$$\sum_i \sum_j r_{ij}^2 = 2R_g^2 = 2S^2$$

where $R_g = S$ is the radius of gyration. Hence

$$P(\theta) = 1 - \frac{16\pi^2}{3 \cdot 2\lambda^2} \sin^2 \frac{\theta}{2} (2S^2)$$

$$= 1 - \frac{16\pi^2}{3\lambda^2} S^2 \sin^2 \frac{\theta}{2}$$

Notice that

$$\frac{1}{1-x} = 1 + x + \cdots$$

So

$$\frac{1}{P(\theta)} = 1 + \frac{16\pi^2}{3\lambda^2} S^2 \sin^2 \frac{\theta}{2} \tag{8.23}$$

We thus have the following expressions, all about the light scattering:

$$\frac{Kc}{R_\theta} = \frac{1}{MP(\theta)} + 2A_2 c + \cdots \qquad \text{(general)} \tag{8.24}$$

$$\lim_{\theta \to 0} \frac{Kc}{R_\theta} = \frac{1}{M} + 2A_2 c + \cdots \qquad \text{(no internal interference)} \tag{8.25}$$

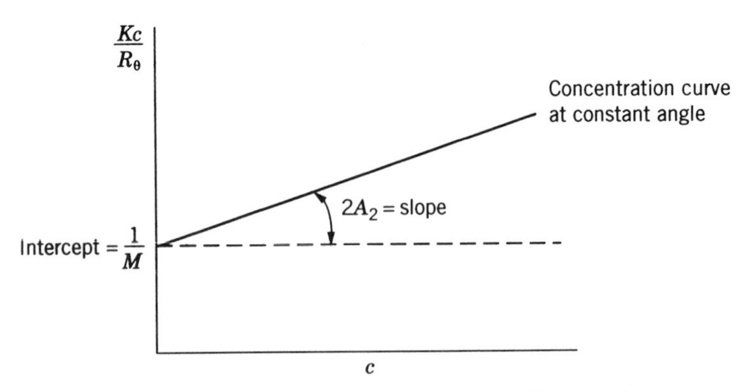

FIGURE 8.6 Determination of the molecular weight and the second virial coefficient.

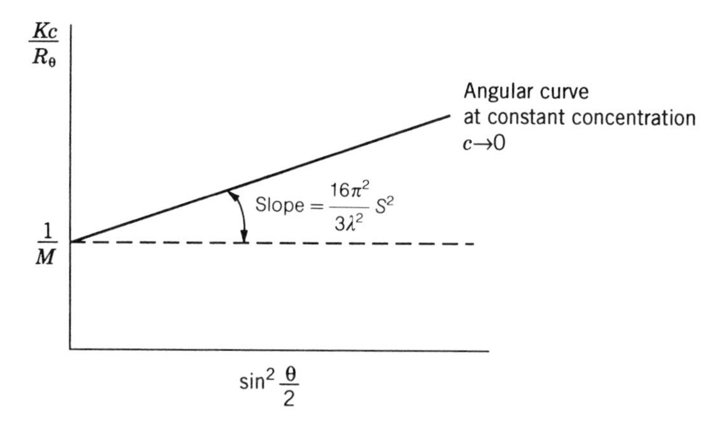

FIGURE 8.7 Determination of the radius of gyration.

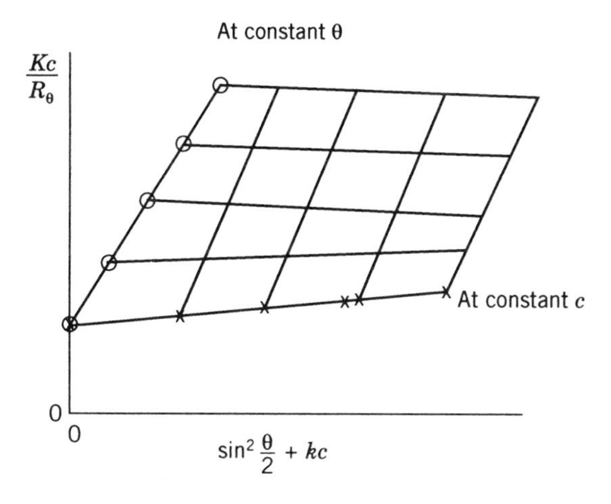

FIGURE 8.8 The Zimm plot.

Equation (8.25) is plotted in Figure 8.6. Equation (8.24) is rewritten as

$$\lim_{c \to 0} \frac{Kc}{R_\theta} = \frac{1}{\mathrm{MP}(\theta)} = \frac{1}{M}\left(1 + \frac{16\pi^2}{3\lambda^2} S^2 \sin^2 \frac{\theta}{2} + \cdots\right) \qquad \text{(with internal interference)}$$

and is plotted in Figure 8.7. The combination of the two plots in Figures 8.6 and 8.7 gives the Zimm plot of Figure 8.8 (see the appendix). In the Zimm plot k is an arbitrary constant to adjust the size of the plot. The lines are defined as follows:

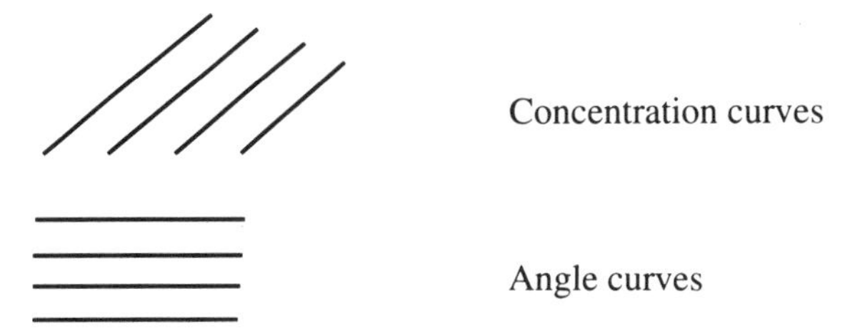

From the plot we obtain

$$\text{Intercept} = \lim_{\substack{c \to 0 \\ \theta \to 0}} = \frac{1}{M}$$

$$\frac{\text{Limiting slope of } c \text{ curve, } c \to 0}{\text{Intercept}} = \frac{16\pi^2}{3\lambda^2} S^2$$

$$\frac{\text{Limiting slope of } \theta \text{ curve, } \theta \to 0}{\text{Intercept}} = 2A_2$$

The radius of gyration, S^2, gives an indication as to whether a chain is compact or extended. If additional information, such as shape, is available, we may estimate the size:

$$S^2 = \frac{3d^2}{20} \qquad \text{for a sphere}$$

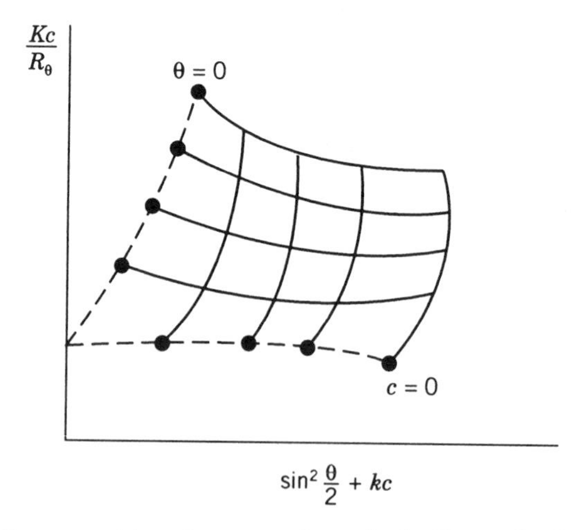

FIGURE 8.9 The external interference as shown in the Zimm plot. In these cases, there is not only internal interference, but also external interference in light scattering. So far there is no simple theory to treat external interferenece in light scattering.

$$S^2 = \frac{l^2}{12} \qquad \text{for a rod}$$

$$S^2 = \frac{\langle R^2 \rangle}{6} \qquad \text{for a coil}$$

Sometimes, the Zimm plot is shown in a distorted form, as in Figure 8.9.

APPENDIX EXPERIMENTAL TECHNIQUES OF THE ZIMM PLOT

Measurement

First a stock solution is prepared, for example, 5.2153×10^{-3} g/mL, and, labeled as 1 unit. The dilution is labeled as $\frac{1}{4}$, $\frac{2}{4}$, and $\frac{3}{4}$ units. Thus, the concentrations are always 1, 2, 3, and 4 (unit: $\frac{1}{4}$ stock concentration, for example, $\frac{1}{4}$ $(5.2153 \times 10^{-3}$ g/mL) or $\frac{1}{4}$, $\frac{1}{2}$, $\frac{3}{4}$, 1 (unit: stock concentration).

Light scattering is then measured: I, the intensity of scattered light. Note that

$$I = R_\theta$$

Recall that

$$K\left(\frac{c}{I}\right) = \frac{1}{M} P^{-1}(\theta) + 2A_2 c$$

where

$$K = \frac{2\pi^2}{\lambda^4 N_A} n^2 \left(\frac{dn}{dc}\right)^2 k'$$

where k' is the instrument constant. For example,

$$k' = \frac{I_B}{R_B}$$

where B is benzene, used as a standard. I_B is the intensity of benzene selected (for example, 0.100, 0.200, 1.00) related to the sensitivity. The values of R_B may be determined experimentally:

$$R_B = \frac{I_{90^\circ}}{I_{0^\circ}} = 16.3 \times 10^{-6} \qquad \lambda = 546 \text{ nm}$$

$$K = \frac{2\pi^2}{\lambda^4 N_A} n^2 \left(\frac{dn}{dc}\right)^2 \frac{I_B}{R_B}$$

$$\left(\frac{2\pi^2}{\lambda^4 N R_B} = 0.2249 \text{ for } \lambda = 546 \text{ nm}\right)$$

The dn/dc values are obtained from the measurement of the solutions (same solutions as for light scattering apparatus) with a differential refractometer (a separate instrument).

Treatment of Data

Data are presented in the following form:

c	c/I			
	30°	45°	60°	⋯
$\frac{1}{4}$	⋯	⋯	⋯	
$\frac{1}{2}$	⋯	⋯	⋯	
$\frac{3}{4}$				
1				

Draw a ruler on graph paper:

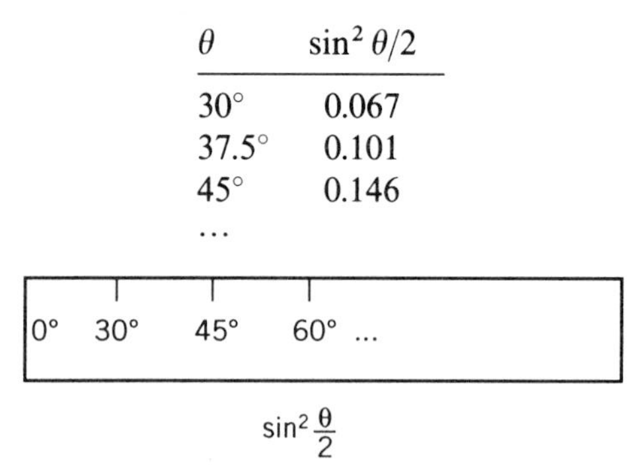

θ	$\sin^2 \theta/2$
30°	0.067
37.5°	0.101
45°	0.146
⋯	

Arbitrarily select 0, $\frac{1}{4}$, $\frac{1}{2}$, $\frac{3}{4}$, 1, units for the absissa: For the $c=\frac{1}{4}$ line, we line up the 0 point of the ruler to the point $\frac{1}{4}$ on the absissa. Plot and then proceed to $c=\frac{1}{2}$, and so forth.

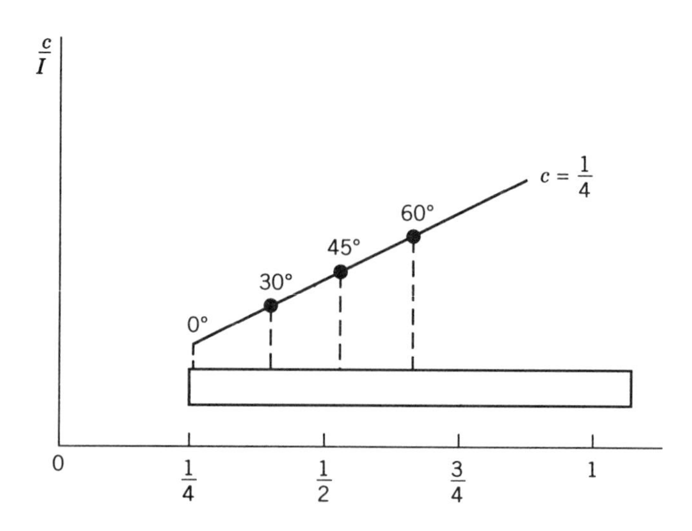

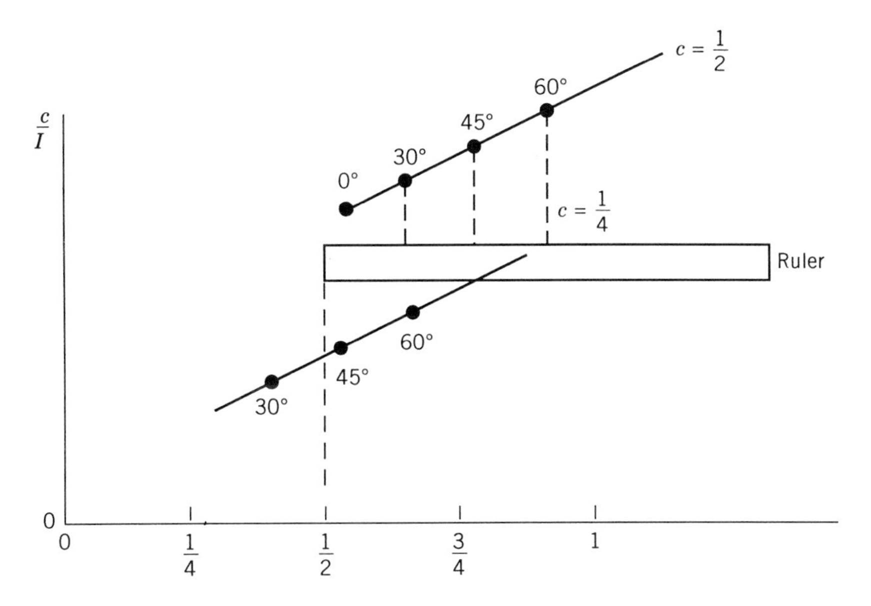

To extrapolate to 45° for example, we first connect the 4 points at 45° from the four lines of c ($\frac{1}{4}, \frac{1}{2}, \frac{3}{4}, 1$) and extend to $c = 0$.

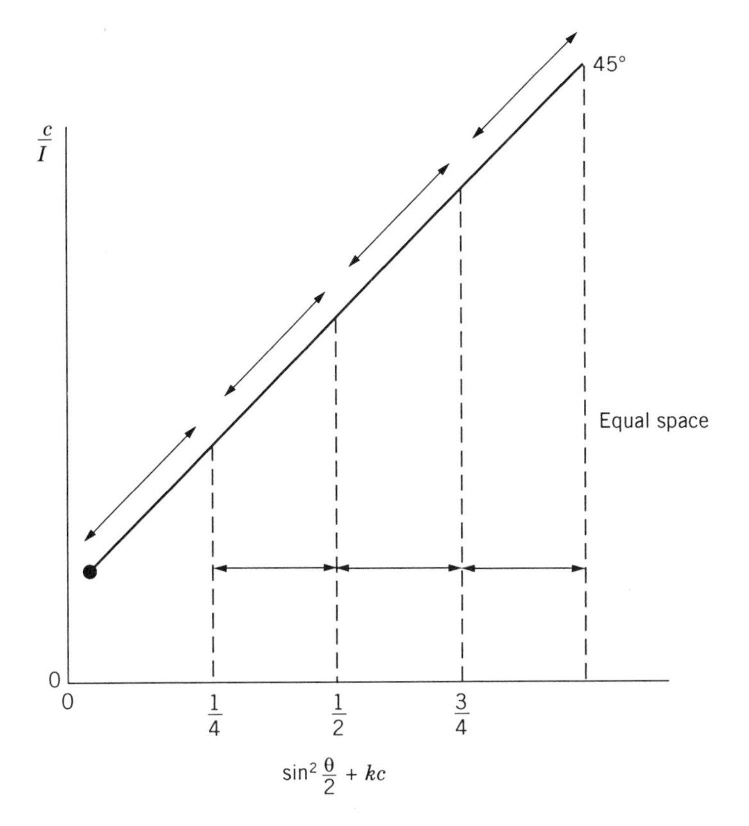

Then extrapolate $c = \frac{1}{4}$. We connect the points at different angles and extend to zero angle.

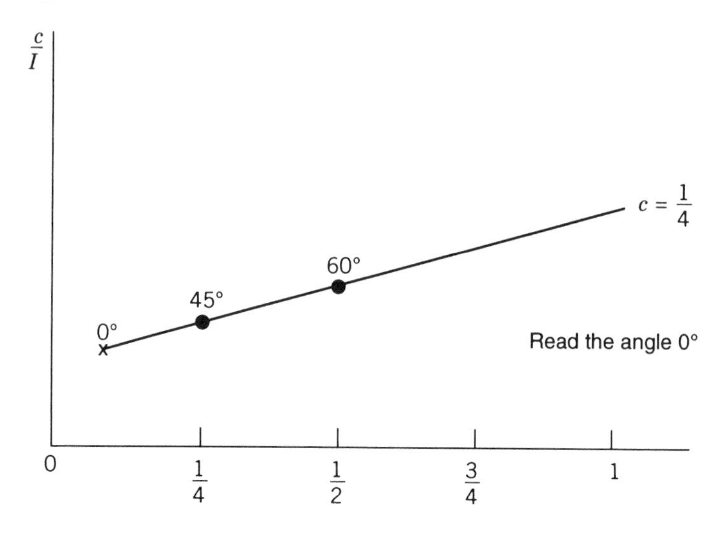

Combining yields

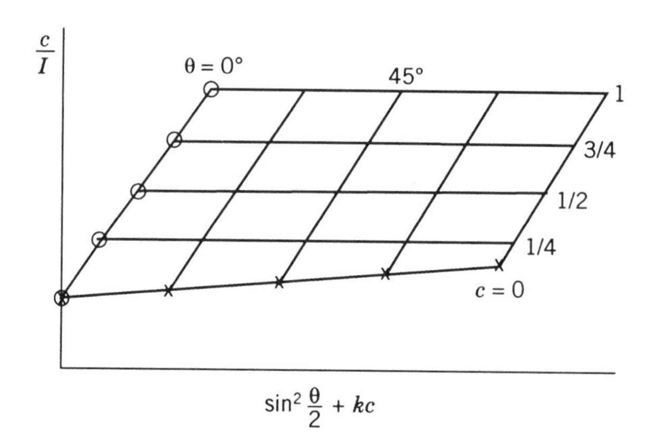

REFERENCES

Bushuk, W., and H. Benoit, *Can. J. Chem.* **36**, 1616 (1958).

Casassa, E. F., and H. Eisenberg, *J. Phys. Chem.* **64**, 753 (1960); **65**, 427 (1961).

Debye, P., *J. Appl. Phys.* **15**, 338 (1944).

Debye, P., *J. Phys. Coll. Chem.* **51**, 18 (1947).

Del Rosario, N. O., and S. F. Sun, *Can. J. Chem.* **51**, 3781 (1973).

Doty, P., and R. F. Steiner, *J. Chem. Phys.* **18**, 1211 (1950).

Edsall, J. T., H. Edelhoch, R. Lontre, and P. R. Morrison, *J. Am. Chem. Soc.* **72**, 4641 (1950).

Eisenberg, H., *J. Chem. Phys.* **36**, 1837 (1962).

Eisenberg, H. and E. F. Cassassa, *J. Polym. Sci.* **47**, 29 (1960).

Inoue, H., and S. N. Timasheff, *J. Am. Chem. Soc.* **90**, 1890 (1968).
Kerker, M., *The Scattering of Light and Other Electromagnetic Radiation*. New York: Academic, 1969.
Krause, S., *J. Phys. Chem.* **65**, 1618 (1961).
Rayleigh, Lord, *Philos. Mag.* **41**, 447 (1871).
Read, B. E., *J. Chem. Soc.* 382 (1960).
Stacey, K. A., *Light Scattering in Physical Chemistry*. London: Butterworths Scientific, 1956.
Stockmayer, W. H., L. D. Moore, Jr., M. Fixman, and B. N. Epstein, *J. Polym. Sci.* **16**, 517 (1955).
Tanford, C., *Physical Chemistry of Macromolecules*. New York: Wiley, 1961.
Tyndall, J., *Phil. Mag.* **37**, 384 (1869); *Proc. Roy. Soc.* (London) **17**, 223 (1869).
Vollmert, B., *Polymer Chemistry*. New York: Springer, 1973.
Zimm, B. H., *J. Chem. Phys.* **16**, 1093 (1948); **16**, 1099 (1948).

PROBLEMS

8.1 The refractive index of dioxane at 25°C is 1.4232 and the refractive index increment of the system poly(methyl methacrylate) in dioxane is 0.065 $cm^3\ g^{-1}$. Given the following turbidities of the solutions of poly(methyl methacrylate) in dioxane with benzene as reference, determined at $\lambda = 5461$ Å, calculate (a) the molecular weight and (b) the radius of gyration of the polymer.

Conc.	C/I										
	30°	37.5°	45°	60°	75°	90°	105°	120°	135°	142.5°	150°
$\frac{1}{4}$	2.37	2.34	2.35	2.46	2.59	2.74	2.89	3.00	3.12	3.18	3.16
$\frac{1}{2}$	3.53	3.52	3.51	3.62	3.79	3.97	4.12	4.20	4.35	4.41	4.35
$\frac{3}{4}$	4.59	4.55	4.55	4.67	4.85	5.05	5.21	5.32	5.46	5.49	5.35
1	5.49	5.43	5.46	5.56	5.68	6.21	6.41	6.58	6.76	6.80	6.71

stock solution = 5.315×10^{-3} g/cc
$I_B = .100$
$R_B = I_{90}/I_0 = 16.3 \times 10^{-6}$ for $\lambda_0 = 5461$ Å

8.2 Light scattering measurements were carried out for bovine serum albumin in 0.1 M KI solution at isoelectric point (pH = 5.17, $z_2 = 0$). The wavelength was set at 546 nm and the experiment was performed at room temperature. The following turbidity data were obtained:

Conc. (g/ml)$\cdot 10^3$	$\tau \cdot 10^4$
0	0.213
0.936	2.88
1.902	5.75
2.801	8.18
3.701	10.59
5.590	15.37

Separate experiments were carried out for the determination of refractive index and differential refractive increment. The value of dn/dc_2 was found to be 0.168 and that of n was 1.3342.

(a) Calculate the Debye factor H.

(b) Plot $HC_2/\Delta\tau$ versus C_2 to determine the molecular weight and the second virial coefficient B.

(c) Suggest a method by which the values of β_{23} and β_{22} could be determined. (*Source*: Edsall et al. (1950); Inoue and Timasheff (1968); and Del Rosario and Sun (1973).)

8.3 A protein dissolved in water ($n_2 = 1.33$) is known to be 1200 Å in diameter. Calculate the degree of dissymmetry in light scattering at $\lambda = 5461$ Å.

8.4 The angular dissymmetry of tobacco mosaic virus is 1.90 with light of $\lambda = 5461$ Å, its density is $2\,\text{g/cm}^3$ and its molecular weight 4.45×10^7. Decide whether the molecule is in the shape of a rod or a sphere.

8.5 Light-scattering measurements at 436 nm of a polystyrene–poly(methylmethacrylate) copolymer in the three different solvents give the following data:

	$\left(\frac{dn}{dc}\right)_0$	$M_{app} \times 10^{-6}$
Butanone	0.184	1.05
1,2-Dichloroethane	0.135	0.97
Toluene	0.075	1.11

The quantity $b = (dn/dc)_A - (dn/dc)_B$ was found to be 0.117 at 436 nm. Calculate the $\bar{M}_w$ of the copolymer.

Hint:

$$\left(\frac{I_\theta}{K'c}\right)_{\substack{c\to 0\\ \theta\to 0}} = \left(\frac{dn}{dc}\right)_0^2 M_{app}$$

$$\left(\frac{I_\theta}{K'c}\right)_{\substack{c\to 0\\ \theta\to 0}} = \left(\frac{dn}{dc}\right)_0^2 \bar{M}_w + 2b\left(\frac{dn}{dc}\right)_0 \langle M\,\Delta X\rangle + b^2\langle M(\Delta X)^2\rangle$$

(*Source*: Krause (1961).)

9

SMALL-ANGLE X-RAY SCATTERING AND NEUTRON SCATTERING

9.1 SMALL-ANGLE X-RAY SCATTERING

Light-scattering apparatus is good for measuring the angle dependence of intensity in the range between 30° and 135°. Below 30° the measurement is usually unreliable. However, by taking extreme precautions and with a special device, the measurement may be carried out down to 10°, which is near the limit. X-ray scattering apparatus, on the other hand, is good for measurements at a very small angle, near 0°. While small-angle x-ray scattering is not as convenient (for example, it is not readily available) as light scattering in determining molecular weight, it is superior for the determination of size ($\sum r_{ij}$) and shape (sphere, rod, and coil) of macromolecules, particularly proteins. This is because the particle size is related inversely to the scattering angle. For proteins with dimensions in the neighborhood of 100 Å, the strong scattering with wavelength of 1.54 Å will be chiefly at angles about 1°.

Originally, small-angle x-ray scattering was a nuisance to the investigators of the x-ray diffraction pattern (which is described in Chapter 14). In the region of small angles the diffraction pattern was completely black due to parasitic scattering. By the early 1930s the use of a crystal monochromater filtered out parasitic scattering, making it possible to resolve the diffraction pattern close to the primary beam. At the same time it also became clear that the parasitic scattering itself, rather than being a nuisance, could actually yield important information on the conformation of macromolecules. Although the work of parasitic scattering (now called small-angle x-ray scattering) is still x-ray diffraction, the laws governing scattering are basically similar to those governing light scattering. For this reason, small-angle x-ray scattering is discussed in this book directly after light scattering, instead of being included with our discussion of x-ray diffraction.

To demonstrate their similarity, small-angle x-ray scattering data may be treated as light-scattering data. In parallel to light scattering, we have

$$\frac{Kc}{I} = \frac{1}{MI_n(h)} + 2A_2T(h)c + \cdots \tag{9.1}$$

where the only two different parameters are $I_n(h)$, which is the single-particle scattering function, and $T(h)$, which is a spatial arrangement of two polymer particles during their mutual approach. The independent variable h is related to the angle θ which we define later. Of course, the constant K is no longer defined in terms of n (the refractive index) and dn/dc (the differential refractive index). Instead, K is defined in terms of other parameters:

$$K = \frac{e^4}{m^2c'^4}\,\frac{P(\Delta z)^2N_A d}{a^2} \tag{9.2}$$

where e is the electrostatic charge, m is the mass of the particle, c' is the velocity of light, P is the total energy per unit time irradiating the sample, Δz is the number of effective moles of electrons per gram, d is the thickness of the sample, N_A is Avogadro's number, and a is the distance between the sample and the plane registration.

9.1.1 Apparatus

A diagram of the small-angle x-ray scattering apparatus is shown in Figure 9.1. The x-ray source is a tube with a water-cooled copper anode, usually using a Kratky camera for measurement. The tube is operated at, for example, 30 kV dc and a current of, for example, 80 mA. The total flux incident on the sample is approximately 10^8 photons/s. Monochromatization was achieved with balanced filters of nickel and cobalt foil which isolate the CuKα line. The x-ray wavelength (the most frequently used CuKα line) is 1.54 Å.

There are two types of detectors: film (which normally anticipates a long exposure time) and counter (for example, Geiger-Muller, scintillation counters,

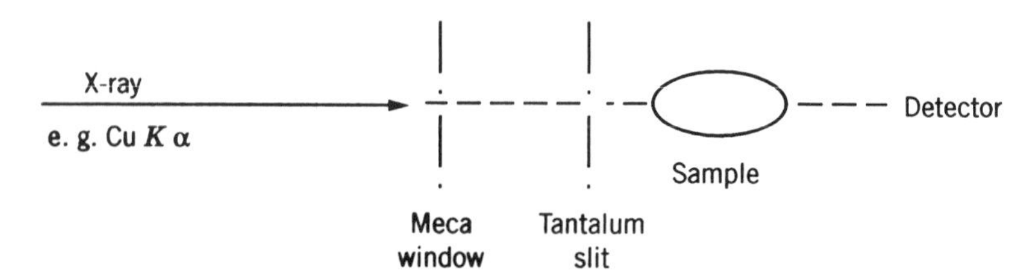

FIGURE 9.1 Diagram of the small-angle x-ray scattering apparatus.

and position-sensitive detector systems). Samples can take the forms of films, fibers, plates, disks, or solutions.

9.1.2 Guinier Plot

In spite of its similarity to light scattering, small-angle x-ray scattering is not usually used to determine M and A_2 as mentioned before; it is used mainly in the determination of the size and shape. The versatility of the experiment for the measurement of size and shape lies on the interpretation of the designated output, $I(h)$.

Intensity I(h) The scattering intensity $I(h)$ is expressed as a function of the scattering angle θ. A scattering curve is shown in Figure 9.2. The variable h can be expressed in terms of magnitude,

$$h = \frac{4\pi}{\lambda} \sin\theta$$

or in terms of vector,

$$\mathbf{h} = \frac{2\pi}{\lambda} (\mathbf{s} - \mathbf{s}_0)$$

In either expression, λ is the wavelength. In the magnitude expressions the angle θ is the scattering angle and h is almost identical to Q:

$$Q = \frac{4\pi}{\lambda} \sin\frac{\theta}{2}$$

In the vector expression, the vector is normal to the plane of atom (see Figure

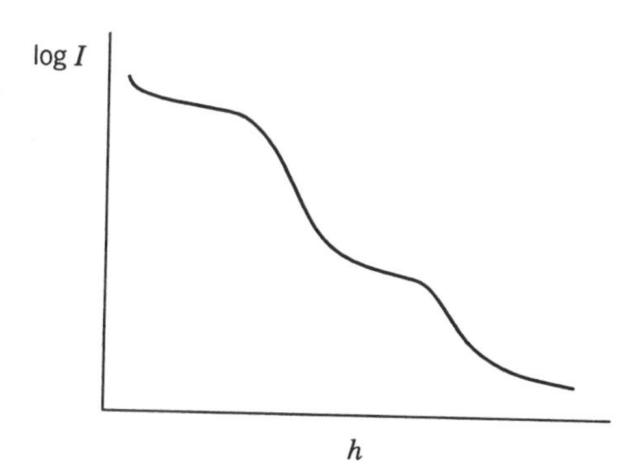

FIGURE 9.2 A scattering curve.

15.11). If the intensity of scattering $I(\mathbf{h})$ is multiplied by the structure factor $\langle F^2(\mathbf{h})\rangle$, we obtain the observed average intensity $\langle I(\mathbf{h})\rangle$:

$$\langle I(\mathbf{h})\rangle = I(\mathbf{h})\langle F^2(\mathbf{h})\rangle$$

It is the structure factor $\langle F^2(\mathbf{h})\rangle$ that correlates the scattering intensity and the configuration of he molecule; it is the center of the study of x-ray scattering. The structure factor can be expressed in the well-known Debye formula (1915):

$$\langle F^2(\mathbf{h})\rangle = \sum_i \sum_j f_i f_j \frac{\sin hr_{ij}}{hr_{ij}}$$

where f is the individual scattering factor of electrons (for example, the ith and jth electrons), and r_{ij} is the distance between the two atoms. The sine function can be expanded in the series,

$$\langle F^2(\mathbf{h})\rangle = \sum_i \sum_j f_i f_j - \frac{h^2}{6} \sum_i \sum_j f_i f_j \langle r_{ij}\rangle^2 + \cdots$$

At the limit $h \to 0$

$$\sum_i \sum_j f_i f_j = F^2(0)$$

and

$$\langle F^2(h)\rangle = n^2 e^{-h^2 R_0^2/3} \tag{9.3}$$

where $n = \sum_i f_i$ is the total number of electrons in the particle and $R_0^2 = \sum \langle r_{ij}\rangle^2$. This is the Guinier equation (1955). The plot of $\ln I$ versus $(2\theta)^2$ is called the Guinier plot. (*Note*: 2θ is the scattering angle. In the small-angle region, $h \sim 2\theta$.)

Analysis of the Data Depending on the way it is plotted, the Guinier equation can provide information about the size of macromolecules in various shapes:

Radius of Gyration of the Whole Particle R Here the equation is

$$I(h) = I(0)e^{-h^2 R^2/3}$$

If the particle is in the shape of a sphere, the linear plot of the experimental data $\ln I(h)$ versus h^2 (or $(2\theta)^2$) will yield the value of R, the radius of gyration of the whole particle. Here R is proportional to the square root of the slope of the tangent in the limit $2\theta \to 0$.

Radius of Gyration of the Cross Section R_c Here the equation is

$$I_c(h) = I_c(0)e^{-h^2 R_c^2/2}$$

If the particle is rodlike, the linear plot of $I(h)\,h$ (or $I(h)\,2\theta$) versus h^2(or $(2\theta)^2$) will yield the value of R_c, the radius of gyration of the cross section.

Radius of Gyration of the Thickness R_t Here the equation is

$$I_t(h) = I_t(0)e^{-h^2R_t^2}$$

If the particle is lamellar (flatlike particles), as in the drawing,

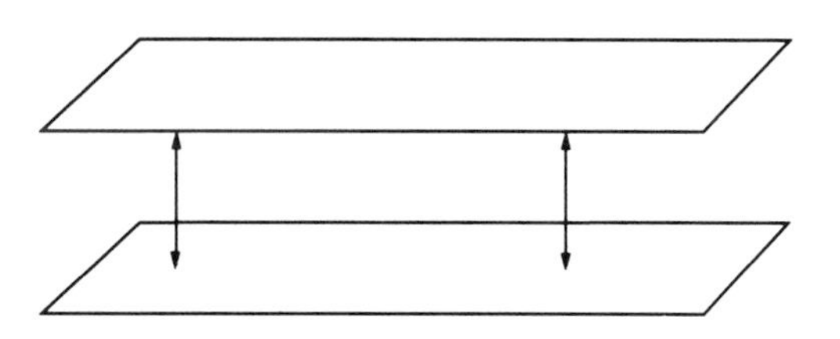

the plot of $I(h)h^2$ (or $I(h)\,(2\theta)^2$) versus h^2 (or $(2\theta)^2$) will yield the information about R_t, the radius of gyration of the thickness of the particle. For particles with uniform electron density distribution, the thickness can be calculated from R_t according to

$$t = R_t\sqrt{12}$$

9.1.3 The Correlation Function

In Guinier's approach, the key term is the scattering factor, which can be analyzed in terms of Debye's formula (the sine function in a power series). The same set of experimental data $I(h)$ can also be analyzed by a different approach. Here, we discuss the correlation function (see the appendix to Chapter 5). Since the evaluation of $\langle r_{ij} \rangle$ is related to statistics we can define a correlation function $\gamma(r)$ as the average of the product of two fluctuations at a distance r:

$$\gamma(r) = \langle \eta(r_1)\eta(r_2) \rangle$$

where

$$r = \langle r_1 - r_2 \rangle$$
$$\eta = \rho - \langle \rho \rangle \qquad \rho \text{ is the density}$$

Then the average square of $\langle \eta^2 \rangle$ may be expressed as

$$\langle \eta^2 \rangle = \langle \rho^2 \rangle - V\langle \rho \rangle^2 = V\gamma(r)$$

where V is the volume of the particle. At the limit $r \to 0$, that is, the two fluctuation points being at the same position, we have

$$V\gamma(0) = \frac{1}{2\pi^2} \int_0^\infty h^2 \, \mathrm{d}h I(h)$$

The integral is known as invariant Q' (due to Porod) (1951); that is,

$$Q' = \int_0^\infty I(h) h^2 \, \mathrm{d}h \tag{9.4}$$

Thus, if the scattering intensity at zero angle is available, Q' can be used to determine the volume of the particle:

$$V = \frac{I(0)}{Q'} 2\pi^2$$

The value of the invariant Q' can be obtained by the evaluation of the integration in Eq. (9.4). The integration is usually carried out numerically with Simpson formula after the intensity data is plotted in the form of $I(h) \cdot h^2$ versus h. The invariant Q' is equal to the area under the curve. If the particle is in rodlike form, Q' can also be used to determine the cross-section area A:

$$A = \frac{I(h)h}{Q'} 2\pi \qquad \text{at } h \to 0$$

It is interesting to note that the correlation function originally was also suggested by Debye and coworker. The correlation function of the particle $\gamma(r)$ represents the probability of a point at a distance r in an arbitrary direction from a given point in the same particle. It can be related to the distance distribution function $p(r)$ by

$$p(r) = \gamma(r) r^2 \tag{9.5}$$

The distribution function itself is related to the shape of the macromolecules:

For spheres

$$p(r) = \frac{1}{2\pi^2} \int_0^\infty I(h) hr \sin hr \, \mathrm{d}h \tag{9.6}$$

For rod like shapes

$$p_c(r) = \frac{1}{2\pi} \int_0^\infty I_c(h)(hr) J_0(hr) \, \mathrm{d}h \tag{9.7}$$

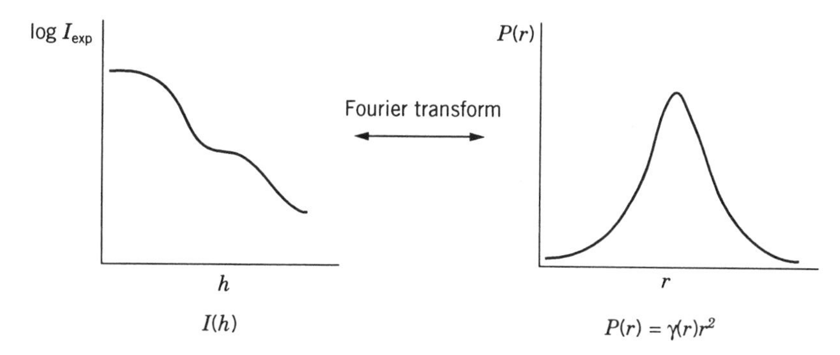

FIGURE 9.3 Fourier transform of $I(h)$ into $P(r)$.

where $J_0(hr)$ is the zero-order Bessel function.
For lamella

$$p_t(r) = \frac{1}{\pi}\int_0^\infty I_t(h)\cos hr\,dh \tag{9.8}$$

All of the above distribution functions, $p(r)$, $p_c(r)$, and $p_t(r)$, are Fourier inversions of $I(h)$, $I_c(h)$, and $I_t(h)$, respectively. Modern experimental techniques can easily transform $I(h)$ into $p(r)$ (Figure 9.3).

Thus by transforming the intensity data to the distance distribution function, we are able to estimate the shape of a macromolecule, particularly a biological polymer.

9.1.4 On Size and Shape of Proteins

With a small modification of Guinier's equation, Luzzati (1960, 1961) suggested that the plot of log (intensity) data versus scattering angle data, as shown in Figure 9.4) enables one to determine the radius of gyration R of protein molecules in solution:

$$\ln j_n(s) = \ln j_n(0) - \tfrac{4}{3}\pi^2 R s^2$$

Here the scattering angle parameter s is slightly different from that of h:

$$s = \frac{2\sin\theta}{\lambda}$$

c.f.

$$h = \frac{4\pi\sin\theta}{\lambda}$$

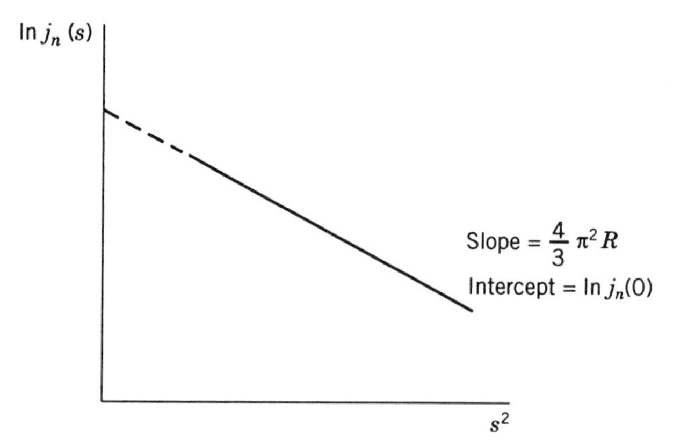

FIGURE 9.4 A modified Guinier plot.

The scattered light intensity parameter $j_n(s)$ is calculated from

$$j_n(s) = \left[\frac{I(s)}{nm\int i_0(s)\,\mathrm{d}s}\right]_{\text{solution}} - \left[\frac{I(s)}{nm\int i_0(s)\,\mathrm{d}s}\right]_{\text{solvent}}$$

where $I(s)$ is the scattered light intensity measured at s, the integral is the energy of the stack of incident beams i_0, n is the number of electrons/cm^2, and m is a constant.

If the internal structure of the protein molecules is stressed, we can plot the experimental results in slightly different way (Figure 9.5):

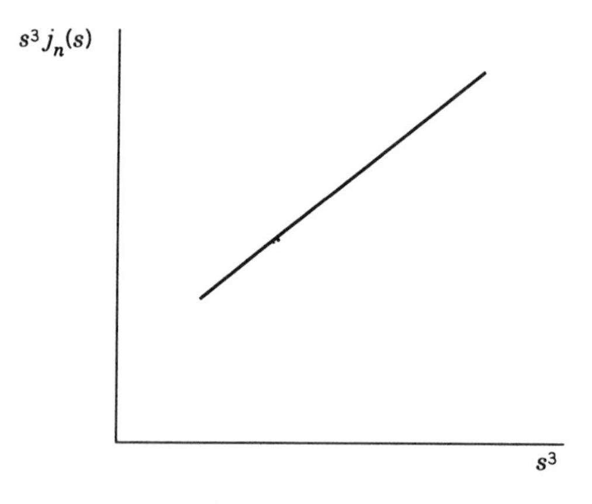

Slope = δ^*

Intercept = $\frac{1}{16\pi^2}\frac{S_1}{v_1}(\rho_1 - \rho_0)\, c_e (1 - \rho_0\bar{v})$

FIGURE 9.5 A Guinier plot for the determination of the internal structure of a protein molecule.

The equation that represents Figure 9.5 is

$$\lim_{s\to\infty} s^3 j_n(s) = \frac{1}{16\pi^2}\frac{S_1}{v_1}(\rho_1 - \rho_0)c_e(1 - \rho_0\bar{v}) + \delta^* s^3$$

where the subscript 0 refers to the solvent and 1 refers to the solute. The slope δ^* is close to the term $j_n(0)$, with, of course, some deviation. The intercept contains two terms related to the internal structures:

$\rho_1 - \rho_0$ = distribution of electron density in the molecule

$\frac{S_1}{v_1}$ = surface dimension of the protein molecule in Å^{-1} which is a characteristic parameter

The symbol ρ refers to the density, S_1, is the surface dimension, V is the volume of the molecule, $\bar{v}$ is the partial specific volume, and c_e is the concentration of electrons. For lysozyme, $c_e = 0.957c\ (1 - 0.55c)^{-1}$; for BSA, $c_e = 0.960c\ (1 - 0.042c)^{-1}$, c being the concentration in g/mL.

From S_1/v_1, the surface to volume ratio, one can calculate the dimension (i.e., the length of the molecule L), using the equation derived by Porod,

$$L_1 = n\frac{v_1}{S_1}\phi_1$$

where ϕ is the volume fraction.

However the two parameters $\rho_1 - \rho_0$ and S_1/v_1 may also be evaluated independently by

$$\rho_1 - \rho_0 \cong \frac{\int_0^\infty 2\pi s j_n^*(s)\,ds}{c_e(1 - \rho_0\bar{v})} + \rho_0 c_e(1 - \rho_0\bar{v})$$

$$\frac{S_1}{v_1} \cong \frac{16\pi^2 \lim_{s\to\infty} s^3 j_n^*(s)}{\int_0^\infty 2\pi s j_n^*(s)\,ds}\left[\rho_1 - \frac{\rho_0 c_e(1 - \rho_0\bar{v})}{\rho_1 - \rho_0}\right]$$

where $j_n^*(s)$ is calculated from

$$j_n^*(s) = j_n(s) - \delta^*$$

The shape of a protein molecule, whether spherical or not, can be determined by plotting $s^4\ I(s)$, (called the Porod term) versus s, as shown in Figure 9.6.

Luzzati (1961) further suggested a set of theoretical equations for estimating the dimension of the equivalent ellipsoid of revolution on the basis of experimental values such as v_1 and S_1:

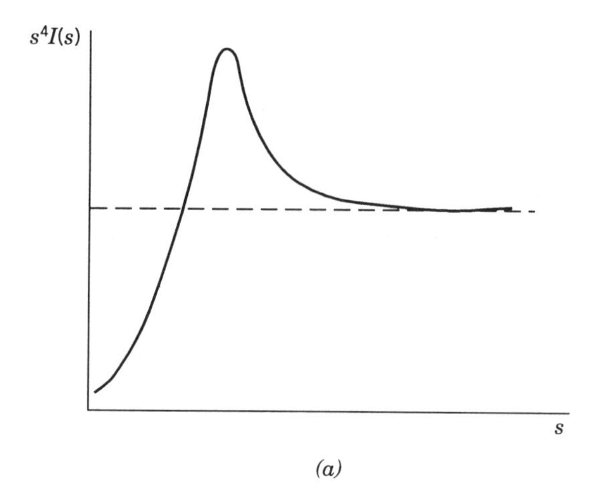

(a)

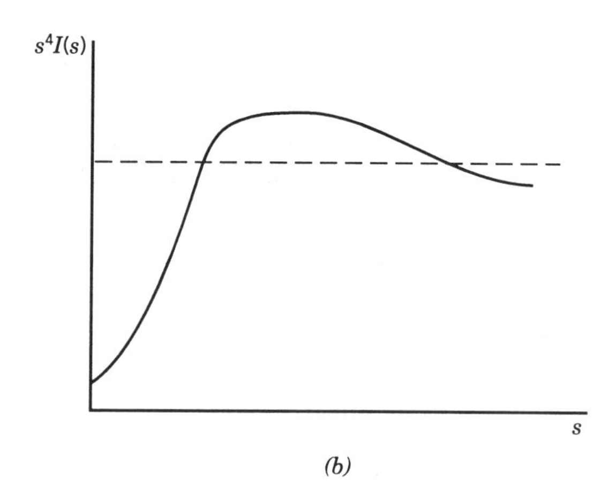

(b)

FIGURE 9.6 Determination of the shape of a protein molecule: (*a*) spherical—a curve that oscillates before setting toward the asymptote; (*b*) nonspherical—a curve that shows little oscilation in the asymptotic approach.

$$v_1 = \frac{4\pi}{3} pb^3$$

$$S_1 = 2\pi b^2\left(1 + \frac{p^2}{\sqrt{p^2-1}} \arcsin \frac{\sqrt{p^2-1}}{p}\right) \qquad p > 1$$

$$= 2\pi b^2\left(1 + \frac{p^2}{\sqrt{1-p^2}} \arctan \sqrt{1-p^2}\right) \qquad p < 1$$

where $p = a/b$, the axial ratio.

TABLE 9.1 Comparison of X-rays with Neutron Beams

	X-rays	Neutrons
Source	Energetic electron beam	Nuclear reactor of pulsed reaction
Wavelength	1.54 Å	2.0–10 Å
Kinetic energy	25–5 keV	0.3–0.0051 eV
Detection	Fluorescence, scintillation counter	Reaction with ^{6}Li or ^{10}Be

9.2 SMALL-ANGLE NEUTRON SCATTERING

Light scattering and small-angle x-ray scattering are based on electromagnetic radiation, while small-angle neutron scattering is based on particle radiation. The interaction between light and the collective electrical charges of a molecule produces the electric field. Neutron beams, on the other hand, interact with the nuclei of atoms via a strong nuclear force (called nuclear scatterings). They also interact with unpaired electron spins of a molecule, if any, via the magnetic dipole (called magnetic scattering). Table 9.1 gives a brief comparison of x-rays with neutron beams.

9.2.1 Six Types of Neutron Scattering

The type of neutron scattering depends on the incident wave frequency ν_0 and the scattered wave frequency ν. If ν_0 is equal to ν, we have coherently scattered radiation; if not, we have incoherently scattered radiation. In coherently scattered radiation, phases of the electric and magnetic fields of incident radiation and those of scattered radiation are in a definite relation to each other.

The scattering events may or may not involve an energy change. If no energy change takes place, the scattering is elastic; if an energy change takes place, the scattering is inelastic. If an energy change is very small and if there is a Doppler shift (that is, a frequency shift), then the scattering is quasielastic. Thus, there are six different types of scatterings: coherently elastic scattering, incoherently elastic scattering, coherently inelastic scattering, incoherently inelastic scattering, coherently quasielastic scattering, and incoherently quasielastic scattering.

Among the six types of scattering, only incoherently elastic scattering seems to have no application in polymer studies. Coherently elastic scattering of neutrons measures the correlations between scattering centers and hence is useful to the study of the conformation of polymers. The coherently inelastic scattering experiment is useful for obtaining information on photon dispersion curves, whereas the incoherently inelastic scattering is useful for determining the vibrational behavior of side groups of the polymer chain (that is, molecular spectroscopy). Neutron quasielastic scattering studies, whether coherent or incoherent, of polymers in bulk and in solution have given much new information on the dynamics of a polymer chain in solution.

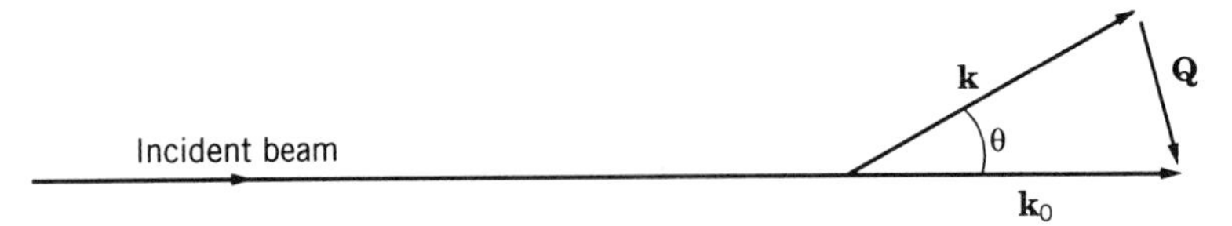

FIGURE 9.7 The neutron scattering phenomenon.

Small-angle neutron scattering, like small-angle x-ray scattering, is used to study the size, shape, and conformation of polymer molecules in bulk and in solution. It is particularly useful to investigate orientation of long polymer chains in solid states. In general, the laws that govern light scattering and small-angle x-ray scattering also govern small-angle neutron scattering. We will compare small-angle neutron scattering with light scattering in detail after we briefly describe the theory.

9.2.2 Theory

The scattering phenomenon is shown in Figure 9.7, where $\mathbf{k}_0$ is the incident wave vector, $\mathbf{k}$ is the scattered wave vector, $\mathbf{Q}$ is the resultant vector in the direction from the scattered beam to the incident beam, that is $\mathbf{Q} = \mathbf{k} - \mathbf{k}_0$, and θ is the scattering angle. In any neutron scattering event energy and momentum transfers are always involved. The energy transfer from the incident neutron to the scattered neutron is given by

$$\Delta E = E - E_0 = \hbar\omega = \frac{1}{2}m(v^2 - v_0^2) = \frac{\hbar^2}{m}(k^2 - k_0^2)$$

where v is the frequency, $\hbar$ is planck's constant ($\hbar = h/2\pi$), m is mass, and the momentum transfer is given by

$$\hbar\mathbf{Q} = \hbar(k^2 + k_0^2 - 2kk_0 \cos\theta)^{1/2}$$

For elastic scattering, $\mathbf{k} = \mathbf{k}_0$ and

$$Q = |Q| = \frac{4\pi}{\lambda}\sin\frac{\theta}{2}$$

where $|k_0| = 2\pi/\lambda$. (Notice the similarities between Q and h.) Scattered intensity is measured as a function of energy and of angle of scatterer. It is expressed in terms of the differential cross section $d^2\delta/d\Omega\, dE$, where δ is the cross section in barns and Ω is the solid angle including θ, which is related to Q. The differential cross section is the probability that neutrons will be scattered by an array of atoms in the sample with energy change dE. It may be used to define a function $S(\mathbf{Q}, \omega)$:

$$S(\mathbf{Q}, \omega) = \frac{k_0}{k} \frac{1}{Nb^2} \frac{d^2\delta}{d\Omega\, dE} \tag{9.9}$$

where b is the scattering length, which is a complex number, and N is the number of incident electrons. The imaginary part of b (in cm) is a measure of the neutron absorption, if any. The function $S(\mathbf{Q}, \omega)$ is called the scattering law by Van Hove.

The differential cross section is a measurable quantity, for example, from the sample count rate f. It can be calculated, using

$$f = nN \frac{d^2\delta}{d\Omega\, d\tau'} \Delta\Omega\tau\varepsilon_{\tau'}$$

where n is the number of atoms/cm^2 in the sample, $\Delta\Omega$ the solid angle subtended by the detector at the sample, $\Delta\tau$ the channel width, and $\varepsilon_{\tau'}$ the efficiency of the detector to neutrons of time of flight τ'.

The scattering law $S(\mathbf{Q},\omega)$ is the double Fourier transform $G(r, t)$ of the space–time correlation function:

$$S(\mathbf{Q}, \omega) = \frac{1}{2\pi\hbar N} \iint \exp[-i(\omega t - \mathbf{Q}\mathbf{r})] G(\mathbf{r}, t)\, d\mathbf{r}\, dt \tag{9.10}$$

The quantity $S(\mathbf{Q}, \omega)$ is usually measured over a limited range of $\hbar\mathbf{Q}$ and $\hbar\omega$. A model correlation function $G(\mathbf{r}, t)$ is then predicted to fit the experimental data.

9.2.3 Dynamics of a Polymer Solution

Dynamics refers to the property that involves the time t. The scattering law $S(\mathbf{Q}, \omega)$ and the intensity $I(\mathbf{Q}, t)$ are related to the correlation function $G(\mathbf{r}, t)$ in the following equations (a Fourier transform pair):

$$I(\mathbf{Q}, t) = \int \exp(i\mathbf{Q}\cdot\mathbf{r}) G(\mathbf{r}, t)\, d\mathbf{r}$$

and

$$S(\mathbf{Q}, \omega) = \int \exp(i\omega t) I(\mathbf{Q}, t)\, dt$$

For a simple liquid, the intensity $I(\mathbf{Q}, t)$ can be described in terms of a self-diffusion coefficient D (see Chapter 11) by

$$I(\mathbf{Q}, t) = \exp(-Q^2 D t)$$

The self-diffusion coefficient is used to describe the center of mass motion for a simple liquid. It can also be used in connection with Rouse–Zimm model to describe the behavior of a long chain. The determination of diffusion coefficient D by the small-angle neutron scattering experiment will be further discussed in a slightly different way in Section 9.2.7.

9.2.4 Coherently Elastic Neutron Scattering

In the study of chain dimensions, we are interested in coherently elastic neutron scattering only, that is, $dE = 0$ and $k = k_0$. The experimental data needed are $d\sigma/d\Omega$ and $S(\mathbf{Q})$, not $d^2\sigma/d\Omega\, dE$ and $S(\mathbf{Q}, \omega)$. The differential cross section $d\sigma/d\Omega$ and the scattering law $S_{coh}(\mathbf{Q})$ are related in a very simplified form:

$$\frac{d\sigma}{d\Omega} = k' S_{coh}(\mathbf{Q})$$

where k' is called the contrast factor (which is discussed further in a later section) and is defined as

$$k' = \left(b_{1_{coh}} - b_{2_{coh}} \frac{\bar{V}_1}{\bar{V}_2} \right)^2$$

The subscripts 1 and 2 refer to one substance dissolved in another substance (including one solid sample mixed with another solid sample) coh refers to the coherent scattering and $\bar{V}$ is the partial molar volume. The scattering law $S(\mathbf{Q})$ here is the Fourier transform of the density fluctuation correlation function (instead of $G(\mathbf{r}, t)$).of the scattering centers:

$$S(\mathbf{Q}) = \sum_{ij} \langle [\exp i(\mathbf{Q}(\mathbf{r}_i - \mathbf{r}_j)]^2 \rangle$$

where $\mathbf{r}_i$ and $\mathbf{r}_j$ are the centers of mass of the scattering points i and j, which correspond to chain segments in a polymer molecule.

9.2.5 Comparison of Small-Angle Neutron Scattering with Light Scattering

The scattering intensity per unit volume $I(\mathbf{Q})$ of a polymer of molecular weight M and concentration c (weight per volume) is

$$I(\mathbf{Q}) = \frac{d\sigma}{d\Omega} \frac{c}{M} N_A$$

$$= k' S_{coh} \frac{c}{M} N_A$$

$$= \frac{k'c}{M} N_A S_{coh}$$

where N_A is Avogadro's number. With some manipulation, we can approximate $S_{coh}(\mathbf{Q})$ to the form

$$S_{coh}(\mathbf{Q}) = \frac{2(M/m)^2}{x - 1 + e^{-X^2}}$$

where M is the molecular weight of the polymer, m is the molecular weight of a segment unit, and

$$x = Q^2 \langle R^2 \rangle$$

$\langle R^2 \rangle$ being the mean-square of the radius of gyration.

If we let $K = k' N_A / m^2$, then we have

$$I(Q) = KCM \frac{2}{x^2}(x - 1 + e^{-X})$$

or

$$\frac{KC}{I} = \frac{1}{M}\left(1 + \frac{\langle R^2 \rangle Q^2}{3} + \cdots\right)$$

Thus, the governing equations of small neutron scattering are also almost identical to those of conventional light scattering. In the analysis of experimental data, we can use the Zimm plot (among others) to obtain the molecular weight and the radius of gyrations of the polymer molecules.

There are, however, some important differences. Because of the involvement of the contrast factor k', in the K term, small-angle neutron scattering can observe the Gaussian behavior of polymer chains in their own bulk, whereas conventional light scattering cannot. The different ranges of Q in small-angle neutron scattering provides more ways than conventional light scattering to obtain molecular parameter of a polymer chain. The four different ranges of Q, as Kratky pointed out, are

1. $Q < (R^2)^{1/2}$ The Guinier domains

$$\frac{K^*C}{I(Q)} = M^{-1}\left(1 + \frac{Q^2 R^2}{3}\right)$$

2. $(R^2)^{1/2} \leqslant Q \leqslant a^{-1}$ The Debye domain where a is the persistent length of the polymer chain

$$\frac{K^*C}{I(Q)} = \frac{M^{-1} Q^2 R^2}{2}$$

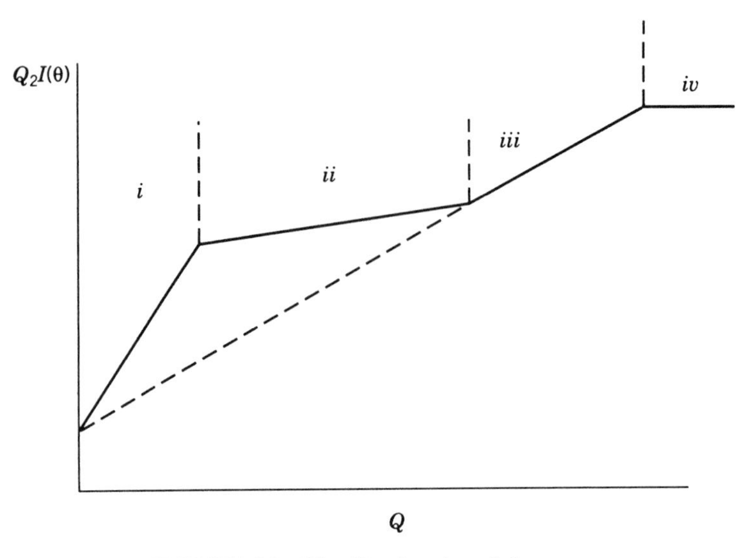

FIGURE 9.8 The Kratky plot of Q ranges.

3. $a^{-1} < Q < l^{-1}$ The rod shape domain where l is the step length of the polymer chain and where n is the number of statistical units in the chain

$$\frac{K^*C}{I(Q)} = \frac{M^{-1}nlQ}{\pi}$$

4. $l^{-1} < Q$ The internal structure domain where no equation similar to that of light scattering can yet be formulated in this range.

Figure 9.8 shows the Kratky plot in an idealized form. Furthermore, small-angle neutron scattering experiments can be carried out on solid solutions, whereas light scattering cannot.

9.2.6 Contrast Factor

The contrast factor k' may be written in a slightly different form:

$$k = |a_D - a_H|^2$$

where a is the coherent scattering length of an atom for the ith nucleus. The values of a for the H atom and the D (deuterium) atom individually, as well as in protonated polystyrene and deuterated polystyrene, are given in Table 9.2. Because of the large difference in coherent scattering lengths between protonated and deuterated monomers, the contrast term, $|a_D - a_H|^2$ is ideal for the study of an isolated chain embedded in its environment. The data for a deuterated polymer and a deuterated matrix containing a small % protonated polymer are sketched in Figure 9.9.

TABLE 9.2 The Values of *a* for H and D

Nucleus	a (10^{-12} cm)
H	−0.374
D	0.667
Polystyrene (C_8H_8)	2.328
d-Polystyrene (C_8D_8)	10.656

Source: Bacon, 1975.

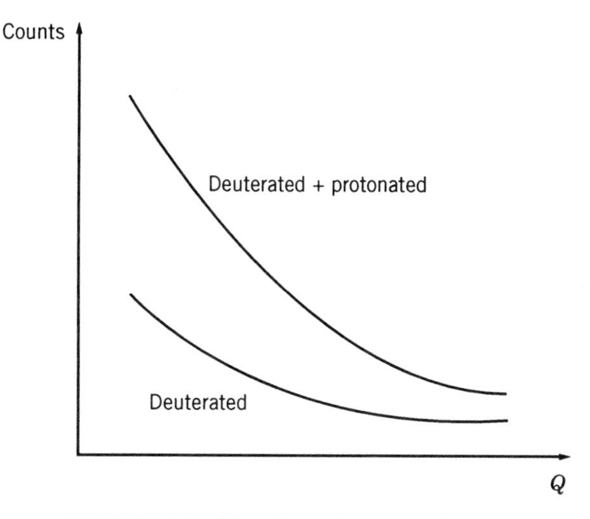

FIGURE 9.9 The effect of contrast factor.

Techniques using the contrast factor have been applied to test the theoretical prediction of the behavior of a polymer in solutions (see Chapter 5):

$$\langle R^2 \rangle = \text{const.} \times M^{2\nu}$$

where ν is a critical exponent for determining the excluded volume in solutions of nonintersecting chain segments. Experiments of small-angle neutron scattering were carried out, for example, on deuterated polystyrene in cyclohexane at 36°C (θ temperature) as well as CS_2 (good solvent) at temperature not equal to θ. It was found that $\nu = 0.5$, as expected, only for chains in a θ condition. The value of k' is the same for the bulk as for a polymer in θ solvent.

9.2.7 Lorentzian Shape

The equation

$$I(Q) = KCM \frac{2}{x^2}(x - 1 + e^{-x})$$

can be modified for the intermediate momentum range to a Lorentzian form:

$$\frac{I(Q)}{KCM} = \frac{1}{Q^2 + \xi^{-2}}$$

where

$$\xi^{-2} = \frac{12cN_A}{M_0 b^2} \phi$$

In the above equations the parameter ξ, as described before, is the screen length, ϕ is the excluded volume per statistical segment, c is the weight fraction concentration of tagged chain (for example, a deuterated polymer embedded in a protonated polymer), and M_0 is the molecular weight of a monomer. Experimentally, ξ^{-2} is equal to a half-width in the plot of the Lorentzian curve (similar to the Gaussian distribution curve).

Thus, the small-angle neutron scattering experiment can be used to determine the screen length ξ. With a small modification the scattering law $S(\mathbf{Q},\omega)$ can be put in a different form of Lorentzian equation:

$$S(\mathbf{Q}, \omega) = \frac{\pi}{2} \frac{DQ^2}{\omega^2 + (DQ^2)^2}$$

where D is the diffusion coefficient and a half-width $\propto DQ^2$. These kinds of experiments have been used to study the dynamics of polymer solutions.

9.2.8 Neutron Spectroscopy

Inelastic neutron scattering is used for the study of transmission or absorption neutron energ spectra, particularly the side-group motion in polymers. All data reported so far for polymers have been concerned with symmetric top molecules. Three spectrometries are available at present: (1) slow neutron spectrometry, which studies slow neutron excitation functions with continuous-energy neutron sources; (2) fast neutron spectrometry, which studies the spectra of neutrons produced in nuclear reactions, and (3) monoenergetic slow neutron spectrometry, which studies the spectra of neutrons corresponding to the inelastic scattering from atoms in solids or fluids.

REFERENCES

Anderegg, J. W., W. W. Beeman, S. Shulman, and P. Kaesberg, *J. Am. Chem. Soc.* **77**, 2927 (1955).

Bacon, G. E., *Neutron Diffraction.* London: Oxford University Press, 1975.

Cotton, J. P., D. Decker, H. Benoit, B. Farnoux, J. Higgins, G. Jannink, R. Ober, C. Picot, and J. des Cloizeaux, *Macromolecules* **7**, 863 (1974).

Daoud, M., J. P. Cotton, B. Farnoux, G. Jannink, G. Sarma, H. Benoit, R. Duplessix, C. Picot, and P. G. de Gennes, *Macromolecules* **8**, 804 (1975).

de Gennes, P. G., *Macromolecules* **9**, 587, 594 (1976).

Debye, P., and A. M. Buche, *J. Appl. Phys.* **20**, 518 (1949).

Debye, P., *Ann. Physik* **46**, 809 (1915).

Glatter, O., and O. Kratky, ed., *Small-Angle X-Ray Scattering*, New York: Academic Press, 1982.

Gupta, A. K., J. P. Cotton, E. Marchal, W. Burchard, and H. Benoit, *Polymer* **17**, 363 (1976).

King, J. K., *Methods Exp. Phys.*, **16A**, 480 (1980).

Kratky, O., and W. Kreutz, *Z Elektrochem.* **64**, 880 (1960).

Kratky, O., *J. Polym. Sci.* **3**, 195 (1948); *Angew. Chem.* **72**, 467 (1960).

Kratky, O., *Pure Appl. Chem.* **12**, 483 (1966).

Luzzati, V., *Acta Crystallogr.* **13**, 939 (1960).

Luzzati, V., A. Nicolaieff, and F. Masson, *J. Mol. Biol.* **3**, 185 (1961).

Luzzati, V., J. Witz, and A. Nicolaieff, *J. Mol. Biol.* **3**, 367, 379 (1961).

Maconnachie, A., and R. W. Richards, *Polymer* **19**, 739 (1978).

Mashall, W., and S. W. Lovessey, *Theory of Thermal Neutron Scattering*. Oxford, UK: Clarendon, 1971.

Ober, R., J. P. Cotton, B. Farnoux, and J. S. Higgens, *Macromolecules* **7**, 634 (1974).

Porod, G., *Kolloid-Z.* **124**, 83 (1951).

Ritland, M. N., P. Kaesberg, and W. W. Beeman, *J. Chem. Phys.* **18**, 1237 (1950).

Safford, G. J., and A. W. Naumann, *Macromol. Rev.* **2**, 1 (1967).

Van Hove, L., *Phys. Rev.* **95**, 249 (1954).

Vonk, C. G., and G. Kortleve, *Kolloid-Z. Polym.* **220**, 19 (1967).

PROBLEMS

9.1 Expand the function (sin hr)/hr in a power series.

9.2 The following experimental data of small-angle x-ray scattering were obtained for lysozyme and bovine serum albumin:

	$\ln I$ (arbitrary units)	
h^2 (radians^2)	Lysozyme	BSA
1×10^{-4}	10.0	4.9
2×10^{-4}	9.6	3.9
3×10^{-4}	9.3	3.0
4×10^{-4}	8.9	1.8

Determine the radius of gyration of each of the proteins. (*Source*: Ritland et al. (1950).)

9.3 Persistence length a is a measure of the rigidity of the conformation of cellulose tricarbanilate in dioxane, which can be determined by a Kratky plot of small-angle neutron scattering. Plot the following data in the Kratky form and determine the persistence length a of the polymer under study:

Iq^2 (arbitrary units)	$q\,(\text{Å}^{-1})$
0.65	0.046
0.55	0.04
0.40	0.03
0.30	0.02
0.22	0.015
0.19	0.01
0.08	0.005

The equation to be used is in the form $a = A/q^*$, where $A = 1.91$ is a constant and q^* is the point at which the wave vector enters a different domain. (*Source*: Gupta et al. (1976).)

9.4 The scattering law $s(q)$ may be expressed as

$$s(q) = \frac{12}{(ql)^{1/\nu} + O(1/N)}$$

where l is the polymer step length, N is the number of links, O refers to the "other" terms, and q is the momentum transfer, which is related to the scattering angle θ and wavelength λ of the neutrons:

$$q = \frac{4\pi}{\lambda}\sin\theta$$

This law can be used experimentally to determine the excluded volume exponent ν. The following data were obtained from the low-angle neutron scattering data for a polystyrene dissolved in CS_2 (good solvent) and cyclohexane (poor solvent), respectively:

Good Solvent		Poor Solvent	
$\log s^{-1}$	$\log q$	$\log s^{-1}$	$\log q$
1.95	6.05	2.60	6.40
2.45	4.65	2.65	5.31
2.65	4.20	3.60	4.20
3.10	2.70	4.15	3.21

The values for s are in arbitrary units, and those for q are $^{10-1}$ Å. Determine graphically the value of ν for the polymer in (a) good solvent and (b) poor solvent and compare the experimental value with that predicted by the scaling law. (*Source*: Cotton et al. (1974).)

9.5 The radii of gyration as a function of concentration in the semidilute regime for a polystyrene with molecular weight 1.14×10^5 in CS_2 were found from small-angle neutron scattering measurements as follows:

c (g/cm^3)	R (Å)
0.00	137
0.03	120
0.06	117
0.10	111
0.15	104
0.20	101
0.33	95
0.50	91
1.06	82

Plot $\log R^2/M$ versus $\log c$ and determine the value of the slope. Show that this experiment confirms the scaling law at the semidilute solution $R^2 \sim Mc^\nu$. (*Source*: Daoud et al. (1975).)

10

DIFFUSION

Diffusion is a process that involves the random motion of particles and the concentration gradient dc/dx in the system. The diffusion coefficient, which is the major concern of this chapter, is a measure of the mass of solute transported in a given period of time under the influence of a known driving force. The driving force is essentially the concentration gradient caused by external forces, such as the gravitational field or centrifugal field. There are two kinds of diffusion: translational and rotational.

10.1 TRANSLATIONAL DIFFUSION

10.1.1 Fick's First and Second Laws

In 1822 Fourier derived an equation for heat conduction. Realizing that the process of transferring heat by induction is analogous to the process of diffusion, and that both are due to random molecular motion, Fick in 1855 adapted the Fourier equation to describe diffusion.

Fick's first law is about the change of concentration (of solute) with respect to coordinates:

$$J = -D\frac{\delta c}{\delta x} \tag{10.1}$$

In Eq. (10.1), J is called the flux (or diffusion flux) or flow (in $kg/m^2\,s$) and is the rate of transfer per unit of section, c is the concentration of solute (or particles) (in g/mL or g/100 mL), x is the space coordinate measured normal to the section where the particles cross, and D is the diffusion coefficient in (cm^2/s). The direction of diffusion is opposite to that of increasing concentration; hence, a negative sign is given in the equation.

Fick's second law describes the change of concentration with time:

$$\frac{\delta c}{\delta t} = -\frac{\delta J}{\delta x} = \frac{\delta}{\delta x}\left(D\frac{\delta c}{\delta x}\right)$$

which leads to

$$\frac{\delta c}{\delta t} = D\frac{\delta^2 c}{\delta x^2} \tag{10.2}$$

if D is a constant. Notice that this is a well-known continuity equation.

10.1.2 Solution to the Continuity Equation

The equation

$$\frac{\delta c}{\delta t} = D\frac{\delta^2 c}{\delta x^2}$$

is difficult to solve, because three variables, c, x, and t, are involved. However, Boltzmann (1894) pointed out that in free diffusion the variables x and t always occur in the ratio $x/\sqrt{t}$. This is true whether or not D depends on the concentration. This suggests a solution:

$$\frac{\delta c}{\delta x} = -\frac{c_0}{\sqrt{4\pi Dt}}\,e^{-x^2/4Dt} \tag{10.3}$$

which satisfies the differential equation (10.2). That this is the solution can be verified by substituting Eq. (10.3) into Eq. (10.2). The great advantages of Eq. (10.3) are that it is identical to the Gaussian equation or error distribution equation, that it is closely related to experimental data of dc/dx and x, and that it is easy to evaluate it numerically to obtain D.

Figure 10.1 shows the graph of dc/dx versus x on the basis of Eq. (10.3). A plot of (Area/Height)^2 versus t will give a straight line of slope $4\pi D$ (Figure 10.2). The dc/dx versus x curve may also be treated in terms of the standard deviation σ (Figure 10.3). From σ we obtain

$$D = \frac{\sigma^2}{2t} \tag{10.4a}$$

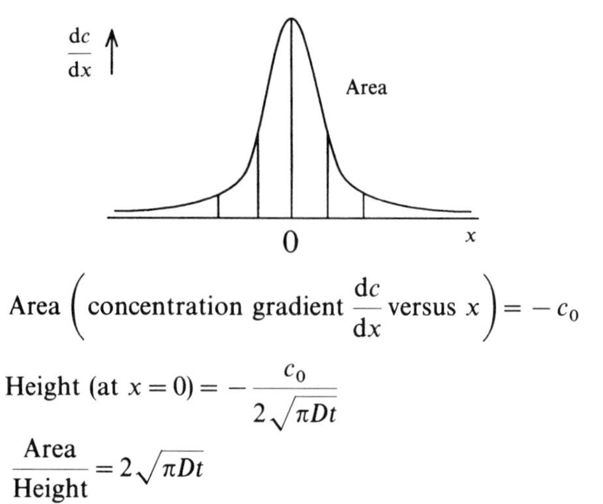

$\frac{dc}{dx}$
Area
0
x
Area $\left(\text{concentration gradient } \frac{dc}{dx} \text{ versus } x\right) = -c_0$
Height (at $x = 0$) $= -\frac{c_0}{2\sqrt{\pi D t}}$
$\frac{\text{Area}}{\text{Height}} = 2\sqrt{\pi D t}$

FIGURE 10.1 Diffusion profile.

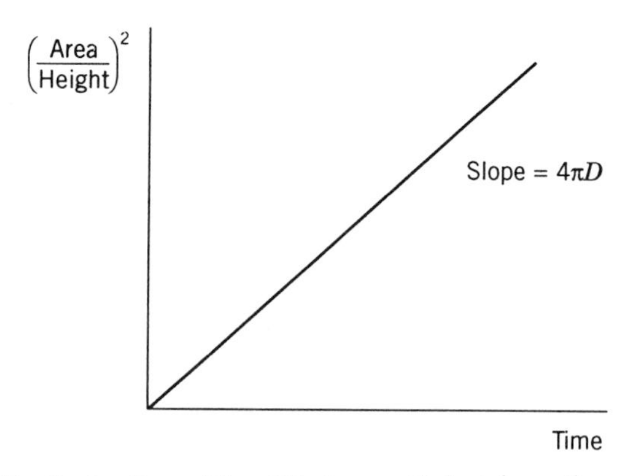

FIGURE 10.2 Evaluation of the diffusion coefficient from the area/height plot.

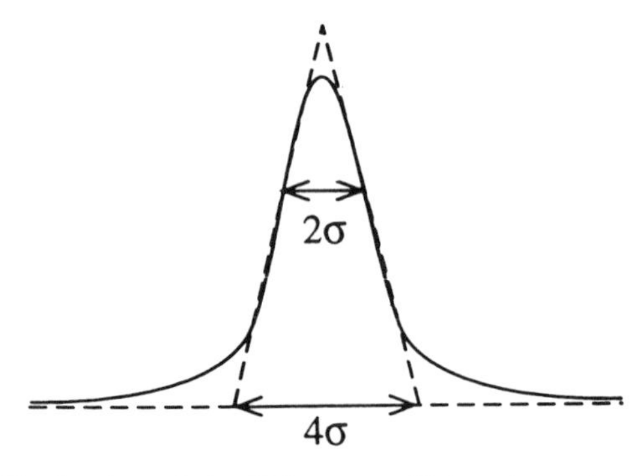

FIGURE 10.3 Evaluation of the diffusion coefficient from the standard deviation.

or

$$D = \frac{\sigma_1^2 - \sigma_2^2}{2(t_1 - t_2)} \tag{10.4b}$$

where t_1 and t_2 refer to the two different times in which the experiment was carried out, and σ_1 and σ_2 represent the standard deviations of the two curves recorded at the two different times. Integrating Eq. (10.3), we obtain

$$c = \frac{c_0}{2}\left(1 - \frac{2}{\sqrt{\pi}}\int_0^{x/2\sqrt{Dt}} e^{-z^2}\, dz\right) \tag{10.5}$$

with $z = x/\sqrt{4Dt}$. The integral term in Eq. (10.5) is the well-known error (or Gaussian) function:

$$\text{erf} = \frac{2}{\sqrt{\pi}}\int_0^{x/\sqrt{4Dt}} e^{-z^2}\, dz$$

Values of the integral are tabulated in most mathematical function books, usually expressed in terms of z:

$$\text{erf}\, z = \frac{1}{\sqrt{\pi}}\int_0^{z} e^{-t^2}\, dt$$

$$\text{erfc}\, z = \frac{2}{\sqrt{\pi}}\int_z^{\infty} e^{-t^2}\, dt = 1 - \text{erf}\, z$$

Both of these equations are called error functions.

10.2 PHYSICAL INTERPRETATION OF DIFFUSION: EINSTEIN'S EQUATION OF DIFFUSION

Einstein interpreted diffusion as being a result of the random thermal motion of molecules. Such a random motion is caused by fluctuations in pressure in a liquid. Thus, diffusion is closely related to Brownian motion. The Brownian motion consists of zigzag motion in all directions. It is a random walk, as discussed in Chapter 5, and is described by the parameter $\langle x^2 \rangle$, the square mean displacement. The equation of motion in one dimension for the Brownian motion of a particle in solution is given by

$$m\frac{d^2x}{dt^2} = -f\frac{dx}{dt} + F' \tag{10.6}$$

where m is the mass of the particle, f is the frictional coefficient, $f(\mathrm{d}x/\mathrm{d}t)$ is the viscous drag, and F' is a force component due to random motion. If we take the average over many particles, then $\langle F' \rangle = 0$. This is because F', being a random force, is as likely to be positive as negative. It eventually vanishes. The equation then becomes

$$m \frac{\mathrm{d}^2 x}{\mathrm{d}t^2} = -f \frac{\mathrm{d}x}{\mathrm{d}t} \tag{10.7}$$

Multiplying Eq. (10.7) by x, we obtain

$$mx \frac{\mathrm{d}^2 x}{\mathrm{d}t^2} = -fx \frac{\mathrm{d}x}{\mathrm{d}t} \tag{10.8}$$

Notice the identities:

$$\frac{\mathrm{d}(x^2)}{\mathrm{d}t} = 2x \frac{\mathrm{d}x}{\mathrm{d}t}$$

$$x \frac{\mathrm{d}x}{\mathrm{d}t} = \frac{1}{2} \frac{\mathrm{d}(x^2)}{\mathrm{d}t}$$

Differentiating $\mathrm{d}(x^2)/\mathrm{d}t$ with respect to t gives

$$\frac{\mathrm{d}^2(x^2)}{\mathrm{d}t^2} = 2\left(\frac{\mathrm{d}x}{\mathrm{d}t}\right)^2 + 2x \frac{\mathrm{d}^2 x}{\mathrm{d}t^2}$$

or

$$x \frac{\mathrm{d}^2 x}{\mathrm{d}t^2} = \frac{1}{2} \frac{\mathrm{d}^2(x^2)}{\mathrm{d}t^2} - \left(\frac{\mathrm{d}x}{\mathrm{d}t}\right)^2$$

Equation (10.8) becomes

$$\frac{m}{2} \frac{\mathrm{d}^2(x^2)}{\mathrm{d}t^2} - m\left(\frac{\mathrm{d}x}{\mathrm{d}t}\right)^2 = -\frac{f}{2} \frac{\mathrm{d}(x^2)}{\mathrm{d}t}$$

Notice also that

$$m\left(\frac{\mathrm{d}x}{\mathrm{d}t}\right)^2 = mv^2$$

The thermal motion assumption provides a justification here to use the principle of equipartition of energy, which states that the kinetic energy of each particle on the average is $\frac{3}{2}kT$; that is,

$$\begin{aligned}\tfrac{1}{2}mv^2 &= \text{the kinetic energy of each particle}\\ &= \tfrac{3}{2}kT\end{aligned}$$

where k is the Boltzmann constant. For one degree of freedom $\frac{1}{2}mv^2 = \frac{1}{2}kT$. If we change the notation x^2 into $\langle x^2 \rangle$ and substitute it, we have

$$\frac{m}{2}\frac{\mathrm{d}^2\langle x^2\rangle}{\mathrm{d}t^2} - kT = -\frac{f}{2}\frac{\mathrm{d}\langle x^2\rangle}{\mathrm{d}t} \tag{10.9}$$

To solve this differential equation, we let

$$y = \frac{\mathrm{d}\langle x^2\rangle}{\mathrm{d}t}$$

Then, Eq. (10.9) is in the form

$$m\frac{\mathrm{d}y}{\mathrm{d}t} + fy = 2kT$$

This is in the standard form for a linear differential equation of first order:

$$\frac{\mathrm{d}y}{\mathrm{d}x} + Py = Q$$

where P and Q are functions of x. The standard solution is

$$y = \mathrm{e}^{-\int P\,\mathrm{d}x}\left(\int \mathrm{e}^{\int P\,\mathrm{d}x} Q\,\mathrm{d}x + c\right)$$

If the integral is from 0 (the origin) to t, the integration constant in our case would be zero. Then

$$\begin{aligned}y &= \mathrm{e}^{-\int (f/m)\,\mathrm{d}t}\left(\int_0^t \mathrm{e}^{\int (f/m)\,\mathrm{d}t}\frac{2kT}{m}\,\mathrm{d}t\right)\\ &= \frac{2kT}{f}\end{aligned}$$

That is,

$$\frac{\mathrm{d}\langle x^2\rangle}{\mathrm{d}t} = \frac{2kT}{f} \tag{10.10}$$

Again, if diffusion starts at $t = 0$, the integration constant obtained from integration of Eq. (10.10) is zero. Thus, we have

$$\frac{\langle x^2 \rangle}{2t} = \frac{kT}{f} \tag{10.11}$$

But the mean-square displacement is proportional to the time $2t$ with the proportionality constant equal to the diffusion coefficient D:

$$\langle x^2 \rangle = D(2t)$$

This leads to

$$D = \frac{\langle x^2 \rangle}{2t} = \frac{kT}{f} \tag{10.12}$$

The expression

$$D = \frac{kT}{f} \tag{10.13}$$

is called the Einstein equation of diffusion. Because of the correlation between D and f, as shown in Eq. (10.13), D becomes an important parameter characterizing macromolecules and relates to other hydrodynamic properties, such as η (viscosity) and S (sedimentation coefficient).

10.3 SIZE, SHAPE, AND MOLECULAR WEIGHT DETERMINATIONS

10.3.1 Size

Kirkwood derived an equation to correlate the diffusion coefficient D with the size of polymer molecule $\sum \langle R_{ij} \rangle$ in the form

$$D = \frac{kT}{N_A \rho}\left[1 + \frac{\rho}{6\pi\eta N_A} \sum_{i=0}^{N_A} \sum_{j=0}^{N_A} (1 - \delta_{ij}) \langle R_{ij}^{-1} \rangle\right] \tag{10.14}$$

where N_A is Avogadro's number, ρ is the density of the solution, $\delta_{ij} = 0$ if $i \neq j$, $\delta_{ij} = 1$ if $i = j$, and R_{ij} is the distance between the polymer segments i and j. The summation of R_{ij} gives the average dimension (size) of the polymer molecule in the solution. Since D is related to f according to the Einstein equation,

Eq. (10.14) can also be expressed in the form

$$f = n\zeta\left(1 + \frac{\zeta}{6\pi n\eta}\sum_{i.j=1.i\neq j}^{n}\left\langle\frac{1}{R_{ij}}\right\rangle\right)^{-1} \tag{10.15}$$

where n is the number of monomer units and ζ is the frictional coefficient of each individual monomer unit.

10.3.2 Shape

If a particle of a suspension (here, we mean a macromolecule) is spherical, then the following equation provides the relationship between the diffusion coefficient D and the radius of the sphere r:

$$D = \frac{kT}{6\pi\eta r} \tag{10.16}$$

where η is the viscosity of the solvent. Equation (10.16) is called the Stokes-Einstein equation. A modification of Eq. (10.16) was proposed by Sutherland in 1905 to take into consideration sliding friction. The diffusion coefficient D is then expressed as

$$D = \frac{kT}{(6\pi\eta r)(1 + 2\eta/\beta r)/(1 + 3\eta/\beta r)} \tag{10.17}$$

where β is the coefficient of sliding friction.

Perrin in 1936 derived equations that relate frictional coefficients to the shape of macromolecules formed as ellipsoids of revolution. The frictional coefficients (f) of prolate and oblate ellipsoids (see chapter 6) are both greater than the frictional coefficients of spheres (f_s or f_0) of equal volume. The difference depends on the ratio of the major to the minor axis. Let $p = b/a$ be the axial ratio where b is the equatorial radius and a is the semiaxis of revolution. For prolate ellipsoids or elongated ellipsoids ($a/b > 1$), we have

$$\frac{D}{D_s} = \frac{f}{f_s} = \frac{f}{f_0} = \frac{\sqrt{1 - p^2}}{p^{2/3}\ln\left[(1 + \sqrt{1 - p})/p\right]} \tag{10.18}$$

For oblate ellipsoids ($a/b < 1$), we have

$$\frac{D}{D_s} = \frac{f}{f_s} = \frac{f}{f_0} = \frac{(p^2 - 1)^{1/2}}{p^{2/3}\tan^{-1}(p^2 - 1)^{1/2}} \tag{10.19}$$

The frictional ratios of irregular shapes, that is, other than the sphere and the ellipsoid, have not yet been investigated.

10.3.3 Molecular Weight

Diffusion and molecular weight are related in three ways:

1. *Einstein-Stokes Relation* From

$$D = \frac{kT}{f}$$

$$f = 6\pi\eta r$$

$$\frac{4}{3}\pi r^3 = \frac{\bar{v}M}{N_A}$$

We obtain

$$M = \frac{4\pi r^3 N_A}{3\bar{v}} \tag{10.20}$$

where $\bar{v}$ is the partial specific volume of the polymer molecule and M is its molecular weight. It should be noted that

$$D = \frac{kT}{f} = \frac{R'T}{N_A f} = \frac{R'T}{6\pi\eta N_A (3M\bar{v}/4\pi N_A)^{1/3}}$$

where R' is the gas constant. Hence, $D \sim 1/M^{1/3}$, that is, D is inversely proportional to the cubic root of molecular weight.

2. *Einstein-Svedberg Relation*

$$M = \frac{SR'T\,.}{D(1 - \bar{v}\rho)} \tag{10.21}$$

where S is the sedimentation coefficient, which is discussed in Chapter 12, and ρ is the density of the solution.

3. *Empirical Relation* For homologous polymers, we have an equation relating D and M analogous to the Mark-Houwink equation for intrinsify viscosity:

$$D = K'M^{-a} \tag{10.22}$$

where K' and a are constants. For example, for polystyrene in 2-butanone at 25°C, we have

$$D = (3.1 \times 10^{-4})M^{0.53}$$

The physical meanings of K' and a are yet to be explored.

10.4 CONCENTRATION DEPENDENCE OF THE DIFFUSION COEFFICIENT

The translational diffusion coefficient is considered to be a constant only for particles in dilute solution. In general, if the solution is not dilute, D is dependent on concentration. This dependence may be expressed as

$$D = D_0(1 + k_D c + \cdots) \tag{10.23}$$

where

$$D_0 = \frac{kT}{f_0}$$

The parameter D_0 is the diffusion coefficient at infinite dilution. The term k_D is the hydrodynamic and thermodynamic combined factor and can be expressed in either of two different ways:

$$k_D = 2A_2M - f' - \frac{N_A V}{M} \tag{10.24}$$

or

$$k_D = \left(\frac{d \ln y}{dc}\right)(\bar{v}_1 \rho) \tag{10.25}$$

In Eqs. (10.24) and (10.25) f' is the first-order frictional coefficient, V is the polymer molecular volume, y is the activity coefficient of the polymer in solution, $\bar{v}_1$ is the partial specific volume of the solvent, and ρ is the density of the solution. Equation (10.24) fits the description for the behavior of synthetic polymers, while Eq. (10.25) fits that for biological polymers, particularly proteins. Substituting Eq. (10.24) into Eq. (10.23), we obtain

$$D = D_0[1 + (2A_2M - f')c + \cdots] \tag{10.26}$$

The relative values of the two terms $2A_2M$ and f' provide a measure of the property of solvent in polymer solutions. For good solvents, the difference between $2A_2M$ and f' is large. For poor solvents, the difference between $2A_2M$ and f' is small.

Substituting Eq. (10.25) in Eq. (10.23), we obtain

$$D = D_0\left(1 + \frac{d \ln y}{dc} c\right)(\bar{v}_1 \rho) \tag{10.27}$$

The quantity $(\bar{v}_1 \rho)$ is approximately equal to unity, except at very high concentrations of solution. This equation is suitable for systems of three or more components with interacting flows, as is the case with protein solutions.

There is a subtle difference in the two interpretations of k_D. In the first interpretation, Eq. (10.26), the driving force dc/dx is identical to a gradient of osmotic pressure; hence, the equation for D involves the second virial coefficient A_2. In the second interpretation, Eq. (10.27), the driving force dc/dx is identical to a gradient of chemical potential; hence, the equation for D involves the activity coefficient y of the polymer. The equation fits into the analysis of an interacting multicomponent system. Of course, it must be remembered that osmotic pressure and chemical potential are closely related, as discussed in Chapter 7.

10.5 THE SCALING RELATION FOR THE TRANSLATIONAL DIFFUSION COEFFICIENT

Scaling relations for quantities that are time-independent; for example, the osmotic pressure π', the mean-square end-to-end distance $\langle R^2 \rangle$, and the screen length ξ are known as static scaling relations. Those for time-dependent quantities, such as the translational diffusion coefficient, are called dynamic scaling relations.

The Rouse model considers the polymer chain as a succession of "beads," $\mathbf{r}_1, \ldots, \mathbf{r}_n \cdot \mathbf{r}_{n+1}$, separated by "springs" along the vectors $\mathbf{a}_1, \ldots, \mathbf{a}_n$ (see Chapter 6). If all the internal forces add up to zero, the equation is reduced to

$$\frac{\partial \mathbf{r}}{\partial t} = \frac{B}{N} \sum_n f_n$$

where f_n is an external force. Since the overall mobility of a single chain is B/N (B is the mobility and N is the number of segments per chain), the corresponding diffusion coefficient can be derived, similarly to Einstein's equation, as

$$D = BkTN^{-1}$$

In scaling relations, we can put the diffusion coefficient in the form (Adler and Freed, 1979)

$$D = \frac{kT}{\eta_0 b} N^{-1/2} f_D(\varepsilon) \tag{10.28}$$

where ε is the argument of the function f_D and is defined as

$$\varepsilon = u e^{-4} N^{(v-d)/2}$$

and b is the Kuhn length (the length of chain segment), u is the excluded volume strength, v is the critical exponent, and d is the dimensionally of the system under consideration.

In a θ solvent where the excluded volume effect vanishes, Eq. (10.28) reduces to Zimm's extension of the Rouse treatment:

$$D = \frac{kT}{\eta_0 b} N^{-1/2} f_D(0) \tag{10.29}$$

where $f_D(0)$ is numerical factor. In a good solvent, the diffusion coefficient can be expressed in a simplified form:

$$D \sim N^{-\nu} \tag{10.30}$$

where $\nu = 0.6$.

10.6 MEASUREMENTS OF THE TRANSLATIONAL DIFFUSION COEFFICIENT

10.6.1 Measurement Based on Fick's First Law

A simple method used by Northrop and Anson (1929) is to measure the diffusion coefficient by a porous disk (not a semipermeable membrane, which was unavailable at that time). Figure 10.4 is a diagram of the experimental setup. Let c be the concentration of macromolecules, with c_{in} as the concentration inside the cell, c_{out} the concentration outside the cell, V the volume of the cell, A the area of the disk, l the thickness of the disk, m the mass, and t the time of measurement. Fisk's first law can be put into a practical form as

$$\mathrm{d}m = -AD \frac{\mathrm{d}c}{\mathrm{d}x} \mathrm{d}t$$

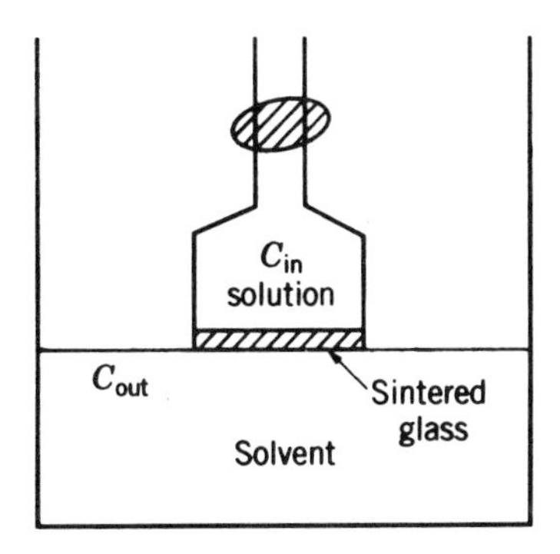

FIGURE 10.4 Northrop and Anson experiment for the determination of diffusion coefficient.

or

$$\Delta m = -AD\frac{\Delta c}{\Delta x}\Delta t \qquad (10.31)$$

The explicit expression to determine D is then

$$D = \frac{-\Delta m}{A(\Delta c/\Delta x)\Delta t}$$

In terms of experimental quantities, we have

$$\Delta m = V(c_{\text{in}} - c_{\text{in}}^0)$$
$$\Delta c = c_{\text{in}}^0 - c_{\text{out}}^0$$
$$\Delta x = 1$$

where the zero superscript refers to the concentration in the reservoir when $t=0$. We then have

$$D = \frac{V(c_{\text{in}} - c_{\text{in}}^0)}{K'(c_{\text{in}}^0 - c_{\text{out}}^0)\Delta t} \qquad (10.32)$$

where $K' = A/l$, a constant known as the cell constant, which can be determined with a substance of known diffusion coefficient. Using a cell constant we can avoid determining A and l.

10.6.2 Measurement Based on Fick's Second Law

Several methods based on Fick's second law are available to measure the diffusion coefficient. Among these are the following three classical methods: the schlieren method, the Gouy interference method, and the Rayleigh interference method. We describe here the Rayleigh interference method, for its application is also found in ultracentrifuge sedimentation. And we use the classical Tiselius electrophoresis cells to illustrate (Figure 10.5).

Typical pictures found in a photographic plate are the displacements of fringes caused by differences in refractive indices in the diffusion region:

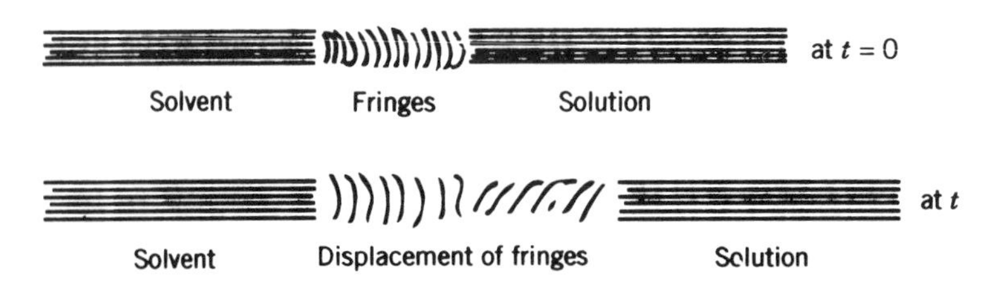

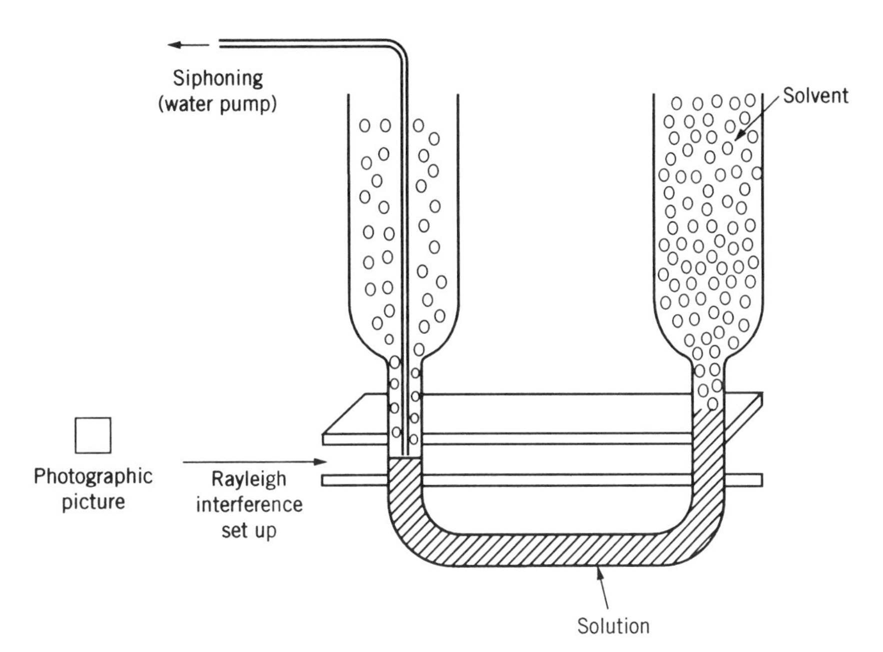

FIGURE 10.5 Rayleigh interference method for the determination of diffusion coefficient.

To aid in our discussion, we enlarge the fringes:

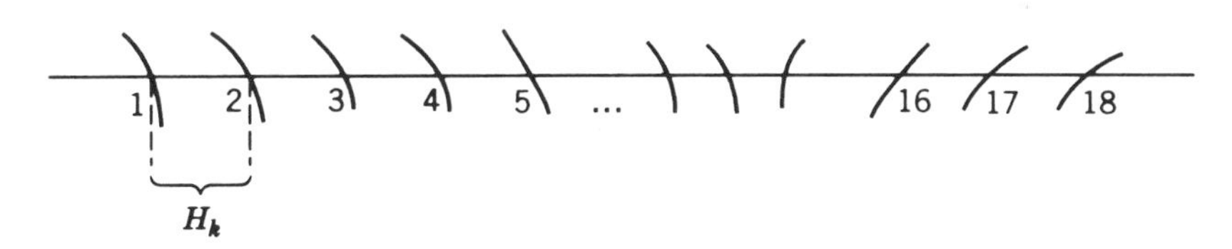

Let $j_k = 1, 2, 3, \ldots$ be the fringe numbers. We then arbitrarily divide these numbers into two sets, j_k and j_l:

Fringes ($J = 18$)

j_k	j_1
2	10
3	11
4	12
5	13
6	14
7	15
8	16
9	17
1(0)	18

Let J be the total number of fringes, for example, 18, and let H_k and H_l be the microcomparator reading for the distance of each j_k and j_l from the reference point, for example, $j_k = 1$. The value of j is related to Δn by*

$$J = \frac{a\,\Delta n}{\lambda}$$

where Δn is the difference in refractive index of the solution and the solvent, a is the cell constant, and λ is the wavelength of the light. Recalling the solution of Fick's second law,

$$\frac{\mathrm{d}c}{\mathrm{d}t} = D\frac{\delta^2 c}{\delta x^2}$$

which is in the form

$$c = \frac{c_0}{2}\left[1 - \underbrace{\frac{2}{\sqrt{\pi}}\int_0^{x/\sqrt{4Dt}} \mathrm{e}^{-x^2/4Dt}\,\mathrm{d}\left(\frac{x}{\sqrt{4Dt}}\right)\Big]}_{j_k^*}$$

This can now be related to the interference fringes by

$$j_k^* = \frac{2j_k - J}{J} = \frac{2}{\sqrt{\pi}}\int_0^z \mathrm{e}^{-z^2}\,\mathrm{d}z$$

With the aid of a table from the *Tables of Probability Functions* (1941) the value of z may then be determined for each fringe. Thus, we have

$$H_k - H_l = \Delta H$$

$$z_k = \frac{h_k}{(4DT)^{1/2}}$$

$$z_l = \frac{h_l}{(4Dt)^{1/2}}$$

* The cell constant a often refers to the cell thickness. It is in the order of 1.2 or 3.0 cm. In a 1.2-cm cell a displacement of one fringe ($j = 1$) corresponds to (assuming $\lambda = 5461 \times 10^{-8}$ cm),

$$\Delta n = \frac{(\lambda)(1)}{1.2} = \frac{5461 \times 10^{-8}\ \mathrm{cm}}{1.2\ \mathrm{cm}} = 4.55 \times 10^{-5}$$

This corresponds to a concentration difference about 0.25 mg/mL. That is, 1 fringe $\cong$ 0.25 mg/mL and 0.03 fringe (measurable) $\cong (0.03)(0.25) = 0.0075$ mg/mL.

$$z_k + z_l = \frac{h}{(4Dt)^{1/2}}$$

$$\frac{\Delta H}{h/(4Dt)^{1/2}} = \text{known value } Q$$

Here we change notation from x to h. The quantity $\Delta H/h$ is the magnification factor for the camera lens that relates distances in the diffusion cell to distances in the photographic plate. From Q, the value of D is determined.

10.7 ROTATIONAL DIFFUSION

A macromolecule may move, not in a transverse direction, but in a rotary motion under a torque, that is, oriented in an angle θ around a reference axis. Let $\rho(\theta)\,\mathrm{d}\theta$ be the number of particles per cubic center of solution, with an orientation between θ and $\theta + \mathrm{d}\theta$. Then, the rotary concentration gradient $\mathrm{d}\rho/\mathrm{d}\theta$ is analogous to the transverse concentration gradient $\mathrm{d}c/\mathrm{d}x$ and the laws governing translational diffusion apply equally to rotational diffusion. Fick's first law is

$$J(\theta) = -\Theta \frac{\delta \rho}{\delta \theta} \tag{10.33}$$

where Θ is the rotational diffusion coefficient. Fick's second law is

$$\frac{\delta \rho}{\delta t} = \Theta \frac{\delta^2 \rho}{\delta \theta^2} \tag{10.34}$$

The rotational frictional coefficient ζ may be defined as

$$\zeta = \frac{T'}{\omega} \tag{10.35}$$

where T' is the torque and ω is the angular velocity. This may be compared to the definition for the translational frictional coefficient f:

$$f = \frac{F}{v}$$

Similar to translational diffusion D, we now have rotational diffusion Θ, which follows the Einstein equation:

$$\Theta = \frac{kT}{\zeta} \tag{10.36}$$

We also have Stokes' law for spherical macromolecules

$$\zeta = 8\pi\eta r^3 \tag{10.37}$$

and Perrin's equation for ellipsoids of revolution

$$\zeta = \frac{16\pi\eta a^3}{3[-1 + 2\ln(2a/b)]} \tag{10.38}$$

with semiaxes of length a and b.

The rotary motion in three dimensions is related to the relaxation time τ:

$$\tau = \frac{1}{2\Theta} = \frac{\zeta}{2kT} \tag{10.39}$$

This equation is important to the design of the experiment to determine Θ. For a sphere, all three τ values in all three dimensions (τ_x, τ_y, and τ_z) are equal, whereas for an ellipsoid of revolution only two of the τ values are equal.

The most common methods for measuring rotational diffusion coefficient Θ are non-Newtonian viscosity, flow birefringence, NMR, dielectric relaxation, fluorescence depolarization, electric birefringence, and polarized light scattering. Here, we describe the methods of flow birefringence and fluorescence depolarization. Currently, the most common and fruitful method is through laser light-scattering measurement, which is discussed in Chapter 11. The laser light-scattering method is used to measure both translational and rotational diffusion coefficients.

10.7.1 Flow Birefringence

The instrument used consists of two cylinders, one fixed and one moving, as shown in Figure 10.6. The sample is placed between these two cylinders. The

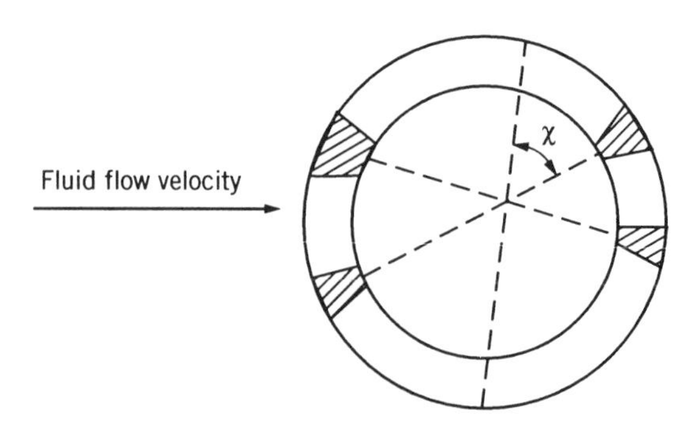

FIGURE 10.6 Flow birefringence.

extinct angle χ is measured where particles orient with respect to flow line. At the shaded part, no light is transmitted; at the empty part, light is transmitted.

The rotational diffusion coefficient Θ can be calculated using

$$\chi = a - \frac{B'}{120} + b\left(\frac{B'}{\Theta}\right)^3 + \cdots \tag{10.40}$$

where a and b are constants and B' is the velocity gradient:

$$B' = \frac{du}{dy}$$

10.7.2 Fluorescence Depolarization

Fluorescence spectroscopy is discussed in Chapter 16. Here we describe the experimental method for determining the rotational diffusion coefficient. The parameter we are interested in is ρ_h, which is the harmonic mean of the two principal relaxation times for the rotation of the ellipsoid. This is based on an equation derived by Perrin (1936):

$$\frac{1}{P} + \frac{1}{3} = \left(\frac{1}{P_0} + \frac{1}{3}\right)\left(1 + 3\frac{\tau_0}{\rho_h}\right) \tag{10.41}$$

Here P is the degree of polarization of fluorescent light emitted at right angles to the direction of the incident light, τ_0 is the lifetime of the excited state of the fluorescence, and P_0 is an empirical constant. The rotational diffusion coefficient Θ_h is obtained from

$$\Theta_h = \frac{1}{2\rho_h} \tag{10.42}$$

REFERENCES

Adler, R. S., and K. F. Freed, *J. Chem. Phys.* **70**, 3119 (1979).

Boltzmann, L., *Ann. Phys. Chem. Wied.* **53**, 959 (1894).

Crank, J., *Mathematics of Diffusion*, 2nd ed. Oxford, UK: Oxford University Press, 1975.

Einstein, A., *Ann. Phys.* **17**, 549 (1905).

Fick, A., *Ann. Phys. Chem.* **170**, 59 (1855).

Fourier, J. B. J., *Theorie Analytique de la Chaleur* (Paris, 1822), translated by A. Freeman, Cambridge England, 1878 and New York, 1955. (See Grattan-Guinness, I., *Joseph Fourier 1768–1830*, Cambridge, Mass: MIT Press, 1972.)

Gosting, L. J., *Adv. Protein Chem.* **11**, 429 (1956).

Kirkwood, J. G., *J. Polym. Sci.* **12**, 1 (1954).

Lamm, O., and P. Polson, *Biochem. J.* **30**, 528 (1936).
Northrop, J. H., and M. L. Anson, *J. Gen. Physiol.* **12**, 543 (1929).
O'Konski, C. T., and A. J. Haltner, *J. Am. Chem. Soc.* **78**, 3604 (1956).
Perrin, F., *J. Phys. Fadium* **1**, 1 (1936).
Schachman, H. K., *Methods Enzymol.* **4**, 78 (1957).
Sutherland, W., *Philos. Mag.* **9**, 781 (1905).
Tables of Probability Functions, Vol. 1. Washington, DC: Government Printing Office, 1941.
Tiselius, A., *Trans. Faraday Soc.* **33**, 524 (1937a).
Tiselius, A., *Biochem. J.* **31**, 1464 (1937b).
Weber, G., and L. B. Young, *J. Biol. Chem.* **239**, 1424 (1964).

PROBLEMS

10.1 Show that Fick's second law (a) can be written in the form

$$\frac{dc}{dx} = \frac{D}{w}\frac{d^2c}{dx^2}$$

where w is the rate of vertical flow (i.e., $w = dx/dt$), and (b) has a solution

$$\frac{c - c_0}{c_{max} - c_0} = \frac{e^{x/x^*} - 1}{e^{x_{max}/x^*} - 1}$$

where c_0 is the concentration at $x = 0$, c_{max} is the asymptotic concentration, x_{max} is the asymptotic depth and $x^* = D/w$.

10.2 A solution was suggested to the diffusion equation

$$c = \alpha t^{-1/2} e^{-x^2/4Dt}$$

where

$$\alpha = \frac{n_0}{2(\pi D)^{1/2}}$$

and n_0 is the number of macromolecules initially present at $x = 0$. Show that this solution satisfies Fick's second law:

$$\frac{\partial c}{\partial t} = D\frac{\partial^2 c}{\partial x^2}$$

10.3 What is the probability that a macromolecule will diffuse a distance x in time t?

10.4 The drawing is a representative Rayleigh pattern from a diffusion experiment on ribonuclease:

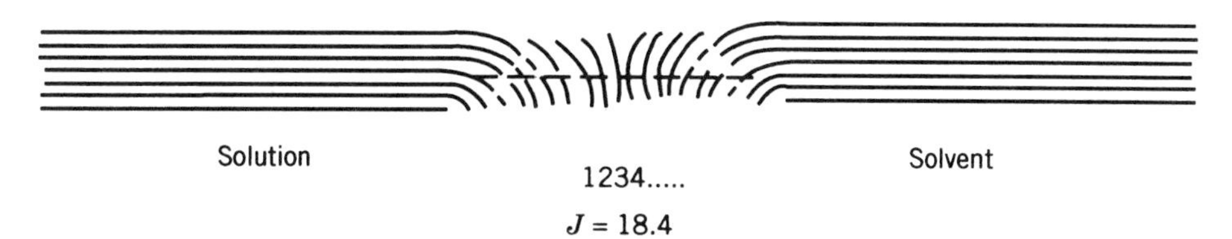

The total number of fringes J was found to be 18.4. The distance H_k between every two neighboring individual fringes j_k was measured on a photographic plate with a microcomparator. The graph was taken 2048 min after the boundary was sharpened. The camera lens magnification factor was $\Delta H/\Delta h = 1.00$. All the pertinent data of this particular graph are given in the following table:

j_k (mm)	j_l (mm)	H_k (mm)	H_1 (mm)
2	10	64.053	58.929
3	11	63.085	58.402
4	12	62.278	57.864
5	13	61.614	57.301
6	14	61.022	56.717
7	15	60.473	56.036
8	16	59.932	55.257
9	17	59.399	54.270

Calculate the apparent diffusion coefficient of ribonuclease. (*Source*: Schachman (1957).)

10.5 The diffusion coefficient of a polymer at infinite dilution is found to be 2.8×10^{-7} cm^2/s. Estimate the hydrodynamic radius of this polymer. Hint: $D_0 = kT/f$.

10.6 The relaxation time τ_i can be determined from the slope of experimentally birefringence decay curves by constructing a tangent to the decay curve at the time t_i, where $i = 1, 2, 3, 4, \ldots$, and t_i is the time required for the birefringence to decay to $\exp(-i)$ of its initial steady-state value. Relaxation times (in m/s) for a tobacco mosaic virus were found to be $\tau_0 = 0.54$, $\tau_1 = 0.53$, $\tau_2 = 0.53$, $\tau_3 = 0.53$, and $\tau_4 = 0.54$. Calculate the rotational diffusion coefficient for this virus. (*Source*: O'Konski and Haltner (1956).)

10.7 The diffusion coefficient of bovine serum albumin at zero concentrations in water at 25°C is 6.75×10^{-7} cm^2/s. The molecular weight is known to be 68 000.

(a) Calculate the frictional factor f.

(b) Calculate the frictional ratio f/f_0.

(c) Using a computer, construct a table for oblong ellipsoids ($p < 1$)

$$\frac{f}{f_0} = \frac{\sqrt{1-p^2}}{p^{2/3} \log(1+\sqrt{1-p^2})/p}$$

and another table for oblate ellipsoids ($p > 1$)

$$\frac{f}{f_0} = \frac{\sqrt{p^2-1}}{p^{3/2} \arctan\sqrt{p^2-1}}$$

The values of $1/p$ or p may be selected from 1.0 to 25. Determine the axial ratio p for the bovine serum albumin and discuss the shape of the molecule.

10.8 At θ temperature and infinite dilution, the diffusion coefficient D_0 of a polymer is found to be empirically related to its molecular weight:

$$D_0 = K'M^{-1/2}$$

where K' is a proportionality constant. Show that at θ temperature, the hydrodynamic volume is proportional to $M^{3/2}$.

11

DYNAMIC LIGHT SCATTERING

In Chapters 8 and 9 we discussed light scattering, small-angle x-ray scattering, and small-angle neutron scattering. In this chapter we discuss dynamic light scattering, also known as laser light scattering. All these types of scattering are basically governed by the same or similar laws. The differences lie in the radiation sources and the techniques used to handle these sources. For light scattering (often called conventional light scattering), the source is light (wavelength 436–546 nm); for small-angle x-ray scattering, the source is x-rays (wavelength 0.05–0.25 nm); for small-angle neutron scattering, the source is neutrons (wavelength 0.05–0.4 nm); and for dynamic light scattering, the source is lasers (wavelength 488–635 nm). The spectrometer for light scattering is relatively simple, while those for the other three are rather complicated.

Dynamic light scattering is characterized by three aspects: (1) The high intensity of the laser source makes it possible to measure weakly scattered light and to observe very small frequency shifts. (2) Time t becomes an important variable in the measurement. The quantity we are interested in, $S(q, t)$, is a function of both the angle θ (included in q) and time t. (3) The correlation function as an experimental device and as a technique in the mathematical analysis of the phenomena is extensively employed.

Like conventional light scattering, dynamic light scattering can be used successfully to determine the molecular weight, size, and shape of macromolecules. But more important, dynamic light scattering is a powerful tool that can measure the diffusion coefficient in a more elegant manner than many other classical tools. For that reason, we place the topic in this chapter after Chapter 10 on diffusion. The extensive use of correlation function techniques in dynamic light scattering also provides methods for studying polymer chain statistics. The serious problem of dust, which causes many spurious signals in light scattering, is greatly minimized in the technique of laser light scattering.

11.1 THE LASER LIGHT-SCATTERING EXPERIMENT

Several well-known experimental techniques are employed in laser light scattering. Here we describe the homodyne and heterodyne methods. Both methods use a laser source for incident radiation. Both use a photo multiplier as a detector, followed by an autocorrelator for computer analysis. The autocorrelator is defined as

$$\langle i(t)i(0)\rangle = B\langle |E(0)|^2 |E(t)|^2\rangle \tag{11.1}$$

where i is the intensity of scattered light, t is the time, E is the scattered electric field, and B is a proportionality constant. The difference between the two methods lies in the substance (scattered light) for analysis. In the homodyne method, we analyze only the scattered light that impinges on the photocathode. The amplitude of the scattered field is proportional to the instantaneous dielectric constant fluctuations. In the heterodyne method, we analyze a small amount of unscattered laser light that is mixed with the scattered light on the photocathode. The heterodyne method, therefore, is also called optical mixing or photon beating method.

11.2 AUTOCORRELATION AND THE POWER SPECTRUM

If two dynamical properties are correlated over a period of time, the function that describes such a correlation is called the correlation function (Chapter 5, Appendix B). Let the dynamical property be $P(t)$. The value of $P(t)$ at t fluctuates about the value at t_0. The average value of $\langle P\rangle$ at average time T is defined as

$$\langle P(t_0, T)\rangle = \frac{1}{T}\int_{t_0}^{t_0+T} P(t)\mathrm{d}t \tag{11.2}$$

The measured bulk property P is simply a time average property $\langle P(t_0, T)\rangle$. At two different times, t and $t+\tau$, the property can have different values:

$$P(t+\tau) \neq P(t)$$

If τ is very small, $P(t+\tau) \equiv P(t)$. If τ is very large, there can be no correlation. Only within a certain range of τ, can there be a correlation between $P(t)$ and $P(t+\tau)$. A measure of this correlation is called the autocorrelation function of

P and is defined as

$$\langle P(0)P(\tau)\rangle = \lim_{T\to\infty}\int_0^T dt\, P(t)P(t+\tau) \tag{11.3}$$

The time τ usually represents the relaxation time.

Experimentally, the autocorrelation function is a measure of the similarity between two noise signals, $P(t)$ and $P(t+\tau)$. If the two signals are completely in phase with each other, then $\langle P(0)P(\tau)\rangle$ is large. If they are out of phase with each other, then the autocorrelation function $\langle P(0)P(\tau)\rangle$ is small. (see Berne and Pecora, 1976.)

The intermediate scattering function $S(g, t)$ is the autocorrelation of density $\rho(t)$, defined as

$$S(q, t) = \langle \rho^*(0)\rho(t)\rangle \tag{11.4}$$

where ρ^* is the complex conjugate of ρ. The density $\rho(t)$ is itself defined as

$$\rho(t) = \sum_j \exp \mathrm{i}\,\mathbf{q}\mathbf{R}_j(t) \tag{11.5}$$

where j sums over all scattering elements $\mathbf{R}$ from an arbitrary origin and $\mathbf{q}$ is the scattering wave function:

$$\mathbf{q} = \frac{4\pi}{\lambda}\sin\frac{\theta}{2} \tag{11.6}$$

The scattering function $S(q, t)$ and the spectra density $I(\omega)$ form a Fourier transform pair:

$$\langle \rho^*(0)\rho(t)\rangle = \int_{-\infty}^{\infty} d\omega\, e^{\mathrm{i}\omega t} I(\omega) \tag{11.7}$$

$$I(\omega) = \frac{1}{2\pi}\int_{-\infty}^{\infty} dt\, e^{-\mathrm{i}\omega t}\langle \rho^*(0)\rho(t)\rangle \tag{11.8}$$

where ω is the angular frequency. In the literature $I(\omega)$ is also called the power spectrum which is the experimental result we are looking for.

11.3 MEASUREMENT OF THE DIFFUSION COEFFICIENT IN GENERAL

Consider a time-dependent correlation function $\langle G(0,0)G(\mathbf{r},t)\rangle$ (first introduced by Van Hove, 1954, see Chapter 9):

$$S(\mathbf{q},\omega) = \frac{1}{2\pi}\int_{-\infty}^{\infty} dt \exp(\mathrm{i}\omega t)\int d\mathbf{r} \exp(\mathrm{i}\mathbf{q}\mathbf{r})\langle G(0,0)G(\mathbf{r},t)\rangle \tag{11.9}$$

By applying the Wiener–Khintchine theorem (Wiener, 1930; Khintchine, 1934) (see appendix to this chapter), the scattering can be transformed into the power spectrum just like the pair by $S(\mathbf{q}, t)$ and $I(\omega)$:

$$I(\omega) = N\frac{1}{2\pi}\int_{-\infty}^{\infty} e^{i(\omega - \omega_0)t}[G_A(\tau)][G_\phi(\tau)]d\tau \quad (11.10)$$

where N is the number of scatterers in a system, ω_0 is the undamped natural angular frequency, $G_A(\tau)$ is the orientation autocorrelation function, and $G_\phi(\tau)$ is the position autocorrelation function. The two autocorrelation functions are defined as

$$G_A(\tau) = \langle A^*(t)A(t-\tau)\rangle$$

$$G_\phi(\tau) = \langle e^{-iqr(t)} e^{-iqr(t+\tau)}\rangle \quad (11.11)$$

If the scatterers are spherical, the orientation autocorrelation function is simplified to

$$G_A(\tau) = |A|^2 \quad (11.12)$$

where A is the amplitude of the wave. The position autocorrelation function, on the other hand, is related to the translational diffusion coefficient D:

$$G_\phi(\tau) = e^{-Dq^2t} \quad (11.13)$$

We now have the scattering intensity expressed more simply as

$$I(\omega) = N|A|^2\frac{1}{2\pi}\int_{-\infty}^{\infty} e^{i(\omega - \omega_0)t} e^{-Dq^2t} dt \quad (11.14)$$

This equation can be further transformed into a Lorentzian equation:

$$I(\omega) = N|A|^2\left[\frac{Dq^2/\pi}{(\omega - \omega_0)^2 + (Dq^2)^2}\right] \quad (11.15)$$

The quantity in brackets is a normalized Lorentzian centered at $\omega = \omega_0$. Thus, the Rayleigh scattered light has a Lorentzian frequency distribution with a half-width:

$$\Delta\omega_{1/2} = Dq^2 = \frac{1}{\tau_0} \quad (11.16)$$

where τ_0 is the correlation time beyond which intensity fluctuations at t are uncorrelated with those at $t + \tau$. The value of D can be determined from $\Delta\omega_{1/2}$ of a plot of experimental data. In experiments, photocurrent $P_i(\omega)$ is usually plotted versus frequency, as shown in the following diagram:

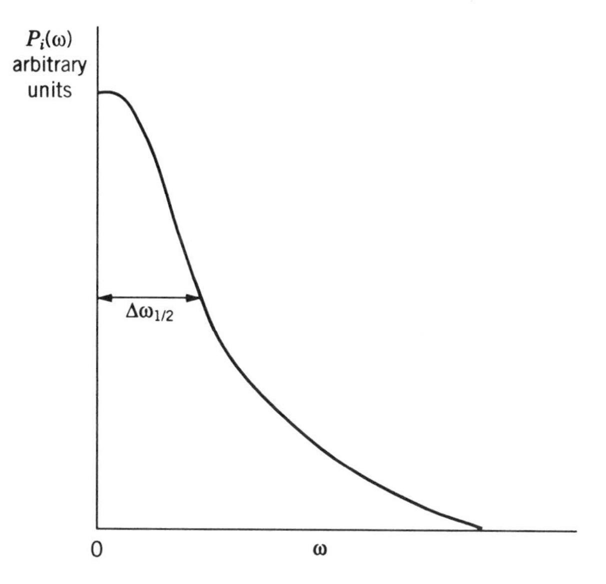

If the macromolecule takes the form of rigid rods, the equation of $\{G_A(\tau)\}$ is more complicated. The power spectrum is now expressed (Wada, 1969) as

$$I(\omega) = \frac{N|A|^2\omega_0^4}{60\pi T^4 R_0^2}(\alpha_z^0 - \alpha_x^0)L(6\theta) \tag{11.17}$$

where T is a parameter related to a diffusion tensor, R is the distance observed from the spectra power density, α is the polarizability, x and z are two Cartesian coordinates, and $L(6\theta)$ is the normalized Lorentzian of half-width $\Delta\omega_{1/2}$. Again, we can determine the value of Θ (the rotational diffusion coefficient) from the graph of $P(\omega)$ versus ω as shown in the determination of D. The equation to be used is now

$$\text{Full width at half-height} = \frac{6\Theta}{\pi} \qquad (\pi = 3.1416) \tag{11.18}$$

11.4 APPLICATION TO THE STUDY OF POLYMERS IN SEMIDILUTE SOLUTIONS

In the previous section, we describe the general method for measuring diffusion coefficients of a polymer solution by laser light scattering. We now list three techniques for measuring diffusion coefficient specifically for polymers in semidilute solutions.

11.4.1 Measurement of Lag Times

According to the reptation theory (Chapter 5), there are two kinds of diffusion of the polymer molecule in the entanglement: the diffusion of individual segments in the tube confined by other chains, and the diffusion of the whole molecule (the polymer chain) out of the tube to assume a new configuration. The former is represented by D_c, the cooperative diffusion coefficient, and the latter is represented by D_s, the self-diffusion coefficient. In the laser light-scattering experiment, if the intensity autocorrelation of the scattered light is measured with two widely different sample times, then the accumulated autocorrelation displays a reasonable portion of two decay times, or e-folds, of an apparent exponential decay. The sample times are new separated into short sample times and long sample times (Amis and Han, 1982). The scattering functions for the two different times can be expressed as follows:

$$S(q,t) = A\,e^{-\tau_c t} + S_c(q) \qquad \text{(Short time)} \tag{11.19}$$

$$S(q,t) = A\,e^{-\tau_s t} + S_s(q) \qquad \text{(Long time)} \tag{11.20}$$

By curve fitting an exponential function to $S(q,t)$ data, we can obtain the decay constant, τ_c or τ_s, either from the slope if the plot is linear or from a second-order cumulant fit (which is discussed later in this section) if the plot is not linear. The diffusion coefficient is obtained by

$$\tau = q^2 D \qquad (\tau = \tau_c \text{ or } \tau_s)$$

From short-time data we obtain D_c and from long-time data we obtain D_s.

11.4.2 Forced Rayleigh Scattering

Forced Rayleigh scattering provides another effective method for measuring D_s. The method makes use of tracer techniques and two laser sources (Leger, Hervet, and Rondelez, 1981). A portion of the sample polymer is labeled with a photochronic probe, such as 1′-(4-iodobutyl)-3′,3′-dimethylindolino-6-nitrobenzospiropyran. The mixture of labeled and nonlabeled chains is dissolved in a solvent. The exciting laser beam (351 nm) is mechanically chopped to give a light pulse of 1 ms to 1 s duration. It passes through another laser light source (633 nm) for scattering measurement. The relaxation of the diffracted intensity is recorded.

For data analysis, the following empirical equation is employed:

$$n(t) = (A\,e^{-t/\tau} + B)^2 + I^2 \tag{11.21}$$

where $n(t)$ is the number of pulses in the pulse counting mode, $A\,e^{-t/\tau}$ represents the amplitudes of the diffracted fields, and B and I are the amplitudes of the scattered fields, which are coherent and incoherent with the signal, respectively. By least-squares curve fitting, with A, B, I, and t as adjustable, the quantity τ is determined, from which we calculate D_s.

11.4.3 Linewidth Analysis

From the power spectrum, we can obtain the linewidth $\Gamma = \Delta\omega_{1/2} = q^2 D$ (Chu and Nose, 1980). If the measurement is carried out for a polymer in a semidilute solution, we can relate asymptotically Γ to D and c (concentration), following the scaling law:

$$\langle \Gamma \rangle \sim D \sim c^{\alpha} \tag{11.22}$$

where

$$D = \frac{kT}{6\pi\eta_0 \xi} \tag{11.23}$$

and $D \sim c^{\alpha}$ represents the scaling law. In Eq. (11.23), k is the Boltzmann constant, T is temperature, and η_0 is the viscosity of the solvent. The quantity ξ is the hydrodynamic length of a polymer in the semidilute solution. The exponent α is a measure of the excluded volume effect v:

$$\alpha = \frac{v}{3v - 1} \tag{11.24}$$

It is the determination of an accurate value of $\langle \Gamma \rangle$ from experimental data that matters. One method to achieve this is first to define a normalized distribution function $G(\Gamma)$:

$$G(\Gamma) = e^{-q^2 Dt} = e^{-\Gamma t} \tag{11.25}$$

Then we calculate

$$\langle \Gamma \rangle = \int_0^{\infty} \Gamma G(\Gamma) d\Gamma \tag{11.26}$$

The evaluation of $G(\Gamma)$ can be performed with the cumulant analysis as follows (Koppel, 1972). We take the logarithm of the normalized correlation function:

$$\ln G(\Gamma) = \ln e^{-q^2 Dt} \tag{11.27}$$

Then we expand the natural log term in series:

$$\ln(1 + x) = x - \tfrac{1}{2}x^2 + \cdots$$

Thus, we have

$$\ln G(\Gamma) = 1 - K_1 t + \frac{1}{2} K_2 t^2 - \frac{1}{3!} K_3 t^3 \cdots \tag{11.28}$$

where

$$K_n = \left[(-1)^n \frac{d^n}{dt^n} \ln G(\Gamma) \right]_{t=0} \tag{11.29}$$

The quantity K_n is the nth cumulant of $G(\Gamma)$. The first few cumulants are

$$\begin{aligned} K_1 &= \langle q^2 D \rangle \equiv q^2 \langle D \rangle_z \quad (= \langle \Gamma \rangle) \\ K_2 &= \langle (q^2 D - \langle q^2 D \rangle)^2 \rangle \\ K_3 &= \langle (q^2 D - \langle q^2 D \rangle)^3 \rangle \\ K_4 &= \langle (q^2 D - \langle q^2 D \rangle)^4 > - 3K_2^2 \end{aligned} \tag{11.30}$$

D_z is the z average diffusion coefficient.

APPENDIX FOURIER SERIES, FOURIER INTEGRAL, AND FOURIER TRANSFORM

In 1807 Fourier presented a paper to the French Academy in which he stated that any function $f(x)$ can be expressed in terms of sines and cosines. The Academy meeting was attended by such great mathematicians as Lagrange, Legendre, and Laplace. Fourier's theorem was unanimously rejected on the basis that it lacked a rigorous proof. To a mathematician, a theorem without a rigorous proof is equivalent to a theory in science without experimental support. It is considered sheer nonsense. Lagrange told Fourier that the same idea had occurred to him as well, but he did not claim it only because he found it difficult to prove. Twenty-two years later (1822), Fourier published a book, *The Theory of Heat Conduction*, in which he again proposed this theorem. His enthusiasm about a function $f(x)$ as expressed in terms of sines and cosines had not been in any way diminished. He reiterated this theory, disregarding early objections by Lagrange and other fellow mathematicians. Unfortunately, he again did not provide any rigorous proof. Eight years later (1830), Fourier died. He must have died a sad man, having never earned distinction in his lifetime among his peers in mathematics.

However, Fourier's position in the history of mathematics has been gradually recognized. Throughout the years, not only has his theorem been partially proved, but he also stimulated a great interest in such mathematicians of the 19th century as Riemann, Presvel, and Weierstrass, in the field of analysis (Bell, 1945). Today, the Fourier series is developed in modern analysis alongside with the rapid growth of automatic computing. The Fourier integral and Fourier transform, which are derived directly from the Fourier series, are involved in all technical fields, including engineering, physics, chemistry, biology, and medicine.

Fourier Series

To present the Fourier series in an exact language (with detailed description), we first introduce three well-known definitions (Hildebrand, 1962):

Definition 1. If $f(x)$ and $g(x)$ are two periodic functions, their *inner product* (f, g) is defined as

$$(f, g) = \int_{-\pi}^{\pi} f(x)g(x)\,\mathrm{d}x \quad \text{or} \quad (f, g) = \int_{-\infty}^{\infty} f(x)g(x)\,\mathrm{d}x$$

Definition 2. Two functions $f(x)$ and $g(x)$ are orthogonal if and only if

$$\int_{-\infty}^{\infty} f(x)g(x)\,\mathrm{d}x = 0$$

Definition 3. A function $f(x)$ is said to be odd if and only if the quantity $f(-x) = -f(x)$ holds for all x's in the given domain; $f(x)$ is an even function if and only if $f(-x) = f(x)$.

Following are some examples. The terms $\cos nx$ and $\sin mx$ form orthogonal functions as seen in their inner product:

$$\int_{-\pi}^{\pi} \cos mx \cos nx\,\mathrm{d}x = \pi\delta_{nm}$$

$$\int_{-\pi}^{\pi} \sin mx \sin nx\,\mathrm{d}x = \pi\delta_{nm}$$

$$\int_{-\pi}^{\pi} \sin mx \cos nx\,\mathrm{d}x = 0$$

where δ_{mn} is the Kronecker delta, defined as

$$\delta_{mn} = 0 \qquad \text{if } m \neq n$$

$$\delta_{mn} = 1 \qquad \text{if } m = n$$

Note that $\sin mx$ are odd functions and $\cos mx$ are even functions. The functions e^x and $\log x$ are neither odd nor even. It can be shown that for an odd function,

$$\int_{-\infty}^{\infty} f(x)\,\mathrm{d}x = 0$$

and for an even function,

$$\int_{-\infty}^{\infty} f(x)\,\mathrm{d}x = 2\int_{0}^{\infty} f(x)\,\mathrm{d}x$$

We are now in a position to describe Fourier series. A suitably restricted solution to a given second-order differential equation

$$\frac{\mathrm{d}^2 y}{\mathrm{d}x^2} + ay = 0$$

may be found in the form of a power series:

$$f(x) = \sum_{n=0}^{\infty} \alpha_n \phi_n(x)$$

where a and α are both coefficients. Fourier proposed that the power series $\sum_{n=0}^{\infty} \alpha_n \phi_n(x)$ can be expressed as a combination of $\sum a_n \cos nx$ and $\sum b_n \sin nx$:

$$f(x) = \sum_{n=0}^{\infty} (a_n \cos nx + b_n \sin nx)$$

The coefficients α_n can be evaluated by the inner product:

$$\alpha_1 = a_1 = (f, \phi_1) = \frac{1}{2\pi}\int_{-\pi}^{\pi} f(x)\,\mathrm{d}x$$

$$\alpha_{2n} = a_n = (f, \phi_{2n}) = \frac{1}{\sqrt{\pi}}\int_{-\pi}^{\pi} f(x)\cos nx\,\mathrm{d}x \qquad n = 1, 2, \ldots$$

$$\alpha_{2n+1} = b_n = (f, \phi_{2n+1}) = \frac{1}{\sqrt{\pi}}\int_{-\pi}^{\pi} f(x)\sin nx\,\mathrm{d}x \qquad n = 1, 2, \ldots$$

A typical example is $f(x) = |x|$, $-\pi \leqslant x \leqslant \pi$. We can calculate the coefficients as follows:

$$a_0 = \frac{1}{\pi}\int_{-\pi}^{\pi} |x|\,\mathrm{d}x = \frac{2}{\pi}\int_{0}^{\pi} x\,\mathrm{d}x = \pi$$

$$a_n = \frac{2}{\pi}\int_{0}^{\pi} |x|\cos nx\,\mathrm{d}x = \frac{-4}{n^2\pi} \qquad n = 1, 3, \ldots$$

$$= 0 \qquad n = 2, 4, \ldots$$

$$b_n = 0$$

Thus, the solution is a Fourier series:

$$f(x) = \frac{\pi}{2} - \frac{4}{\pi}\left(\frac{\cos x}{1^2} + \frac{\cos 3x}{3^2} + \frac{\cos 5x}{5^2} + \cdots\right)$$

This indicates that a Fourier series does not always contain both sine and cosine terms. If $f(x)$ is even and $b_n = 0$, a Fourier series becomes a pure cosine series:

$$f(x) = \frac{1}{2}a_0 + \sum_{n=1}^{\infty} a_n \cos nx$$

where

$$a_n = \frac{2}{\pi}\int_0^{\pi} f(x)\cos nx \, dx \qquad n = 0, 1, 2, \ldots$$

If $f(x)$ is odd and $a_n = 0$, a Fourier series becomes a pure sine series:

$$f(x) = \frac{1}{2}a_0 + \sum_{n=1}^{\infty} b_n \sin nx$$

where

$$b_n = \frac{2}{\pi}\int_0^{\pi} f(x)\sin nx \, dx \qquad n = 1, 2, \ldots$$

In practice, for example, the interferogram of FT-IR (Fourier transform–infrared spectroscopy, Chapter 16) is defined as the sum of cosine waves of all the frequencies present in the source:

$$I(\delta) = \sum_{v_i} \beta(v_i)\cos(2\pi\delta v_i)$$

where $I(\delta)$ is the intensity of the detector signal, $\beta(v)$ is the intensity of brightness of the source, δ is the optical difference, and v is the frequency.

The superimposition of Fourier coefficients c_n can be calculated (Rao, 1986) by

$$c_n = (a_n^2 + b_n^2)^{1/2}$$

The plot of c_n versus v gives a typical harmonic graph, called the frequency spectrum, which represents a harmonic oscillator or any system involving oscillation, as shown in Figure 11.1. The phase angle ϕ_n can be calculated by

$$\phi_n = \tan^{-1}\left(\frac{b_n}{a_n}\right)$$

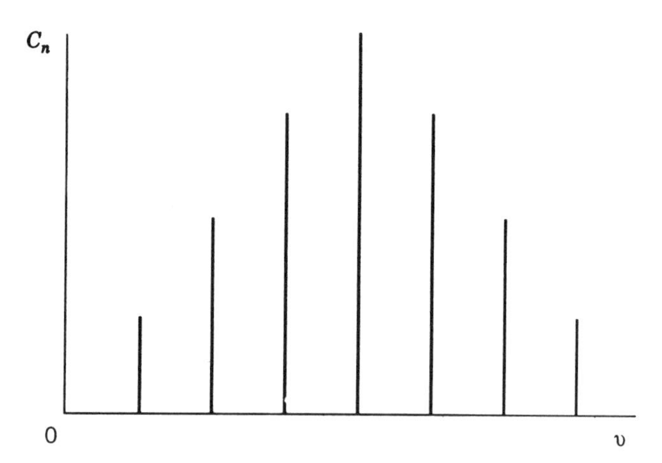

FIGURE 11.1 Frequency spectrum form 1.

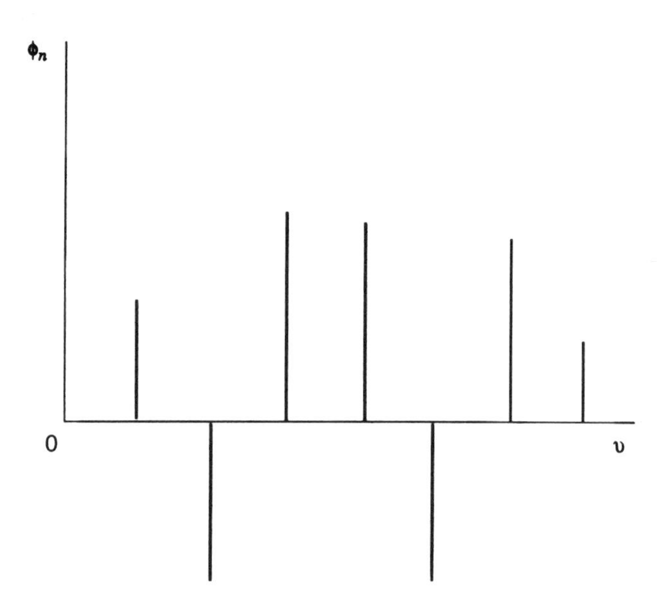

FIGURE 11.2 Frequency spectrum form 2.

The plot of ϕ_n versus v is another form of frequency spectrum (Figure 11.2). The graph separates the data into two opposite directions, providing a choice of information.

The Fourier Integral

Fourier series $f(x)$ is a discontinuous function. The combination of harmonic functions $a_n \cos nx$ and $b_n \sin nx$ is obtained by summation $\sum$. The periodicity of $f(x)$ is limited to the interval $(-\pi, \pi)$. The purpose of the Fourier integral is to make $f(x)$ a continuous function, to combine cosine and sine terms by integration, and to extend the limit to $(-\infty, \infty)$. This requires that certain manipulations be performed (Hildebrand, 1962; Olmsted, 1959).

Let us first define $f(x)$ on $[0, \pi]$ only:

$$f(0) = f(\pi) = 0$$

Then the domain can be extended from $[-\pi, \pi]$ to $[-\infty, \infty]$ by periodicity (Bracewell, 1990). The contribution of $\sin nx$ to $f(x)$ can be put in the form $A(n) \sin nx$. The complete representation of $f(x)$ is now obtained by superimposing these contributions by means of a semi-infinite integral:

$$f(x) = \int_0^\infty A(n) \sin nx \, dn$$

We multiply both sides by $\sin n_0 x$, where n_0 is any positive value of n:

$$f(x) \sin n_0 x = \sin n_0 x \int_0^\infty A(n) \sin nx \, dn$$

Then we integrate both sides with respect to x:

$$\int_0^\infty f(x) \sin n_0 x \, dx = \int_0^\infty \sin n_0 x \int_0^\infty A(n) \sin nx \, dn \, dx$$

Since the order of integration can be assumed to be immaterial, we have

$$\int_0^\infty f(x) \sin n_0 x \, dx = \int_0^\infty A(n) \left(\int_0^\infty \sin n_0 x \sin nx \, dx \right) dn$$

The integral in parentheses equals $\pi/2$, if $n_0 = n$. Hence,

$$A(n) = \frac{2}{\pi} \int_0^\infty f(x) \sin nx \, dx$$

The function

$$f(x) = \frac{2}{\pi} \int_0^\infty \sin nx \int_0^\infty f(t) nt \, dt \, dn$$

is called the Fourier sine integral. Similarly, we can obtain the Fourier cosine integral:

$$f(x) = \frac{2}{\pi} \int_0^\infty \cos nx \int_0^\infty f(t) \cos nt \, dt \, dn$$

For the complete Fourier integral we can write

$$f(x) = \int_{0}^{\infty} [A(n)\cos nx + B(n)\sin nx]\,dn \qquad -\infty < x < \infty$$

where

$$A(n) = \frac{1}{\pi}\int_{-\infty}^{\infty} f(t)\cos nt\,dt$$

$$B(n) = \frac{1}{\pi}\int_{-\infty}^{\infty} f(t)\sin nt\,dt$$

If we substitute the two integrals $A(n)$ and $B(n)$ into $f(x)$, we obtain

$$f(x) = \frac{1}{2\pi}\int_{-\infty}^{\infty}\int_{-\infty}^{\infty} f(t)\cos n(t-x)\,dt\,dn \qquad -\infty < x < \infty$$

This double integral leads to our next topic, the Fourier transform.

The Fourier Transform

Using the identity

$$e^{ix} = \cos x + i\sin x$$

we can write the cosine term in the double integral as

$$\cos n(t-x) = \tfrac{1}{2}e^{in(t-x)} + \tfrac{1}{2}e^{-in(t-x)}$$

where $i = \sqrt{-1}$. The Fourier integral can now be written in the form

$$f(x) = \frac{1}{2\pi}\int_{-\infty}^{\infty}\int_{-\infty}^{\infty} f(t)e^{-in(t-x)}\,dt\,dn$$

$$= \frac{1}{2\pi}\int_{-\infty}^{\infty} e^{inx}\underbrace{\int_{-\infty}^{\infty} e^{-int}f(t)\,dt}_{F(n)}\,dn \qquad -\infty < x < \infty$$

Denoting the inner product in the above equation as $F(n)$ and changing the dummy variable x to t, we obtain the following pair of equations:

$$F(n) = \int_{-\infty}^{\infty} e^{-int}f(t)\,dt \qquad -\infty < n < \infty$$

$$f(t) = \frac{1}{2\pi}\int_{-\infty}^{\infty} e^{int}F(n)\,dn \qquad -\infty < t < \infty$$

The pair is called the Fourier transform pair. The function $F(n)$ is called the Fourier transform of $f(t)$, and the function $f(t)$ is the inverse Fourier transform of $F(n)$.

In a similar way, we can have Fourier sine and cosine transforms of $f(t)$:

$$F(n) = \int_0^\infty f(t) \sin nt \, dt \qquad 0 < n < \infty$$

$$f(t) = \int_0^\infty F(n) \sin nt \, dn \qquad 0 < t < \infty$$

and

$$F(n) = \int_0^\infty f(t) \cos nt \, dt \qquad 0 < n < \infty$$

$$f(t) = \int_0^\infty F(n) \cos nt \, dn \qquad 0 < t < \infty$$

That is, Fourier transform is not limited to complex numbers only; it may be used in real numbers. The following are the Fourier transforms most well-known to chemists (see Chapter 16).

FT – IR

$$F(x) = \int_0^\infty A(v) \cos(2\pi v x) \, dx$$

$$A(v) = 2 \int_0^\infty F(x) \cos(2\pi v x) \, dv$$

where $A(v)$ is the interferogram for a polychromatic source, and $F(x)$ is the actual interferogram.

NMR All the information of an NMR spectrum is contained in the frequency domain, $F(\omega)$. The Fourier transform between the time domain $f(t)$ and the frequency domain $F(\omega)$ is given by

$$F(\omega) = \int_{-\infty}^\infty f(t) e^{-i\omega t} \, dt$$

Since there is an identity

$$e^{-ix} = \cos x - i \sin x$$

we can decompose $F(\omega)$ into sine $S(\omega)$ and cosine transforms $C(\omega)$:

$$S(\omega) = 2\int_0^\infty f(t)\sin\omega t\,\mathrm{d}t$$

$$C(\omega) = 2\int_0^\infty f(t)\cos\omega t\,\mathrm{d}t$$

The time domain function, which is the time decay of signal intensity, can be expressed as

$$f(t) = A\,\mathrm{e}^{-t/\tau}\cos(\omega - \omega_\mathrm{i})t \qquad t > 0$$

where A is the coefficient and τ is the relaxation time. The coefficient A itself is given by

$$A_f = \sum_{t=0}^{N-1} x_t \exp\left(-\frac{2\pi\nu t}{N}\right) \qquad f = 0, 1, 2, \ldots, N-1$$

where x_t is the value of the time decay at time t. The equation of A_f is a discrete Fourier transform, which will be discussed later.

X-Ray Crystallography (see Chapter 15)

$$F_{hkl} = \int_V \rho(x, y, z)\exp\left[2\pi\mathrm{i}\left(\frac{hx}{a} + \frac{ky}{b} + \frac{lz}{c}\right)\mathrm{d}v\right.$$

$$\rho(x, y, z) = \frac{1}{V}\sum_h\sum_k\sum_l F_{hkl}\exp\left[-2\pi\mathrm{i}\left(\frac{hx}{a} + \frac{ky}{b} + \frac{lz}{c}\right)\right]$$

where h, k, l are Miller indices. The values of structure factor F_{hkl} were obtained for all the reflections when an x-ray hits a molecule. They become the coefficients of a Fourier series for the electron density function $\rho(x, y, z)$, from which an electron density distribution can be constructed.

Fourier Transform and Computers

Given a function $f(x)$, Fourier transform is an operation O such that

$$Of(x) = F(n)$$

where O is the operation $\mathrm{e}^{-2\pi nx}$ and n is the transform variable. This is the transformation of one point x in one domain to another point n in another domain. The transform of the whole domain of x to another domain of n

requires another operation, integration, with respect to x from the limit $-\infty$ to the limit ∞. This is based on the assumption that x is continuous. The transformation, as mentioned before, is called the (continuous) Fourier transform:

$$F(n) = \int_{-\infty}^{\infty} f(x)\,\mathrm{e}^{-\mathrm{i}2\pi nx}\,\mathrm{d}x$$

The term "continuous" is often omitted. In modern technology, the calculation of Fourier transform is often performed by a digital computer, and for convenience the Fourier transform has to be converted to the discrete form:

$$F(n) = \frac{1}{n}\sum_{x=0}^{n-1} f(x)\,\mathrm{e}^{-\mathrm{i}2\pi nx/n}$$

This is called the discrete Fourier transform. The variable n refers to the data points. Customarily, the capital letter N, instead of n, is used in the literature (Bracewell, 1990). In 1965, Cooley and Tukey proposed an algorithm in which N is factored to some power, $N = 2^p$ or $N = 4^q, p$ and q being the powers or exponents. This algorithm not only simplifies the number of computer calculations required, but also saves money. It became the basis of what is now called the fast Fourier transform (FFT), which is available in numerous software packages.

The transformation of one domain, say time, to a corresponding domain, say frequency, is very important in science and technology. The transformation is shown in Figure 11.3. Often the experimental curve $f(t)$ has no apparent physical significance. After transforming to a frequency domain, $F(v)$, the peak immediately reveals to us the hidden meaning of our observation.

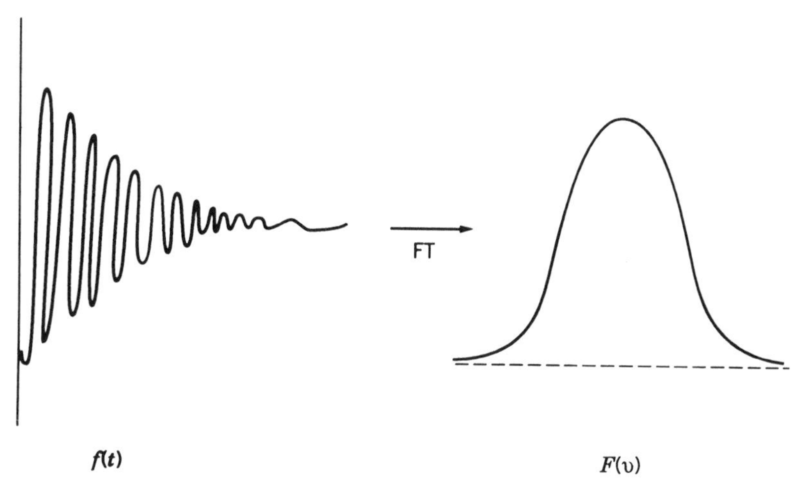

FIGURE 11.3 Transformation of a time domain to a frequency domain.

With some modification, Fourier transform can be converted to an apodization function. This could remove false sidelobes in the spectra. Furthermore, it can be used to delay the input to an amplifier or a vibratory mechanical system.

Extension of Fourier Series and Fourier Transform

We further describe two subjects that are an extension of Fourier series and Fourier transform: the Lorentz line shape and the autocorrelation function. These are both employed in the photon beat technique as well as small-angle neutron scattering. A knowledge of their mathematical backgrounds will facilitate an understanding of the experimental device and the treatment of experimental data.

The Lorentz Line Shape In optics the general solution of the differential equation of oscillators is given in terms of a Fourier series:

$$x(t) = e^{-(1/2)\Gamma t}(A \sin \omega t + B \cos \omega t)$$

where Γ is a damping constant and ω is the angular frequency (Crawford, 1968). Using the formulas of definite integrals

$$\int_0^{\infty} e^{-ax} \sin bx = \frac{b}{b^2 + a^2}$$

$$\int_0^{\infty} e^{-ax} \cos bx = \frac{a}{b^2 + a^2}$$

the Fourier coefficients are given:

$$A \sim \frac{(\frac{1}{2}\Gamma)^2}{(\omega_0 - \omega)^2 + (\frac{1}{2}\Gamma^2)}$$

$$B \sim \frac{(\omega_0 - \omega)^2}{(\omega_0 - \omega)^2 + (\frac{1}{2}\Gamma)^2}$$

where A is the absolute amplitude, B is the elastic amplitude, and the term on the right

$$L \sim \frac{(\frac{1}{2}\Gamma)^2}{(\omega_0 - \omega)^2 + (\frac{1}{2}\Gamma^2)}$$

is the Lorentz function L which describes the Lorentz line shape curve as in Figure 11.4. The quantity $\Gamma = \Delta\omega$ is called the linewidth of the frequency spectrum of the Fourier superposition. The Lorentz line shape is the basis of

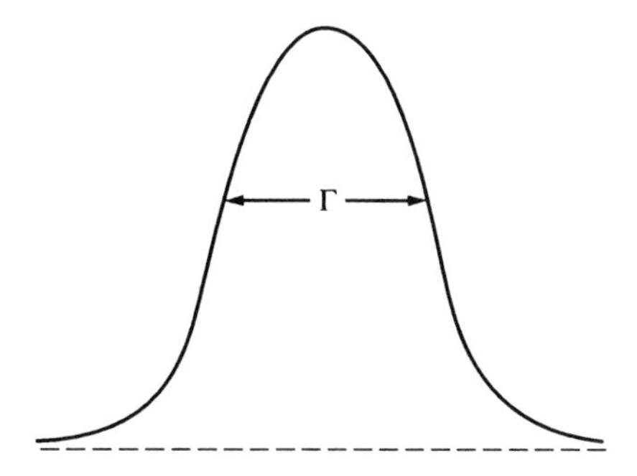

FIGURE 11.4 Lorentz line shape.

the photon beat technique for the determination of translational and rotational diffusion coefficients.

Correlation Function The inner product of the two functions of time $f(t)$ and $f(t+\tau)$ form a correlation function (see Chapter 5) $\langle f(t)f(t+\tau)\rangle$ defined as

$$\langle f(t)f(t+\tau)\rangle = \lim_{T\to\infty}\frac{1}{2T}\int_{-T}^{T} f(t)f(t+\tau)\,\mathrm{d}t$$

where T is the period of length. This is the Fourier integral in a slightly different form. The Fourier transform pairs are

$$\langle f(t)f(t+\tau)\rangle = \int_{-\infty}^{\infty} s(\omega)\mathrm{e}^{-\mathrm{i}\omega t}\mathrm{d}\omega$$

$$s(\omega) = \frac{1}{2\pi}\int_{-\infty}^{\infty} \langle f(t)f(t+\tau)\rangle \mathrm{e}^{\mathrm{i}\omega t}\mathrm{d}t$$

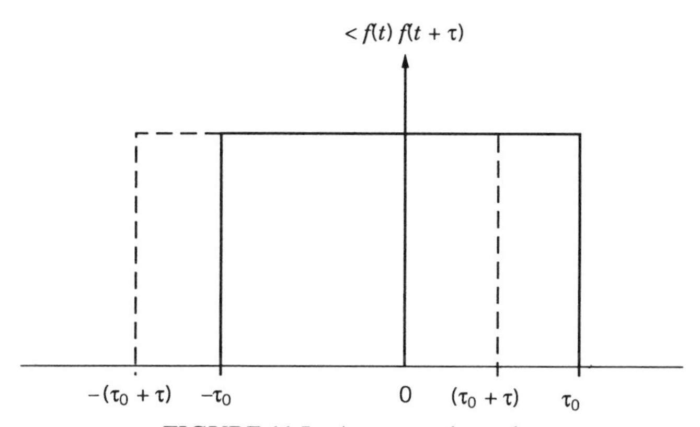

FIGURE 11.5 A rectangular pulse.

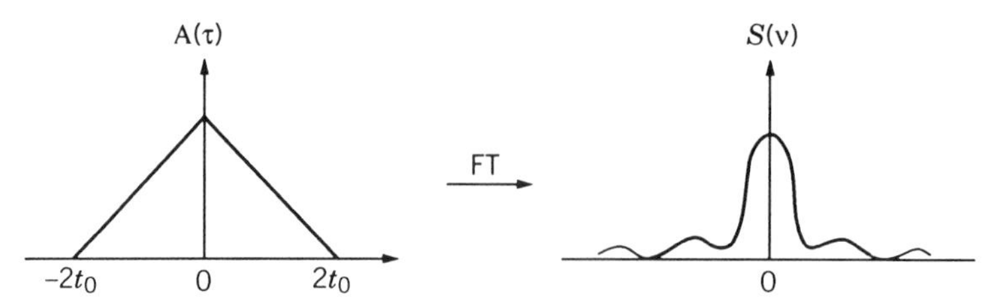

FIGURE 11.6 Graphical representation of an autocorrelation function to a spectral density function.

where $\langle f(t)f(t+\tau)\rangle$ is the correlation function and $s(\omega)$ is the power special density. Either of the equation of the pair is called the Wiener-Khinechine theory. Experimentally, $s(\omega)$ is the photocurrent density and $\langle f(t)f(t+\tau)\rangle$ is the current–density correlation function.

A typical example is the rectangular pulse shown in Figure 11.5. The autocorrelation function in the diagram is

$$A(\tau) = \langle f(t)f(t+\tau)\rangle = x_0^2\left(1 - \frac{\pi}{2t_0}\right)$$

The spectral density function $s(\nu)$ is the Fourier transform of $A(\tau)$:

$$S(\nu) = 4\int_0^\infty A(\tau)\cos 2\pi\nu t\, d\tau = 4x_0^2 t_0 \frac{\sin 2\pi\nu t_0}{2\pi\nu t_0}$$

In graphical representation, we see the transformation shown in Figure 11.6.

REFERENCES

Akcasu, A. Z., M. Benmouna, and C. C. Han, *Polymer* **21**, 866 (1980).

Amis, E. J., and C. C. Han, *Polymer* **23**, 1403 (1982).

Bell, E. T., *The Development of Mathematics*. New York: McGraw-Hill, 1945.

Berne, B. J., and R. Pecora, *Dynamic Light Scattering*. New York: Wiley, 1976.

Bracewell, R. N., *Science* **248**, 697 (1990).

Chu, B., *Laser Light Scattering Basic Principles and Practice*, 2d ed. New York: Academic, 1991.

Chu, B., and T. Nose, *Macromolecules* **13**, 122 (1980).

Cooley, J. W., and J. W. Tukey, *Math. Comput.* **19**, 297 (1965).

Crawford, F. S., Jr., *Waves, Berkeley Physics Course*, Vol. 3. New York: McGraw-Hill, 1968.

Cummins, H. Z., F. D. Carlson, T. J. Herbert, and G. Woods, *Biophys. J.* **9**, 518 (1969).

Dubin, S. B., J. H. Lunacek, and G. B. Benedek, *Proc. Natl. Acad. Sci. U.S.A.* **57**, 1164 (1967).

Hildebrand, F. B., *Advanced Calculus for Applications.* Englewood Cliffs, NJ: Prentice-Hall, 1962.
Khintchine, A., *Math. Ann.* **109**, 604 (1934).
Koppel, D. E., *J. Chem. Phys.* **57**, 4814 (1972).
Leger, L., H. Hervet, and F. Rondelez, *Macromolecules* **14**, 1732 (1981).
Olmsted, J. M. H., *Real Variables.* New York: Appleton-Century-Crofts, 1959.
Rao, S. S., *Mechanical Vibrations.* Reading, MA: Addison-Wesley, 1986.
Van Hove, L., *Phys. Rev.* **95**, 249 (1954).
Wada, A., N. Suda, T. Tsuda, and K. Soda, *J. Chem. Phys.* **50**, 31 (1969).
Wiener, N., *Acta Math.* **55**, 117 (1930).

PROBLEMS

11.1 The time-correlation function is defined as

$$\langle x(t+\tau)x(t)\rangle = \langle x(\tau)x(0)\rangle$$

and the Fourier transform pairs are

$$x_{2T}(t) = \int_{-\infty}^{\infty} \hat{x}_{2T}(\omega)\exp(-i\omega t)\,d\omega$$

$$\hat{x}_{2T}(\omega) = \frac{1}{2\pi}\int_{-\infty}^{\infty} x_{2T}(t)\exp(-i\omega t)\,dt$$

$$2\pi\delta(\omega-\omega') = \int_{-\infty}^{\infty} \exp[i(\omega-\omega')t]\,dt$$

Show that

$$\langle x(t+\tau)x(t)\rangle = \int_{-\infty}^{\infty} S_x(\omega)\exp(-i\omega\tau)\,d\omega$$

$$S_x(\omega) = \frac{1}{2\pi}\int_{-\infty}^{\infty} \langle x(t+\tau)x(t)\rangle\exp(i\omega\tau)\,d\tau$$

where $S_x(\omega)$ is the power spectral density. Note: These two equations comprise the Wiener-Khintchine theorem.

11.2 Let τ be the correlation time beyond which intensity fluctuations at t are uncorrelated with those at $t+\tau$. The time correlation function of $x(t)$ can be defined as

$$\langle x(t)x(t+\tau)\rangle = \lim_{T\to\infty}\frac{1}{2T}\int_{-\infty}^{\infty} x(t)x(t+\tau)\,dt$$

where T is the length of the time. The power spectral density $S_x(\omega)$ is the time Fourier transform of the correlation function:

$$S_x(\omega) = \frac{1}{2\pi} \int_{-\infty}^{\infty} \langle x(t+\tau)x(t) \rangle \exp(i\omega\tau)\, dt$$

$$\langle x(t+\tau)x(t) \rangle = \int_{-\infty}^{\infty} S_x(\omega) \exp(i\omega t)\, d\omega$$

Show that the real part of the Fourier transform of the time correlation function is Lorentzian:

$$\int_{-\infty}^{\infty} \exp(i\omega\tau)\exp(-\Gamma\tau)\, d\tau = 2\frac{\Gamma}{\Gamma^2 + \omega^2}$$

11.3 The following data of the self-beating experiment were obtained for a protein at $\theta = 90°$, $T = 25°C$, and concentration 40 mg/mL:

$s(\omega)$ (arbitrary unit)	ω ($\times 10^{-3}, s^{-1}$)
69	0
59	10
44	20
31	30
24	40
18	50
14	60

The wavelength of the photons is 6328 Å and the refractive index of the solution is 1.453. Calculate the diffusion coefficient of the protein.

11.4 Show that the Fourier transform of an exponential $e^{-a|t|}$ is

$$\pi^{-1} \operatorname{Re} \exp \frac{1}{i\omega + \alpha} = \pi^{-1}\left(\frac{\alpha}{\omega^2 + \alpha^2}\right)$$

11.5 Show that the equilibrium mean-square value of the property A, as involved in the time correlation function

$$\langle A(0)A(t) \rangle = \int_{-\infty}^{\infty} d\omega\, e^{i\omega t} I(\omega)$$

is

$$\langle |A|^2 \rangle = \int_{-\infty}^{\infty} d\omega I(\omega)$$

12

SEDIMENTATION

Like intrinsic viscosity and diffusion, sedimentation is a frictional property. It is the transport of mass from the surface toward the bottom. In ultracentrifuge sedimentation, which is our major subject in this chapter, macromolecules in solution are forced to sediment by a centrifugal force that is 100 000 times greater than gravitational force. Sedimentation can be described by Fick's two laws with some modification, since sedimentation is often accompanied by diffusion. The modified equation of Fick's first law is

$$\underset{\text{Flow of the solute}}{J} = \underbrace{cS\omega^2 r}_{\text{Sedimentation}} - \underbrace{D(\partial c/\partial r)}_{\text{Diffusion}} \tag{12.1}$$

and that of the second law is

$$\frac{\partial c}{\partial t} = \frac{1}{r}\frac{\partial}{\partial c}\left[\left(cS\omega^2 r - D\frac{\partial c}{\partial r}\right)r\right] \tag{12.2}$$

where c is the concentration of the solute in solution, r is the distance from the axis of rotation, S is the sedimentation coefficient, ω is the number of rotations per minute (angular velocity), D is the diffusion coefficient, and t is time.

The first law is more important than the second law in treating sedimentation data. Using Eq. (12.1) two basic methods have been developed in ultracentrifuge sedimentation studies:

1. The sedimentation velocity method, in which diffusion is negligible. This is done by setting the rotor at a very high speed, for example, 60 000 rpm (rotations per minutes), which is equivalent to a pressure of 420 bars.
2. The sedimentation equilibrium method, in which diffusion plays an equally important role. This is done at a relatively low speed, for example, 12 000 rpm.

In between, there are two techniques that are extensively used to study biological polymers:

1. The approach to equilibrium method, in which sedimentation plays an important role, but diffusion is not completely neglected. This method sets up a speed that is neither too fast nor too slow. During the run, macromolecules do not diffuse into the region of the meniscus, nor do they diffuse onto the bottom. Even at a relatively early stage of sedimentation, equilibrium reaches the regions of the meniscus and bottom, whereas the concentration in the region between the meniscus and bottom still changes with time. Thus, analysis need be focused only on the meniscus and bottom regions.
2. Density gradient at equilibrium, in which a density gradient is created by adding materials such as CsCl and CsBr to the system. The macromolecules float between the meniscus and the bottom and remain in a definite position when the force exerted on the macromolecules from the meniscus is balanced by that from the bottom.

12.1 APPARATUS

The ultracentrifuge apparatus comprises three major components: cell coordinates (Figure 12.1), rotor (Figure 12.2), and optical systems. There are three

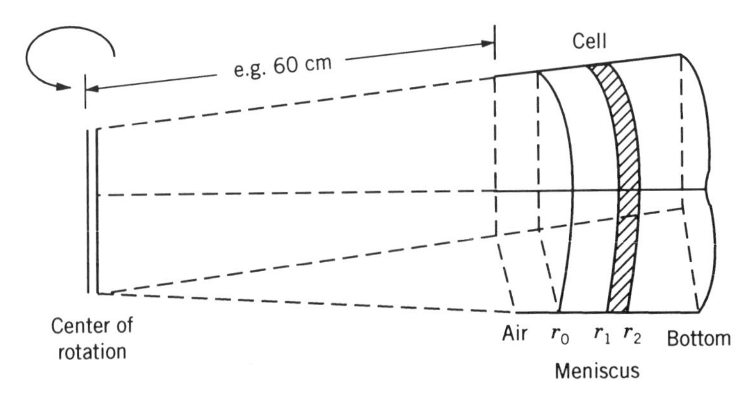

FIGURE 12.1 Coordinates of the ultracentrifuge.

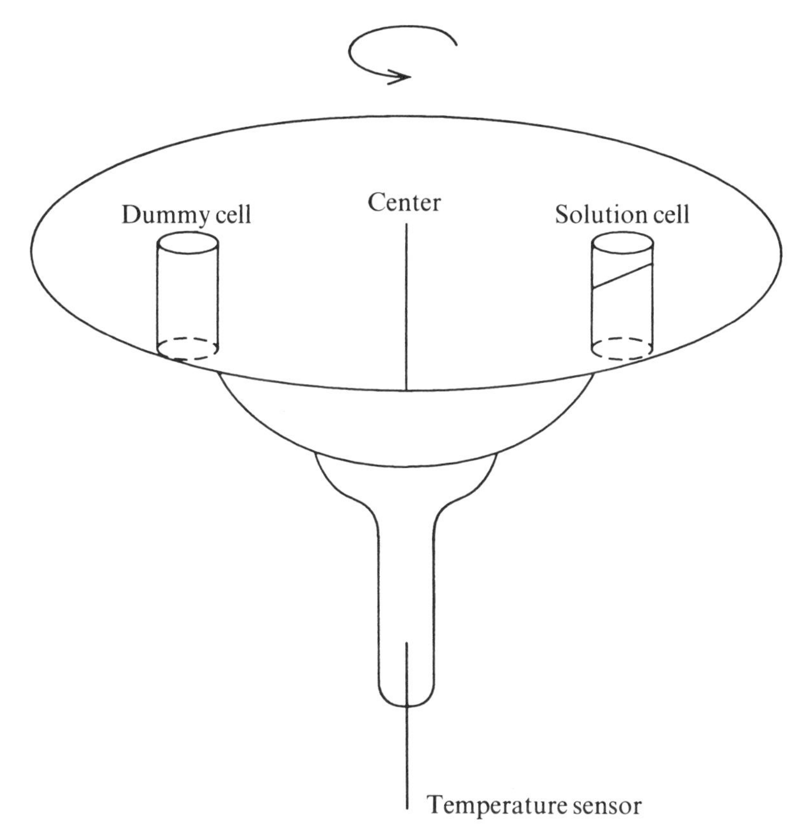

FIGURE 12.2 A rotor of the ultracentrifuge.

different optical systems: Schlieren, Rayleigh interferometric, and absorption. The schlieren system is a basic unit of the ultracentrifuge instrument. It is commonly used in both sedimentation velocity and sedimentation equilibrium experiments. The system's only drawback is that the concentration of sample required is 5–50 mg/mL, which is too high and too costly for biological polymers (for example, enzymes). The Rayleigh interferometric system measures the difference in refractive index Δn between the solution and a reference column of the solvent. This difference is measured through the displacement of interference fringes formed by slits placed behind the two columns. As mentioned before (Chapter 10), a displacement of one fringe ($j = 1$) corresponds to a concentration difference of about 0.25 mg/mL for most proteins and it is not difficult to measure displacement to about 0.02 fringes. The absorption system needs a scanner and a monochromator light source as accessories. The advantage of the absorption system is its need for even smaller amounts of sample for measurement, that is, 20 μg (20×10^{-6} g).

12.2 SEDIMENTATION VELOCITY

Since it is usually included with the Spinco E analytical ultracentrifuge (a commercially available instrument), we will first analyze the graph obtained with the schlieren system.

12.2.1 Measurement of Sedimentation Coefficients: Moving Boundary Methods

Assuming that the boundary is sharp and no diffusion occurs (which is usually the case for a homogeneous system), we have

$$J = cS\omega^2 r$$

But at r,

$$J = c\frac{\mathrm{d}r}{\mathrm{d}t}$$

We thus have

$$c\frac{\mathrm{d}r}{\mathrm{d}t} = cS\omega^2 r$$

and

$$S = \frac{1}{\omega^2 r}\frac{\mathrm{d}r}{\mathrm{d}t} = \frac{1}{\omega^2}\frac{\mathrm{d}\ln r}{\mathrm{d}t} \tag{12.3}$$

This is the equation used in the experimental determination of the sedimentation

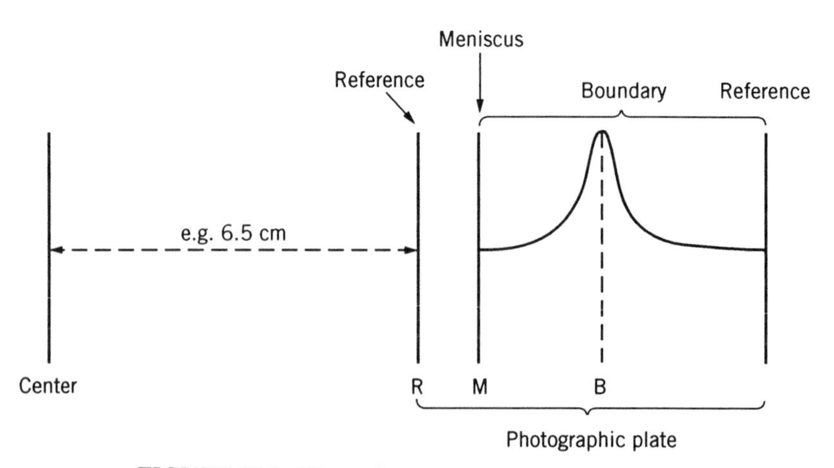

FIGURE 12.3 The sedimentation velocity experiment.

coefficient S. Figure 12.3 describes the experiment. Let R be the reference point, M the meniscus, and B the boundary. All of these points are measured from the center of rotation. Then we have

$$\Delta R = B - R$$

where ΔR is the distance of the boundary from the reference point that was read from the plate. The value of ΔR may be converted to that of Δr:

$$\Delta r = \Delta R \frac{1}{F} = (\Delta R) \quad (0.462)\,\text{cm}$$

where Δr is the corrected distance of the boundary from the reference point and F is the magnification of the camera, which depends on the instrument. For example, $1/F = 0.462$ for a given instrument. The value of Δr is further converted to r_b,

$$r_b = \Delta r + 6.5$$

where r_b is the distance of the boundary from the center of rotation. The value 6.5 cm is measured for a given rotor. It varies from rotor to rotor but usually is in the neighborhood of 6.5 cm.

Now we divide the values of r_b for the second, third, fourth, and fifth pictures (usually 5 pictures in a photographic plate taken at certain time intervals, for example, 16 min) by the value of r_b for the first picture to obtain the values

$$\frac{r_b(2)}{r_b(1)} \qquad \frac{r_b(3)}{r_b(1)} \qquad \frac{r_b(4)}{r_b(1)} \qquad \frac{r_b(5)}{r_b(1)}$$

If we set the time of the first picture at t_1, then we have

$$\Delta t = t - t_1$$

where Δt is the time interval between the two pictures (for example, 16, 32, 48, and 64 min). The rotor speed (rpm) is changed into the angular velocity ω:

$$\omega = \frac{\text{rpm}}{60} 2\pi$$

With all the pertinent values available we can now calculate the sedimentation velocity coefficient directly by using Eq. (12.3):

$$S = \ln \frac{r_b/r_b(1)}{\omega^2 \Delta t}$$

The unit of S is sec or svedbergs (in honor of Theodor Svedberg, who was a pioneer in developing the untracentrifuge). One svedberg $= 1 \times 10^{-13}$ sec.

To facilitate comparison with other biological polymers, the experimental value of the sedimentation coefficient is usually corrected to a standard basis corresponding to a reference solvent having the viscosity and density of water at 20°C. The equation for correction is

$$S_{20,\mathrm{w}} = S_{t,\mathrm{b}} \frac{(1 - \bar{v}\rho)_{20,\mathrm{w}}}{(1 - \bar{v}\rho)_{t,\mathrm{b}}} \frac{\eta_{t,\mathrm{b}}}{\eta_{20,\mathrm{w}}} \cong S_{t,\mathrm{b}} \frac{\eta_{\mathrm{t}}}{\eta_{20}}$$

where $\bar{v}$ is the partial specific volume, ρ is the solution density, and η is the viscosity. The subscripts w and b refer to water and the buffer used (if any). Usually, the major effect is the change of water viscosity with temperature; data for making this correction are readily available in the literature.

Two important factors have to be considered to report accurate values of sedimentation coefficient. First is the possible dependence of the sedimentation coefficient on concentration. Although the sedimentation coefficient is supposed to be a constant under specified conditions, it may vary with concentration. A correction formula is given as follows:

$$S = \frac{S_0}{1 + kc}$$

where S_0 is the value of the sedimentation coefficient at infinite dilution, k is a constant that expresses the dependence of the sedimentation coefficient on concentration, and S is the sedimentation coefficient at the concentration c. A plot of $1/S$ versus c is expected to give a straight line, thereby enabling an extrapolation to infinite dilution. If k is small, that is, if the dependence of S on c is slight, then the equation is in the form

$$S = S_0(1 - kc)$$

Note:

$$\frac{1}{1 + kc} = 1 - kc + \cdots$$

The second factor is the Johnston-Ogston effect, which is related to the multi-component system. If a solution contains the fast and slow components and if the mixture is diluted, the relative peak area of the fast components increases. Thus, in the region of the fast component boundary there is a decrease in the concentration of the slow component, which causes a decrease in the height of schlieren peak. The slow component often appears to be greater than its true concentration, whereas the faster component appears to be less than its true concentration. Consequently, there may be an error.

12.2.2 Svedberg Equation

If the rotor turns with an angular velocity ω (radius per second), n molecules suspended in a liquid will experience a centrifugal force F_c which is balanced by a frictional force F_f. The centrifugal force is expressed by

$$F_c = nv(\rho_m - \rho)\omega^2 r$$

where v is the volume of a single molecule, ρ_m is the density of the molecule, ρ is the density of the solution, ω is the angular velocity, and r is the distance from the center of rotation. If n is taken as Avogadro's number, then

$$F_c = M(1 - \bar{v}\rho)\omega^2 r$$

where $\bar{v}$ is now the partial specific volume. The parameter $1 - \bar{v}\rho$ is a correction of the buoyancy factor. The frictional force may be expressed by

$$f_f = f\frac{dr}{dt}$$

If we equate the two forces

$$M(1 - \bar{v}\rho)\omega^2 r = f\frac{dr}{dt}$$

and rearrange the terms, we obtain

$$\frac{M(1 - \bar{v}\rho)}{f} = \frac{dr/dt}{\omega^2 r} = S$$

This is the Svedberg equation. It is also the definition of the sedimentation coefficient S.

12.2.3 Application of the Sedimentation Coefficient

Molecular Weight Determination Combining Svedberg's equation

$$S = \frac{M(1 - \bar{v}\rho)}{f}$$

with Einstein's equation

$$D = \frac{R'T}{f}$$

we have

$$M = \frac{R'TS}{D(1 - \bar{v}\rho)}$$

This equation was mentioned in Chapter 10. Here, D and S are measurable quantities and $\bar{v}$ and ρ are usually available in the literature. Thus, from the two separate experiments (sedimentation and diffusion), we can determine the molecular weight of a polymer (particularly biological polymers).

Characterization of Molecular Species The sedimentation coefficient can be used to evaluate the purity of a material. A sharp boundary is usually (though not always) an indication of the homogeneity of the particles. A broad boundary (asymmetrical) or more than one boundary grouped together is a clear indication that there is a heterogeneity of particles. The number of fast-moving molecules (for example, dimers) can be estimated. Caution must be exercised, however, since the most serious consequence of the dependence of the sedimentation coefficient on concentration is a possible distortion of the boundary. Thus, before we draw any conclusions, the degree of dependence of S upon c must be assessed from a series of experiments performed at various concentrations.

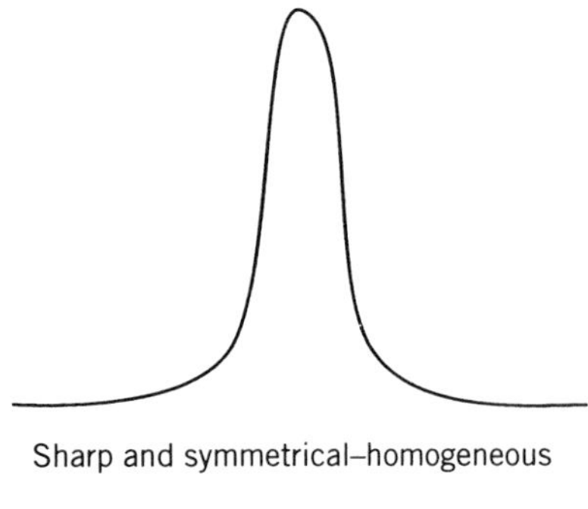

Sharp and symmetrical–homogeneous

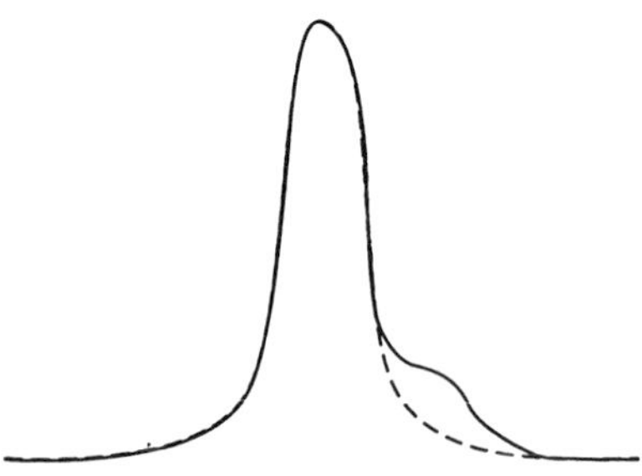

Broad and asymmetrical–heretogeneous

In biochemistry different species of the same kind of molecules are often labeled in terms of sedimentation coefficients. As an example, *Escherichia coli* ribosome has been separated into four different species, 30S, 50S, 70S, and 100S. The larger the number of S, the larger the size of the molecule.

12.3 SEDIMENTATION EQUILIBRIUM

Recall Eq. (12.1):

$$J = cS\omega^2 r - D\frac{\mathrm{d}c}{\mathrm{d}r}$$

When equilibrium is reached, the force of diffusion equals the force of sedimentation, that is, $J = 0$. We then have

$$0 = cS\omega^2 r - D\frac{\mathrm{d}c}{\mathrm{d}r}$$

Substituting Svedberg's equation for S,

$$S = \frac{M(1 - \bar{v}\rho)}{f}$$

and Einstein's equation for D,

$$D = \frac{R'T}{f}$$

we obtain

$$0 = \frac{M(1 - \bar{v}\rho)\omega^2 rc}{f} - \frac{R'T}{f}\frac{\mathrm{d}c}{\mathrm{d}r}$$

If we divide the equation by the factor c, we then have

$$M = \frac{2R'T}{(1 - \bar{v}\rho)\omega^2}\frac{\mathrm{d}\ln c}{\mathrm{d}r^2} \tag{12.4}$$

Equation (12.4) was first derived by Goldberg in 1953 on the assumption that the total potential of any component in the centrifugal solution is constant at equilibrium.

Three experimental methods are used to evaluate the term $\mathrm{d}\ln c/\mathrm{d}r^2$ in Eq. (12.4): the Archibald method, the van Holde and Baldwin method, and the Yphantis method. All are designed to determine the molecular weight of polymers (particularly biological polymers).

12.3.1 The Archibald Method

At the beginning of the run, the material enters the column of solution at the meniscus and the bottom. It soon becomes saturated, so that the material can no longer transport to the meniscus and bottom. A simple criterion is the plot of the experimental data in the form of $(1/rc)(dc/dr)$ versus r. If the plot produces a horizontal line, then equilibrium is attained.

Archibald suggested that we need focus only on the top and the bottom of the cell during the process of equilibrium sedimentation to determine molecular weight; we do not have to wait for equilibrium to be reached. Hence, the method is also called approach to equilibrium. Equation (12.4) can be rearranged for meniscus (m) and bottom (b) as follows:

$$M_{\mathrm{m}} = \frac{R'T}{\omega^2(1-\bar{v}\rho)} \frac{(\partial c/\partial r)_{\mathrm{m}}}{c_{\mathrm{m}} r_{\mathrm{m}}}$$

$$M_{\mathrm{b}} = \frac{R'T}{\omega^2(1-\bar{v}\rho)} \frac{(\partial c/\partial r)_{\mathrm{b}}}{c_{\mathrm{b}} r_{\mathrm{b}}}$$

where

$$c_{\mathrm{m}} = c_0 - \frac{1}{r_{\mathrm{m}}^2} \int_{r_{\mathrm{m}}}^{r_{\mathrm{p}}} r^2 \frac{\partial c}{\partial r} \, \mathrm{d}r$$

$$c_{\mathrm{b}} = c_0 + \frac{1}{r_{\mathrm{b}}^2} \int_{r_{\mathrm{p}}}^{r_{\mathrm{b}}} r^2 \frac{\partial c}{\partial r} \, \mathrm{d}r$$

The value of c_0 is obtained from a synthetic boundary cell run, which we describe later. The term r_{p} refers to the position at plateau region.

Let us consider the equilibrium at meniscus. To determine the value of M, we need the values of four paramenters (in addition to other constants): $(\partial c/\partial r)_{\mathrm{m}}$, r_{m}, c_{m}, and c_0. Using the schlieren system, a typical graph in the photographic plate is obtained in the following form:

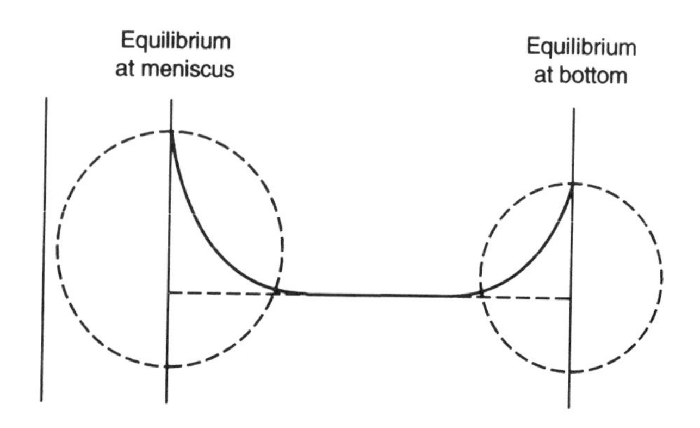

The four parameters may then be evaluated by using data read from the plate:

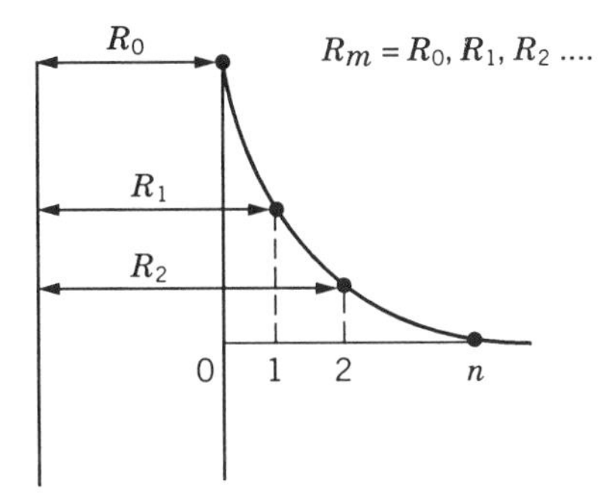

The division from 0 to n is arbitrary. The length of the vertical line gives dc/dr, whereas the length of the division is R:

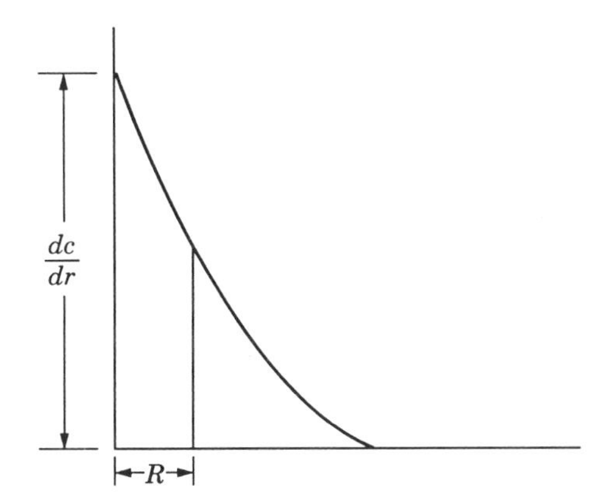

The values of R should be converted to r as follows:

$$r_m = \frac{R_m}{F} + k \qquad m = 0, 1, 2, \ldots$$

where F is the magnification factor (relating distances in the cell to distances on the enlargement), and the value k is the distance of the reference point in the plate from the center of rotation. The value of k is measured on the rotor and depends on the rotor used, for example, 6.5 cm.

The value of c_0 is usually obtained from an independent synthetic boundary experiment that is relatively simple. A layer of pure solvent (for example, 6.45 mL) is placed over a solution (for example, 0.15 mL) of known concentration in a synthetic boundary cell, and a sedimentation velocity (8000 rpm) is run to obtain a graph:

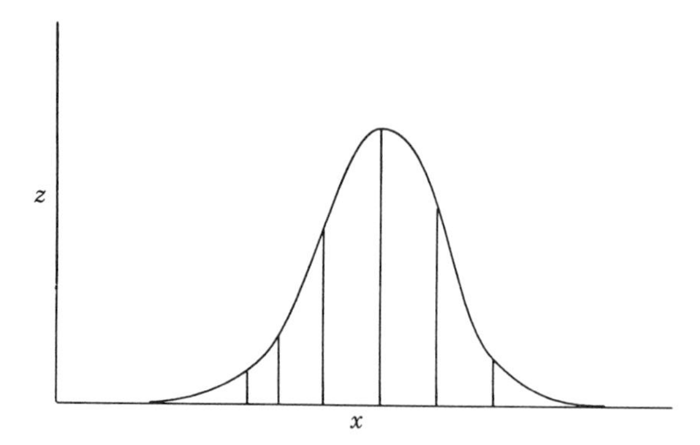

The abscissa x is divided into intervals of 0.1. From the summation

$$\sum x_n z_n = (0.1) \sum z_n$$

we obtain

$$c_0 = \frac{0.1 \sum x_n}{F} = \text{Area}$$

Note that the concentration unit is not g/100 mL; instead, it is in arbitrary units. The value of c_{m} can then be evaluated from the integral:

$$c_{\mathrm{m}} = c_0 - \frac{1}{r_{\mathrm{m}}^2} \sum_{n=0}^{q} r_n^2 \left(\frac{\mathrm{d}c}{\mathrm{d}r} \right)_n \frac{\Delta r}{F}$$

where Δr may be chosen to be 0.1 cm, and q is the number of divisions in the horizontal coordinate. Since we have the values of c_0, c_{m}, r_{m}, $(\mathrm{d}c/\mathrm{d}r)_{\mathrm{m}}$, ω, T, $\bar{v}$, and ρ, we can calculate the molecular weight.

Sedimentation equilibrium experiments do not have to be run with the schlieren system. They can also be run with the interference or absorption system. The only difference is in the expression of concentration in the photographic plate. The concentrations are expressed experimentally in terms of arbitrary units: square centimeters for schlieren optics, number of fringes for interference optics, and optical density for absorption optics.

12.3.2 The van Holde and Baldwin (Low-Speed) Method (rotor velocity 10 000–14 000 rpm)

Equation (12.4) may be rewritten in the form

$$\frac{1}{c} \frac{\mathrm{d}c}{\mathrm{d}r} = \frac{M(1 - \bar{v}\rho)\omega^2 r}{R'T}$$

Multiplying both sides by the factor $c\,dr$ and integrating from the limit m (the meniscus) to the limit b (the bottom), we have

$$\int_{c_m}^{c_b} dc = \frac{M(1-\bar{v}\rho)\omega^2}{R'T} \int_{r_m}^{r_b} rc\,dr$$

The two definite integrals may be evaluated as follows:

$$\int_{c_m}^{c_b} dc = c_b - c_m$$

$$\int_{r_m}^{r_b} rc\,dr = c_0 \int_{r_m}^{r_b} r\,dr = c_0 \frac{r_b^2 - r_m^2}{2}$$

The equation of sedimentation equilibrium then becomes

$$\frac{c_b - c_m}{c_0} = \frac{M(1-\bar{v}\rho)}{2R'T}(r_b^2 - r_m^2)$$

Both terms $(c_b - c_m)/c_0$ and $r_b^2 - r_m^2$ can be experimentally determined, using either schlieren optics or the Rayleigh interference system. Here, we illustrate with the Rayleigh interference system.

First, we run a synthetic boundary experiment to determine c_0. The following drawing is an example:

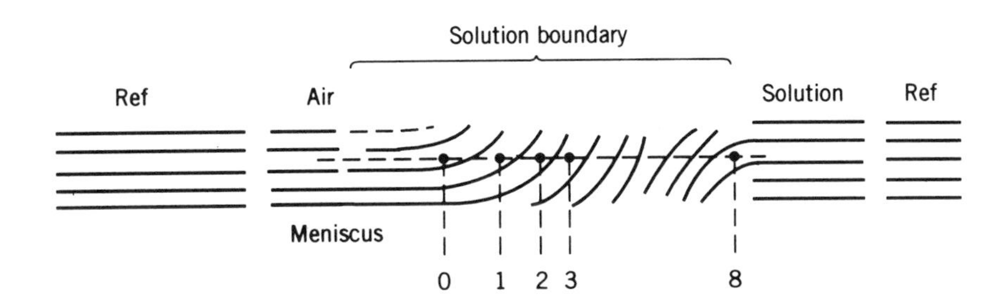

We count the number of fringes. Let Δj_{sb} be the total number of fringes crossed from the meniscus to the bottom, where the subscript sb denotes synthetic boundary. We then have

$$c_0 = k\Delta j_{sb}$$

where k is the proportionality constant which need not be evaluated, as we will see later.

Second, we run an equilibrium experiment for the sample. As an example of the moving boundary, consider the following drawing:

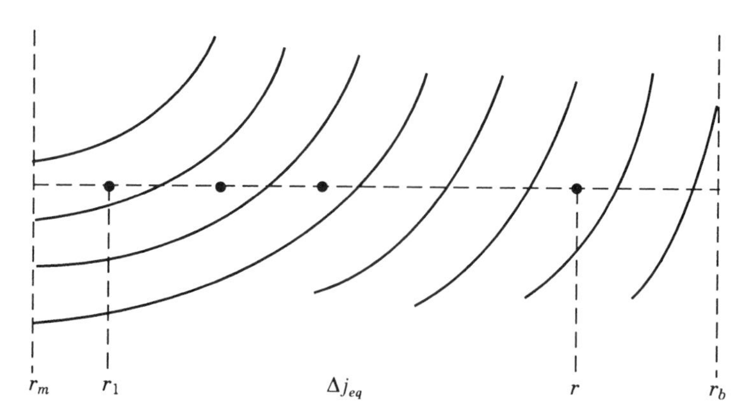

Δj_{eq} is the total number of fringes, Δj_{eq} where the subscript eq refers to the equilibrium. Then,

$$c_{\text{b}} - c_{\text{m}} = k \Delta j_{\text{eq}}$$

where k is the same constant that appeared in the evaluation of c_0.

The values of r_{m} and r_{b} are measured by reading on the plate, while the value of the distance of the reference from the center is available as was described previously:

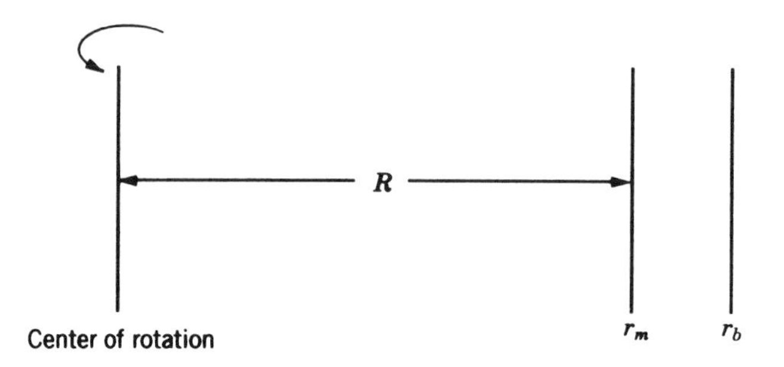

We thus can determine the molecular weight of the macromolecule:

$$M = \frac{2R'T}{\omega^2(1 - \bar{v}\rho)(r_{\text{b}}^2 - r_{\text{m}}^2)} \frac{\Delta j_{\text{eq}}}{\Delta j_{\text{sb}}}$$

The above equation is based on the assumption that the solution under study is an ideal solution, that is, there is no interaction between solute molecules. If it is not an ideal solution, the term of a second virial coefficient, A_2, should be included in the equations:

$$\frac{M(1 - \bar{v}\rho)\omega^2}{(1 + 2A_2 Mc)R'T} = \frac{(c_{\text{b}} - c_{\text{m}})/2}{c_0(r_{\text{b}}^2 - r_{\text{m}}^2)}$$

or

$$M = \frac{2R'T(1 + 2A_2Mc)}{\omega^2(1 - \bar{v}\rho)(r_b^2 - r_m^2)} \frac{\Delta j_{eq}}{\Delta j_{sb}}$$

12.3.3 Yphantis (High-Speed) Method (rotor velocity 28 000–32 000 rpm)

Equation (12.4) may be remodified as

$$M = \frac{2R'T}{(1 - \bar{v}\rho)\omega^2} \frac{d \ln c}{dr^2}$$

$$= \frac{2R'T}{(1 - \bar{v}\rho)\omega^2} \frac{d \ln(y - y_0)}{dr^2}$$

where y refers to the fringe displacement. The graph of a high-speed run is sketched as follows:

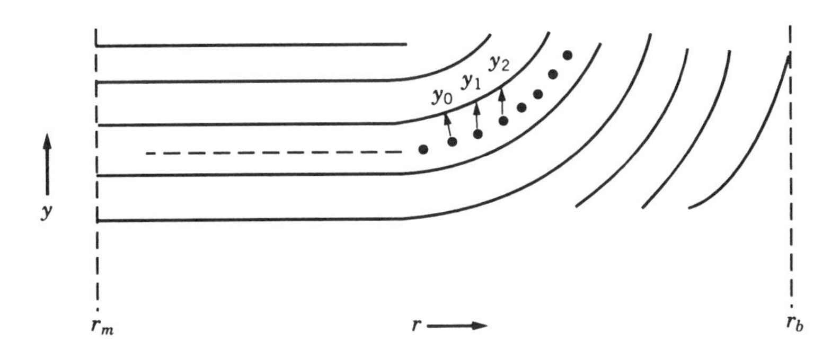

The crucial part in analyzing the data is the reading of r versus y of the curve in the photographic plate. The enlarged curve is in the following form:

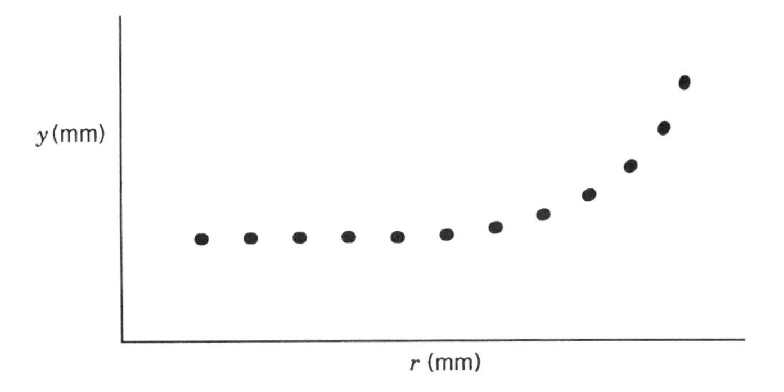

With the data of y, y_0, and r now available, we plot the final results:

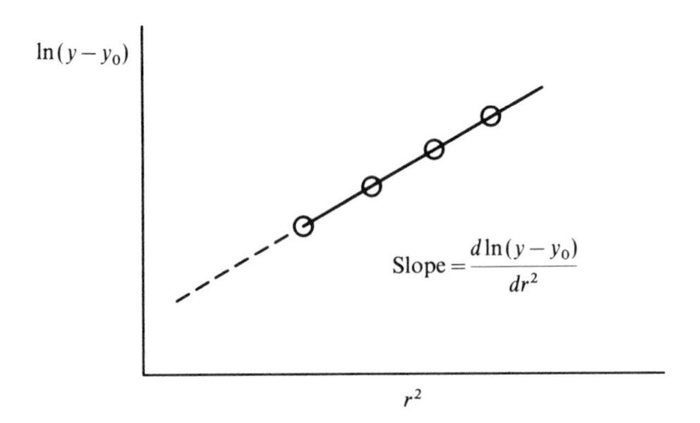

The experimental points are expected to be on a straight line and the slope is equal to $[d\ln(y-y_0)]/dr^2$. From the slope of the plot we can thus calculate molecular weight.

The following table (van Holde, 1967), compares the low-speed method to the high-speed method:

Method	Approximate Concentration of Sample (mg/mL)	Approximate Solution Volume (mL)	Total Weight of Solute (mg)
Low speed	2	0.10 (for run) 0.15 (for synthetic boundary)	0.5
High speed	0.5	0.1	0.05

The high-speed method seems more economical and convenient than the low-speed method. The low-speed method requires an additional experiment, the synthetic boundary, whereas the high-speed method does not.

12.3.4 Absorption System

Equation (12.4) is rearranged in the form

$$\ln\frac{c_r}{c_m} = \frac{M(1-\bar{v}\rho)\omega^2}{R'T}\frac{r^2-r_m^2}{2}$$

In the absorption system, we measure c_r and c_m directly. The measurement is of the absorption of light and on the development of the scanner, which allows direct photoelectric recording of the optical density versus the r curve, as in the following sketch:

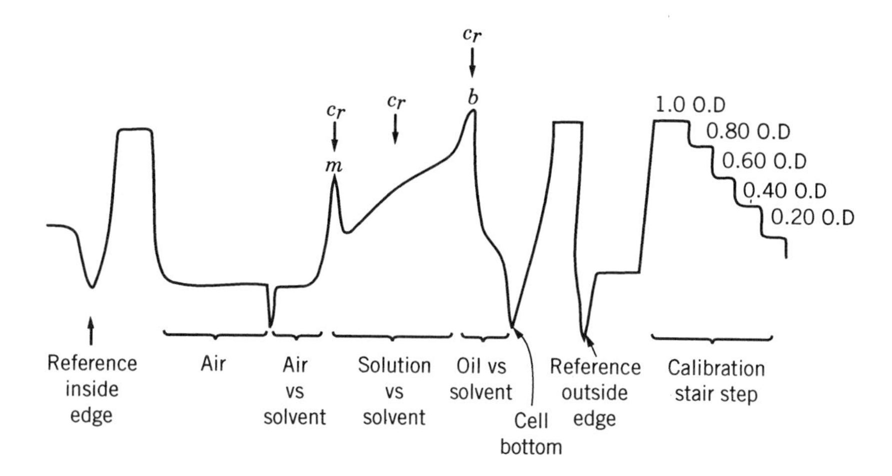

The coordinates are the absorption o.d. and the distance r. Since the o.d. is directly proportional to the concentration (see Chapter 16), we can measure all four parameters used in the equation to calculate the molecular weight. They are c_r, c_m, r^2 (the distance of the position from the center of rotation), and r_m^2 (the distance of the meniscus from the center of rotation).

12.4 DENSITY GRADIENT SEDIMENTATION EQUILIBRIUM

When materials with very high densities, such as CsCl and CsBr, are introduced into the macromolecular solution, a new density gradient is created. The centrifugal force sediments the macromolecules down to the bottom; but the heavy density from the sedimentation of CsCl or CsBr in the bottom forces the macromolecule to float in the region between the bottom and the meniscus. At equilibrium, the macromolecule rests at a definite position in the density gradient. The position that represents this density is called the buoyancy density, ρ_0, of the macromolecule. The main purpose of the experiment for density gradient sedimentation equilibrium is to obtain the value of ρ_0. Important information in characterizing a macromolecule may be obtained by analyzing ρ_0 and the gradient density $d\rho/dr$, where r is the distance of the point in a gradient from the center of rotation. Most density gradient experiments so far have been carried out wth nucleic acids and proteins. Our discussion is therefore focused on biological polymers.

The buoyancy density of a protein, ρ_0, is related to the net hydration of the protein salt complex, and the net hydration of the salt-free protein. Let the number 1 represent water, 2 the protein, and 3 the salt. Then

$$\rho_0 = \frac{1 + \Gamma'_1}{\bar{v}_3 + \Gamma'_1 \bar{v}_1}$$

where Γ'_1 is the net hydration and $\bar{v}$ the partial specific volume. The concentration gradient $d\rho/dr$ is related to rotor speed and the physical properties of CsCl. If we use the activity of the solute a (here, protein) instead of the concentration of the solute c, Eq. (12.4) could be put in the form

$$\frac{da}{dr} = \frac{M(1-\bar{v}\rho)\omega^2 r}{R'T} a$$

We now relate the gradient density to the above equation by a simple manipulation:

$$\frac{d\rho}{dr} = \frac{d\rho}{da} \frac{da}{dr} = \frac{d\rho}{da} \frac{M(1-\bar{v}\rho)\omega^2 r}{R'T} a$$

$$= \frac{d\rho}{d\ln a} \frac{M(1-\bar{v}\rho)\omega^2 r}{R'T}$$

We define a parameter β by

$$\beta = \frac{R'T}{(1-\bar{v}\rho)M} \frac{d\ln a}{d\rho}$$

Then the density gradient is related to β in the form

$$\frac{d\rho}{dr} = \frac{\omega^2 r}{\beta}$$

From the photograph of the experiment for gradient density in equilibrium we can locate not only ρ_0, but also the values for r_m, r_b, and r_0, where r_0 is the radial position of the center of band or bands. Let us define a new term r_e, the isoconcentration point in the cell:

$$r_e = \sqrt{\frac{r_b^2 + r_m^2}{2}}$$

At position r_e, the density of the solution is the same as that of the initial solution, ρ_e. The isoconcentration is the same as the isodensity.

If the photographic image of the macromolecular band is a Gaussian distribution curve, then we have

$$c = c_0 e^{-(r-r_s)^2/2\sigma^2}$$

and

$$\sigma^2 = \frac{R'T}{M\bar{v}(\mathrm{d}\rho/\mathrm{d}r)_{r_s}\omega^2 r_s}$$

Thus, the graphic determination of the standard deviation σ enables us to calculate the molecular weight of the sample.

The density gradient in equilibrium is an elegant method for determining molecular weight and for binding salt and water to the proteins. However, the experimental run is complicated and the analysis of data is tedious. For these reasons, this method is never extensively used in macromolecular chemistry, except for proteins and some nucleic acids.

12.5 SCALING THEORY

Except for water-soluble polymers, most synthetic polymers dissolve only in organic solvents, which are usually volatile and thus more difficult to deal with in the ultracentrifuge. Furthermore, unlike proteins whose polydispersity is near unity, polymer (synthetic) solutions are often heterogeneous. For this reason, sedimentation experiments are used less frequently in (synthetic) polymer chemistry. Investigators usually avoid using the ultracentrifuge for the determination of molecular weight.

However, while the ultracentrifuge is no longer heavily used in biochemistry (because of the development of SDS—polyacrylamide gel electrophoresis for the estimation of molecular weight, see Chapter 17), it has become an important tool for the study of dimensions of synthetic polymers in solution.

In the semidilute range, we may imagine the solution as a continuum formed by entangled macromolecules, which can be divided into spheres (or blobs) of radius equal to the screen length ζ. If we write the Svedberg equation in its approximate form,

$$S \sim \frac{M}{f}$$

and make use of two other approximations

$$M \sim c\zeta^3$$

$$f \sim \zeta$$

we can then obtain the approximate relation

$$S \sim c\zeta^2$$

In good solvents, it can be shown that

$$\zeta^2 \sim c^{-3/2}$$

Hence

$$S \sim c^{-1/2} \tag{12.5}$$

Equation (12.5) predicts that in the semidilute range of concentration ($c \sim c^*$), the sedimentation coefficient should be independent of the molecular weight of the polymer and the plot of $\log S$ versus $\log c$ should give the slope -0.50.

This theory has been partially confirmed by sedimentation experiment (Langevin and Rondelez, 1978). The value of the slope so far found was -0.50 ± 0.10. We now have some evidence to believe that in the semidilute range of polymer solution, the solvent is forced through in orderly fashion around the blob of radius ζ, but still cannot penetrate the interior of the blob. Note that this theory is reminiscent of the pearl necklace model and the hydrodynamic equivalent sphere.

Sedimentation experiments may also be used to test the scaling theory in another way. A small number of inert spheres with diameters of $2R \sim 100$–200 Å (for example, bovine serum albumin) is added to an aqueous solution of synthetic polymer (such as polyethylene oxide). If $2R > \zeta$, the spheres should move easily and f (the frictional coefficient) is related to η_0 (the viscosity of solvent). If $2R < \zeta$, the spheres will be trapped and f is related to η (the viscosity of entangled solution). To measure η (the absolute viscosity) as a function of concentration is relatively difficult. It is much more convenient to test through the sedimentation experiment, since the sedimentation coefficient, as shown in Svedberg's equation, is also closely related to f. The equation to be used is

$$\frac{f}{f_0} = \frac{S_0 - S_p}{S - S_p}$$

where S_0 is the sedimentation coefficient of the inert sphere alone, S_p that of the polymer alone, and S that of the inert sphere in the polymer solution. According to a scaling law

$$\frac{f}{f_0} = \psi\left(\frac{R}{\xi}\right)$$

we have

$$\psi\left(\frac{R}{\xi}\right) \sim 1 \qquad \text{for } R < \xi$$

$$\psi\left(\frac{R}{\xi}\right) \sim \frac{f_0}{f} \qquad \text{for } R \gg \xi$$

Thus, we can have

$$\frac{S}{S_0} \sim \exp(-c^{0.50})$$

By plotting $\ln(S/S_0)$ versus c, we can test whether the exponent is close to 0.50.

REFERENCES

Archibald, W. J., *J. Phys. Colloid Chem.* **51**, 1204 (1947).

Baldwin, R. L., *Biochem. J.* **65**, 503 (1957).

Chervenka, C. H., *A Manual of Methods for the Analytical Ultracentrifuge*. Palo Alto, CA. Spinco Division of Beckman Instruments, Inc., 1969.

Goldberg, R. J., *J. Phys. Chem.* **57**, 194 (1953).

Johnston, J. P., and A. G. Ogston, *Trans. Faraday Soc.* **42**, 789 (1946).

Langevin, D., and F. Rondelez, *Polymer* **19**, 875 (1978).

Meselson, M., F. W. Stahl, and J. Vinograd, *Proc. Natl. Acad. Sci. USA* **43**, 581 (1957).

Schachman, H. K., in *Methods in Enzymology*, Vol. 4, S. P. Colowick and N. O. Kaplan (Eds.). New York: Academic, 1957.

Studier, F. W., *J. Mol. Biol.* **11**, 373 (1965).

Svedberg, T., and K. O. Pedersen, *The Ultracentrifuge*. Oxford, UK: Clarendon Press, 1940.

van Holde, K. E., and R. L. Baldwin, *J. Phys. Chem.* **62**, 734 (1958).

van Holde, K. E., *Fractions* (Spinco Division), **1**, 1 (1967).

Vinograd, J., *Methods Enzymol.* **6**, 854 (1963).

Vinograd, J., and J. E. Hearst, *Fortschr. Chem. Org. Naturstoffe* **20**, 372 (1962).

Williams, J. W., K. E., van Holde, R. L. Baldwin, and H. Fujita, *Chem. Rev.* **58**, 715 (1958).

Yphantis, D. A., *Biochemistry*, **3**, 297 (1964).

PROBLEMS

12.1 The speed of the ultracentrifuge (rpm) is often expressed in terms of g, the gravitational constants; for example, 45 000 rpm = 100 000 g. Express the following in terms of g: 12 000, 18 000, 24 000, 57 890, 60 000 rpm.

12.2 A protein was dissolved in buffer at pH 3.2 and was studied in the ultracentrifuge sedimentation velocity experiment. After the boundary was developed, five pictures were taken at intervals of 16 min. A typical picture with labels in a photographic plate is shown in the following sketch:

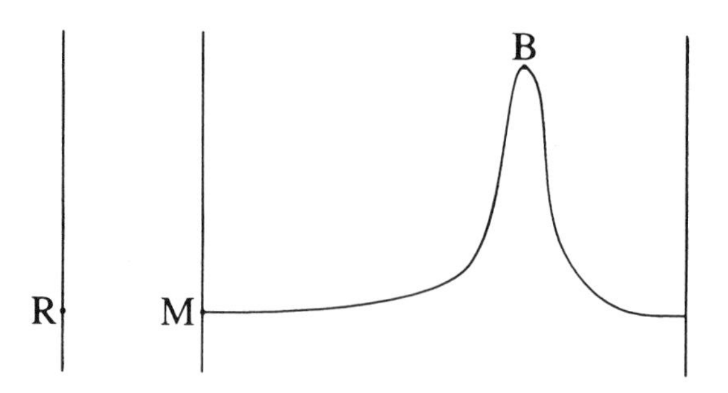

where R is the reference point, M is the meniscus, and B is the boundary. The plate was measured with a comparator and data are given as follows ($\Delta t = 16$ min, $1/F = 0.428$, $R = 5.765$ cm, $T = 298°$K, rpm = 59 780):

R (in)	M (in)	B (in)
9.000	9.037	11.076
48.958	55.321	59.203
9.859	16.213	21.96
57.099	63.466	71.12
46.88	40.49	30.91

Calculate the sedimentation coefficient of the protein.

12.3 In a study of the sedimentation of bovine serum albumin, S was determined at a number of concentrations. When corrected to 25.0°C, the data are as follows:

c_0	$S_{25}(10^{-13}$ sec)
0.29	4.927
0.59	4.839
0.87	4.772
1.17	4.663
1.17	4.662
1.76	4.475

(a) From these data determine the value of S at $c_0 = 0$.

(b) Using this value and the current best value for D_0 at 25°C (6.97×10^{-7} cm^2/s) and $\bar{v}$(0.734 mL/g), calculate the molecular weight.

12.4 Two ultracentrifuge sedimentation equilibrium experiments were carried out with ribonuclease in aqueous solution. The first was run in a conventional 12-mm cell, and the second was performed in a synthetic boundary cell. The schematic diagrams of the ultracentrifuge patterns are shown in the following two figures:

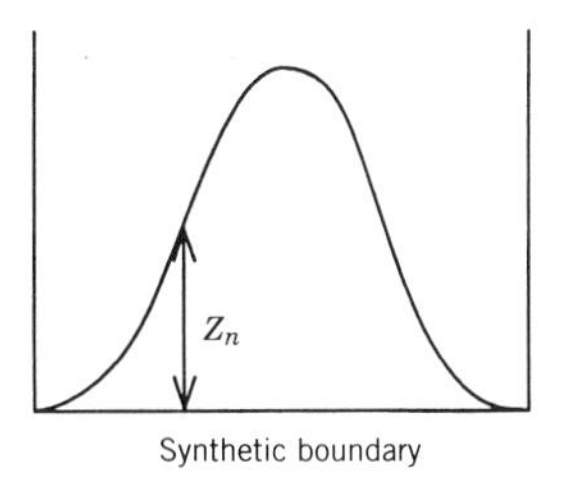

Synthetic boundary

Sedimentation equilibrium pattern

All the pertinent data are given in the following table for the sedimentation equilibrium of ribonuclease calculation at the meniscus (time = 38 min, $F = 12.19$, $\bar{v} = 0.709$ mL/g, $T = 298.9$°K, $\omega^2 = 1.3633 \times 10^6$):

m	R_n (cm)	z_n (cm)
0	2.2	1.41
1	2.3	1.30
2	2.4	1.20
3	2.5	1.08
4	2.6	0.99
5	2.7	0.88
6	2.8	0.77
7	2.9	0.66
8	3.0	0.57
9	3.1	0.50
10	3.2	0.42
11	3.3	0.35
12	3.4	0.25
13	3.5	0.19
14	3.6	0.13
15	3.7	0.10
16	3.8	0.08
17	3.9	0.04
18	4.0	0.02

The value of c_0 was found from the synthetic boundary measurement to be 1.2167. Calculate the molecular weight of ribonuclease.

12.5 Following is a diagram of the fringe displacements as a function of the radial distance from the center of rotation typically found on a photographic plate:

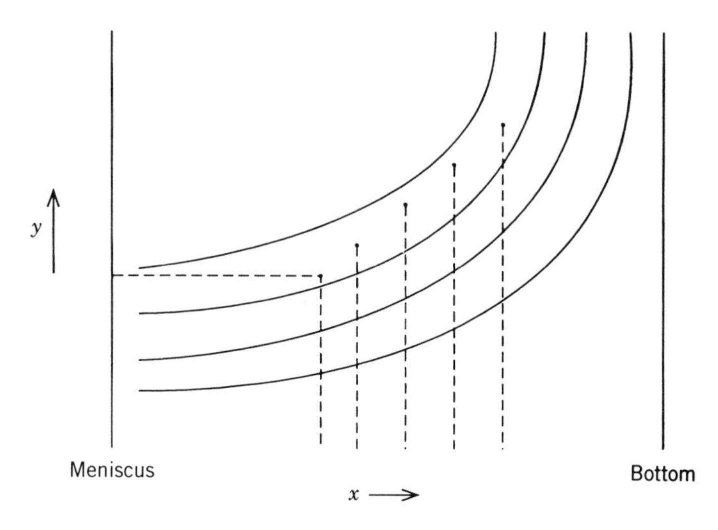

Experimental data actually recorded on the sedimentation equilibrium run for a protein, sodium dioxidase, in an acetate buffer solution at pH 5, 25°C, and rotor speed 2.42×10^3, rotation per minute, are given as follows:

x (mm)	y (mm)	x (mm)	y (mm)
28.00	18.68	40.80	19.64
29.00	18.68	41.00	19.71
30.00	18.68	41.20	19.79
31.00	18.71	41.40	19.89
32.00	18.71	41.60	19.93
33.00	18.71	41.80	20.08
34.00	18.75	42.00	20.21
35.00	18.78	42.20	20.33
36.00	18.85	42.40	20.47
37.00	18.93	42.60	20.65
38.00	19.04	42.75	20.81
39.00	19.20	42.90	20.95
40.00	19.41	43.00	21.10
40.30	19.50	43.10	21.25
40.50	19.55		

Plot $\log(y - y_0)$ versus r^2 and determine the molecular weight of the protein. The values of r may be calculated from the following practical equation:

$$r = 5.64 + \frac{x - 10.00}{(2.15)(10)}$$

where 5.64 cm is the distance of the center of rotation of a rotor, the value 10.00 in the numerator refers to an arbitrary setup in the comparator which was used for reading the plate, and that 10 in the denomination refers to the conversion factor 10 mm/cm.

12.6 To illustrate the scaling law, sedimentation coefficients were determined for poly(ethylene oxide) (mol wt 3×10^5) in aqueous solution as a function of the concentration. Data are as follows:

Sedimentation Coefficient S (10^{-13} s)	Concentration c (g/cm^3)
1.24	0.99
0.82	2.98
0.56	4.78
0.41	6.62

Show that $S \sim c^{-\alpha}$. Determine α. (*Source*: Langevin and Rondelez (1978).)

13

OPTICAL ROTATORY DISPERSION AND CIRCULAR DICHROISM

Optical rotatory dispersion (ORD) and circular dichroism (CD) are useful for the study of the molecular structure, particularly the secondary structure, (helices), of biological polymers. The instrument used is called a spectropolarimeter, and is easy to operate. Information is abundant in the literature for comparison and interpretation of new experimental results.

13.1 POLARIZED LIGHT

A light that oscillates in a single plane is called plane polarized light, as shown in Figure 13.1. The magnetic field M is perpendicular to the electric field E. The direction is from left to right. It is the electric field that is related to ORD and CD. Plane polarized light can be decomposed into circularly polarized components: right-handed E_R and left-handed E_L. Conversely, the two circularly polarized components may combine to form plane polarized light. If the two components, E_R and E_L, are not equal in amplitude, then we have elliptically polarized light (Figure 13.2).

13.2 OPTICAL ROTATORY DISPERSION

As plane polarized light passes through a substance, the velocities of the two circularly polarized components are reduced. If they are reduced to the same extent, the substance is optically inactive; if not, the substance is optically active. After passing through an optically inactive substance, recombination of the two

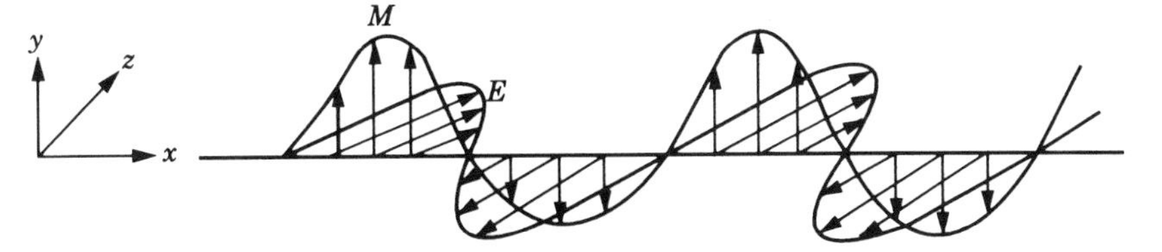

FIGURE 13.1 Plane-polarized propagation in the x direction.

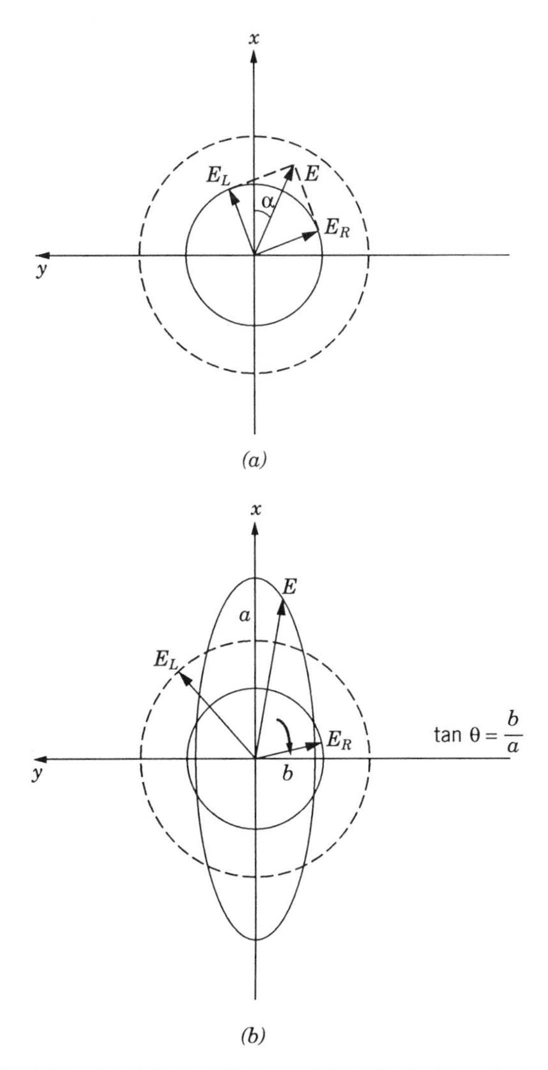

FIGURE 13.2 (*a*) Polarized light and (*b*) elliptically polarized light.

circularly polarized components does not create a phase difference; it emerges as an outgoing plane polarized wave. On the other hand, after passing through an optically active substance, there is a phase difference and there will be a rotation of the plane of polarization, designated α.

Experimentally, the velocity of light in a medium is characterized by the refractive index of the medium. The different velocities of the two circularly polarized components are thus expressed in terms of their different refractive indices. An optically active substance is one that has different refractive indices for its left and the right circularly polarized lights, n_L and n_R. The optical rotation α at a given wavelength of incident light λ is directly proportional to the difference between the refractive indices of the two circularly polarized components:

$$\alpha = k(n_L - n_R)$$

where k is the proportionality constant

$$k = \frac{180l}{\lambda}$$

and l is the path length in the medium.

Customarily, the optical rotation of an optically active substance is expressed in terms of the specific rotation $[\alpha]_\lambda$ and the molar rotation $[M]_\lambda$:

$$[\alpha]_\lambda = \frac{\alpha}{l'c}$$

$$[M]_\lambda = \alpha \frac{M}{100l'c}$$

where α (in degrees) is the observed rotation, λ (in cm) is the wavelength of incident light, l' (in dm) is the light path, c (in g/mL) is the concentration of the optically active substance, and M (in g/mole) is its molecular weight. Molar rotation is used to compare substances of different molecular weight.

In the study of macromolecules, such as proteins and nucleic acids, optical rotation is expressed in mean residual rotation $[m]_\lambda$:

$$[m]_\lambda = \alpha \frac{M_0}{100lc}$$

where M_0 is the mean residual molecular weight (for example, for most globular proteins, $M_0 = 117$ or 115), and c is the concentration of the sample in g/100 mL.

If the refractive index of solvent n is included for correction, the equation of $[m]_\lambda$ is

$$[m]_\lambda = \frac{3}{n^2 + 2} \frac{\alpha M_0}{100lc}$$

where $n^2 = 1 + a\lambda^2/(\lambda^2 - \lambda_0^2)$, and a and λ_0 are both constants.

Optical rotatory disperson is the change of optical rotation with wavelength. The data for optical rotatory dispersion are often analyzed in terms of the Drude (1900) equation:

$$[\alpha'] = \sum_i \frac{K_i}{\lambda^2 - \lambda_i^2}$$

where $[\alpha']$ is the observed rotation corrected for the refractive index of the solvent, K_i is a constant, and λ_i the wavelength of the ith transition. It has been suggested that K_i is not really a constant. It is related to λ_i by another constant A_i, such that

$$K_i = A_i \lambda_i^2$$

where A_i is related to the rotational strength of the ith transition R_i:

$$R_i = \frac{hc}{96\pi n} A_i$$

The rotational strength is the dot product of vectors $\boldsymbol{\mu}$ and $\mathbf{m}$:

$$R_i = \boldsymbol{\mu}_i \cdot \mathbf{m}_i$$

where $\boldsymbol{\mu}$ is the charge transition and $\mathbf{m}$ is the magnetic transition. Both $\boldsymbol{\mu}$ and $\mathbf{m}$ result from the interaction of electromagnetic radiation (light) with an optically active matter.

In many cases (such as protein) the Drude equation is used with only one term:

$$[\alpha'] = \frac{A_0 \lambda^2}{\lambda^2 - \lambda_0^2}$$

However, the one-term Drude equation has three drawbacks: (1) The equation is valid only at wavelengths far from the absorption band, (2) the analysis is informative only when the molecule has low α-helix content, and (3) the constants A_0 and λ_0 do not provide physical meaning.

Multiterm Drude equations are an improvement on the one-term Drude equation. There is no a priori way to demonstrate which multiterm Drude equation is best for analysis of any particular experimental data; whichever fits the data is the best.

The Moffit–Yang (1956) equation provides a different way to analyze experimental data, and is given in the form

$$[m']_\lambda = \frac{a_0 \lambda_0^2}{\lambda^2 - \lambda_0^2} + \frac{b_0 \lambda_0^4}{(\lambda^2 - \lambda_0^2)^2}$$

where $[m']$ is called the effective residue rotation, and a_0 and b_0 are constants. The parameter a_0 seems to have no physical meaning, but b_0 does. The plot of $[m'](\lambda^2 - \lambda_0^2)$ or $[\alpha](\lambda^2 - \lambda_0^2)$ versus $1/(\lambda^2 - \lambda_0^2)$ is expected to give a straight line from which b_0 can be determined. The terms b_0 and λ_0 are principally functions of the helical backbone, independent of side chains and environment. The value of λ_0 is assumed to be 212 nm for the wavelength range between 600 and 350 nm, and 216 nm for the wavelength range between 280 and 240 nm. The value of b_0, which is of primary importance in ORD measurements, is a measure of the helical content of a macromolecule. On the basis of experimental studies of some polypetides, the maximum value for b_0 is 630. The sign of the value for b_0 indicates the direction of the helix: minus for a right-handed helix and plus for a left-handed helix. Thus, there are three extreme values for b_0: -630 for a 100% right-handed helix; 0 for no helix at all; $+630$ for a 100 left-handed helix. In between these three values, the molecule is supposed to have partial helix in its secondary structure. The Moffit-Yang equation has been used extensively in the literature for the estimation of the α-helix content of polypeptides and proteins.

Figure 13.3 shows a typical rotatory dispersion curve, where there is no maximum or minimum. Figure 13.4 shows a plot of $[m']$ versus λ, in which a protein in denatured form is compared with the same protein in native form. The plot of the Moffit-Yang equation is shown in Figure 13.5. If a synthetic polypeptide is in a random conformation, the rotatory dispersion, which is simple, may be adequately expressed by a one-term Drude equation. If a synthetic polypeptide is in an α-helical conformation, the Moffitt-Yang plot may be employed to describe the rotatory dispersion.

In assessing the optical rotatory change of proteins, the sequence and molecular weight are unimportant. The important paramaters are the composition

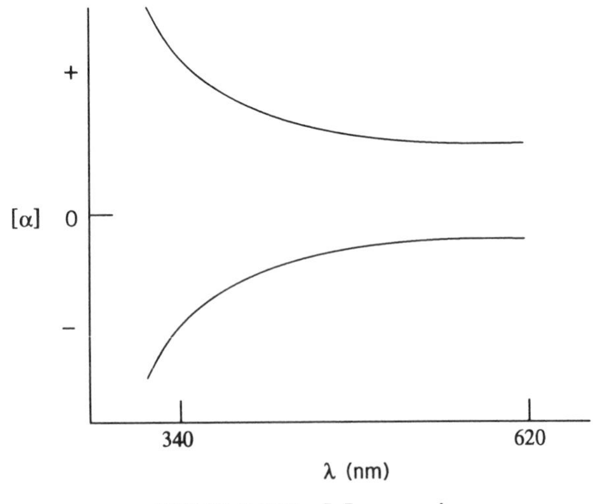

FIGURE 13.3 $[\alpha]$ versus λ.

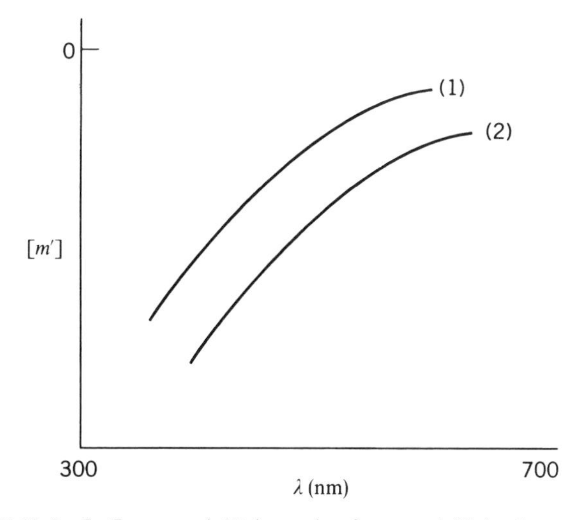

FIGURE 13.4 $[m']$ versus λ (1) in native form and (2) in denatured form.

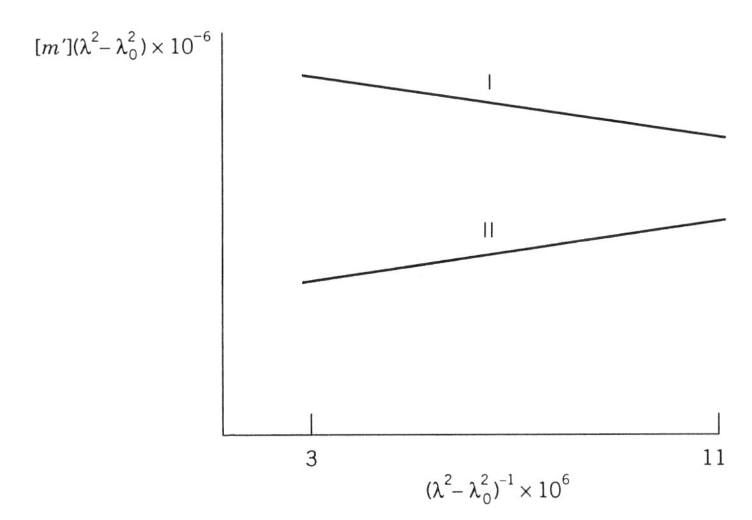

FIGURE 13.5 Plot of the Moffit-Yang equation.

and conformation, particularly the latter. Any change in environment (such as pH or salt concentration) could cause a change in conformation. Quantitatively, the numerical values of $[\alpha]$ or $[m']$, λ_0, and b_0 provide information on the extent of change.

13.3 CIRCULAR DICHROISM

If the intensity of absorption (not the refractive index n) is used as a function of the orientation of the plane of polarization, we have a phenomenon called circular dichroism. As in the case of refractive indices, for an optically inactive substance, the intensities of the left and right circularly polarized components are equal. For an optically active substance, they are not equal. The difference $\Delta\varepsilon$ is expressed in the equation

$$\Delta\varepsilon = \varepsilon_{\mathrm{L}} - \varepsilon_{\mathrm{R}}$$

where ε is the molar absorptivity, and L and R refer to the left- and right-circularly polarized components, respectively. $\Delta\varepsilon$ varies with wavelength and can be positive or negative. The values of $\varepsilon_{\mathrm{L}}, \varepsilon_{\mathrm{R}}$, and $\Delta\varepsilon$ are in a narrow region in the absorption maximum of the compound. The plot of $\Delta\varepsilon$ versus λ for a simple optical transition is bell-shaped, similar to the ordinary absorption curve in spectroscopy.

Circular dichroism is usually measured as a dichroic ratio, which is the ratio of the optical densities of an absorption band in the direction of the polarized light (parallel over perpendicular) to a specified direction in the sample. More specifically, dichroic ratios depend on the angle θ, as shown in Figure 13.2b. The tangent of θ is the ratio of the minor axis b to the major axis a of the ellipse that is defined as

$$\theta = \tan^{-1}\frac{b}{a}$$

The angle θ, written as $[\theta]$, is called the molar ellipticity. The relationship between $\Delta\varepsilon$ and $[\theta]$ is expressed in the equation

$$[\theta] = (3300^\circ)(\Delta\varepsilon)$$

The dimension of $\Delta\varepsilon$ is in (L/cm) mol^{-1}, while that of $[\theta]$ is in $\mathrm{deg \cdot cm^2/d\,mol}$. The molar ellipticity $[\theta]$ is often considered to consist of possible electronic transitions in the molecules:

$$[\theta] = \sum_i [\theta_i]$$

where $[\theta_i]$ is the ellipticity for the ith transition. $[\theta_i]$ is characterized by three parameters:

λ_i^0 the wavelength at which the maximum ellipticity is located
Δ_i^0 the half-width of the dichroism band
R_i the rotational strength of the ith transition

All three parameters could easily be evaluated as shown in Figure 13.6. R_i can be calculated using the following equation:

$$R_i = \frac{3hc'}{8\pi^3 N_1}\int_0^\infty \frac{\theta_k(\lambda)}{\lambda}\,d\lambda$$

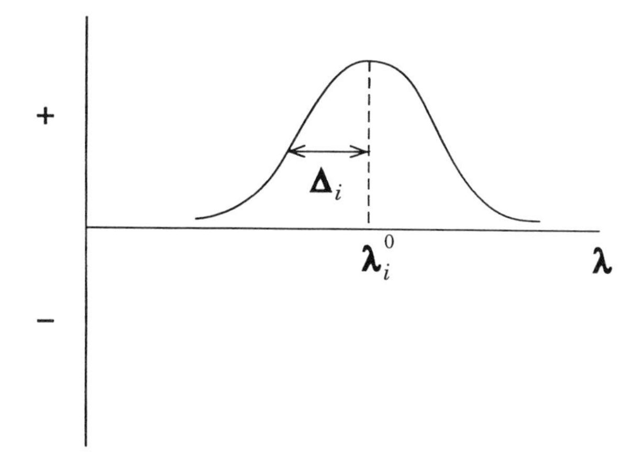

FIGURE 13.6 A dichroic band.

where $\theta_k(\lambda)$ is the partial ellipticity for the kth transition, h is Planck's constant, c' is the velocity of light, and N_1 is the number of absorbing molecules per cubic centimeter. The integral is the area under the dichroic band. The rotational strength is related to the induced electric and magnetic dipole moments by $R_i = \mu_l^i \mu_m^i$, as mentioned before.

The dichroic bands are not always neatly separated as in Figure 13.7a; they may overlap and be located in positive and negative directions as in Figure 13.7b. When they overlap, each band should be resolved. Provided that the dichroic bands are Gaussian, the three parameters λ_i^0, Δ_i^0, and R_i can be evaluated as a single separate dichroic band.

Circular dichroism is extensively employed in the structural study of biological polymers, but not in the study of synthetic polymers. This is because most biological polymers are optically active, whereas most synthetic polymers are not. However, if the optically active vinyl or vinylidene monomers are incorporated in a synthetic polymer, the synthetic polymer could become optically active.

13.4 COTTON EFFECT

An anomalous curve may appear within the optical rotatory spectrum, showing one or more maxima or minima in the neighborhood of the absorption band. Such an anomalous curve is attributed to a special combination of inequal refractive indices known as the Cotton effect. The maxima are called peaks and the minima are called troughs (Figure 13.8). If the peak is at the longest wavelength (that is, right of the trough), there is a positive Cotton effect. Otherwise, the Cotton effect is negative.

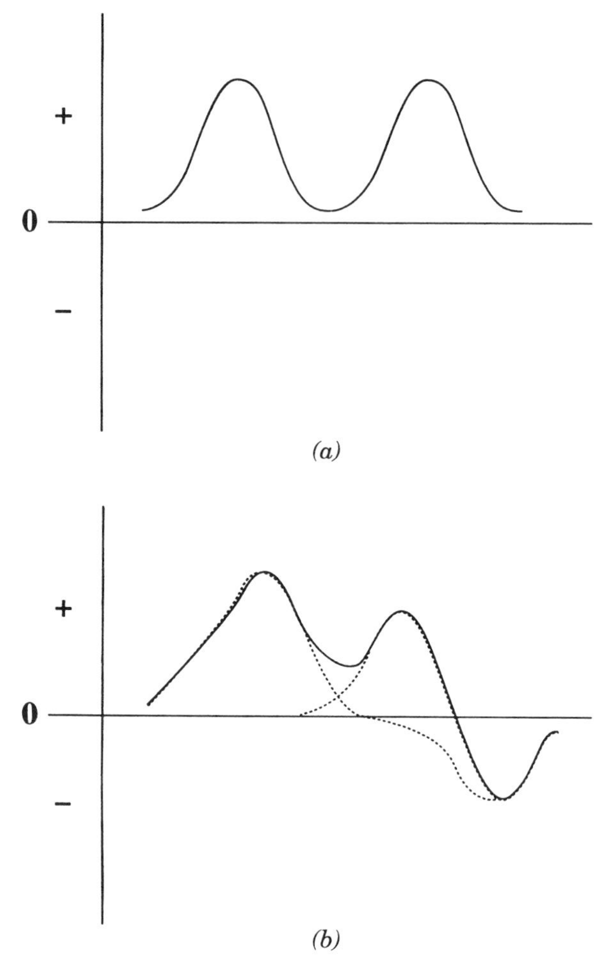

FIGURE 13.7 Dichroic bands.

The Cotton effect can also appear in circular dichroism, but there is only a positive or a negative maximum, not peaks and troughs. The positive maximum in CD corresponds to the inflection of the positive Cotton effect and the negative maximum corresponds to the negative Cotton effect (see Figure 13.9). Since the two extremes, peak and trough, are so obvious, the Cotton effect is more easily visualized in optical rotatory dispersion than in circular dichroism; but CD is much more sensitive than ORD because the trailing parts of the ORD curves representing Cotton effects often overlap.

The Cotton effect curve is very sensitive to conformational alterations of proteins and polypeptides. It is a method that can also be used to assess α-helical content. In a dissymmetric environment, the α-helical conformation is characterized

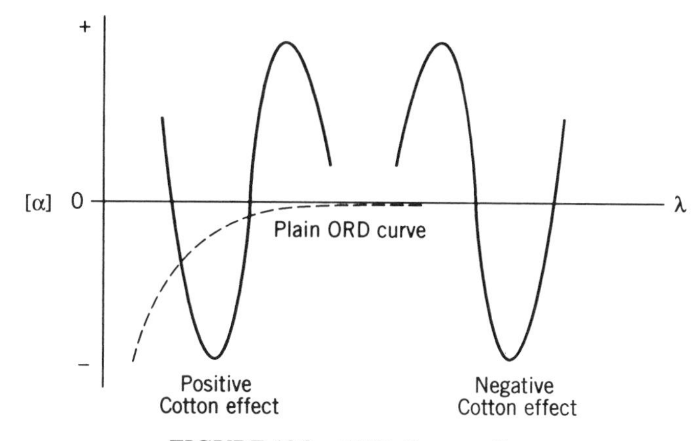

FIGURE 13.8 ORD Cotton effect.

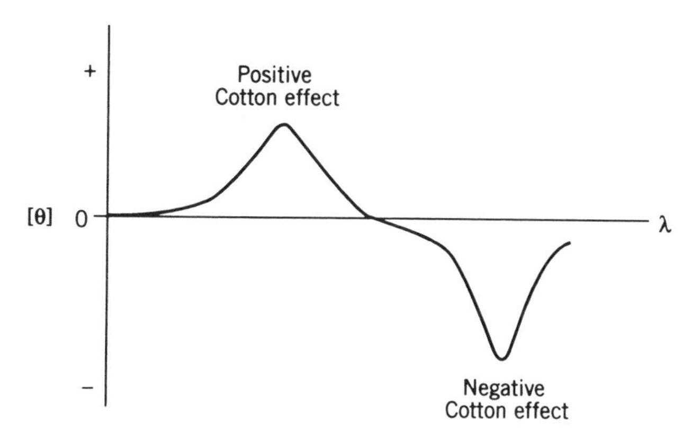

FIGURE 13.9 CD Cotton effect.

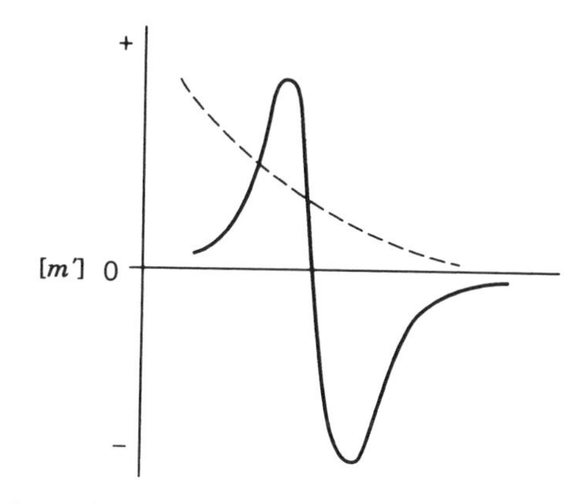

FIGURE 13.10 Loss of Cotton effect: —, Cotton effect; – –, loss of Cotton effect.

in ORD by a large negative Cotton effect with a trough at 233 nm and an inflection at about 225 nm. Quantitatively, the magnitude of the trough at 233 nm provides an estimation of α-helix content. When the helix is destroyed, the Cotton effect is lost (Figure 13.10). For mononucleosides and mononucleotides, a single Cotton effect appears above 220–240 nm and a crossover appears near 260 nm (Figure 13.11). For synthetic polynucleotides, such as poly A and poly U, the multiple Cotton effects all have a large peak at 282–286 nm and a trough at 252–260 nm, followed by a small peak near 230–240 nm (Figure 13.12).

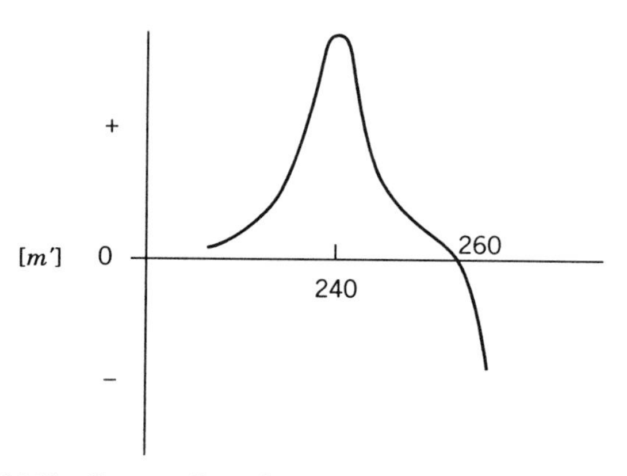

FIGURE 13.11 Cotton effect of mononucleosides and mononucleotides.

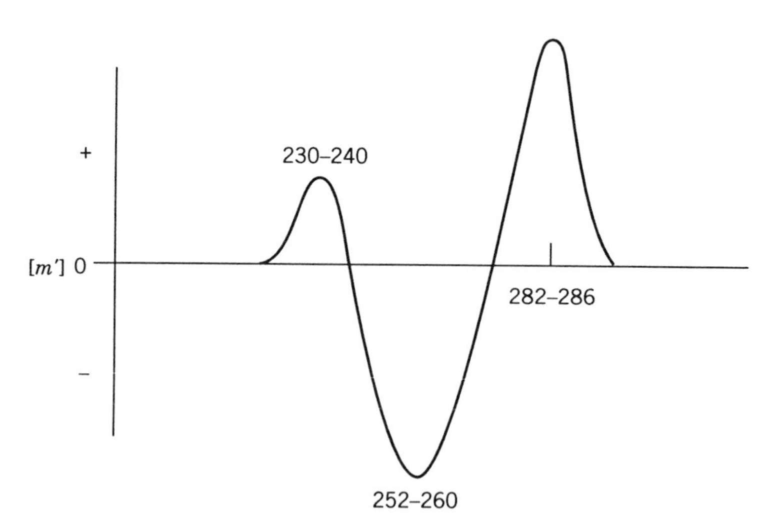

FIGURE 13.12 Cotton effect of synthetic polynucleotides.

13.5 CORRELATION BETWEEN ORD AND CD

Both ORD and CD are sensitive to conformational changes and chemical transformation. ORD has the following advantages over CD: (1) It is easier to visualize the Cotton effect with ORD because of the three distinct points in the ORD curve: the peak, the crossover, and the trough (in that order or in reverse order). (2) An optically active compound that does not show the band in the wavelength

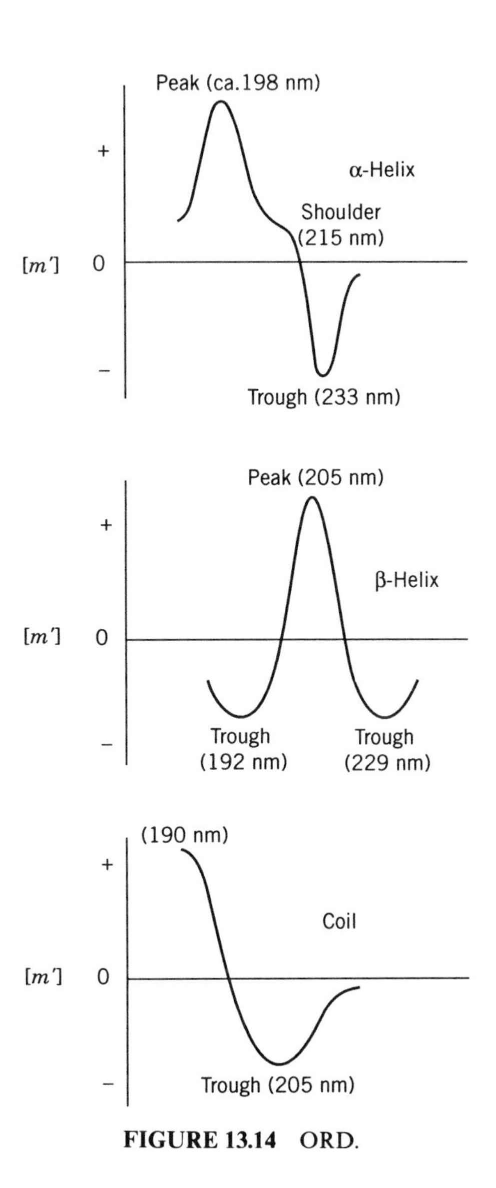

FIGURE 13.14 ORD.

range of interest in the absorption spectrum will not show a CD curve, but will show a plain ORD curve. CD, on the other hand, possesses an intrinsic discreteness and is a more sensitive tool in examining the environmental effect on the conformation of macromolecules.

A reciprocal relationship exists between rotatory dispersion and circular

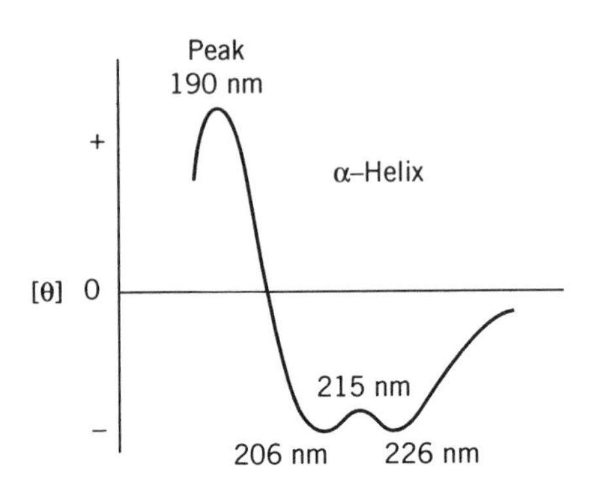

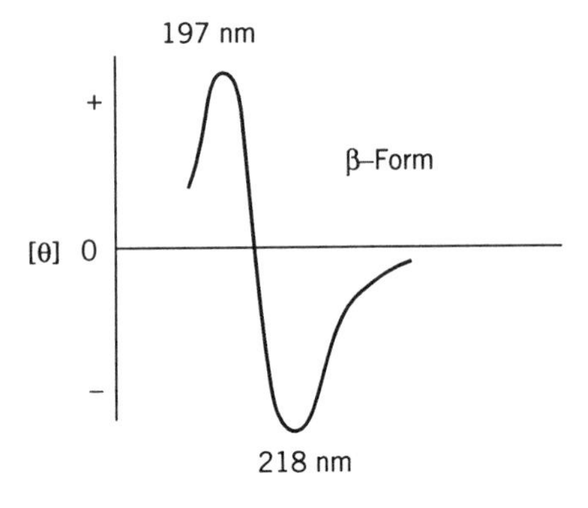

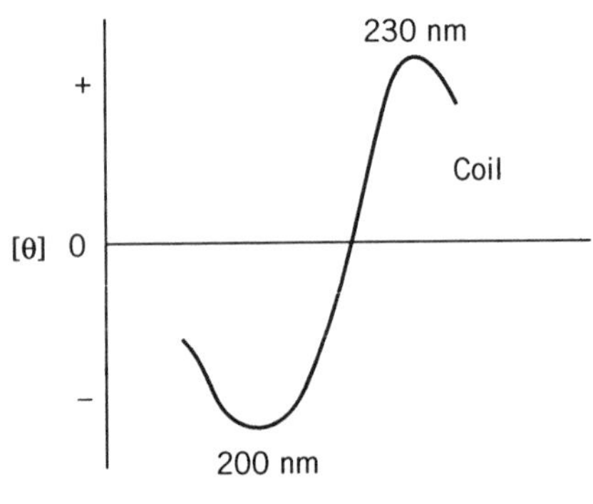

FIGURE 13.15 CD.

dichroism, using Kronig–Kramers (Kronig, 1926; Kramers, 1927) transform:

$$[m_i(\lambda)] = \frac{2}{\pi}\int_0^\infty [\theta_i(\lambda')]\frac{\lambda'}{\lambda^2 - \lambda'^2}\,d\lambda'$$

$$[\theta(\lambda)] = -\frac{2}{\pi\lambda}\int_0^\infty [m(\lambda')]\frac{\lambda'^2}{\lambda^2 - \lambda'^2}\,d\lambda'$$

The prime (′) represents the assigned vacuum values (see Moscowitz 1960). The condition attached to this transform is that the rotatory dispersion curve must be Gaussian. Thus, from the circular dichroism curves, we can calculate the rotatory dispersion and vice versa.

13.6 COMPARISON OF ORD AND CD

In what follows, we compare ORD and CD by sketching the typical curves for proteins and polypeptides in α-helical, β-helical, and random coil forms. We first give a few theoretical remarks to provide background. It is now well-known that the spectral region of the amide involves π°–π^- and n_1–π^- transitions. According to Moffit, the electronic dipole transition moments and the optical activity of helical molecules are coupled. The electronic transition π°–π^- occurs at around 190 nm, while that of n_1–π^- occurs at around 225 nm. The actual

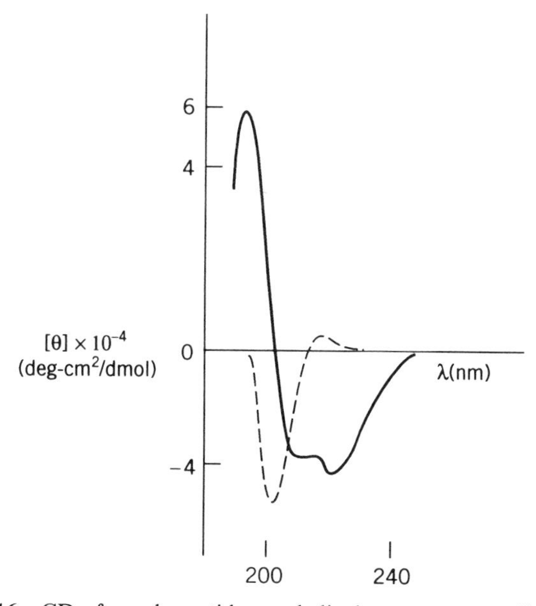

FIGURE 13.16 CD of a polypeptide: —, helical structure; – –, disordered state.

values, of course, depend on the structure of the molecules at specific conditions. They are not exactly at 190 and 225 nm, respectively. Still, it is worthwhile to focus on the curve near these two wavelengths.

In general, an ORD curve of α-helical polypeptides or proteins shows a trough at $\sim$233 nm, a crossover at $\sim$225 nm, a shoulder at $\sim$215, and a peak at 198 nm. At random conformation, the Cotton effect at $\sim$225 is lost (Figure 13.14). In the study of synthetic polypeptides and proteins, CD is found in the region between 190 and 250 nm, where absorption is basically due to peptide linkage. The value of molar ellipticity has been suggested to the estimate of α-helical content (Figure 13.15).

Figure 13.16 shows CD of a typical polypeptide. There are three bands in the accessible region: 191, 206 (or 209), and 227 nm. The two dichroic peaks (222 and 209) are negative; they are clearly separated by a distinct notch at 215 nm. The dichroic band at 191 nm is positive. Whenever there is a conformational change, for example, from α-helical to disordered, there is a change in the three bands. In certain cases the change is strong; in others it is weak.

REFERENCES

Beychok, S., *Science* **154**, 1288 (1966).

Beychok, S., and G. D. Fasman, *Biochemistry* **3**, 1675 (1964).

Blout, E. R., I. Schnier, and N. S. Simmons, *J. Am. Chem. Soc.* **84**, 3193 (1962).

Brahms, J., *J. Am. Chem. Soc:* **85**, 3298 (1963).

Chen, Y. H., and J. T. Yang, *Biochem. Biophys. Res. Commun.* **44**, 1285 (1971).

Cotton, A., *Ann. Chim. Phys.* **8**, 347 (1896).

Djerassi, C., *Optical Rotatory Dispersion.* New York: McGraw-Hill, 1960.

Doty, P., A. Wada, J. T. Yang, and E. R. Blout, *J. Polym. Sci.* **23**, 831 (1957).

Drude, P., *Lehrbuch der Optik*, Leipzig: Hirzel, 1900.

Greenfield, N., and G. D. Fasman, *Biochemistry* **8**, 4108 (1969).

Holzwarth, G., W. B. Gratzer, and P. Doty, *J. Am. Chem. Soc.* **84**, 3194 (1962).

Holzwarth, G., and P. Doty, *J. Am. Chem. Soc.* **87**, 218 (1965).

Iizuka, E., and J. T. Yang, *Proc. Natl. Acad. Sci. USA* **55**, 1175 (1966).

Kronig, R. de. L., *J. Opt. Soc. Am.* **12**, 547 (1926).

Kramers, H. A., *Atticongr. Intern. fisici, como* **2**, 545 (1927).

Moffitt, W., and J. T. Yang, *Proc. Natl. Acad. Sci. USA* **42**, 596 (1956).

Moscowitz, A. in *Optical Rotatory Dispersion*, C. Djerassi, Ed., New York: McGraw-Hill, 1960.

Shechter, E., and E. R. Blout, *Proc. Nat. Acad. Sci. USA* **51**, 695 (1964).

Simmons, N. S., C. Cohen, A. G. Szent-Gyorgyi, D. B. Wetlaufer, and E. R. Blout, *J. Am. Chem. Soc.* **83**, 4766 (1961).

Tinoco, I., and C. Cantor, *Methods Bio-chem. Anal.* **18**, 81–203 (1970).

Yang, J. T., and T. Sameijima, *Progr. Nucl. Acids Res.* **9**, 223 (1969).

PROBLEMS

13.1 In a certain environment, the residual rotation value $[R']_{233}$ band was found to be $-12\,600$ for poly-L-glutamic acid, $-10\,400$ for poly-L-methionine, $-12\,800$ for poly-γ-benzyl-L-glutamate. Calculate the percentage helix of each of the polypeptides, assuming a mean value of $b_0 = -630$ for 100% helix and $b_0 = 0$ for 0% helix. (*Source*: Simons et al. (1961).)

13.2 Three bands were found for poly-L-tyrosine in 0.2 M NaCl, pH 11.2, at 24°C in the CD. The maximum ellipticity $[\theta]^\circ$ and the half-width Δ° for each band's λ° are given as follows:

λ° (nm)	$[\theta]^\circ$	Δ°
224	-1.51×10^4	8
248	6.70×10^3	8
270	2.00×10^3	10

Calculate $[m']$ and plot to obtain the ORD curves. (*Source*: Beychok and Fasman (1964).)

13.3 The optical rotatory dispersion of poly-γ-benzyl-L-glutamate in chloroform does not fit a simple Drude equation. Such dispersion results may be plotted in the general form proposed by Moffit. The data are as follows:

λ (nm)	$[a]$ (deg-cm^2/dg)
340	-40
420	9
500	12
580	8
600	10

(a) Give a Moffit plot ($[\alpha](\lambda^2 - \lambda_0^2)$ versus $1/(\lambda^2 - \lambda_0^2)$), assuming that $\lambda_0 = 212 \pm 5\ \mu$m.

(b) Determine the value of b_0.

(c) Interpret the results.

(*Source*: Doty et al. (1957).)

13.4 The following are the experimental data for circular dichroism of polyuridylic acid in 0.01 M $MgCl_2$, pH 6.8, at 1°C. Concentrations ranged from 0.008 to 0.0013%.

λ (nm)	$\varepsilon_L - \varepsilon_R$ (cm^2/mol)
230	-3
240	-5

λ(nm)	$\varepsilon_L - \varepsilon_R$(cm^2/nol)
243	−6
250	−2
255	6
260	14
265	17
270	14
275	10
280	7
290	2
295	0

(a) Plot $\varepsilon_L - \varepsilon_R$ versus λ.

(b) Determine λ_m, λ_c and $\Delta\nu\,(\text{cm}^{-1})$ respectively, where λ_m is the wavelength where the maximum occurs, λ_c is the wavelength at which the curve is crossing, and $\Delta\nu$ is the half-width of the band.

(c) Calculate the rotational strength R_{ba}, using

$$R_{ba} = \frac{3hc \times 10^8 \ln 10}{32\pi^3 N} \int \left| \frac{\varepsilon_L - \varepsilon_R}{\nu} \right| d\nu$$

(*Source*: Brahms (1963).)

13.5 Circular dichroism possesses the advantage of intrinsic discreteness and can be transformed to rotatory dispersion. Holzwarth et al. (1962) measured circular dichroism of poly-L-glutamic acid in 0.1 M NaF at concentrations from 0.03 to 0.4%. At pH 4.3, poly-L-glutamic acid exists in the helical form, whereas at pH 7.3, in the disordered form. The data are as follows:

pH 4.3		pH 7.3	
λ(nm)	$[\theta] \times 10^{-4}$	λ(nm)	$[\theta] \times 10^{-4}$
185	4.8	198	−2.0
190	7.5	200	−4.0
192	8.0	205	−4.5
195	7.3	210	0.01
200	2.0	215	0
205	0.2	220	0
210	0.4		
215	−4.4		
220	−4.2		
225	−3.0		
230	−2.0		
240	−0.2		

(a) Plot the circular dichroism of poly-L-glutamic acid in 0.1 M NaF.
(b) Express each curve in Gaussian form.
(c) Calculate the rotational strength of each solution.
(d) Calculate the rotational dispersion curves (as $[m]$, the mean residue rotation) from the CD data, using the Kronig-Kramers transformation:

$$[m_K] = \frac{2[\theta_K^0]}{\sqrt{\pi}}\left(\int_0^{\lambda - \lambda_K/\Delta_K} e^{x^2}\,dx - \frac{\Delta_K}{\lambda + \lambda_K}\right)$$

(*Source*: Holzwarth et al. (1962).)

14

NUCLEAR MAGNETIC RESONANCE

Since Purcell and Bloch in 1946 announced the observation of the phenomenon in bulk matter, nuclear magnetic resonance has become an indispensable tool in chemistry for the study of molecular structure and behavior. In this chapter we first describe the basic theory that underlies the nuclear magnetic resonance phenomenon. Then we discuss the techniques involved in its spectroscopy. Finally we illustrate the spectra of some well-known synthetic and biological polymers.

14.1 MAGNETIC FIELD (H) AND RESONANCE FREQUENCY (ν)

The spin of a nucleus may be described in terms of two quantum numbers: the spin quantum number or spin I and the spin magnetic quantum number or projection m_I. The two numbers are related as shown in the equation

$$m_I = 0, \pm 1, \pm 2, \ldots, \pm I$$

Since, for a nucleus, the number I can take only two values, 1 or $\frac{1}{2}$, the number m_I can take only values 0, $\pm\frac{1}{2}$, and ± 1. From I and m_I, two quantities, $\mathbf{I}$ and $\mathbf{I}_z$, are defined:

$$\mathbf{I} = \sqrt{I(I+1)}\,\hbar$$

and

$$\mathbf{I}_z = m_I \hbar$$

where $\hbar = h/2\pi$ (h is Planck's constant and $\pi = 3.1416$, an irrational number). The quantity I is called the nuclear spin momentum and the quantity I_z is called the z component of the nuclear spin momentum.

In a magnetic field H, the energy of a nucleus with a spin quantum number I is expressed as

$$E_{m_I} = -g\mu m_I H$$

where g is the nuclear g factor, which varies with different nuclei, and μ is the nuclear magnetic moment. The following are some examples of g values:

Nucleus	^1H	^2H	^{13}C
I	$\frac{1}{2}$	1	$\frac{1}{2}$
g	5.5854	0.85738	1.4043

The energy of a nucleus with the same g factor splits into different levels depending on the values of I (Figure 14.1). The splitting of nuclear energy levels in a magnetic field is called Zeeman splitting or the Zeeman effect, a phenomenon that also occurs with electron energy. The basis of nuclear magnetic resonance (NMR) instrument is to induce transitions between energy levels of a nucleus through the absorption or emission of energy quanta. The instrument applies an electromagnetic radiation with frequency ν to the nuclei of the sample in a magnetic field H. At a certain frequency, known as the resonance frequency, this transition occurs. The relationship between H and ν is described by

$$\begin{aligned}\Delta E &= E_2 - E_1 = \mu H m_I - (-\mu H m_I)\\ &= 2\mu H m_I = h\nu\end{aligned}$$

or

$$\nu = \frac{2\mu H m_I}{h} \tag{14.1}$$

An NMR experiment is, therefore, performed either by sweeping the magnetic

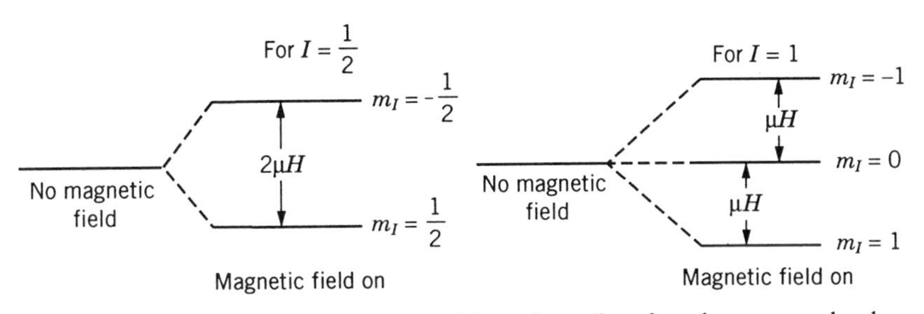

FIGURE 14.1 The effect of values of I on the splits of nuclear energy levels.

field H at constant frequency ν or by sweeping the frequency ν at constant magnetic field H to match the resonance frequency of the nuclei of the sample.

The unit of magnetic field is the tesla (T), also called the weber per square meter (Wb/m^2). In the SI system, $1\,T = 1\,N\,C^{-1}\,M^{-1}\,s = 1\,kg\,s^{-1}\,C^{-1}$ (where N is newton, C is columb, and s is second). Chemists commonly use Gaussian units (G, gauss):

$$1\,G = 1\,\text{dyn stat}\,C^{-1} = 1\,g^{1/2}\,s^{-1}\,cm^{-1/2}$$

$$1\,T = 10\,000\,G$$

The unit of frequency is hertz (Hz) or cycles per second (cps). In a magnetic field of 1.5 T (15 kG), the resonance frequency of protons is 60 MHz; in a field of 7.0 T (70 kG), the frequency is 300 MHz. In general, the higher the frequency, the higher is the signal intensity.

14.2 CHEMICAL SHIFT (δ) AND SPIN–SPIN COUPLING CONSTANT (J)

When a molecule is placed in a magnetic field H, orbital currents are induced in the electron clouds, which create a small local magnetic field, H_{loc}. Each nucleus is shielded by its surrounding electrons. The effective magnetic field felt by the nucleus is not the same as the applied field. The nucleus now requires a slightly higher values of H to achieve resonance:

$$H = H_{loc} + \sigma H$$

or

$$H_{loc} = H(1 - \sigma)$$

where σ is the screening constant. The parameter σ is independent of H, but

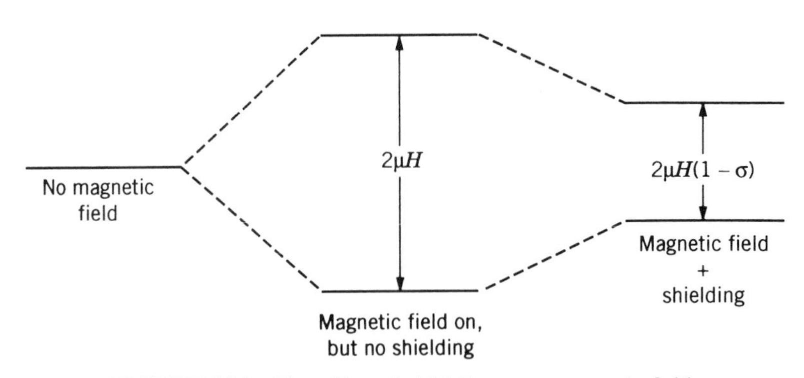

FIGURE 14.2 The effect of shielding on a magnetic field.

dependent on the chemical (electronic) environment. The effect of shielding is shown in Figure 14.2.

The term $H\sigma$ (chemical shift) is the displacement of a signal in different chemical environments due to variations in screening constants. Since the screening constant is of the order of 10^{-5} to 10^{-6} (that is, an external field of 1 T stirs up an extra local field of about 1 μT), the chemical shift is expressed in parts per million (ppm).

In this practice, chemical shift is not expressed in terms of the electron shielding term $H\sigma$. It is expressed in terms of the paramater δ, which is defined in several different ways:

$$\delta = (\sigma_{\text{ref}} - \sigma_{\text{s}}) \times 10^6$$

$$\delta = \frac{H_{\text{r}} - H_{\text{s}}}{H_{\text{r}}} \times 10^6$$

$$\delta = \frac{\Delta\nu}{\text{Oscillator frequency (cps)}} \times 10^6$$

where $\Delta\nu = \nu_{\text{s}} - \nu_{\text{r}}$. The subscript r refers to the reference and s refers to the sample. Oscillator frequency refers to the characteristic frequency of the instrument. For example, a 60-MHz instrument for protons has an oscillator frequency of 60×10^6 cps. The standard reference chosen for proton NMR is usually TMS ($Si(CH_3)_4$, tetramethylsilane). It is chemically inert and magnetically isotropic, and it gives a single sharp line.

Chemical shift (cps) when expressed as a δ value is said to be on the δ scale. The value of δ could be positive or negative with respect to TMS (the δ value for TMS being arbitrarily chosen as 0). To avoid a negative value for δ, another scale is often used, which is called the τ scale. The relationship between δ and τ is expressed by

$$\tau = 10 - \delta$$

The value of τ is always positive. The larger the value of τ, the greater is the magnetic shielding of the nucleus. Figure 14.3 shows the relationship between the two scales.

The resolution of nuclear magnetic resonance depends on both the magnetic field strength and the resonance frequency of the instrument. In the early years (ca. 1953), for example, NMR instruments could resolve only three proton resonances pertaining to ethyl alcohol. Modern instruments can show the multiplicity of the line shape, revealing more details of the nuclear environment. This multiplicity arises due to spin–spin coupling, another important NMR spectrum parameter.

Spin–spin coupling is the result of three types of interactions: (1) interactions of the electrons as moving charged particles in a magnetic field of nuclei; (2) dipole–dipole interactions between nuclear magnetic moments and electron

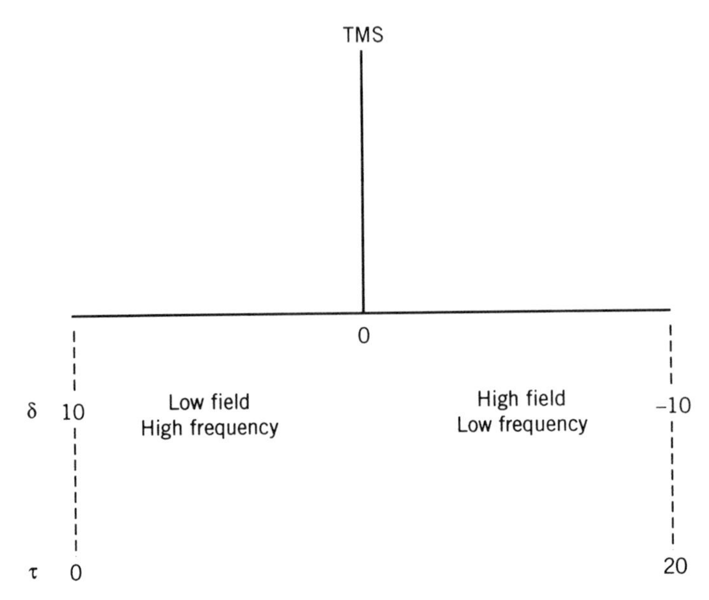

FIGURE 14.3 The two scales of a chemical shift.

magnetic moments, and (3) contact potential interactions. Spin–spin coupling in the NMR spectrum is often expressed as a numerical value, called the spin–spin coupling constant J, whose unit is Hz. The values of J vary from 0.05 to 6000 Hz.

A classical example of chemical shifts and spin–spin coupling is the NMR spectrum of ethyl alcohol, which is shown in Figure 14.4. Often an NMR spectrum is displayed in terms of intensity in arbitrary units versus chemical shifts in ppm or Hz. The two quantities, ppm and Hz, are interchangeable:

$$\text{ppm} = \frac{\nu_{\text{sample}} - \nu_{\text{reference}}}{\nu_{\text{reference}}} \times 10^6$$

For example, $\nu_{\text{reference}} = 60 \times 10^6$ cps for a 60-MHz instrument.

The chemical shift values depend on its applied field, but the value of the spin–spin coupling constant is independent of its applied field. Figure 14.5 is a schematic diagram of the spin resonances of two nuclei, A and X, of the same type (for example, both are protons), a different positions with or without coupling. It also shows the change in J_{AX} values in a different field ($H_0 + H_{01}$).

The value of the spin–spin coupling constant in proton NMR also varies according to the number of intervening bonds (that is, geometry). For geminal protons, H–C–H, $J = 12$ Hz in saturated systems; for vicinal protons, H–C–C–H, $J = 7$ or 8 Hz. If the two protons are separated by four or more single bonds, $J = 0$. Because of the σ–π configuration interaction, the J value can be increased if

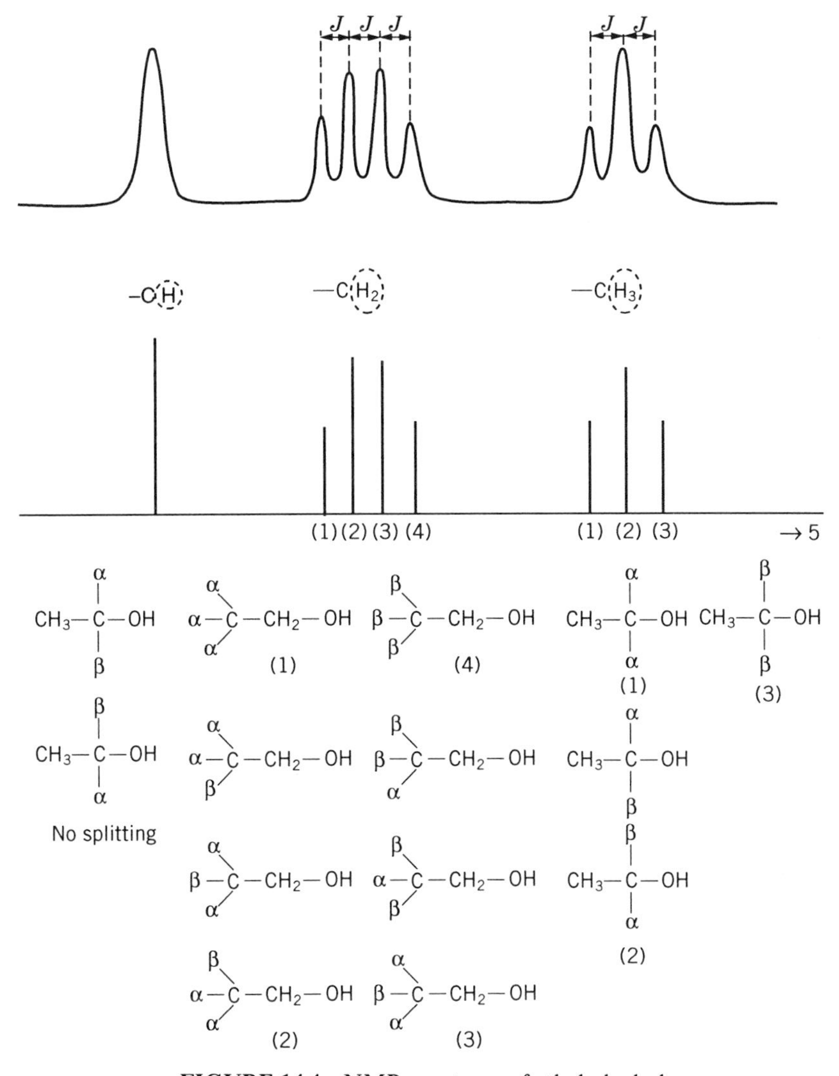

FIGURE 14.4 NMR spectrum of ethyl alcohol.

there is an unsaturated bond in the coupling path. Furthermore, Karplus (1959) found that the values of the vicinal interproton coupling constants depend on the dihedral angle ϕ between the carbon–hydrogen bonds:

H
C ϕ
H

This relationship may be expressed by

$$J_{\text{vic}} = 4.22 - 0.5\cos\phi + 4.5\cos^2\phi$$

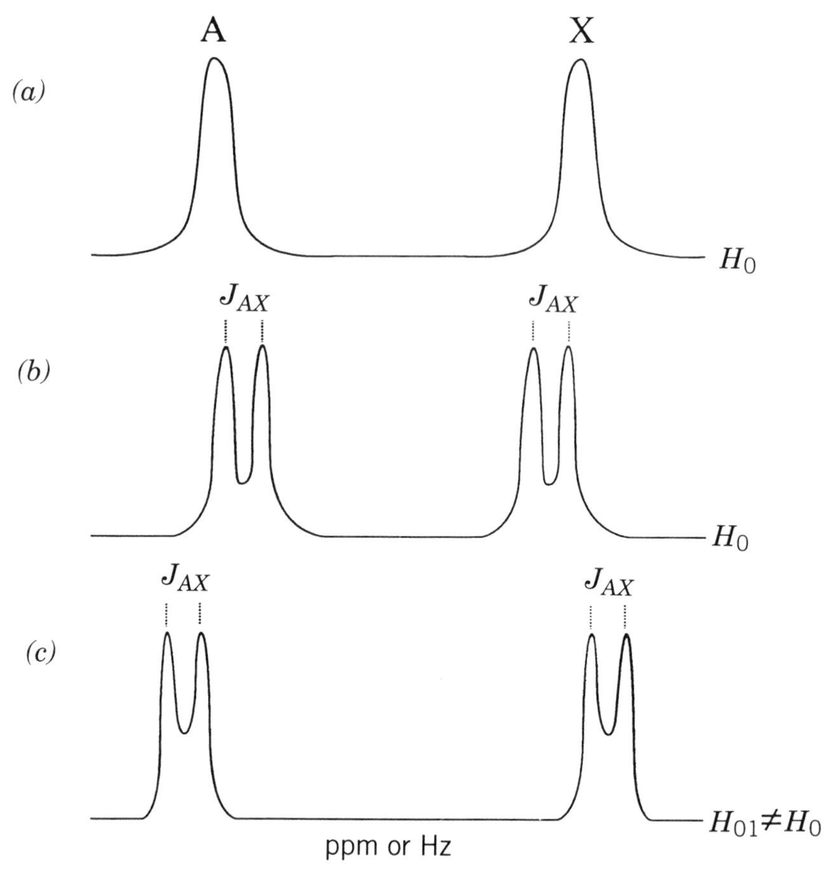

FIGURE 14.5 Schematic diagram of the spin resonances of two nuclei, A and X: (a) spin A and spin X without coupling; (b) spin A and spin X with coupling J_{AX}, split of spin A as well as of spin X; (c) an applied magnetic field H_{01} different from that in (a) and (b).

14.3 RELAXATION PROCESSES

In the absence of magnetic field, the populations of the two nuclear spin states (n_+, the spin-up number and n_-, the spin-down number per unit volume) are equal. When a magnetic field is applied, there is a slight excess of nuclei in the lower spin state. The populations follow the Boltzmann distribution law:

$$\frac{n_+}{n_-} = \exp\left(\frac{2\mu H}{kT}\right)$$

where k is the Boltzmann constant. For protons at 300° K in a field of 10 000 G

(60 MHz),

$$\exp\left(\frac{2\mu H}{kT}\right) \cong 1 + 6.0 \times 10^{-6}$$

As the nuclei continually absorb electromagnetic radiation, this small excess soon diminishes. Fortunately, for molecules that are undergoing Brownian motion, neighboring spins directly influence each molecule's magnetic field. This causes fluctuating magnetic fields, thereby inducing transition between energy states. This radiationless transition, i.e., the relaxation process, means that the induced population difference will soon return to equilibrium.

The relaxation rate can be described in terms of characteristic time τ. If τ is small, relaxation is fast; if τ is large, relaxation is slow. There are two types of relaxation: spin–lattice and spin–spin. The latter is unrelated to the spin–spin coupling constant: Spin–spin relaxation is a kinetic concept, while the spin–spin coupling constant is a thermodynamic constant.

Spin–lattice relaxation is also called longitudinal relaxation and is designated by relaxation time τ_1. The term lattice refers to the framework of molecules that includes the sample and its medium (the neighboring nuclei) with the same meaning as in Chapter 4. Spin–lattice relaxation has a component of random intensity and phase at the nuclear resonant frequency. This involves interchange of nuclear energy with the rest of the system (the surroundings).

In solids, there is no Brownian motion and no random relaxing field; τ_1 is very long. In both solids and liquids, the magnitude of τ_1 is determined by the concentration of paramagnetic impurities in the lattice. The uncertainty principle,

$$\Delta E\,\Delta t \cong \hbar$$

dictates that

$$\Delta v \cong \frac{1}{2\pi\Delta t} \sim \frac{1}{\Delta t} \qquad (\Delta E = h\,\Delta v)$$

Thus, the linewidth measured on a frequency scale (Δv) is roughly equal to $1/\tau_1$ ($\sim 1/\Delta t$). That is, τ_1 broadens the peak in the spectrum. In terms of the transition of a spin from one energy state to another, the quantity $1/\tau_1$ may be defined as

$$\frac{1}{\tau_1} = p_\uparrow + p_\downarrow$$

where $p_\downarrow$ is the probability of a downward transition per unit time of a spin from the higher to the lower magnetic level, and $p_\uparrow$ is the probability of an upward transition per unit time of a spin from the lower to the higher magnetic level.

Spin–spin relaxation is also called the transverse relaxation and is designated by a relaxation time τ_2. It occurs when there is an exchange of spins by two recessing nuclei in close proximity to one another. τ_2 may be defined as

$$\frac{1}{\tau_2} = \pi\,\delta\nu$$

where δ is Dirac delta (here, δ is not chemical shift), which possesses the following property:

$$\begin{aligned} \delta(E_\downarrow - E_\uparrow - h\nu) &= 1 \qquad \text{if } h\nu = E_\downarrow - E_\uparrow \\ &= 0 \qquad \text{if } h\nu \neq E_\downarrow - E_\uparrow \end{aligned}$$

While τ_2 is short in solids (often in the range of microseconds), it is about equal to τ_1 in liquids.

Both τ_1 and τ_2 are responsible for resonance signal shapes and intensities. There is a relationship between the linewidth Δ and τ_2. The linewidth Δ is shown in Figure 14.6 (the distribution of frequency $f(\omega)$ around its resonance ω_0). Note that $\omega = 2\pi\nu$. The equation that relates τ_2 and ω is

$$\int_0^\infty \nu(\omega_k - \omega_0)\cos(\omega_k - \omega_0)t\,\mathrm{d}\omega = \mathrm{e}^{-t/\tau_2}$$

The solution to this integral is

$$\nu = \frac{\tau_2}{1 + \tau_2^2(\omega_k - \omega_0)^2}$$

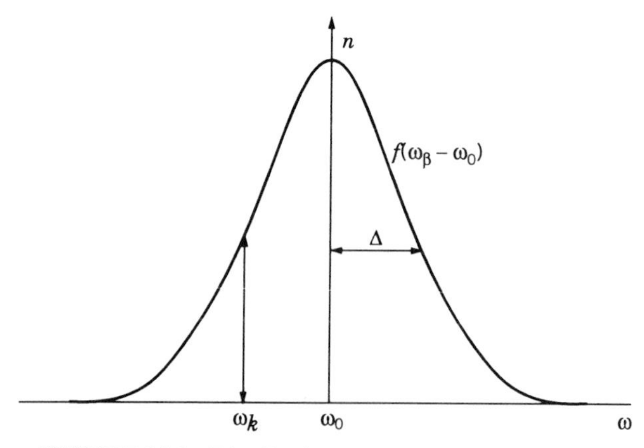

FIGURE 14.6 Distribution of frequency of resonance.

where v is the absorption mode of the Lorentzian line shape. (Note that the Lorenzian line shape is similar to Gaussian distribution but less symmetrical.)

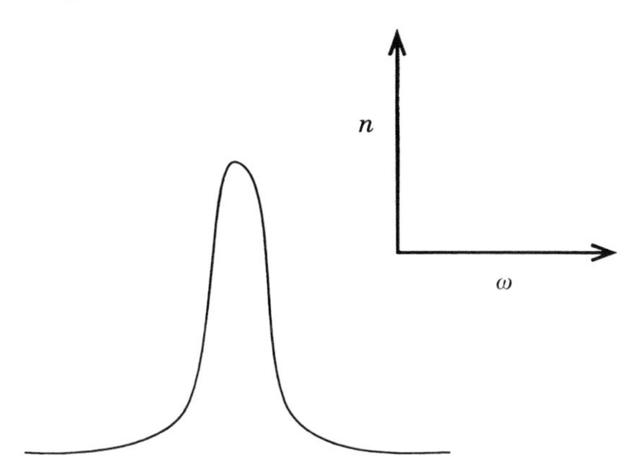

The quantity $(\omega_k - \omega_0)^2$ gives the value of linewidth Δ.

In addition to the two major relaxation processes, τ_1 and τ_2, two other processes important to the study of nuclear properties are nuclear quadrupole relaxation and the Overhauser effect. Nuclear quadrupole relaxation occurs only during nuclear spins with $I = 1$, not with $I = \frac{1}{2}$. For nuclei with $I = \frac{1}{2}$, there is a spherical distribution of the nuclear charge that is unaffected by the electric environment. For nuclei with $I > \frac{1}{2}$, the charge distribution in the nucleus is not spherical, but ellipsoidal in revolution (prolate or oblate), which gives the nucleus a quadrupole moment Q. The relationship between I and Q may be expressed by

$$\frac{1}{\tau_1} = \frac{1}{\tau_2} \sim \frac{(2I+3)Q^2}{I^2(2I-1)} \left(\frac{d^2H}{dz^2} \right)^2 \tau_c$$

where H is the field, $-d^2H/dz^2$ is the maximum field gradient in the z direction, and τ_c is the correlation time (the average time for a molecule to diffuse a distance equal to its own dimension or to rotate through one radian.)

The Overhauser effect is closely related to the spin–lattice relaxation of nuclei by unpaired electrons. It is a nuclear polarization; that is, it involves direct magnetic coupling between nuclei (dipolar coupling). This effect is seen as a change in the intensity of the NMR resonance when transitions of another nucleus are perturbed. Studying the Overhauser effect provides information about internuclear distances and molecular motion.

14.4 NMR SPECTROSCOPY

With the discovery that nuclear precession frequency depends on the chemical environment of the nuclei, chemists immediately recognized that NMR would be a powerful tool for investigating the structure and properties of molecules. Early attention was focused on proton nuclear magnetic resonance. During the last two decades, the resolving power of NMR instruments has improved in both its magnetic resolution and magnetic field strength. After an NMR of 60 MHz was achieved, NMRs of 100, 200, 220, 270, 300, 360, and 500 MHz became availble. Current spectrometers can reach over 750 MHz.

While the proton has been the major focus in NMR studies, investigators have also turned to nuclei other than hydrogen. Instruments for fluorine-19, carbon-13, nitrogen-15, deuterium, and sodium-29 are all commercially available. Different types of nuclear magnetic resonance spectroscopy often serve different purposes and may compliment each other in providing information about a particular compound. For example, fluorine-19 NMR has a larger chemical shift than proton NMR, and carbon-13 is superior to proton NMR in tackling complicated systems.

Advances in NMR research have led to an important application in medical science: imaging. The technique is based on the superimposition of a gradient on the main field, causing different parts of the sample to experience different magnetic fields and to resonate at different frequencies. As a result, the NMR signals provide a linear profile of the distribution of magnetic nuclei across the sample. If gradients are used in different directions, a two- or three-dimensional image of the sample can be constructed. Hence, imaging is primarily a function of proton densities and relaxation times.

14.4.1 Resonance

NMR spectroscopy is different from other forms of spectroscopy in that it does not necessarily measure the absorption or emission of energy directly associated with a particular transition in a molecule (see Chapter 16). Instead, it measures signals that are a response to the change in energy. These are signals of resonance.

Let a vector represent the spin of a nucleus with μ as the magnetic moment.

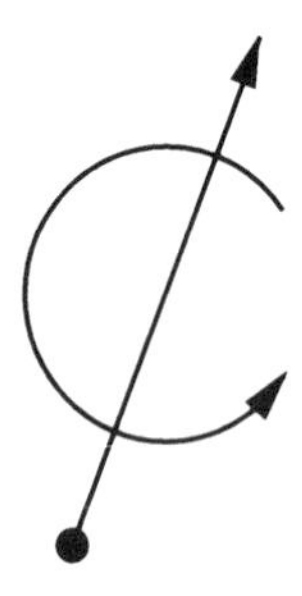

If a magnetic field H_0 is applied to the spin system, we observe a phenomenon known as the Larmor precession:

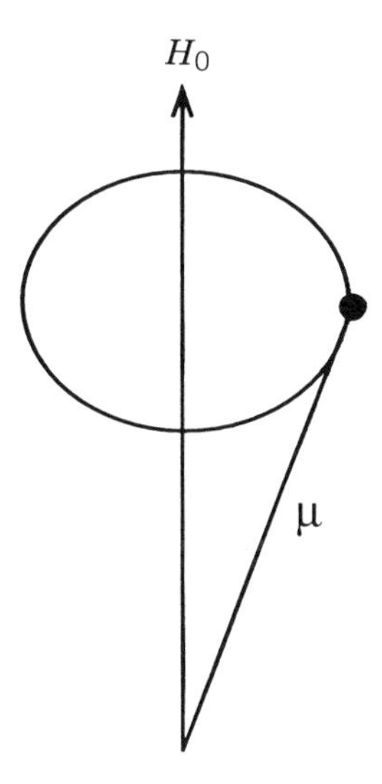

The Larmor precession frequency ω_0 is given by

$$\omega_0 = \gamma H_0$$

or

$$v = \frac{\gamma H_0}{2\pi}$$

where γ is the magnetogyric ratio, which is related to μ by

$$\gamma = \frac{2\pi\mu}{Ih}$$

If the Larmor precession frequency induced by the magnetic field matches the natural frequency of the spin system, then resonance occurs. Thus, the signal of resonance to be observed depends on the values of two factors, γ and H_0, which are related by μ and γ. The NMR spectrometer is also specified by these two parameters.

Table 14.1 gives the values of μ and γ for some important nuclei that are currently used in NMR spectroscopy. From these values we can easily calculate H_0 and γ for some well-known NMR spectrometers. For example, for protons, we have

γ (MHz)	H_0 (T)
60	1.41
100	2.34
200	4.70
220	5.16
500	11.75

TABLE 14.1 Nuclei Magnetic Constants

Isotopes	Spin	μ (Nuclear Magneton)	$\gamma \times 10^{-8}$	Natural Abundance (%)
^{1}H	$\frac{1}{2}$	2.7927	2.6752	99.98
^{2}H	1	0.8574	0.4107	0.0016
^{13}C	$\frac{1}{2}$	0.7022	0.6726	1.1
^{14}N	1	0.4036	0.1933	99.6
^{15}N	$\frac{1}{2}$	−0.2830	−0.2711	0.37
^{19}F	$\frac{1}{2}$	2.6273	2.5167	100
^{31}P	$\frac{1}{2}$	1.1305	1.0829	100

For ^{13}C, we have

γ (MHz)	H_0 (T)
25.2	2.35
45	4.20
50	4.70

Notice that for the same strength magnetic field, 4.7 T, the resonance signal (Larmor frequency) occurs at 200 MHz for proton, but at 50 MHz for ^{13}C.

14.4.2 Continuous-Wave Method and Pulsed Fourier Transform Method

For many years the continuous-wave (cw) method was used to detect resonance, but now most measurements of resonance rely on the pulsed Fourier transform (FT) method. In the continuous-wave method, only a single frequency which is attributed to a magnetically equivalent group (for example, a methyl group) is excited and detected at any one moment. This is accomplished by monitoring the disturbance of the magnetization of a spin system with either the radio frequency varied or the applied field swept. In the Fourier transform method, the entire spectrum of frequencies is stimulated by a pulse of radio frequency energy and the response of the system is observed as a function of time. Using the Fourier transform method the time domain data are then transformed into the classical frequency domain spectrum.

The pulsed Fourier transform method itself has been developed from a one-pulse method to a multiple-phase method. The technique is becoming more and more advanced and complicated. Correspondingly, NMR as a tool has become more versatile in obtaining information about the structures and properties of molecules.

In the one-pulse method, the spin system after being magnetized in a magnetic field is perturbed from equilibrium with a pulse. The response of the system is monitored and the nuclear magnetic resonance is generated. The one-pulse method provides the same information as the cw method, but with better resolution.

By contrast, the multiple-phase method is divided into three separate time intervals: preparation, evolution, and detection. In the preparation period, the system is perturbed from equilibrium by a pulse or pulses. The system then evolves by some combination of precession and relaxation. We do not observe the time evolution directly. Instead, we determine evolution by a manipulation of the signal through an inversion–recovery experiment, which involves spin–lattice relaxation, and a spin-echo experiment, which involves spin–spin relaxation. In the detection period, the classical spectrum is obtained by rotating the frame frequencies, which include the effects of chemical shifts and of all the associated complex constants. The multiple-phase method is capable of handling complicated molecules.

14.4.3 Two-Dimensional NMR

Two-dimensional NMR is obtained by an advanced multiple-phase Fourier transform method. It uses a second Fourier transformation to convert the time dependence of the evolution into a second frequency. The double transformations may be described as

$$S(t_1, t_2) \xrightarrow[\text{FID}]{\text{FT over } t_2} S(t_1, f_2) \xrightarrow{\text{FT over } t_1} S(f_1, f_2)$$

Time domain Mixed domain Frequency domain

where S refers to a time-domain signal, FT refers to Fourier transformation, FID refers to the process of free induction decay, t_1 is the first time variable in pulse sequences, t_2 is the elapsed time of recording, f_1 is the frequency expected in the first time period, and f_2 is the frequency that is produced in a series of spectra manipulated by a variety of techniques. The first time variable t_1 is related to the evolution period, whereas the second time variable is related to the detection period:

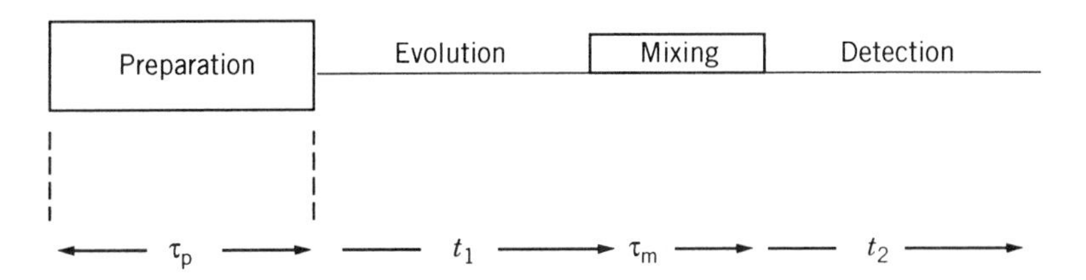

where τ_p and τ_m are relaxation times known to the operator of the spectrometer.

A two-dimensional NMR spectrum is often displayed in the form of a map. One axis, f_2, contains the conventional chemical shifts. The other axis, f_1 or

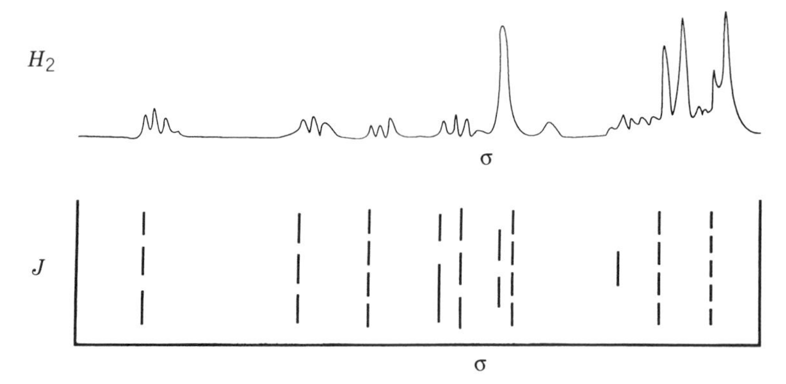

FIGURE 14.7 A 2D J spectrum. The 1D proton spectrum is overlaid on a 2D J spectrum.

t_1, may contain chemical shift or J-coupling information, or both. The position of a signal on the map tells us the chemical shift of that signal and the effect on that signal due to the presence of other nuclei.

There are two types of 2D NMR spectra: a J-resolved spectrum and a chemical shift correlated spectrum. In a J-resolved spectrum the complex and overlapping NMR signals are resolved into individual shifts and coupling patterns. This is obtained by J couplings that modulate the nucleus signal during t_1 operation. The spectrum is displaced as depicted in the form of J versus σ, shown in Figure 14.7. In a correlated spectrum, one dimension is the chemical shift of any type of spin nuclei, such as ^{13}C, ^{15}N, ^{31}P, and the other dimension is usually the chemical shift of the coupled spin proton. Each peak on a contour map gives the shifts of a carbon and its coupled counterpart proton. The correlated spectrum identifies spin-coupled pairs of nuclei in a molecule. An example is discussed in the next section about the polypeptide polybenzyl-γ-glutamate.

14.5 NMR SPECTRA OF MACROMOLECULES

In this section we first discuss two synthetic polymers, poly(methyl methacrylate) and polyproplyne, and one biological polymer lysozyme, to illustrate the proton and ^{13}C NMR spectroscopy of macromolecules. All three polymers have been well-investigated in other areas. Whenever a new research technique is developed, these polymers will always be targets for investigation. We then describe deuterium NMR spectroscopy by illustrating the characteristic features of the motion in the polyethylene molecule. The motion is caused by a change in thermal energy. Finally we describe polybenzyl-γ-glutamate as an example of two-dimensional NMR spectroscopy.

14.5.1 Poly(methyl methacrylate)

Below is the formula of the polymer molecule:

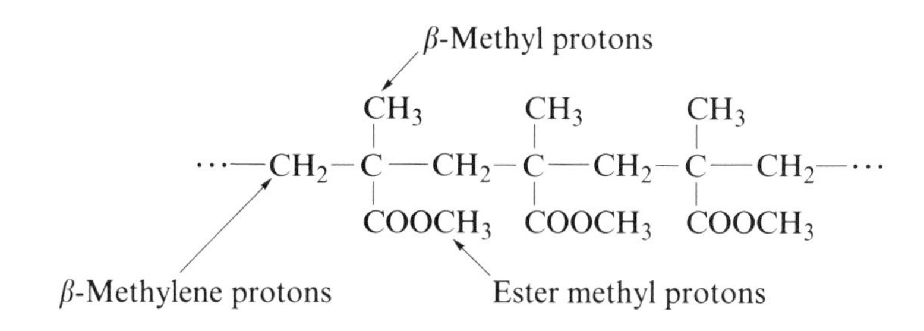

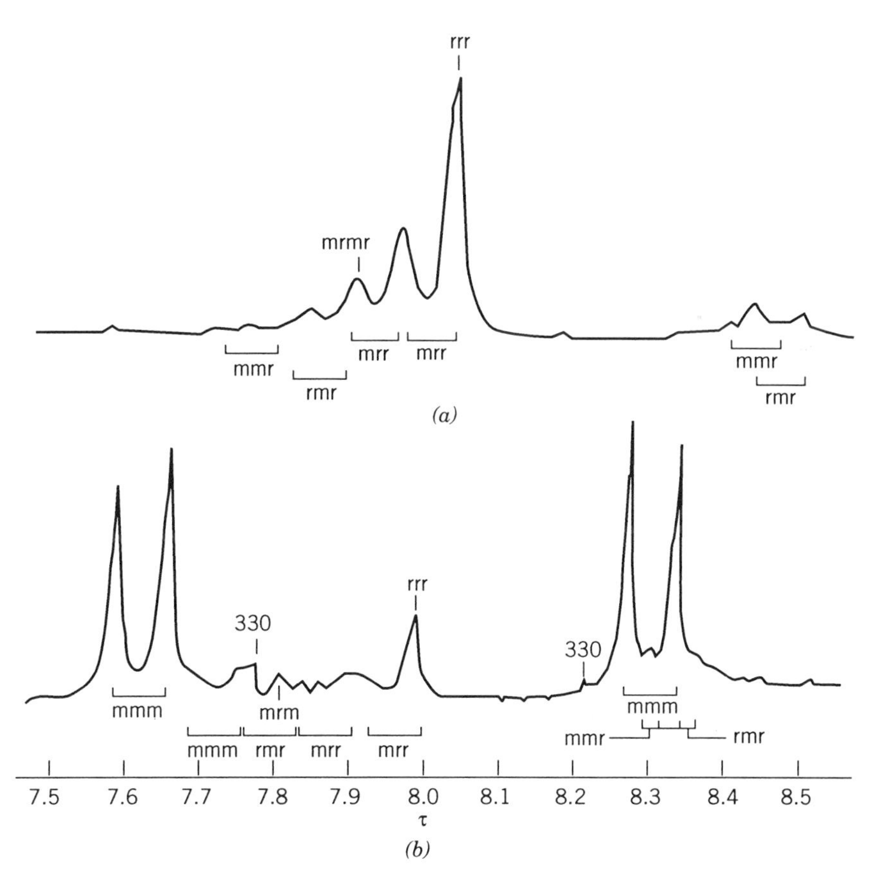

FIGURE 14.8 220-MHz β-methylene proton spectra of poly(methyl methacrylate) in chlorobenzene at 135°C: (a) syndiotactic; (b) isotactic. (From Bovey (1972) by permission of Dr. Bovey and Academic Press.)

Since there is no J coupling between the CH_2 and α-CH_3 protons the phenomenon for poly(methyl methacrylate) is relatively simple. In the proton NMR spectrum, the ester methyl protons appear near 6.5τ, the β-methylene protons appear near 8.0τ, and the β-methyl protons appear between 8.5 and 9.0τ. Figure 14.8 shows the β-methylene proton spectrum; Figure 14.9 shows the β-methyl proton spectrum. Both were observed at 220 MHz. In both spectra the ester methyl protons appear near 6.5τ, but, in order to show the details of β-methylene protons and β-methyl protons, they were not shown in the two figures. In both

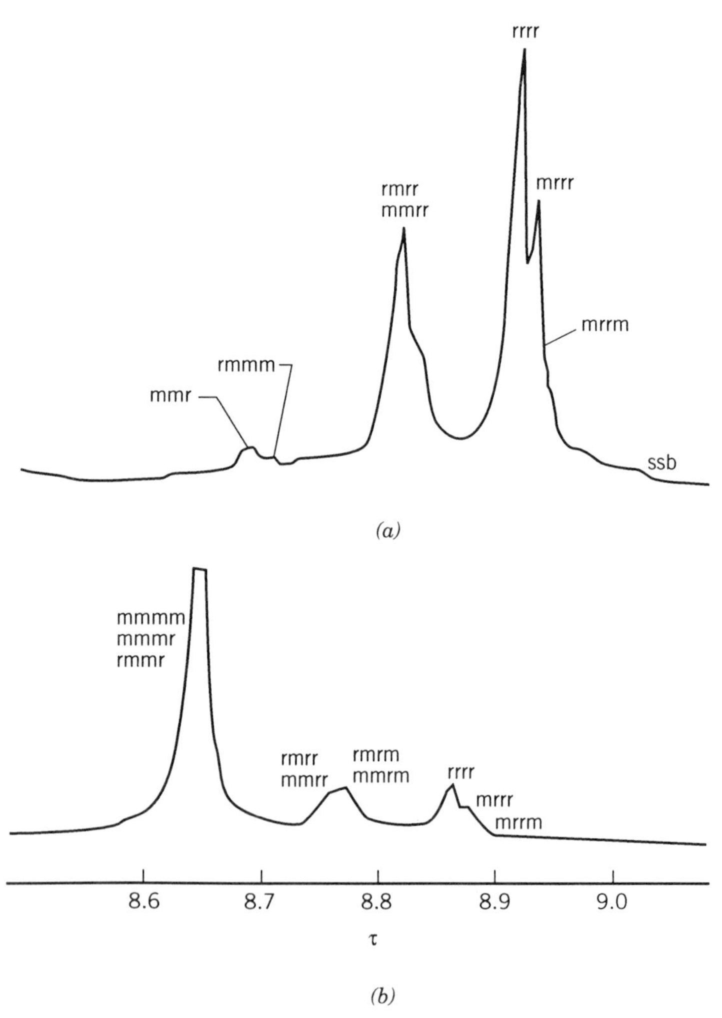

FIGURE 14.9 220-MHz ester methyl proton spectra of poly(methyl methacrylate) in chlorobenzene at 135°C: (a) syndiotactic; (b) isotactic. (From Bovey (1972) by permission of Dr. Bovey and Academic Press.)

cases the sample was prepared by dissolving poly(methyl methacrylate) in 10–15% chlorobenzene, with the temperature of the NMR measurement at 135°C.

The methylene spectrum of the syndiotactic isomer (Figure 14.8a) shows three peaks in the range between 7.5 and 8.5τ (only one being shown in the 60-MHz spectrum). The isotactic isomer (Figure 14.8b) in the same range shows five distinct peaks. The three peaks of the syndiotactic isomer are concentrated at 7.8–8.1τ; there are no signals between 7.5 and 7.8τ or between 8.1 and 8.5τ. The five distinct peaks of isotactic isomer spread from 7.5 to 8.5τ.

By comparison, the spectrum of β-methyl proton (Figure 14.9) spreads wider for the syndiotactic isomer than for the isotactic isomer. In the range between

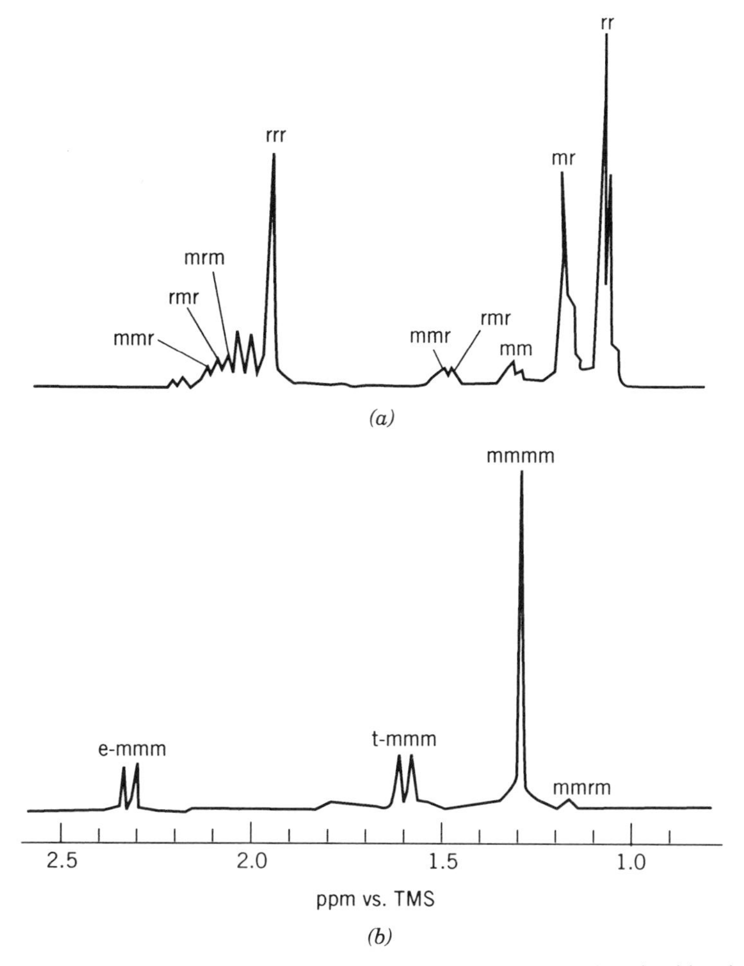

FIGURE 14.10 500-MHz proton spectra of poly(methyl methacrylate) in chlorobenzene at 100 °C: (a) syndiotactic; (b) isotactic. (From Bovey (1988) by permission of Dr. Bovey and Academic Press.)

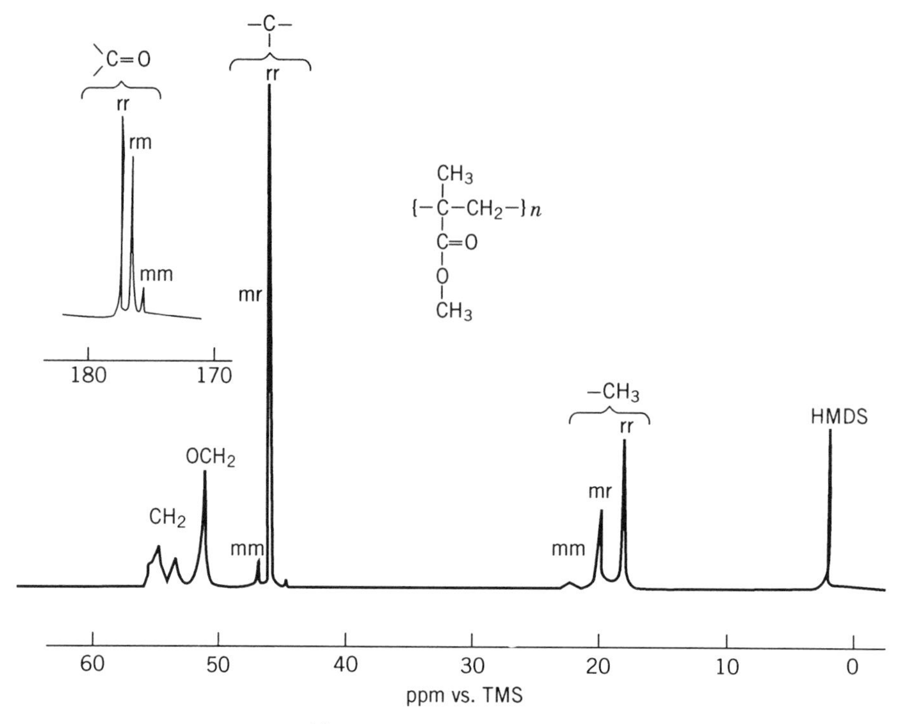

FIGURE 14.11 25.2-MHz ^{13}C spectrum of poly(methyl methacrylate) in 1,2,4-trichlorobenzene at 120°C. (From Randall, (1977) by permission of Academic Press.)

8.7 and 9.0τ, the syndiotactic isomer shows two gigantic peaks, whereas the isotactic isomer shows only two baby peaks. These comparisons clearly show the structural differences between the two isomers. The symbols m and r, which refer to the meso mixture and the racemic mixture, are indicated on each peak in the graphs. A mixture of isomers is said to be racemic if one form is indistinguishable from the other by NMR alone. Their enantiomers are dissymmetric and magnetic equivalent. If the macromolecule involves symmetry and their enantiomers are magnetic nonequivalent, the mixture of isomers is said to be meso. The nuclei of meso polymers have different chemical shifts and couplings, while the nuclei of the racemic isomers have identical chemical shifts and couplings.

Following is an example of meso and racemic mixture of a triad:

isotactic	atactic	syndistactic
C C C; R R R; m m	r; R; C C C; R R; r	r m; R R; C C C; R

Figure 14.10 is similar to Figures 14.8 and 14.9 except that a more powerful magnetic field has been applied and the resonance frequency changes from 220 to 500 MHz. The scales are also different. As expected the basic spectrum is the same, but more structural details are displayed.

Figure 14.11 shows the ^{13}C NMR spectrum of poly(methyl methacrylate) in 1,2,4-trichlorobenzene at 25.2 MHz and 120°C. Here the signals are of C instead of protons. The spectrum shows the stereochemistry of the main chain carbons

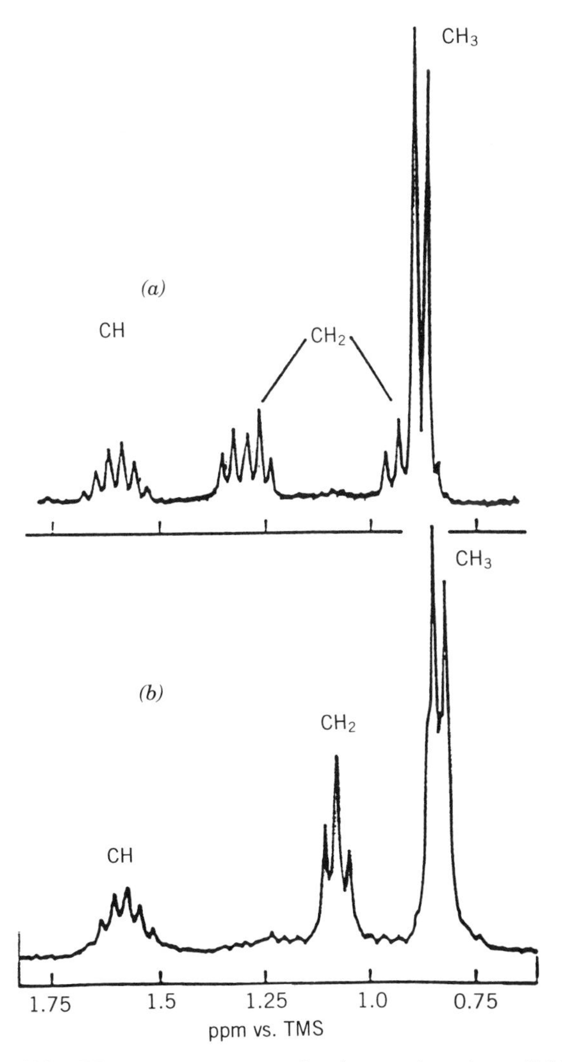

FIGURE 14.12 220-MHz proton spectra of polypropylene in *o*-dichlorobenzene at 165°C: (a) isotactic; (b) syndiotactic. (From Ferguson (1967) by permission of the New York Academic of Science.)

(CH_2 and $-\overset{|}{\underset{|}{C}}-$) and the chain C–$CH_3$ and OCH_3. The signals of $>C=O$ is off the scale and is shown in the left corner of Fig. 14.11.

14.5.2 Polypropylene

As seen in the formula of polypropylene,

$$\left[\begin{array}{cc} \overset{\displaystyle CH_3}{\overset{|}{CH}} - CH_2 \\ \alpha \quad \beta \end{array} \right]$$

the proton spectrum is complicated due to the vicinal coupling between α and β protons and between α and methyl protons, together with the geminal methyl

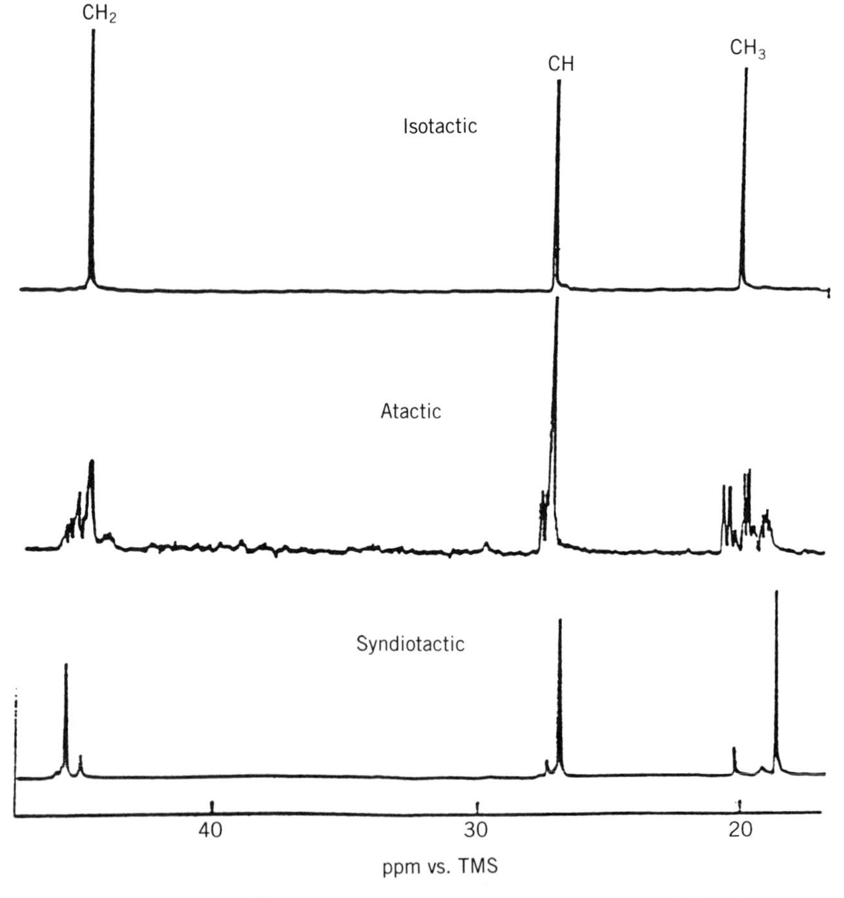

FIGURE 14.13 25-MHz ^{13}C spectra of polypropylene in 1,2,4-trichlorobenzene at 140°C. (From Bovey (1988) by permission of Dr. Bovey and Academic Press.)

proton coupling in the case of the isotactic polymer. Figure 14.12 shows the 220-MHz proton spectra of isotactic and syndiotactic polypropylene in *o*-dichlorobenzene at 165°C. The β protons of the isotactic polymer are more widely spread than those of the syndiotactic polymer. The proton spectrum of atactic polypropylene is not shown, because of a complex of overlapping multiplets. The situation is different in the case of ^{13}C spectra, where we can have three distinct spectra of all three forms of the polymer. Figure 14.13 shows

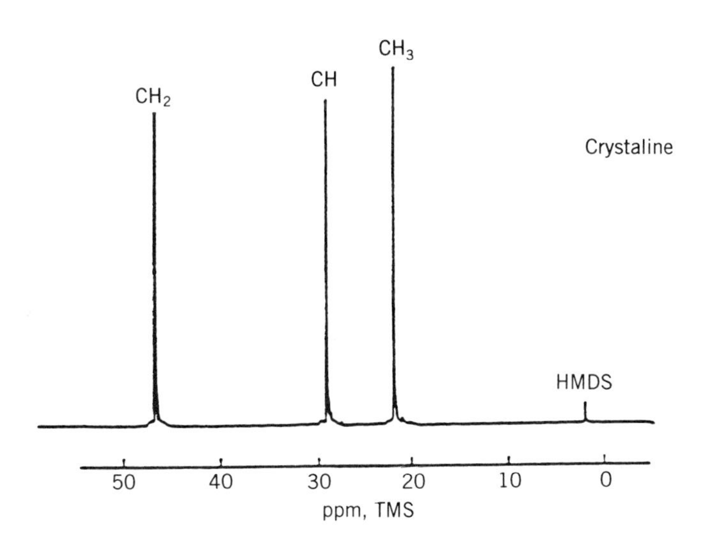

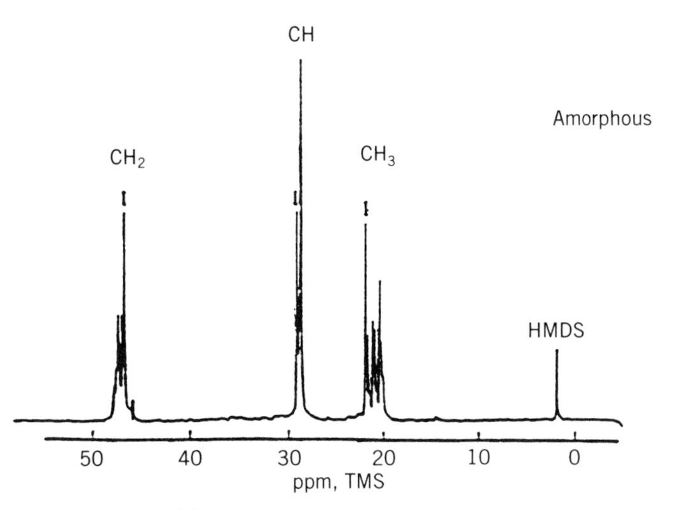

FIGURE 14.14 25.2-MHz ^{13}C spectra of polypropylene in 1,2,4-trichlorobenzene at 120°C. (From Randall (1977) by permission of Academic Press.)

25-MHz ^{13}C NMR spectra for comparing the structural differences. ^{13}C NMR spectra can also show the difference between crystalline polypropylene from the amorphous polypropylene, as in Figure 14.14.

14.5.3 Lysozyme

Proteins are more complicated than synthetic polymers. The NMR spectra of native proteins are very sensitive to perturbation by the environment, for example, temperature, pH, and ionic strength. A large number of individual resonances of protons overlap to form an envelop, thereby severely limiting resolution. Nevertheless, as with other types of spectroscopy, the characteristic band of individual amino acids can be reasonably determined from standard under specific conditions. For example, the resonance triangle positions of protons from internal DDS (sodium 2,2-dimethyl-2-silapentane-5-sulfornate) dissolved in D_2O at pD 7 were found as follows: valine β-CH, 495 Hz; glutamic acid β-CH_2, 435 Hz; glutamine β-CH_2, 455 Hz; methionine CH_3, 454 Hz; aspartic acid β-CH_2, 590 Hz; crysteine β-CH_3, 665 Hz; tryptophan β-CH_2, 745 Hz; and histidine imidazole C-2, 1740 Hz. These amino acids constitute the side chains of lysozyme in contact with solvent under random coil condition. In Figure 14.15, the 220-MHz proton spectrum of lysozyme in D_2O at 79°C, we see significant peaks in the regions near 454, 610, 665, and 704 Hz. The peak at 454 Hz may be assigned to methionine methyl protons, that at 610 Hz to aspartic acid, which has been shifted from 590 Hz as expected, that at 665 Hz to the cysteine β protons, and that at 704 Hz to arginine δ-CH_2 which may also have been contributed from tryptophan β protons. The resonances expected to occur for valine, glutamic acid, and glutamine protons are not observed, nor are histidine imidazole C-2. These peaks have been either broadened or shifted to other regions.

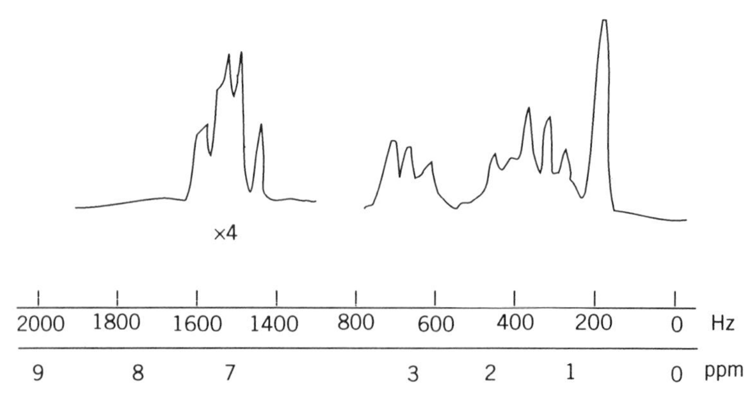

FIGURE 14.15 220-MHz proton NMR spectrum of lysozyme in D_2O at 80°C (random coil). From McDonald and Phillips (1969) by permission of the American Chemical Society.)

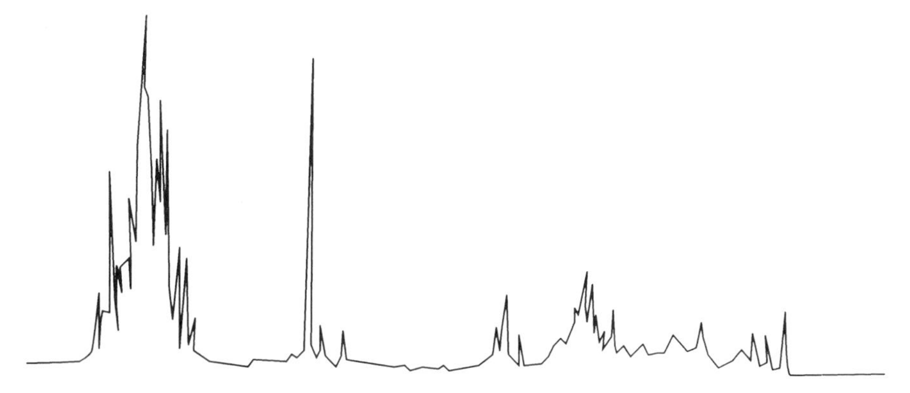

FIGURE 14.16 45-MHz ^{13}C spectrum of native lysozyme. (From Levy, et al. (1980). Reproduced by permission of John Wiley.)

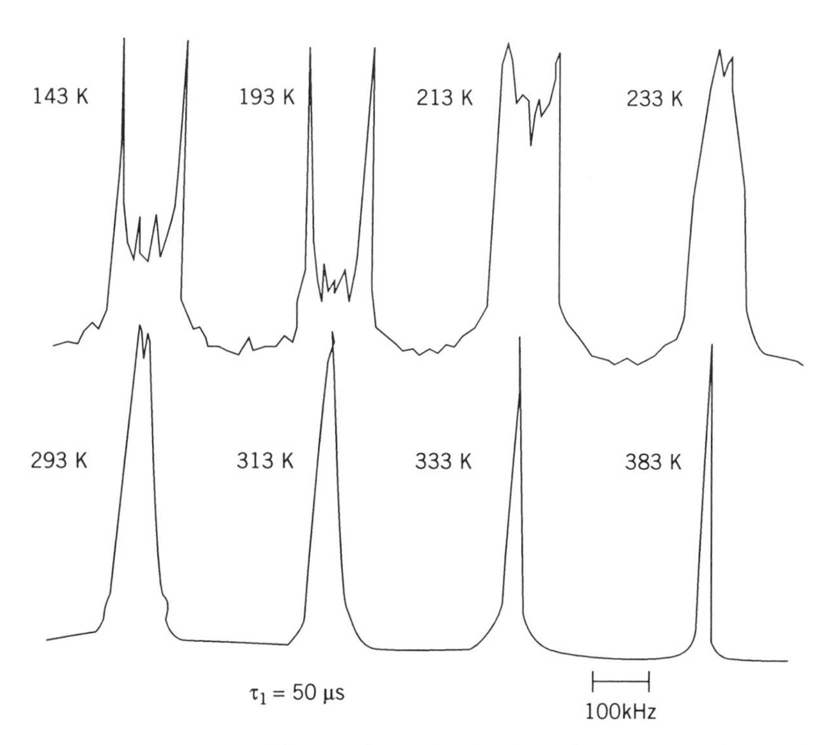

FIGURE 14.17 Chain mobility as shown in the deuterium spectra of amorphous regions of linear polyethylene at various temperatures. (From Speiss (1985). Reproduced by permission of the publisher Springer-Verlag.)

The ^{13}C NMR spectrum appears to be much superior in revealing the resonance triangular positions of amino acids, as shown in Figure 14.16. Individual amino acid residues are resolved for several carbons.

14.5.4 Deuterium NMR Spectra of Chain Mobility in Polyethylene

The deuterium NMR spectrum is basically the observation of the quadrupole coupling, $e^2 q_{\alpha\beta} Q/h$, where e is the electric charge, $q_{\alpha\beta}$ is the electric field gradient tensor defined as

$$q_{\alpha\beta} = \left(\frac{\partial^2 v}{\partial r_\alpha \, \partial r_\beta} \right)_{r=0}$$

Q is the quadrupole moment, and h is the Planck's constant. The magnitude of the splitting may be described by

$$\delta v = \frac{3}{2} \left(\frac{e^2 q_{\alpha\beta} Q}{h} \right) \left(C_1 S + \frac{3}{2} C_2 K \right)$$

where C_1 and C_2 are related coordinates, and S and K are ordering parameters which give the orientation of the molecules in the magnetic field. The nucleus ^2H is well suited for studying molecular order and mobility in polymers. The intermolecular interaction is usually measured in terms of the spin–lattice relaxation time τ_1, whereas the mobility is measured through the change of line shapes. Figure 14.17 shows the chain mobility in the amorphous regions of linear polyethylene, which is compared to the chain motion of polystyrene in the vicinity of the glass transition.

14.5.5 Two-Dimensional NMR Spectra of Poly-γ-benzyl-L-glutamate

The polypeptide, poly-γ-benzyl-L-glutamate, may be labeled as follows:

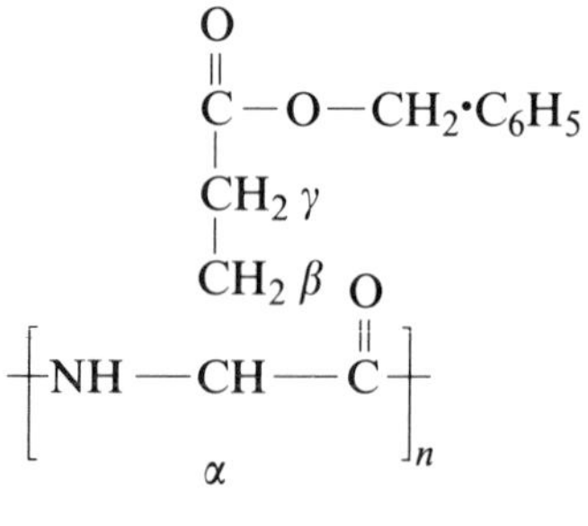

```
        O
        ||
        C—O—CH2·C6H5
        |
        CH2 γ
        |
        CH2 β  O
        |      ||
 [—NH —— CH —— C—]n
         α
```

Figure 14.18 shows a one- and a two-dimensional NMR spectrum of a polypeptide.

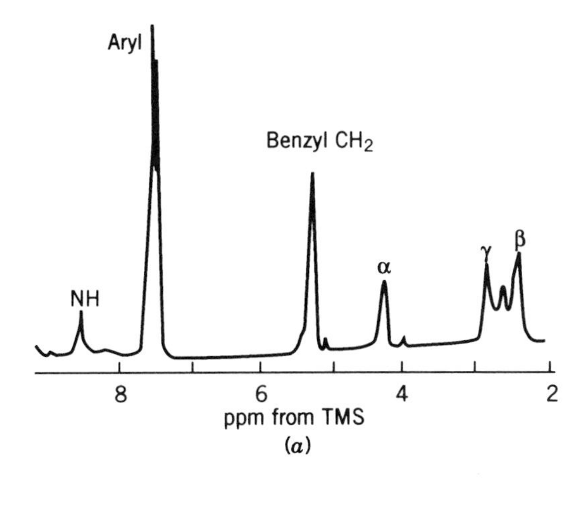

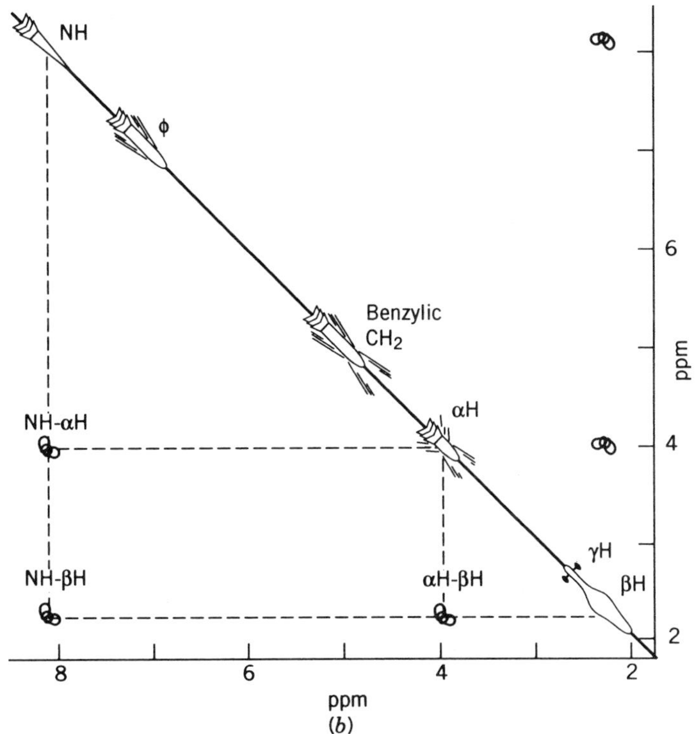

FIGURE 14.18 500-MHz proton spectra of poly-γ-benzyl-L-glutamate 20-mer in α-helical form in 95.5 chloroform trifluoroacetic acid at 120°C: (a) one dimensional; (b) two dimensional. (From Mirau and Bovey (1986) and Bovey (1988). Reproduced by permission of Dr. Bovey and the American Chemical Society as well as Academic Press.)

REFERENCES

Atkins, P. W., *Phys. Chem.* 3rd ed. New York: Freeman, 1986.

Bovey, F. A., *High Resolution Nuclear Magnetic Resonance of Macromolecules.* New York: Academic, 1972.

Bovey, F. A., *Nuclear Magnetic Resonance Spectroscopy*, 2d ed. New York: Academic, 1988.

Breitmaier, E., and W. Voelter, *^{13}C NMR Spectroscopy: Methods and Applications in Organic Chemistry.* Weinheim: Verlag Chemie, 1978.

Croasmun, W. R., and R. M. K. Carlson (Eds.), *Two-Dimensional NMR Spectroscopy: Application for Chemists and Biochemists.* New York: VCH, 1987.

Derome, A. E., *Modern NMR Techniques for Chemistry Research.* New York: Pergamon, 1987.

Ernst, R. R., G. Bodenhausen, and A. Wokaun, *Principles of Nuclear Magnetic Resonance in One and Two Dimensions.* Oxford, UK: Clarendon, 1987.

Everhart, C. H., and C. S. Johnson, *J. Magn. Reson.* **48**, 466 (1982).

Ferguson, R. C., *Trans. N.Y. Acad. Sci.* **29**, 495 (1967).

Harris, R. K., *Nuclear Magnetic Resonance Spectroscopy.* London: Pitman, 1983.

James, T. L., *Nuclear Magnetic Resonance in Biochemistry.* New York: Academic, 1975.

Karplus, M., *J. Chem. Phys.* **30**, 11 (1959).

Levy, G. C. (Ed.), *Topics in Carbon-13 Nuclear Magnetic Resonance*, Vol. 1. New York: Wiley–Interscience, 1974.

Levy, G. C., *NMR Spectroscopy: New Methods and Applications*, ACS Symposium Series 191. Washington, DC: American Chemical Society, 1982.

Levy, G. C., R. L. Lichter, and G. L. Nelson, *Carbon-13 Nuclear Magnetic Resonance Spectroscopy.* New York: Wiley, 1980.

Lyerla, J. R., *Methods Exp. Phys.* **16A**, 241 (1980).

McDonald, C. C., and W. D. Phillips, *J. Am. Chem. Soc.* **91**, 1513 (1969).

Mirau, P. A., and F. A. Bovey, *J. Am. Chem. Soc.* **108**, 5130 (1986).

Overhauser, A. W., *Phys. Rev.* **91**, 476 (1953).

Randcell, J. C., *Polymer Sequence Determination: Carbon-13 NMR Method.* New York: Academic, 1977.

Sanders, J. K. M., and B. K. Hunter, *Modern NMR Spectroscopy: A Guide for Chemists.* Oxford, UK: Oxford University Press, 1987.

Spiess, H. W., *Adv. Polym. Sci.* **66**, 23 (1985).

PROBLEMS

14.1 Polymer chain conformation can often be deduced from the coupling constant ΔJ, which is defined as

$$\Delta J = J_{AX} - J_{BX}$$

where J_{AX} and J_{BX} are vicinal ^{1}H–^{1}H coupling constants, as shown in the diagram:

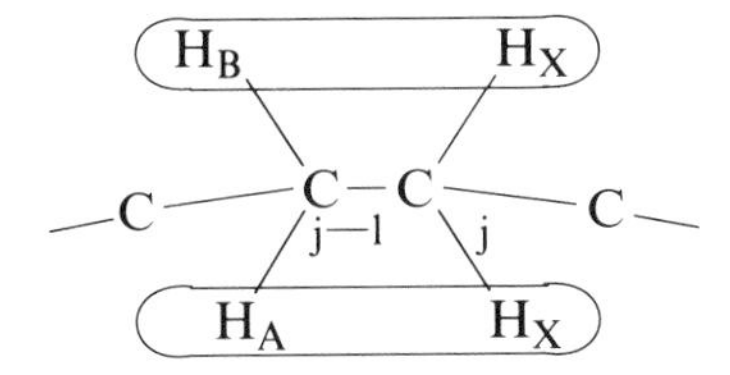

For a two-state conformational system (t, *trans*; g, *gauche*) we have the following equations to calculate:

$$J_{AX} = P_{t,j}(J_{AX,t} - J_{AX,g}) + J_{AX,g}$$

$$J_{BX} = P_{t,j}(\Delta J_t - \Delta J_g) + \Delta J_g$$

where

$$\Delta J_t = J_{AX,t} - J_{BX,t}$$

$$\Delta J_g = J_{AX,g} - J_{BX,g}$$

Assume that

$$J_{vic} = 4.22 - 0.5\cos\phi + 4.5\cos^2\phi \qquad (0 \leqslant \phi \leqslant 180^\circ)$$

Show that $\Delta J = 2\Delta J_t(2P_{t,j} - 1)$ and calculate ΔJ for isotactic poly(methyl acrylate). (*Source*: Lyerla (1980).)

14.2 The Fourier pair relationship between the time (t) and frequency (ω) is expressed in the form

$$F(\omega) = \int_{-\infty}^{\infty} f(t)\,e^{-i\omega t}\,dt$$

The shape of a spectral line $J(\omega)$, which is essentially the time dependence of the oscillations responsible for absorption, is the real part of the integral:

$$J(\omega) \sim \int_{0}^{\infty} f(t)\,e^{i\omega t}\,dt$$

Let $f(t)$ be the oscillating, decaying function:

$$f(t) = \cos\omega_0 t\,e^{-t/\tau}$$

where τ is the time constant for the decay. Plot the line shape for different values of τ. (*Source*: Atkins (1986).)

14.3 The pulsed field gradient NMR method can be used to determine the diffusion coefficients D for protein by

$$\ln R = -\gamma^2 \delta G^2 \left(\Delta - \frac{\delta}{3} \right) D$$

where R is the ratio of the amplitude of the spin echo in the presence of the gradient pulses to the amplitude in the absence of the gradient pulses, γ is the gyromagnetic ratio of the proton nuclei, and G is the amplitude of the field gradient pulse (G/cm). Experimental data were obtained for hemoglobin in buffer at 25°C as follows:

R	G^2
1.31	0.775
1.29	0.866
1.28	0.938
1.26	1.00
1.24	1.14

Experimental parameters are $\delta = 1.75$ m/s, $\Delta = 8.10$ m/s, $\tau = 6.0$ m/s.

(a) Plot $\ln R$ versus G^2.

(b) Determine the translation diffusion coefficient D for the protein.

(*Source*: Everhart and Johnson (1982).)

15

X-RAY CRYSTALLOGRAPHY

15.1 X-RAY DIFFRACTION

X-rays are produced by bombarding a metal with high-energy electrons. They can be diffracted when passed through a crystal (first suggested by von Laue in 1912). X-ray diffraction depends on the angle of the incident beam, as shown in Figure 15.1. This observation was described in the Bragg equation, in the form

$$n\lambda = 2d \sin \theta \tag{15.1}$$

where λ is the wavelength of the incident x-ray (in the range between 0.001 and 50 Å), d is the interplanar spacing of the crystal, and n is an integer ($n = 1, 2, \ldots$).

Analysis of x-ray diffraction gives numerical values of two important parameters: the interplanar spacing and the intensity of diffraction. Interplanar spacing is characteristic of the pattern of the crystal from which one learns the packing of the repeating units, while the intensities of a certain number of diffractions can provide information on the structure of a crystal.

For the past several decades x-ray studies have provided a genuine elegance to macromolecular chemistry. Investigators have relied much on x-ray crystallography to develop a sense of how synthetic and biological polymers are shaped. Since the molecule under study must be crystallized, we begin our description with crystals. The structure of a crystal is the symmetrical arrangement of one or more species of atoms in three directions at certain angles, including right angles. A modern x-ray analysis of a crystal involves the determination of the quantity $F(hkl)$ which is a measure of the intensity of the x-beam, for each

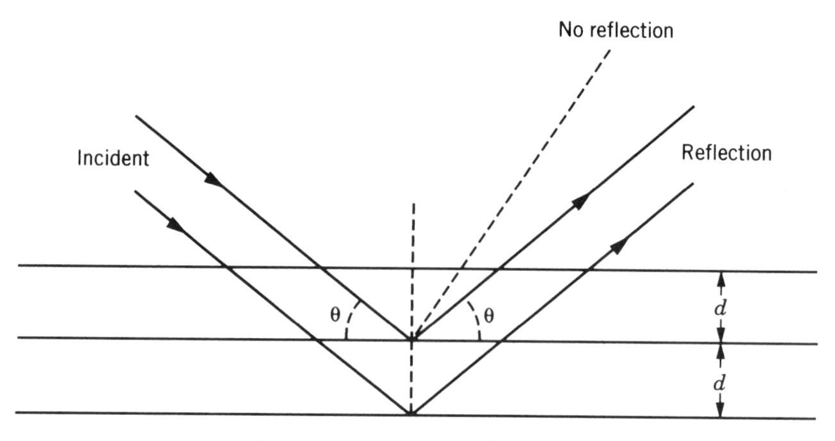

FIGURE 15.1 X-ray diffraction.

order (hkl), scattered by the whole unit of a pattern. The final result is the detailed information about the geometry of a macromolecule including the specific places where the atoms or the groups of atoms (e.g., amino acids in protein) are packed.

15.2 CRYSTALS

A crystal consists of periodic repetitions of some molecular arrangement, known as a "unit cell," throughout space. The array of points that constitutes the regular arrangement is represented by lattices.

15.2.1 Miller Indices, *hkl*

Consider two-dimensional lattices (the array of points in Figure 15.2). The four lattices in two dimensions can be represented by four sets of numbers (x, y):

$$(1, 1) \qquad (-1, 1) \qquad (3, 2) \qquad (\infty, 1)$$

For three dimensions we just add another number to each set (x, y, z). If the plane cuts through the z axis, as in Figure 15.3 (or parallel to z), that is, if z intercepts at ∞, then we have four sets of trios:

$$(1, 1, \infty) \qquad (-1, 1, \infty) \qquad (3, 2, \infty) \qquad (\infty, 1, \infty)$$

These trios are expressed in what is known as the Weiss system. Miller indices are formed by taking two additional steps. First, the reciprocal of the three

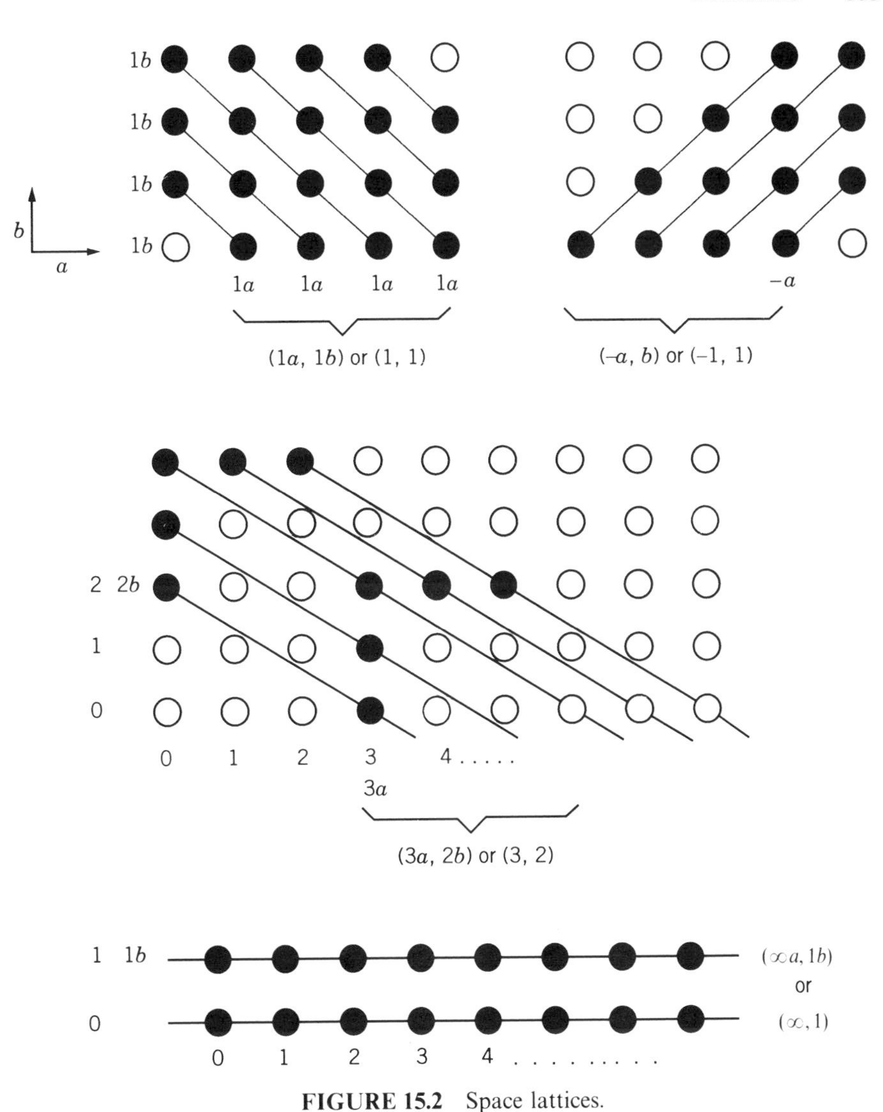

FIGURE 15.2 Space lattices.

numbers in parentheses is taken:

$$\left(\frac{1}{1}, \frac{1}{1}, \frac{1}{\infty}\right) \quad \left(\frac{1}{-1}, \frac{1}{1}, \frac{1}{\infty}\right) \quad \left(\frac{1}{3}, \frac{1}{2}, \frac{1}{\infty}\right) \quad \left(\frac{1}{\infty}, \frac{1}{1}, \frac{1}{\infty}\right)$$

Second, the denominator is removed by multiplying each number with the least common factor so that the trios are made of integers:

$$(1, 1, 0) \quad (\bar{1}, 1, 0) \quad (2, 3, 0) \quad (0, 1, 0)$$

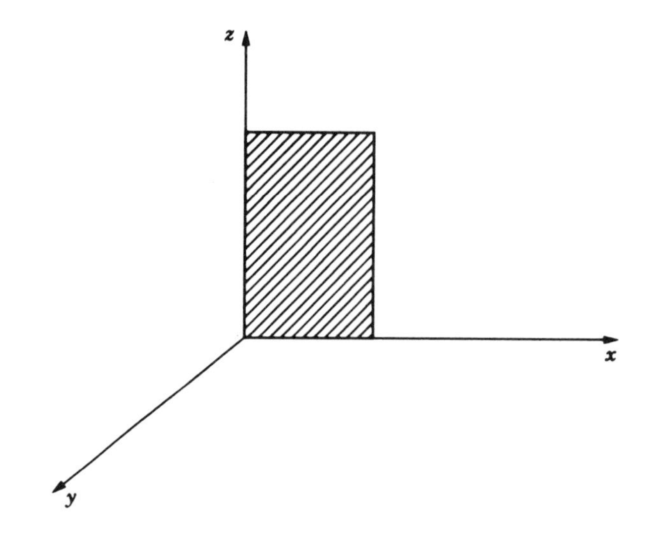

FIGURE 15.3 A representative plane cutting through the z axis.

The integers in each trios are called Miller indices and are expressed in terms of (hkl).

Figure 15.4 illustrates four of the simplest planes expressed in terms of Miller indices hkl. Planes with indices, for example, (200) in dash lines, have half the spacing of planes with indices (100). A very important property of crystals is that for each crystal face there exists three integers h, k, and l such that

$$hx + ky + lz = \text{constant} \tag{15.2}$$

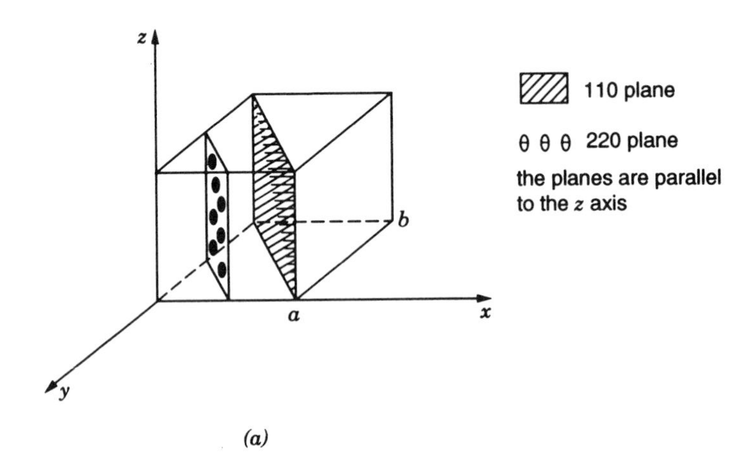

FIGURE 15.4 Four representative planes and Miller indices: hatchmarks, 110 plane; circles, 220 plane. The planes are parallel to the (a) z axis; (b) y and z axes; (c) x and z axes; (d) x and y axes.

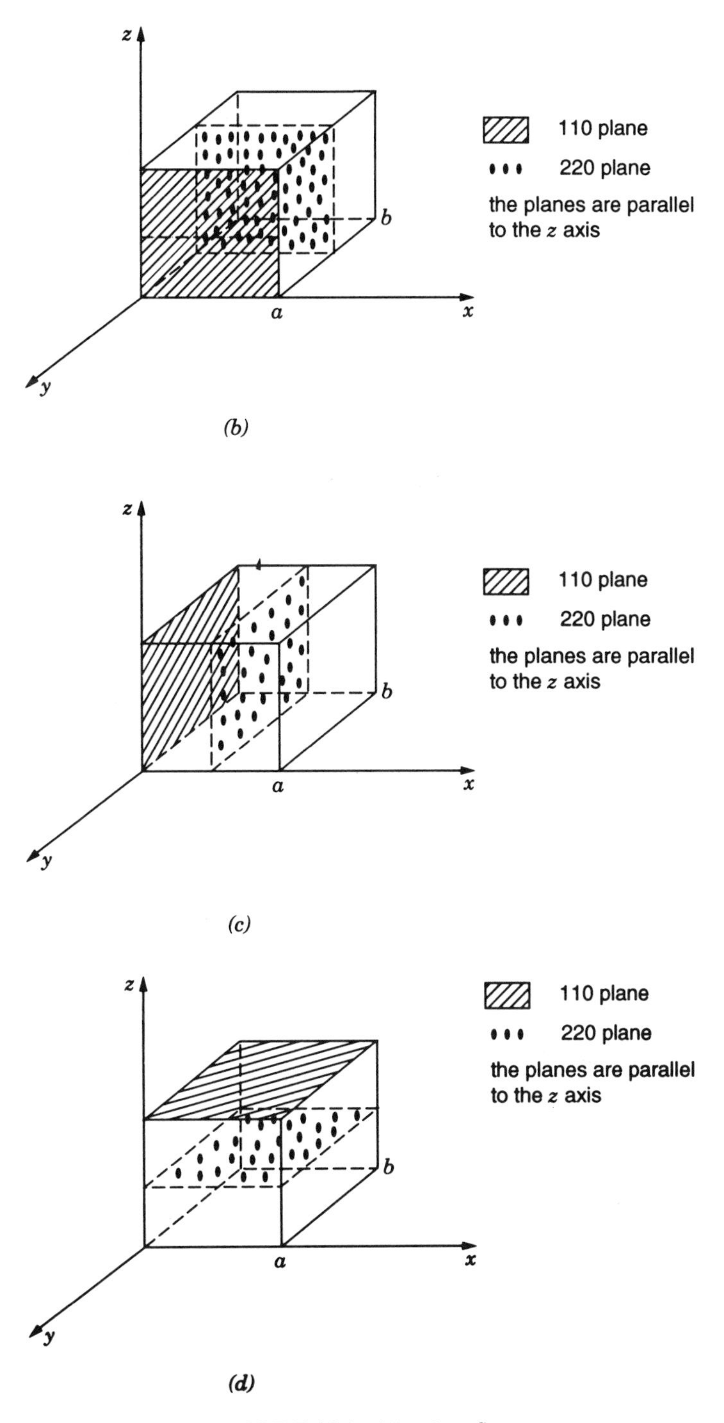

FIGURE 15.4 (*Continued*)

15.2.2 Unit Cells or Crystal Systems

The unit cell is the repeating unit in a crystal system. The coordinate system used to describe the unit cell is x, y, and z. The unit of each coordinate is a, b, and c, respectively (Figure 15.5). However, x, y, and z do not have to be Cartesian, that is, α, β, and γ could be any angle, and not necessarily $\pi/2$ (Figure 15.6). For the coordinates x, y, and z and the angles α, β, and γ there are seven different kinds of unit cells, referred to as crystal systems:

I. Cubic	$\alpha = \beta = \gamma = 90°$	$a = b = c$
II. Tetragonal	$\alpha = \beta = \gamma = 90°$	$a = b \neq c$
III. Hexagonal	$\alpha = \beta = 90°, \gamma = 120°$	$a = b \neq c$
IV. Rhombohedral	$\alpha = \beta = \gamma \neq 90°$	$a = b = c$
V. Orthorhombic	$\alpha = \beta = \gamma = 90°$	$a \neq b \neq c$
VI. Monoclinic	$\alpha = \gamma = 90°, \beta \neq 90°$	$a \neq b \neq c$
VII. Triclinic	$\alpha \neq \beta \neq \gamma \neq 90°$	$a \neq b \neq c$

Since angle and distance are the two parameters that characterize the repeating units, the seven crystal systems can further be classified into 14 lattices, called Bravais lattices. The classification is based on the position of lattice point(s) carried in the unit cell, namely, P (primitive), I (body centered), F (face-centered) and C (end-centered). The 14 lattices are:

1. Simple (or primitive) cubic (I P)
2. Body-centered cubic (I I)

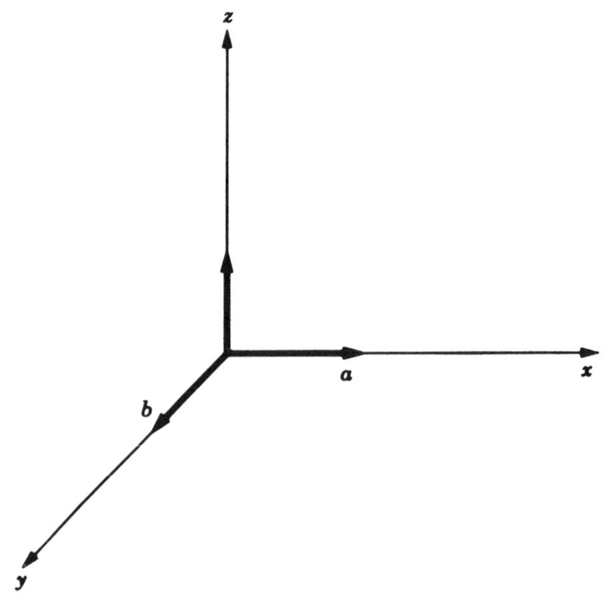

FIGURE 15.5 Crystal coordinates.

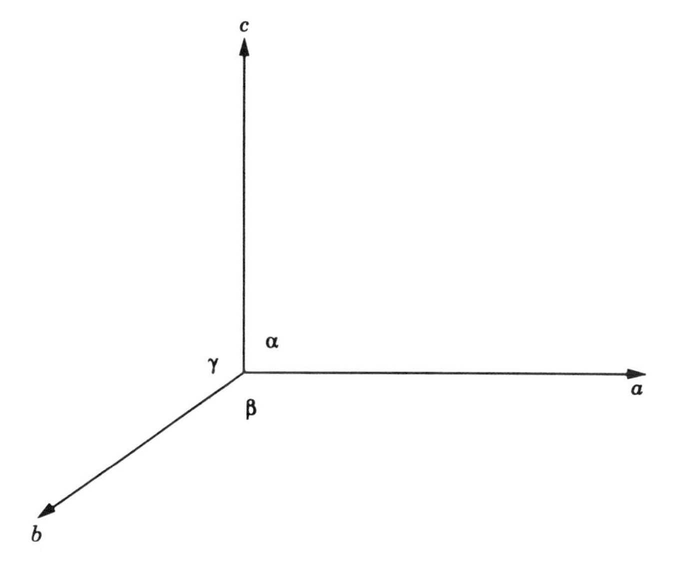

FIGURE 15.6 Crystal angles.

3. Face-centered cubic (I F)
4. Simple tetragonal (II P)
5. Body-centered tetragonal (II I)
6. Hexagonal (III P)
7. Rhombohedral (trigonal) (IV P)
8. Simple orthorhombic (V P)
9. End-centered orthorhombic (V C)
10. Body-centered orthorhombic (V I)
11. Face-centered orthorhombic (V F)
12. Simple monoclinic (VI P)
13. End-centered monoclinic (VI C)
14. Triclinic (VII P)

These 14 lattices are shown in Figure 15.7. In addition, if a unit cell contains only one repeating unit, it is said to be primitive; if a unit cell contains more than one repeating unit, it is said to be centered.

For purposes of data analysis, let us first review the two-dimensional lattice. In Figure 15.8, d_{hk} is the interplanar spacing of the crystal and a is the length of the unit cell. According to the Pythagorean theorem, d_{hk} and a are related by the simple equation

$$d_{hk} = \frac{a}{(h^2 + k^2)^{1/2}}$$

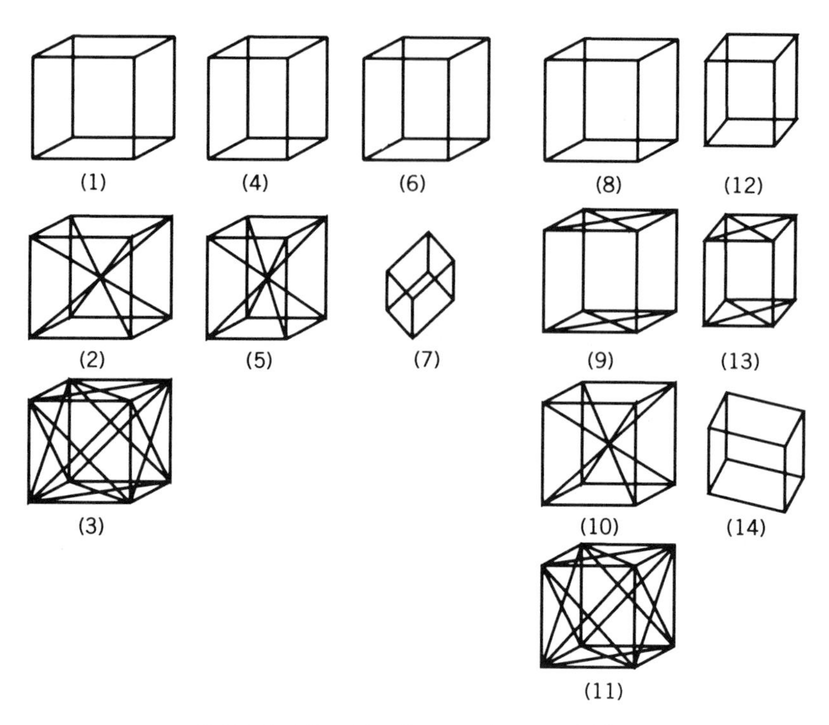

FIGURE 15.7 The fourteen Bravais lattices.

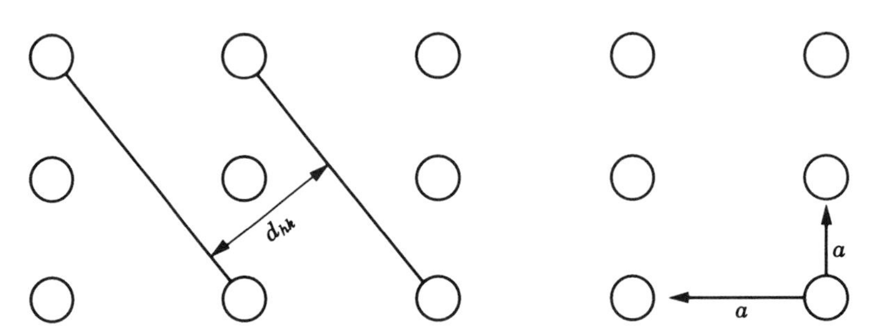

FIGURE 15.8 The interplanar spacing of a crystal.

By extending to three dimensions, we have

$$d_{hkl} = \frac{a}{(h^2 + k^2 + l^2)^{1/2}} = \frac{a}{M} \tag{15.3}$$

where

$$M = \sqrt{h^2 + k^2 + l^2}$$

TABLE 15.1 Possible Values of M^2 for a Cubic Lattice

hkl	M	M^2	Simple Cubic	Body-Centered Cubic	Face-Centered Cubic
100	1.0000	1	×		
110	1.4142	2	×	×	
111	1.7321	3	×		×
200	2.000	4	×	×	×
210	2.2361	5	×		
211	2.4495	6	×	×	
There is no hkl value that makes M^2 equal to the integer 7.					
220	2.8284	8	×	×	×
—	⋮	⋮			
—	⋮	no 15			
—	⋮	⋮			

Since the experimental values of θ, n, and λ (the wavelength of the x-ray) are known, we can calculate d_{hkl} using Bragg's equation:

$$2d_{hkl}\sin\theta = n\lambda \tag{15.4}$$

Once the d_{hkl} value is known, we can calculate the values a, h, k, and l. This is done by rearranging Eq. (15.3) in a slightly different form:

$$\left(\frac{1}{d_{hkl}}\right)^2 = \left(\frac{1}{a}\right)^2 M^2 = \left(\frac{1}{a}\right)^2 (h^2 + k^2 + l^2) \tag{15.5}$$

The crucial part is the analysis of the values of M^2. Table 15.1 gives some possible values of M^2 calculated from Miller indices. For example, let us take the values of M^2 to be 1, 2, and 3 and calculate a. If the three calculated values of a are in good agreement, we can immediately conclude with reference to Table 15.1 that the crystal is a simple cubic with the three reflections (100), (110), and (111). If the three calculated values of a are not in agreement, we can try another set of M^2 values: 2, 4, and 6 corresponding to the three reflections (110), (200), and (211) to see whether the crystal is body-centered. We proceed in this fashion until we find calculated values of a that are in agreement, which tells us both the type of crystal lattice and the length of the unit cell of the cubic simultaneously.

15.2.3 Symmetry of Crystals

Crystals exhibit six symmetry elements, also called the six operators:

1. *Inversion centers*
   Symbol: i

2. *Rotation axis*
   Symbols: 1 (identity), 2 (twofold rotation), 3 (threefold rotation), up to 6 (sixfold rotation)
3. *Mirror planes*
   Symbol: m
4. *Rotation–inversion axes* (combination of rotation and inversion)
   Symbols: $\bar{1}$ (a simple center of conversion), $\bar{2}$ (a 180° rotation followed by an inversion, $\bar{2} = m$), $\bar{3}$ $(= 3 \cdot \bar{1})$, $\bar{6}$ $(= 3 \cdot m)$
5. *Screw axes*
   Symbols: 2_1 (twofold screw rotation), 3_1 (threefold screw rotation),

   $4_1, 4_2, 4_3$

   (Note: 4_1 and 4_3 are enantiomorphs, that is, mirror images)
6. *Glide plane* (a combination of a translation and a reflection across a plane)
   Symbols: a, b, and c

From these symmetry elements, crystallographic point groups and space groups are constructed. Crystallographic point groups, or simply point groups, are also known as crystal classes. There are 32 point groups, each of which is either one or a combination of the following elements:

$$1, 2, 3, 4, 6, \bar{1}, \bar{2}, \bar{3}, \bar{4}, \bar{6} \quad \text{(Note: 5 and } \bar{5} \text{ are forbidden)}$$

Those without bars, $(1, 2, 3, 4, 6)$, are axes of symmetry of order (rotation), while those with bars, $(\bar{1}, \bar{2}, \bar{3}, \bar{4}, \bar{6})$, are inversion of axes. The two sets together form a 10-point group: $\{1, 2, 3, 4, 6, \bar{1}, \bar{2}, \bar{3}, \bar{4}, \bar{6}\}$. The element $\bar{2}$ denotes a plane of symmetry and is identical to the symbol m: $\bar{2} = m$. To combine a plane of symmetry, or construct a twofold axis of symmetry, we add $/m$ or m to an element, as in the following examples:

$$
\begin{array}{ll}
1 + /m = 1/m & \equiv m \equiv \bar{2} \\
2 + /m = 2/m & \\
3 + /m = 3m & \equiv \bar{6} \\
4 + /m = 4/m & \\
5 + /m = 5/m & \text{This does not exist.} \\
6 + /m = 6/m &
\end{array}
$$

and

$$2 + m = 2m$$

Similarly, we have

$$
\begin{array}{l}
2m, 3m, 4mm, 6mm \\
\bar{3}m, \bar{4}m, (= \bar{4}2m = \bar{4}m2) \\
\bar{6}m^2
\end{array}
$$

In addition, we can also add 2 or $/mm$ to an element to form

$$222,\ 32\ (\text{note: no. }23),\ 42,\ 62$$
$$2/mm(=mmm),\ 4/mmm,\ 6/mmm$$

Note that among $1/m$, $2/m$, $3/m$, $4/m$, and $6/m$ only three are new: $2/m$, $4/m$, and $6/m$.

Adding $1/m$ is the adding of symmetry at a right angle; adding m is the adding of symmetry vertically. The operation of adding a symmetry element results in the creation of another symmetry such as 22, 32. Also one can add more than one plane of symmetry, that is, a plane can be added through the axis and a plane can also be added at right angles through the crystal. For example, $4mm$ and $4/mmm$. Figure 15.9 shows diagrams of the 32 point groups, and Figure 15.10 illustrates that certain operations are identical, for example $\bar{1}/m = 2/m$, $\bar{2}/m = m$.

Space groups consist of a symmetry operation of the point groups and a translation. The translation is the result of a screw operation and a glide operation. A screw axis is a clockwise rotation about an n-fold axis during translation. The symbols for a screw axis are as follows:

2_1	a twofold screw axis which causes a 180° rotation and a $\frac{1}{2}$ unit cell translation
$3_1, 3_2$	where the subscript indicates the translation in unit cell length, for example, 1/3, 2/3 units

Other examples are

$$4_1, 4_2, 4_3$$
$$6_1, 6_2, 6_3, 6_5$$

A glide operation involves (1) translation along the unit cell axes with $a/2$, $b/2$, and $c/2$ motion, labeled a, b, and c, respectively; (2) translations of $(a+b)/2$, $(b+c)/2$, and $(c+a)/2$, labeled n; and (3) translations along the diagonal of the net for one-fourth the total distance, labeled d. There are 230 different space groups.

The 32 point groups, 230 space groups, with Bravais lattices, are listed in Table 15.2. The point groups are classified into seven crystal systems. There are five crystal systems with principal axes of order 1 (triclinic), 2 (monoclinic), 3 (trigonal), 4 (tetragonal), and 6 (hexagonal). The other two systems are orthorhombic, which has three twofold axes ($2mm$, 222, mmm), and cubic, which has four threefold axes ($\bar{4}3m$, $m3$, 43, $m3m$). The point group $m3m$, however, describes not only cubic but also the regular octahedron and the rhombic dodecahedron. Space groups can be read from Table 15.2 as follows:

$Im3m$	body-centered cubic
$Fm3m$	face-centered cubic

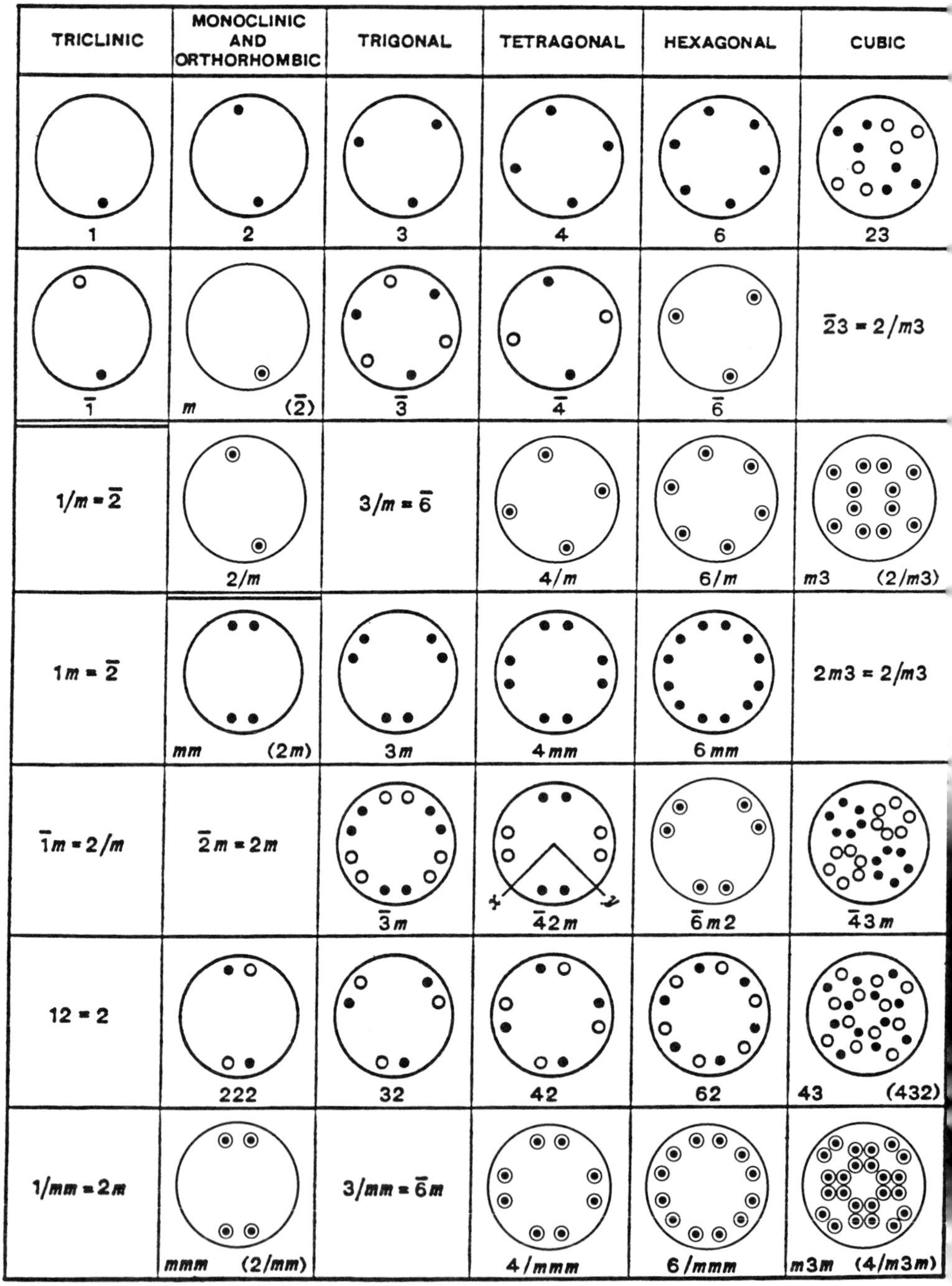

TRICLINIC	MONOCLINIC AND ORTHORHOMBIC	TRIGONAL	TETRAGONAL	HEXAGONAL	CUBIC
1	2	3	4	6	23
$\bar{1}$	m $(\bar{2})$	$\bar{3}$	$\bar{4}$	$\bar{6}$	$\bar{2}3 = 2/m3$
$1/m = \bar{2}$	$2/m$	$3/m = \bar{6}$	$4/m$	$6/m$	$m3$ $(2/m3)$
$1m = \bar{2}$	mm $(2m)$	$3m$	$4mm$	$6mm$	$2m3 = 2/m3$
$\bar{1}m = 2/m$	$\bar{2}m = 2m$	$\bar{3}m$	$\bar{4}2m$	$\bar{6}m2$	$\bar{4}3m$
$12 = 2$	222	32	42	62	43 (432)
$1/mm = 2m$	mmm $(2/mm)$	$3/mm = \bar{6}m$	$4/mmm$	$6/mmm$	$m3m$ $(4/m3m)$

FIGURE 15.9 The 32 point groups. (From Phillips (1971). Reproduced by permission of John Wiley & Sons.)

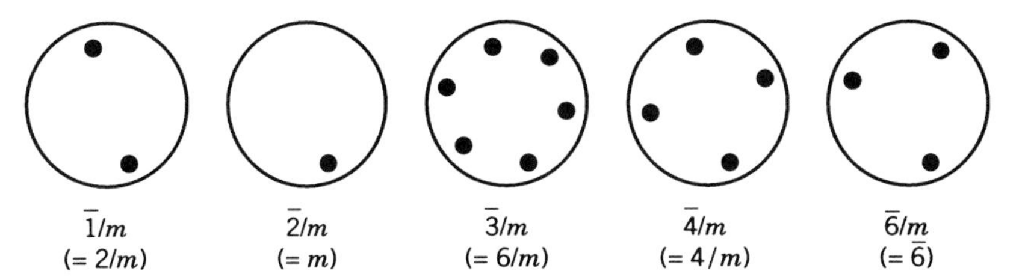

FIGURE 15.10 Illustration of the operation of symmetry elements. Compare with the second row ($\bar{1}$, $\bar{2}$, $\bar{3}$, $\bar{4}$, $\bar{6}$) in Figure 15.9.

*Fd*3*m* glide face-centered cubic

*P*2/*m* a primitive monoclinic space lattice with a twofold axis parallel to the *b* axis and a mirror plane *m* in perpendicular to this axis.

Cmc2_1 a *c*-centered orthorhombic lattice with a mirror plane perpendicular to the *a* axis, a *c* glide plane perpendicular to the *b* axis, and a 2_1 screw axis parallel to the *c* axis.

15.3 FOURIER SYNTHESIS

Once we have determined the symmetry, the space group, and the unit cell dimensions of the crystal, we can focus our attention on the intensities that describe the positions of the atoms. These calculations make use of the Fourier transforms as an essential technique.

15.3.1 The Atomic Scattering Factor

If we consider a volume element, dV, in which the electrons of an atom can be found, we find that a phase difference occurs between the wave scattered by the center of the atom and that scattered by the volume element. This effect is due to the interference of electrons. The atomic scattering factor, which represents the amplitude scattered by the electrons in the atom, is obtained by integrating over the volume elements:

$$f = \int \rho(\mathbf{r}) \exp\left[\frac{2\pi i}{\lambda}(\mathbf{s} - \mathbf{s}_0)\cdot\mathbf{r}\right] \mathrm{d}V \tag{15.6}$$

where ρ is the density of electrons. The coordinates of the scattering are described in Figure 15.11. The vector $\mathbf{s} - \mathbf{s}_0$ is normal to the reflecting plane of the crystal.

TABLE 15.2 Space Groups

Point Groups	Space Groups						
TRICLINIC							
1	$P1$						
$\bar{1}$	$P\bar{1}$						
MONOCLINIC							
2	$P2$	$P2_1$	$C2$				
m	Pm	Pc	Cm	Cc			
$2/m$	$P2/m$	$P2_1/m$	$C2/m$	$P2/c$	$P2_1/c$	$C2/c$	
ORTHORHOMBIC							
222	$P222$	$P222_1$	$P2_12_12$	$P2_12_12_1$	$C222_1$	$C222$	$F222$
	1222	$12_12_12_1$					
$mm2$	$Pmm2$	$Pmc2_1$	$Pcc2$	$Pma2$	$Pca2_1$	$Pnc2$	$Pmn2_1$
	$Pba2$	$Pna2_1$	$Pnn2$	$Cmm2$	$Cmc2_1$	$Ccc2$	$Amm2$
	$Abm2$	$Ama2$	$Aba2$	$Fmm2$	$Fdd2$	$Imm2$	$Iba2$
	$Ima2$						
mmm	$Pmmm$	$Pnnn$	$Pccm$	$Pban$	$Pmma$	$Pnna$	$Pmna$
	$Pcca$	$Pbam$	$Pccn$	$Pbcm$	$Pnnm$	$Pmmn$	$Pbcn$
	$Pbca$	$Pnma$	$Cmcm$	$Cmca$	$Cmmm$	$Cccm$	$Cmma$
	$Ccca$	$Fmmm$	$Fddd$	$Immm$	$Ibam$	$Ibca$	$Imma$
TETRAGONAL							
4	$P4$	$P4_1$	$\mathrm{P}4_2$	$P4_3$	$I4$	$I4_1$	
$\bar{4}$	$P\bar{4}$	$I\bar{4}$					
$4/m$	$P4/m$	$P4_2/m$	$P4/n$	$P4_2/n$	$I4/m$	$\mathrm{I}4_1/a$	
422	$P422$	$P42_12$	$\mathrm{P}4_122$	$P4_12_12$	$P4_222$	$P4_24_12$	$P4_322$
	$P4_32_12$	$I422$	$I4_122$				
$4mm$	$P4mm$	$P4bm$	$P4_2cm$	$P4_2nm$	$P4cc$	$P4nc$	
	$P4_2bc$	$I4mm$	$I4cm$	$I4_1md$	$I4_1cd$		
$\bar{4}2m$	$P\bar{4}2m$	$P\bar{4}2c$	$P\bar{4}2_1m$	$P\bar{4}2_1c$	$P\bar{4}m2$	$P\bar{4}c2$	$P\bar{4}b2$
	$P\bar{4}n2$	$I\bar{4}m2$	$I\bar{4}c2$	$I\bar{4}2m$	$I\bar{4}2d$		
$4/mmm$	$P4/mmm$	$P4/mcc$	$P4/nbm$	$P4/nnc$	$P4/mbm$	$P4/mnc$	$P4/nmm$
	$P4/ncc$	$P4_2/mmc$	$P4_2/mcm$	$P4_2/nbc$	$P4_2/nnm$	$P4_2/mbc$	$P4_2/mnm$
	$P4_2/nmc$	$P4_2/ncm$	$I4/mmm$	$I4/mcm$	$I4_1/amd$	$I4_1/acd$	
TRIGONAL							
3	$P3$	$P3_1$	$P3_2$	$R3$			
$\bar{3}$	$P\bar{3}$	$R\bar{3}$					
32	$P312$	$P321$	$P3_112$	$P3_121$	$P3_212$	$P3_221$	$R32$
$3m$	$P3m1$	$P31m$	$P3c1$	$P31c$	$R3m$	$R3c$	
$\bar{3}m$	$P\bar{3}/m$	$P\bar{3}1c$	$P\bar{3}m1$	$P\bar{3}c1$	$R\bar{3}m$	$R\bar{3}c$	
HEXAGONAL							
6	$P6$	$P6_1$	$P6_5$	$P6_2$	$P6_4$	$P6_3$	
$\bar{6}$	$P\bar{6}$						
$6/m$	$P6/m$	$P6_3/m$					
622	$P622$	$P6_122$	$P6_522$	$P6_222$	$P6_422$	$P6_322$	
$6mm$	$P6mm$	$P6cc$	$P6_3cm$	$P6_3mc$			
$\bar{6}m2$	$P\bar{6}m2$	$P\bar{6}c2$	$P\bar{6}2m$	$P\bar{6}2c$			
$6/mmm$	$P6/mmm$	$P6/mcc$	$P6_3/mcm$	$P6_3/mmc$			

TABLE 15.2 (*Continued*)

Point Groups	Space Groups						
CUBIC							
23	$P23$	$F23$	$I23$	$P2_13$	$I2_13$		
$m3$	$Pm3$	$Pn3$	$Fm3$	$Fd3$	$Im3$	$Pa3$	$Ia3$
432	$P432$	$P4_232$	$F432$	$F4_132$	$I432$	$P4_332$	$P4_132$
	$I4_132$						
$\bar{4}3m$	$P\bar{4}3m$	$F\bar{4}3m$	$I\bar{4}3m$	$P\bar{4}3n$	$F\bar{4}3c$	$I\bar{4}3d$	
$m3m$	$Pm3m$	$Pn3n$	$Pm3n$	$Pn3m$	$Fm3m$	$Fm3c$	$Fd3m$
	$Fd3c$	$Im3m$	$Ia3d$				

Symbols: *P*, primitive; *C*, end centered; *F*, face centered; *I*, body centered; *R*, primitive in rhombohedral axes; *a*, *b*, *c*, glide planes; *n*, number of rotation axes. For symmetry elements (e.g., 1, 2/*m*), see Figures 15.9 and 15.10.

Source: *International Tables for X-ray Crystallography*, Vol. 1, *Symmetry Groups*, Section 4.3, "The 230 Three-Dimensional Space Groups: Equivalent Positions, Symmetry and Possible Reflections." Rearranged in the form of Table 15.2 (in this book), which is taken from Wunderlich (1973). With permission from Dr. Wunderlich and Academic Press.

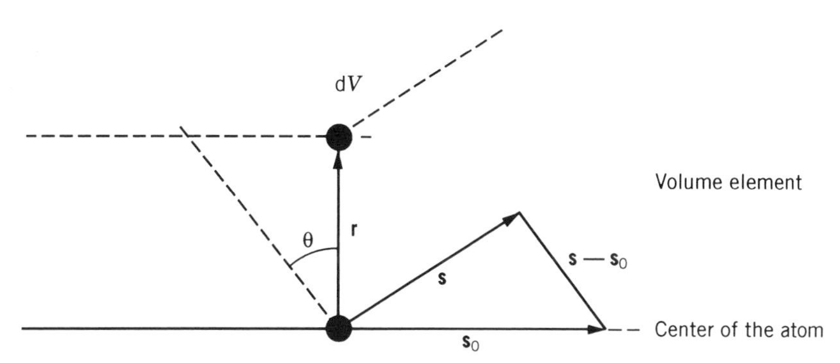

FIGURE 15.11 The coordinates of atomic scattering.

15.3.2 The Structure Factor

By carrying out a summation over all the atoms in a unit cell, we obtain the structure factor $F(hkl)$ of the crystal system:

$$F(hkl) = \sum_i f_i \exp\left[\frac{2\pi i}{\lambda}(\mathbf{s} - \mathbf{s}_0)\cdot\mathbf{r}_i\right]$$

where f_i is the atomic scattering factor of the ith atom. The vector $\mathbf{r}$ is now expressed in terms of the crystal coordinates:

$$\mathbf{r}_i = x_i\mathbf{a} + y_i\mathbf{b} + z_i\mathbf{c}$$

where **a**, **b**, and **c** are coordinates of the unit length, and x_i, y_i, and z_i are the

Cartesian coordinates at which the (hkl) planes intersect. On the basis of the Laue equations,

$$(\mathbf{s}-\mathbf{s}_0)\cdot\mathbf{a} = h\lambda$$
$$(\mathbf{s}-\mathbf{s}_0)\cdot\mathbf{b} = k\lambda$$
$$(\mathbf{s}-\mathbf{s}_0)\cdot\mathbf{c} = l\lambda$$

we can make the vector dot product into the form

$$\mathbf{r}_1\cdot(\mathbf{s}-\mathbf{s}_0) = \lambda(hx_i + ky_i + lz_i)$$

The structure factor now correlates to the Miller indices:

$$F(hkl) = \sum_i f_i \exp[2\pi i(hx_i + ky_i + lz_i)] = \sum_i f_i \exp(i\phi_i) \qquad (15.7)$$

where

$$\phi_i = 2\pi(hx_i + ky_i + lz_i) \qquad (15.8)$$

However, at this point we still have no knowledge of the phase difference ϕ_i, which we discuss in the next section.

15.3.3 Fourier Synthesis of Electron Density

The electron density $\rho(x, y, z)$ of the whole molecule, existing in the form of a crystal, is a periodic function in three dimensions. For this reason we can write the density as a Fourier series in summation (not integral) form for the convenience of computer calculations:

$$\rho(x, y, z) = \frac{1}{V}\sum_h\sum_k\sum_l F(hkl)\exp[-2\pi \mathrm{i}(hx + ky + lz)] \qquad (15.9)$$

The quantities $(1/V)F(hkl)$ are the Fourier coefficients. Here V is the volume of the unit cell. The summation is extended over all diffraction maxima.

15.4 THE PHASE PROBLEM

If the values of the structure factor $F(hkl)$ were all known, we would calculate the values of ρ and thereby determine the locations of the atoms in a crystal. However, in an actual recorded diffraction pattern, only the intensities and, consequently, the amplitudes of the diffracted rays may be measured. We do not have specific information on the crystal's phase. Since all atoms are not at the same place, the scattering wave from one atom may be in phase or out of phase with that of its neighbor. A value of ϕ has to be assigned to each value of $F(hkl)$. This poses a thorny problem, which has been called the *phase problem*. At one time it was generally believed that crystal structures could not be determined from diffraction intensities alone. In this section we describe two ingenious and successful approaches to this problem.

15.4.1 The Patterson Synthesis

In 1934, Patternson suggested that instead of using the structure factors themselves, $F(h,k,l)$, we may use the square modulus of the structure factors, $|F(h,k,l)|^2$, which are the intensities of the diffraction. Equation (15.9) is now modified to the form

$$P(u,v,w) = \frac{1}{V^2} \sum_h \sum_{k=-\infty}^{\infty} \sum_l |F(hkl)|^2 \exp[-i2\pi(hu + kv + lw)] \tag{15.10}$$

where (u,v,w) refers to all possible vectors between the atoms in the cell, and not just those peaks (x,y,z) representing atomic position. The function $P(u,v,w)$ is called the Patternson function. The maxima of a Patterson function give the interatomic factor in a crystal structure.

In an x-ray diffraction experiment, if heavy atoms are introduced as part of the molecule, the scattering from these heavy atoms tends to dominate over the scattering from the light atoms. The summation in a Patterson function may be taken only over the heavy atom positions. The light atoms are considered to be arranged in the form of a random walk around the heavy atoms. Frequently, some light atoms can be recognized in a diffraction pattern. As a result, a map can be constructed to elucidate the structure of a whole molecule.

15.4.2 The Direct Method (Karle-Hauptmann Approach)

The direct method is a mathematical process in which different phases and amplitudes of the molecular transform are related to the contribution of individual atoms. This method is based on statistical procedures used to extract the phase information. The missing phase information is believed to be present in the statistical intensity distribution. One of the most well-known statistical proposals was offered by Karle and Hauptmann (1986). These authors assumed that the Fourier transform of the intensity represents the probability of finding interatomic distances in a molecule, and that the transform is nonnegative.

The intensity data collected from an experiment can be transformed into normalized structure factor magnitudes by

$$|E_{\mathbf{h}}| = \frac{|F_{\mathbf{h}}|}{(\varepsilon \sum_{j=1}^{N} f_{j\mathbf{h}}^2)^{1/2}} \tag{15.11}$$

where $E_{\mathbf{h}}$ is the normalized structure factor and $F_{\mathbf{h}}$ is equivalent to $F(hkl)$. The vector **h** (as well as **k** and **l**) has three integer components: h, k, l (the Miller indices). The symbol ε represents the statistical weighing factor, and the symbol $f_{j\mathbf{h}}$ represents the jth atomic scattering factor, as tabulated in the *International Tables for X-ray Crystallography*.

A formula may be derived for centrosymmetric crystals:

$$sE_{\mathbf{h}} \simeq s\sum_{\mathbf{k}_r} E_{\mathbf{k}} E_{\mathbf{h}-\mathbf{k}} \tag{15.12}$$

The symbol s means "sign of" and the vector $\mathbf{k}_r$ gives the restriction that $|E_{\mathbf{k}}|$ and $|E_{\mathbf{h}-\mathbf{k}}|$ must have large values. If the sign is plus, then the phase is zero; if the sign is minus, then the phase is π. These two values, σ and π, are the only values possible for centrosymmetric crystals.

Two formulas may also be derived for noncentrosymmetric crystals:

$$\phi_{\mathbf{h}} \cong \langle \phi_{\mathbf{k}} + \phi_{\mathbf{h}-\mathbf{k}} \rangle_{\mathbf{k}_r} \tag{15.13}$$

$$\tan \Phi_{\mathbf{h}} \cong \frac{\sum_{\mathbf{k}} |E_{\mathbf{k}} E_{\mathbf{h}-\mathbf{k}}| \sin(\Phi_{\mathbf{k}} + \Phi_{\mathbf{h}-\mathbf{k}})}{\sum_{\mathbf{k}} |E_{\mathbf{k}} E_{\mathbf{h}-\mathbf{k}}| \cos(\phi_{\mathbf{k}} + \phi_{\mathbf{h}-\mathbf{k}})} \tag{15.14}$$

The angle $\phi_{\mathbf{h}}$ is the phase associated with the structure factor $F_{\mathbf{h}}$. Equation (15.14) is called the tangent formula. For noncentrosymmetric crystals, certain phases are specified to establish an origin in the crystal. Certain symbols are also assigned for later evaluation, which is done by use of the theory of invariants and semi-invariants. The procedure continues in a stepwise fashion until a sufficiently reliable set of phase values is built up to solve the structure problem.

15.5 REFINEMENT

The calculated structure factor, whether based on the Patterson synthesis or on the direct method, must be in agreement with the observed structure factor. The uncertainty involved in assigning the positional and thermal parameters in the calculation of the structure factor must be minimized so that the results will be close to the structure factor amplitudes $F_0(\mathbf{h})$ for each reflection **h** observed. The minimization procedure is called the *refinement*, and is based on Legendre's (1806) classical suggestion. If we have a set of unknown parameters $x_1, x_2, \ldots, x_n$ that are related to a set of m observable values of $O_1, O_2, \ldots, O_m$ by

$$\begin{aligned} O_1 &= a_{11}x_1 + a_{12}x_2 + \cdots + a_{1n}x_n \\ &\vdots \\ O_m &= a_{m1}x_1 + a_{m2}x_2 + \cdots + a_{mn}x_n \end{aligned}$$

then the error E between the set of values x and the set of values O can be expressed by their differences:

$$E_1 = a_{11}x_1 + a_{12}x_2 + \cdots + a_{1n}x_n - O_1$$
$$\vdots$$
$$E_m = a_{m1}x_1 + a_{m2}x_2 + \cdots + a_{mn}x_n - O_m$$

The best values of x are those for which $\sum_{i=1}^{m} E_i^2$ is at a minimum:

$$\frac{\partial \sum_{i=1}^{m} E_i^2}{\partial x_1} = \frac{\partial \sum_{i=1}^{m} E_i^2}{\partial x_2} = \cdots = \frac{\partial \sum_{i=1}^{m} E_i^2}{\partial x_n} = 0$$

We now have n equations for the n unknowns (x's) to solve. These x's will give the least errors for our interpretation of the results. In a crystallographic problem, Legendre's procedure may be expressed in the equation

$$F_o(\mathbf{h}) - F_c(\mathbf{h}) = \sum_{j=1}^{n} \partial x_j \frac{\partial F_c}{\partial x_j}$$

where $F_o(\mathbf{h})$ is the observed factor and $F_c(\mathbf{h})$ is the calculated factor. The refinement is often expressed as a quantity R, called the reliability index:

$$R = \frac{\sum_h |F_o(\mathbf{h}) - F_c(\mathbf{h})|}{\sum_h F_o(\mathbf{h})}$$

Experimentally, modern crystallographic study of the structure of macromolecules may be summarized into three steps:

1. The growth of a single crystal of the macromolecule under study.
2. The preparation of isomorphous heavy atom derivatives. If only one isomorphous derivative is prepared, the method is known as a single isomorphous replacement; if more than one derivative is prepared the method is known as multiple isomorphous replacement.
3. Data analysis, which includes calculating phases, determining electron density maps, and refinement.

15.6 CRYSTAL STRUCTURE OF MACROMOLECULES

15.6.1 Synthetic Polymers

Not all polymer chains can be crystallized; only those with stereoregularity can. For example, polyethylene can be crystallized because there is a regular configuration inherent in the monomer. Polypropylene, on the other hand, can be crystallized only under certain conditions. The crystal structures of these two polymers are also quite different. The former is packed in a zigzag form, whereas the latter has helical content (not, of course, 100%). We select these two polymers to illustrate the crystal structure, if any, of synthetic polymers.

Polyethylene The C atoms form a plane of zigzag chain:

$$\diagup CH_2 \diagdown CH_2 \diagup CH_2 \diagdown CH_2 \diagup$$

The distance between the two carbons C–C is 1.52 Å, the angle —CH_2—is 114°. The unit cell is orthorhombic, with $a = 7.40$ Å, $b = 4.93$ Å, and $c = 2.53$ Å. Taking its position from the c axis, the structure is illustrated in Figure 15.12. The motif, CH_2, has the point symmetry $mm2$. The chain axis (c coordinate) is a 2_1 screw axis. The space group is *Pnam*, that is, the unit cell is primitive with an n glide plane perpendicular to the a axis, an a glide plane perpendicular to the b axis, and a mirror plane perpendicular to the c axis. The four carbon atoms are at positions: $x, -y, \frac{1}{4}$; $-x, y, \frac{3}{4}$; $-x + \frac{1}{2}, -y + \frac{1}{2}, \frac{3}{4}$; $x + \frac{1}{2}, y + \frac{1}{2}, \frac{1}{4}$ where $x = 0.038a$ and $y = 0.065b$. The helix of polyethylene is represented by the symbol $1 * 3/1$. The first number, 1, signifies the class of the chain, the 3 signifies the number of motifs, and the last 1 signifies the number of turns. Thus, polyethylene

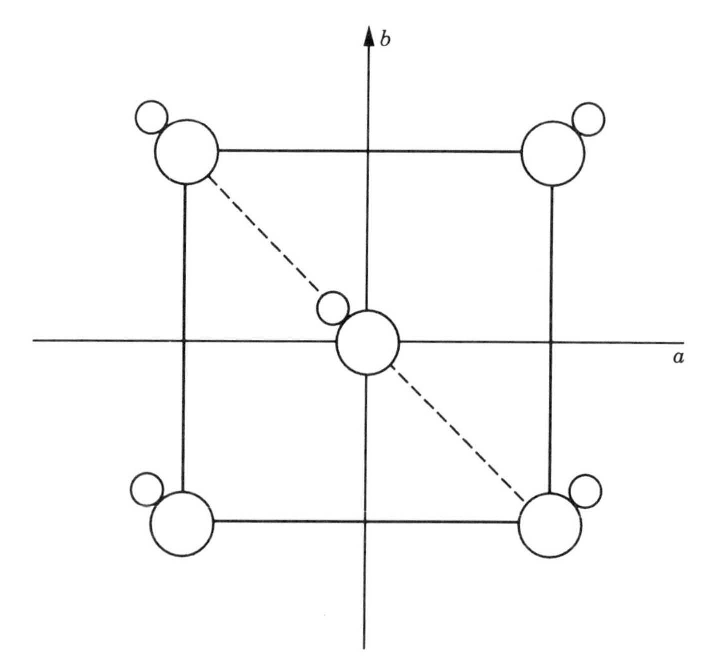

FIGURE 15.12 The structure of polyethylene.

has three CH_2 groups per turn. The width of the Bragg reflections shows polyethylene "crystallites" to be only 100–300 Å in size; hence, the polymer is not completely crystallized. A chain contains crystallites as well as an amorphous part.

Polypropylene Unlike polyethylene, which has only one configuration, polypropylene has three configurations: atactic, isotactic, and syndiotactic. We describe syndiotactic and isotactic polypropylene here because their configurations have been studied extensively in crystallography. Syndiotactic polypropylene has an orthorhombic unit cell with $a = 14.50$ Å, $b = 5.60$ Å, and $c = 7.40$ Å. Its space group is $C222_1$. The projection of the unit cell on the 001 plane is shown in Figure 15.13. There are eight monoclinic lattices in the unit cell. The chain has the linear repetition group $S(2/1)2$ symmetry. The model of the chain is in the form shown in Figure 15.14. The chain requires three repeating units for one turn and so it is a $2*3/1$ helix. The isotactic polypropylene has an end-centered monoclinic unit cell with $a = 6.66$ Å, $b = 20.78$ Å, $c = 6.49$ Å, and the angle $\beta = 99.62°$. There are 12 units in the cell. The chain has a $4*2/1$ helix. Figure 15.15 compares the helical structure of isotactic polypropylene with that of syndiotactic polypropylene.

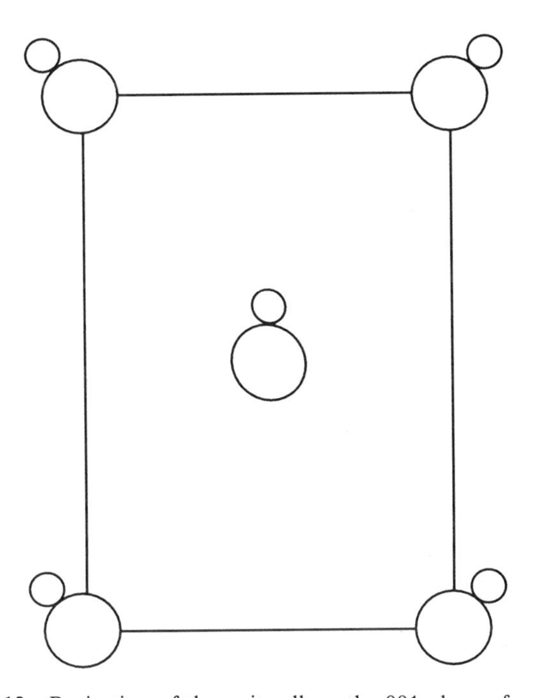

FIGURE 15.13 Projection of the unit cell on the 001 plane of polypropylene.

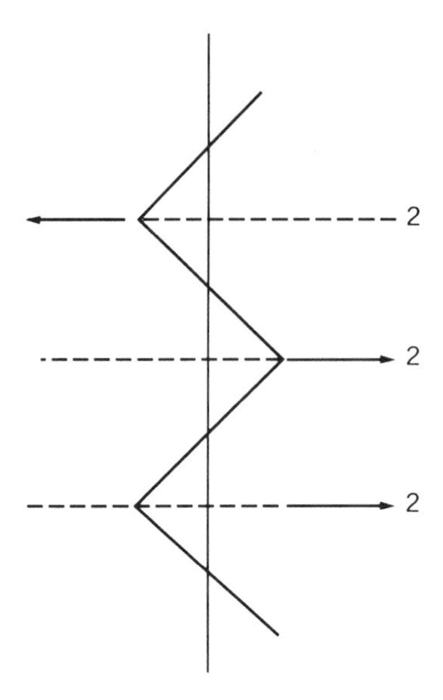

FIGURE 15.14 The linear repetition group $S(2/1)2$ symmetry.

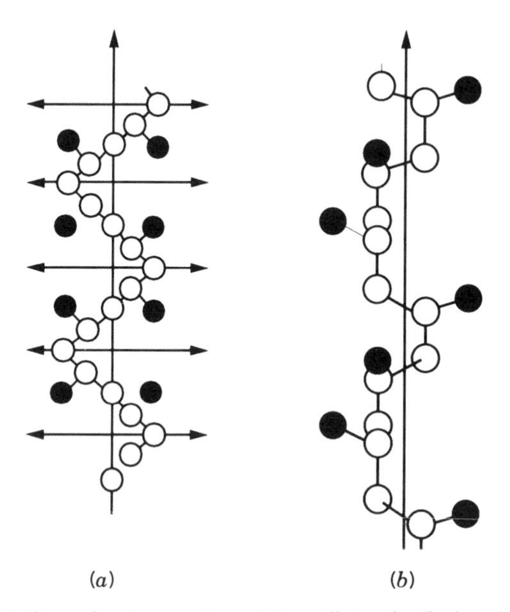

FIGURE 15.15 Helices of polypropylene: (a) syndiotactic; (b) isotactic. Adapted from G. Natta et al (1960) and Wunderlich (1973). With permission from Dr. Wunderlich and Academic Press.

15.6.2 Proteins

Among the landmarks in the elucidation of crystal structure is the crystallography work on myoglobin. Since 1961, most of the well-known globular proteins, such as ribonuclease, insulin, hemoglobin, lysozyme, β-lactorglobulin, chymotrysipnogen, and carboxypeptidase, have been investigated. Studies on myoglobin have established the general observation on protein structure that nearly all the polar and charged groups lie on the surface, with van der Waal's contact between side groups on the interior of the molecule. Figure 15.16 is a diagram of the conformation of sperm whale myoglobin.

The space group of myoglobin is $P2$, with unit cell axes $a = 64.5$ Å, $b = 30.9$ Å,

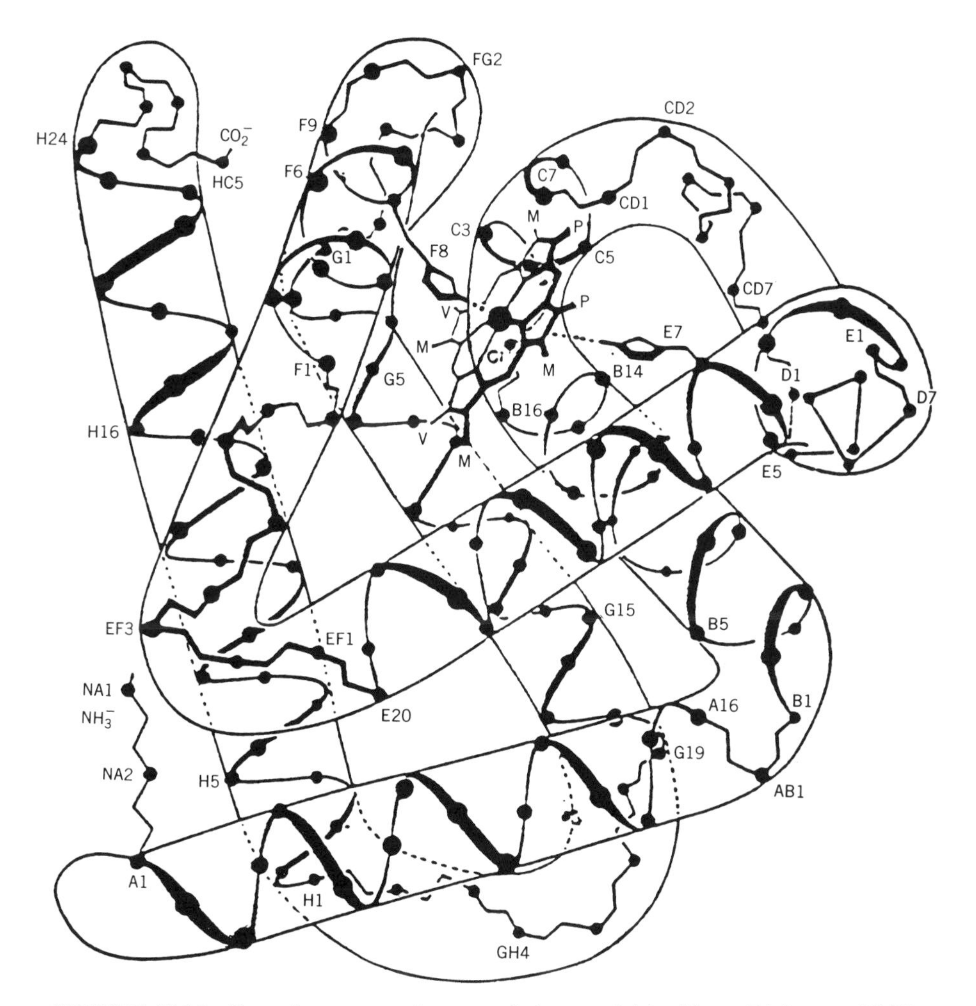

FIGURE 15.16 Crystal structure of sperm whale myoglobin. (From Dickerson (1964). Reproduced by permission of Dr. Dickerson and Academic Press.)

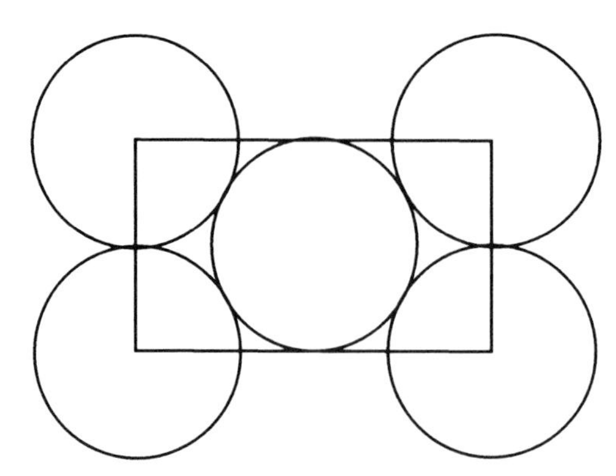

FIGURE 15.17 Projection of the *ab* plane of a DNA crystal.

and $c = 34.7$ Å. There are two molecules in a unit cell. Within its folding the molecule contains 20% water. The 153 amino acids and one heme group have all been identified. Among these 153 amino acid residues, 132 form the right-handed α-helices. They constitute a series of straight rods roughly 5 Å in diameter and 20–40 Å long.

15.6.3 DNA

The DNA molecule, which contains a large percentage of water (up to 40%), exists in three configurations, *A*, *B*, and *C*. The three configurations are of the same structure and differ only in the number of nucleotides per helix turn in helix pitch and in nucleotide conformation. The DNA crystal is monoclinic with the space group $C2$. The molecules occupy special positions $(0, 0, 0)$ $(\frac{1}{2}, \frac{1}{2}, 0)$. Figure 15.17 shows the projection of the *ab* plane with respect to the arrangement of DNA in the unit cell. The cell axes are $a = 22.2$ Å, $b = 40.6$ Å, and $c = 28.2$ Å and the cell angles are $\alpha = \gamma = 90°$ and $\beta = 97°$. The molecule possesses a dyad axis that passes through the helix axis of each molecule and is perpendicular to the helix axis. The symmetry is described almost exactly as in the well-known Watson-Crick model.

It has been suggested that the sodium salt of DNA is in the A form with space group $C2$, whereas the lithium salt of DNA is in the B form with the space group $P2_1\,2_1\,2_1$. Form A belongs to the monoclinic crystal class; form B belongs to the orthorhombic crystal class.

APPENDIX NEUTRON DIFFRACTION

Neutron diffraction and x-ray diffraction are fundamentally similar. Strong reflections occur when Bragg conditions are satisfied. The experimental data of the wavelength λ and scattering angle θ are needed to calculate the interplanar distance d; and the intensity of the reflection is useful in determining molecular structure. The difference lies in the fact that in x-ray diffraction, electrons are the principal cause of scattering, whereas in neutron diffraction, the scattering involves nuclear rather than electronic processes. For this reason, neutron diffraction can easily distinguish between atoms such as carbon and nitrogen in C=N or nitrogen and oxygen in N=O; x-ray diffraction cannot. Neutron diffraction can even distinguish between isotopes. It can be used to observe magnons and to study the dynamics of the spin state of magnetic crystals.

NEUTRON CRYSTALLOGRAPHY

The basic equations that apply in the x-ray diffraction pattern also apply in neutron diffraction:

$$\lambda = 2\frac{d_{hkl}}{n}\sin\theta \qquad \text{(Bragg equation)}$$

$$\frac{1}{d_{hkl}} = f(a, b, c, \delta, \beta, \gamma)$$

where, as mentioned before, a, b, and c are the periods of the unit cell, while α, β, and γ are the angles between the coordinate axes.

The Fourier transform for x-ray form factor $F_x(q)$ is slightly different from that for neutron form factor $F_{\text{neutron}}(q)$, as shown in the following equations:

$$F_{\text{x-ray}}(q) = \int_\tau e^{iqr}\{N_\uparrow|\psi_\uparrow(r)|^2 + N_\downarrow|\psi_\downarrow(r)|^2\}\,dr$$

$$F_{\text{neutron}}(q) = \int_\tau e^{iqr}\{N_\uparrow|\psi_\uparrow(r)|^2 - N_\downarrow|\psi_\downarrow(r)|^2\}\,dr$$

where

$$q = \frac{4\pi}{\lambda}\sin\theta$$

The arrows refer to electron spin: $\uparrow$ in the upward direction and $\downarrow$ in the downward direction. The term N is the number of electrons of an atom and ψ is the wave function of an electron. The radical density of neutron diffraction is in the form

$$U(r) = \frac{2r}{\pi} \int_0^\infty qF(q) \sin(qr) \, dq$$

which corresponds to the equation of electron-density synthesis in x-rays:

$$\rho(x, y, z) = \frac{1}{V} \sum_h^\infty \sum_k^\infty \sum_l^\infty F_{hkl} \exp[-2\pi i(hx + ky + lz)]$$

Because of the cost of neutron-diffraction experiment, so far it has not been used much for polymers, but its potential for the study of polymer structure is obvious.

REFERENCES

Atkins, P. W., *Phys. Chem.* 3rd ed. New York: W. H. Freeman and Co. 1986.

Blundell, T. L., and L. N. Johnson, *Protein Crystallography*. London: Academic, 1976.

Bragg, W. L., *Proc. R. Soc. London Ser. A* **123**, 537 (1929).

Buerger, M. J., *Elementary Crystallography*. New York: Wiley, 1963.

Bunn, C. W., *Trans. Faraday Soc.* **35**, 482 (1939).

Bunn, C. W., *Proc. R. Soc. London Ser. A* **189**, 39 (1947).

Carradini, P., G. Natta, P. Ganis, and P. A. Temussi, *J. Polym. Sci. C* **162**, 477 (1967).

Dickerson, R. E., in *The Proteins*, Vol. 2, H. Neurath (Ed.). New York: Academic, 1964, Chapter 11.

Dickerson, R. E., J. C. Kendrew, and B. E. Strandberg, *Acta Crystallogr.* **14**, 1188 (1961).

Fuller, W., M. H. F. Wilkins, H. R. Wilson, and L. D. Hamilton, *J. Mol. Biol.* **12**, 60 (1965).

Green, D. W., V. M. Ingram, and M. F. Perutz, *Proc. R. Soc. London Ser. A* **225**, 287 (1954).

Hauptman, H., *Science* **233**, 178 (1986).

Hauptman, H., and J. Karle, *The Solution of the Phase Problem, 1: The Centrosymmetric Crystal*, ACA Monograph No. 3. Ann Arbor, MI: Edwards Brothers, 1953.

Izyumov, Y. A., and R. P. Ozerov, *Magnetic Neutron Diffraction*. New York: Plenum, 1970.

Karle, J., *Science* **232**, 837 (1986).

Karle, J., and H. Hauptman, *Acta Crystallogr.* **9**, 635 (1956).

Kavesh, S., and J. M. Schultz, *J. Polym. Sci.* **8**, 243 (1970).

Kendrew, J. C., R. E. Dickerson, B. E. Strandberg, R. G. Hart, D. R. Davis, D. C. Phillips, and V. C. Shore, *Nature (London)* **185**, 442 (1960).

Kendrew, J. C., H. C. Watson, B. E. Strandberg, R. E. Dickerson, D. C. Phillips, and V. C. Shore, *Nature (London)* **190**, 666 (1961).

Lipscomb, W. N., and R. A. Jacobson, in *Techniques of Chemistry*, III, Vol. 1, A. Weissberger and B. W. Rossiter (Eds.). New York: Wiley–Interscience, 1972.

Natta, G., P. Corradini, and P. Gans, *Makromol. Chem.* **39**, 238 (1960).

Patterson, A. L., *Phys. Rev.* **48**, 372 (1934).

Phillips, F. C., *An Introduction to Crystallography*. London: Longmans, 1963.

Phillips, F. C., *An Introduction to Crystallography*, 4th ed. New York: Wiley, 1971.

Spruiell, J. E., and E. S. Clark, *Methods Exp. Phys.* **16B**, 1 (1980).

von Laue, M., *Muench. Sitzungsber.*, p. 363 (1912).

Wunderlich, B., *Macromolecular Physics*, Vol. 1. New York: Academic, 1973.

Wyckoff, H. W., C. H. Hiss, and S. N. Timasheff (Eds.), *Methods Enzylmol.* **115** (1985).

PROBLEMS

15.1 A unit cell has two atoms. Their coordinates (x, y, z) are $(0.10, 0.10, 0.10)$ and $(0.20, 0.75, 0.40)$ and their atomic scattering factors are 3.10 and 4.20. Calculate F_{hkl} for the (121) plane.

15.2 The unit cell constants for human serum albumin are $a = b = 186.5$ Å and $c = 81.0$ Å. What is the spacing of the (120) and (112) planes?

15.3 The structure factors for $(h00)$ planes of a crystal were found as follows:

h	0	1	2	3	4	5	6	7	8	9	10	11	12	13	14	15
F_h	16	−10	2	−1	7	−10	8	−3	2	−3	6	−5	3	−2	2	−3

Construct a plot of the electron density along the X axis of the unit cell. (*Source*: Atkins (1986).)

15.4 Describe the space groups $Im3m$, $Fm3m$, and $Fd3m$, and the five cubic point groups $F23$, $Fm3$, $F\bar{4}3m$, $F432$, and $Fm3m$. Note that $Fm3m$ is both a space group and a point group.

15.5 Polyorthomethylstyrene forms a crystal with a body-centered lattice. The structure factor at $(\frac{1}{2}, \frac{1}{2}, \frac{1}{2})$ is given by

$$F_{hkl} = \left\{ \sum_{m}^{M} f_m \exp[2\pi i(hx_m + ky_m + lz_m)] \right\} \times \left\{ 1 + \exp\left[2\pi i\left(\frac{h}{2} + \frac{k}{2} + \frac{l}{2} \right) \right] \right\}$$

where m refers to the atoms that compose the motif at each lattice point. Show that if $h + k + l = 2n + 1$, where n is any integer, then

$$F_{hkl} = 0$$

that is, there is no intensity. If $h + k + l = 2n$, then

$$F_{hkl} = \sum_m f_m \exp[2\pi i(hx_m + ky_m + lz_m)]$$

that is, they are still functions of the relative positions of the atoms in the motif. (*Source*: Spruiell and Clark (1980).)

15.6 Following is the diagram of the helical structure of a crystal:

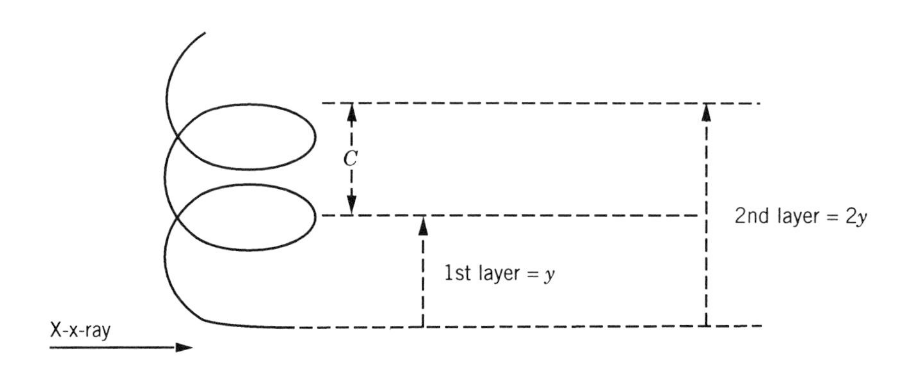

For polyoxymethylene, the intense upper layer line has $l = 5$ at a $2y$ value of 29 mm. The cylindrical camera diameter ($2D$) was 57.3 mm and $\lambda = 1.54$ Å. Calculate the interplanar distance d and the repeat helix distance c. Hint:

$$d = \frac{\lambda}{2\sin[\frac{1}{2}\tan^{-1}(2x/2d)]}$$

$$c = \frac{l\lambda}{\sin[\tan^{-1}(y/D)]}$$

(*Source*: Spruiell and Clark (1980).)

16

ELECTRONIC AND INFRARED SPECTROSCOPY

Spectra are observed as the result of the interchange of energy between a substance and electromagnetic radiation. If the energy is absorbed by a substance from the radiation field, we have absorption spectra; if the energy is added to the radiation field, we have emission spectra. An electromagnetic radiation may be characterized by the frequency ν, the wavelength λ or the wave number $\bar{\nu}$. They are all related as shown in the following equations:

$$\nu\lambda = c'$$

$$\frac{1}{\lambda} = \bar{\nu}$$

where c' is the velocity of light. The units are hertz (or s^{-1}) for frequency, m^{-1} or cm^{-1} for wave number, and m or cm for wavelength. Table 16.1 shows the relationship between some of these units. Table 16.2 shows the wavelength of spectroscopic regions of the electromagnetic spectrum.

In this chapter we describe the three most important types of optical spectroscopy: ultraviolet (UV) and visible, fluorescence, and infrared. UV and visible spectra can be obtained through the same spectrometer. They differ only in the selection of the wavelength of the incident light: UV at 180–400 nm and visible at 400–760 nm. Both UV (and visible) and fluorescence spectra describe the phenomenon of electron excitation; namely, a valence electron of a molecule is excited upon absorbing energy from the electromagnetic radiation, and is thereby transferred from one energy level to another energy level. The spectra are electronic.

TABLE 16.1 Examples of the Relationship Between Frequency, Wave Number, and Wavelength

Frequency (s^{-1})	Wave Number (m^{-1})	Wavelength
3×10^8 (300 MHz)	1	1 m
4.3×10^{14}	1.4×10^8	700 nm (blue)
7.5×10^{14}	2.5×10^6	400 nm (red)
3×10^{18}	1.0×10^{10}	0.1 nm (γ)

TABLE 16.2 Wavelength of Spectroscopic Regions of the Electromagnetic Spectrum

X-rays	Vacuum UV	UV	Visible	Near infrared	Far infrared	Microwave		Radio frequency

1.0 200 400 750 (nm) 1.0 50 (μm) 1.0 30 (mm) 100 (cm)

By contrast, infrared spectra describe the vibration of the atoms (not electrons) around a chemical bond. When the frequency of the incident light (radiation) coincides with the characteristic frequency of the vibration of a chemical bond, a band (or peak) appears. Infrared spectra are not electronic spectra.

16.1 ULTRAVIOLET (AND VISIBLE) ABSORPTION SPECTRA

When a valence *s* electron of one atom moves into the same orbital as a valence *s* electron of another atom, the two atoms may form either a bond or an antibond, depending on whether the resultant bond energy is stable. The antibond is also called an excited state. The bond, which is stable, is designated as a σ bond; the antibond, which is not stable, is designated as a σ^* bond. If a valence *s* electron of one atom shares its orbital with a valence *p* electron of another atom or if a valence *p* electron of an atom shares its orbital with another valence *p* electron of another atom, then we have either a π bond or a π^* antibond, which again depends on whether the resultant bond energy is stable. In some atoms, such as N, O, and S, or molecules, such as $>C{=}O$ and $-ONO$, there are lone pair electrons. In Lewis dot symbols, for example, >C::Ö:, two lone pair electons of the oxygen atom do not form bonds. These electrons are designated as n (nonbonding) electrons.

Figure 16.1 shows four types of electronic transition, $n \rightarrow \sigma^*$ $n \rightarrow \pi^*$, $\pi \rightarrow \pi^*$, and $\sigma \rightarrow \sigma^*$. Among the four types of electronic transitions in the UV (and visible) regions, the $\sigma \rightarrow \sigma^*$ transition requires the highest input of energy. For this reason, compounds where all the valence shell electrons are involved in single bond (σ bond) formations, such as saturated hydrocarbons (C_nH_{n+2}), do

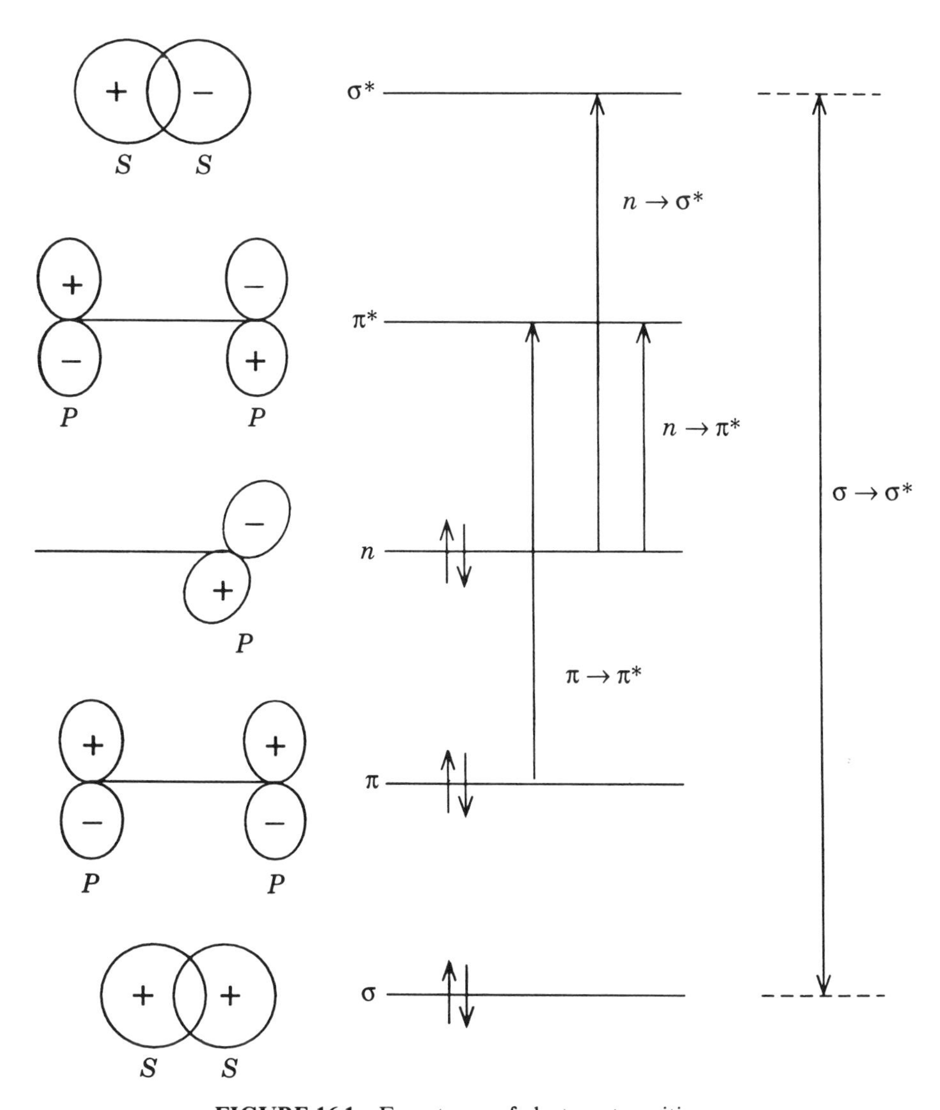

FIGURE 16.1 Four types of electron transitions.

not show absorption in the UV region. Most $\sigma \rightarrow \sigma^*$ transitions for individual bonds take place below 200 nm and a compound containing only σ bonds is transparent (that is, near zero absorption) in the near UV/visible region. The energy required for an $n \rightarrow \sigma^*$ transition is much lower than that required for a $\sigma \rightarrow \sigma^*$ transition. As a result, molecules containing a nonbonding electron, such as oxygen, nitrogen, sulfur, or the halogens, often exhibit absorption in the UV region. The $n \rightarrow \pi^*$ and $\pi \rightarrow \pi^*$ transitions require even lower inputs of energy. These are the two most important types of transitions for organic molecules and are commonly found in aldehydes and ketones. For an aldehyde or a ketone, two peaks often appear: one at 285 nm, which is attributed to an $n \rightarrow \pi^*$ transition, and the other at 180 nm, which is attributed to a $\pi \rightarrow \pi^*$ transition.

16.1.1 Lambert-Beer Law

In UV absorption spectroscopy, one important law frequently used is the Lambert-Beer law, which is a combination of Lambert's law and Beer's law. Lambert's law states that as radiation of a given frequency passes through a medium (which contains the sample molecules), its intensity is reduced exponentially in the form

$$I = I_0 e^{-\alpha x}$$

or

$$\ln \frac{I_0}{I} = \alpha x$$

where I_0 is the intensity of the incident light, I is the intensity after passing through a medium of thickness x, and α is a constant known as the absorption coefficient. If we use common logarithm (base ten), the equation is converted to

$$I = I_0 10^{-Kx}$$

or

$$\log \frac{I_0}{I} = Kx$$

where $K = \alpha/2.303$ is the extinction coefficient. Beer's law states that the absorption is proportional to the concentration of the substance, that is, $K = \varepsilon c$ where ε is a constant equal to the extinction coefficient for unit concentration. The combination of these two laws gives

$$\log \frac{I_0}{I} = \varepsilon c x \quad \text{or} \quad \log \frac{I_0}{I} = \varepsilon c b \tag{16.1}$$

which is the well-known Lambert-Beer law. If c is in moles/liter, ε is called the molar extinction coefficient. Customarily, x is kept constant and is renamed b, which is identical to the cell path length (for example, 1 cm) of the spectrometer. The logarithm term,

$$\log \frac{I}{I_0} \equiv A$$

is called the absorbance or optical density. Its range is often recorded between 0 and 2. In the UV spectrum, absorbance is plotted versus the wavelength λ, or more commonly as ε versus λ or $\log \varepsilon$ versus λ.

16.1.2 Terminology

Chromophores Electronic transitions in complex molecules, such as in polymer molecules, are regarded as being localized around certain chemical groups or bonds that are present in the molecule. It has been observed that certain series of related molecules that all contain a particular chemical group often produce absorption spectra of similar appearance. Such groups are called chromophores. If the chromophore contains π electrons, the absorption band appears in both

TABLE 16.3 Some Characteristic Chromophores[a]

Chromophore	λ_{max} (nm)	ε
$>C=C<$	171–177	15 530–12 600
$-C\equiv C-$	178–223	10 000–160
$>C=O$	160–279	16 000–15
$-C(=O)-CH$	208	32
$-C(=O)-Cl$	220	100
$-C(=O)-NH_2$	178–220	9500–63
$>C=C(-)-C(-)=C<$	217	20 900
$C=C-C\equiv C$	219	6500

[a]All chromophores have $\pi \rightarrow \pi^*$ electronic transitions. The chromophore $>C=O$ may also have a $n \rightarrow \pi^*$ transition.

the visible and the UV regions; if it contains σ electrons the band will appear only in the far ultraviolet region. Table 16.3 gives the absorption of chromophores for some well-known compounds.

Auxochromes Groups such as $-OH$, $-NH_2$, and the halogens, which all possess n electrons, are not chromophores themselves. But they can cause normal chromophoric absorptions to occur at longer wavelengths and with an increase in the value of molar absorptivity. These groups are called auxochromes. Auxochromes are responsible for the electronic transitions $n \rightarrow \sigma^*$ and $n \rightarrow \pi^*$. Table 16.4 shows the effect of auxochromes on the absorption of benzene.

Isobestic Point In the plot of ε versus λ, the absorbance curve generally shifts when the concentration of the solution is changed, as shown in Figure 16.2. At one particular wavelength, λ', however, the concentration does not affect the extinction coefficient. This wavelength is called the isobestic point. When experiments are run at λ', the interference caused by the concentration in the spectrum is avoided.

Red Shift and Blue Shift The shift of an absorption maximum toward a longer wavelength is called a red shift; that toward a shorter wavelength is called a blue shift. The shift of an absorption maximum is caused by either the change of the medium or the presence of an auxochrome. For example, in an aniline molecule the conjugation of the lone pair of electrons on the nitrogen atom gives an absorption at 230 nm ($\varepsilon = 8600$), but when the molecule is placed in an acid solution, the occurrence of the main peak is almost identical to that of benzene, that is, at 203 nm ($\varepsilon = 7500$). In this example a blue shift has occurred.

Hypochromism Hypochromism H is defined as

$$H \equiv 1 - \frac{A_p}{A_m}$$

TABLE 16.4 Auxochromes of Some Monosubstituted Benzenes

Auxochrome	λ_{max} (nm)	ε
$-H$	203.5	7400
$-CH_3$	206.5	7000
$-I$	207	7000
$-Cl$	209.5	7400
$-Br$	210	7900
$-OH$	210.5	6200
$-CN$	224	8700
$-COOH$	230	11600
$-NH_2$	230	8600
$-NO_2$	268.5	7800

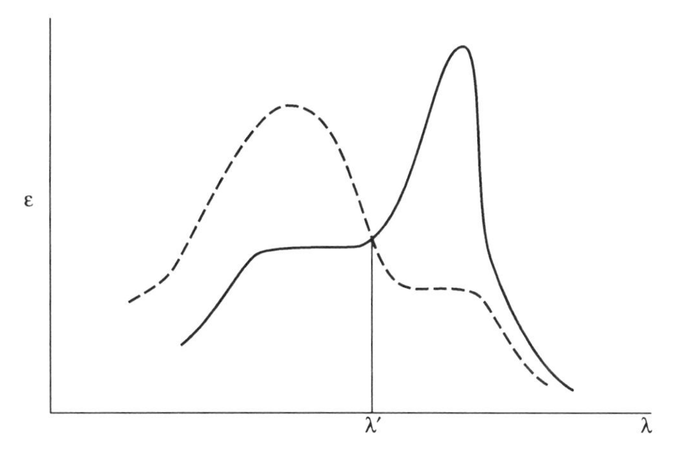

FIGURE 16.2 Isobestic point λ'.

where the subscripts p and m indicate polymer and monomer, respectively, and A refers to area. The values of A_p or A_m are obtained by computing the area under the curve of the polymer or monomer extinction coefficient ε versus frequency ν:

$$A \equiv \int \varepsilon \, d\nu$$

Hypochromism is a measure of the decrease in absorption intensity due to the mixture of a monomer with a polymer. It is often an indication of a change in the geometry of the polymer segments due to their interaction with other segments (or with the free monomers, if present).

16.1.3 Synthetic Polymers

Aromatic polymers, such as polystyrene, and nonaromatic polymers with certain functional groups, such as C–N, can have their basic units excited to produce spectra in the UV region. Following are some of the UV absorption ranges for the major groups of polymers:

Group (Bond)	Absorption Range (nm)
C—N (amid)	250–310
C=O	187; 280–320
C—C	195; 230–250
O—H	230
C=C	180

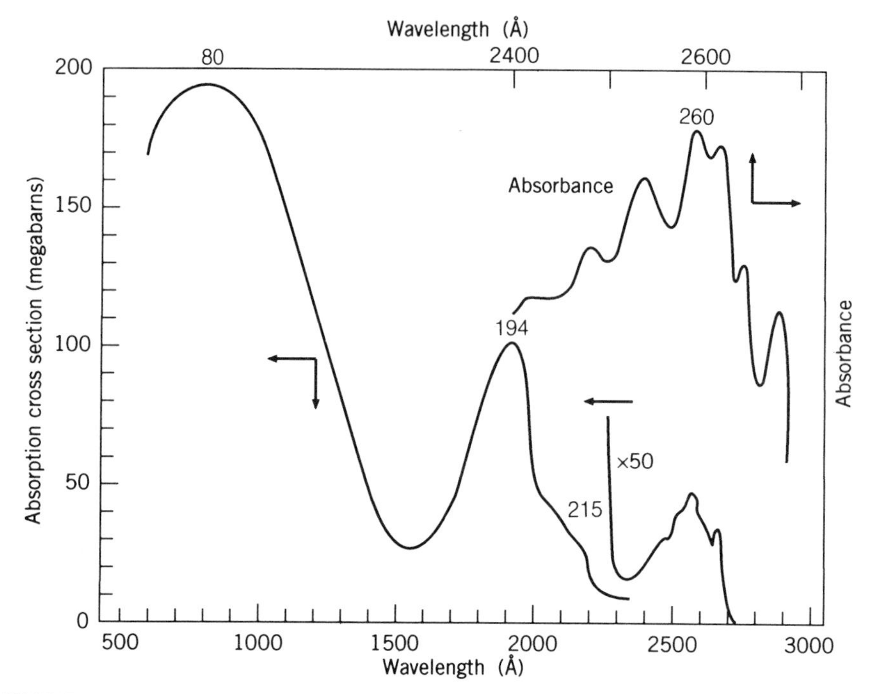

FIGURE 16.3 UV absorption spectrum of polystyrene. (Adopted from *Encyclopedia of Polymer Science and Technology* Vol. 13 (1970). With permission from John Wiley.)

Figure 16.3 shows the UV absorption spectrum of polystyrene. In the spectrum, absorption peaks for polystyrene are found at 260, 215, 194, and 80 nm. The first three peaks are the result of the transition of π electrons, whereas the 80-nm peak involves σ electrons.

16.1.4 Proteins

UV spectroscopy plays a very important role in the study of biological polymers both as an analytical tool and as a probe to determine their structural properties. In the spectroscopy of biological polymers an absorption band at about 190 nm is attributed to the amides. This band is less intense for the helical polypeptides and helical proteins than for the same polymers in the random coil conformation. For this reason the measurement of the extinct coefficient for a biological polymer near 190 nm can be used to estimate the helical content of the biological polymer sample in much the same way that optical activity was used in Chapter 13. The measurement of this band also reflects the hypochromism of the biological compound. Another important band occurs near 280 nm, and is generally attributed to the presence of either tyrosine, tryptophan, phenylalanine,

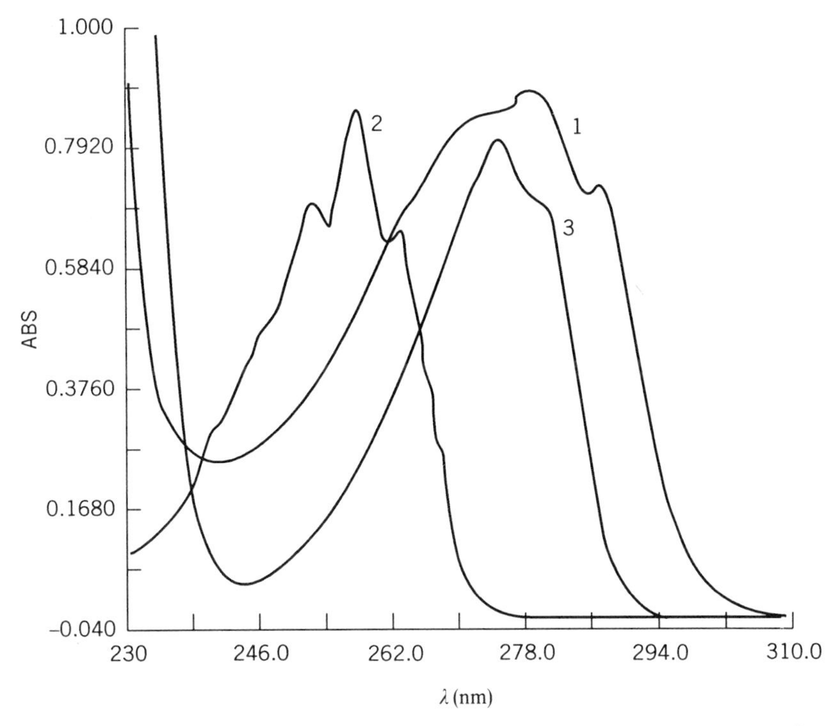

FIGURE 16.4 UV absorption spectra of the amino acids: 1, L-tryptophan; 2, L-phenylalanine; 3, D-tyrosine.

or even arginine. The band at 280 nm is so strong that it often obscures the band at 190 nm.

Figure 16.4 shows the UV absorption spectra for the three amino acids: tyrosine, tryptophan, and phenylalanine. It is clear that near 280 nm, tyrosine and tryptophan are stronger absorbers than phenylalanine. The UV absorption of tyrosine is often used to characterize the structure of the whole protein, as we discuss further. The following is a description of some of the applications of UV absorption spectroscopy in protein chemistry.

Elucidation of Protein Structure UV absorption has been used to study the intra- and intermolecular interactions of biological chromophores in different environments. These interactions are often manifested in the shapes of macromolecules. For example, poly-L-lysine hydrochloride exists in three different forms in aqueous solution, depending on the pH and temperature of the solution: random coil, pH 6.0, 25°C; helix, pH 10.8, 25°C; and β-form, pH 10.8, 52°C. The UV absorption spectra of poly-L-lysine hydrochloride for these three forms are shown in Figure 16.5.

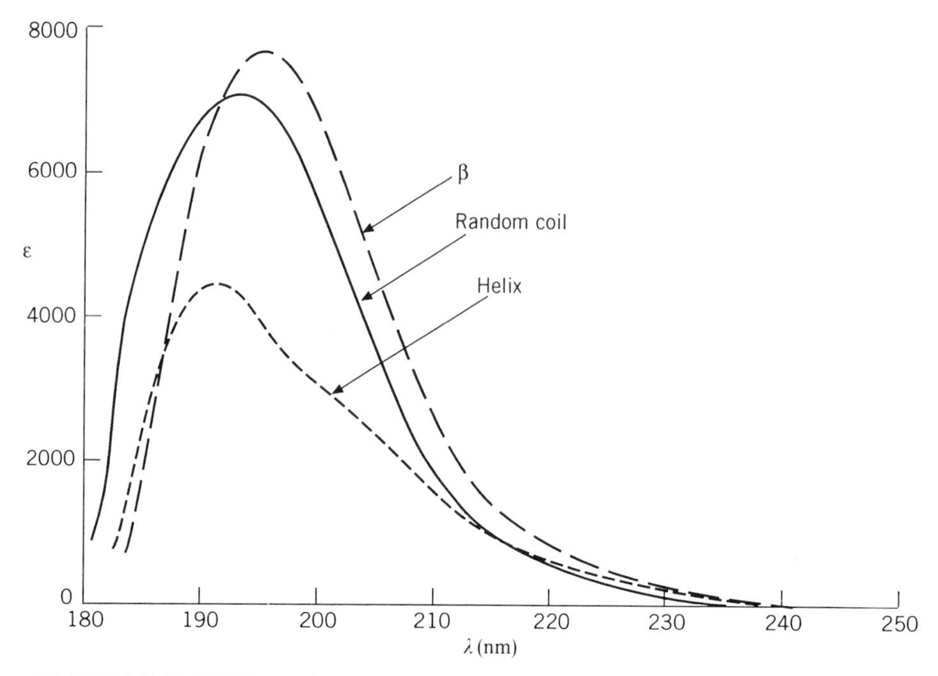

FIGURE 16.5 UV absorption spectra of poly-L-lysine hydrochloride in aqueous solution. (From Rosenbeck and Doty (1961) with permission of Dr. Doty.)

Determination of the Concentration Equation (16.1) can be expressed in the form

$$\varepsilon = \frac{A}{bC}$$

Let $E_{1\,\text{cm}}^{1\%}$ or $A_{1\,\text{cm}}^{1\%}$ be defined as the absorbance of a 1% solution of the substance in a 1-cm cell. Then $E_{1\,\text{cm}}^{1\%}$ is related to ε by

$$10\varepsilon = (E_{1\,\text{cm}}^{1\%}) \quad (M)$$

where M is the molecular weight of the substance. The concentration c in (g/100 mL) can now be determined by the measurement of its absorbance A:

$$c = \frac{A}{E_{1\,\text{cm}}^{1\%}}$$

Both A and E must be measured at the same wavelength λ, which is usually in the region of 210–290 nm. For example, the $E_{1\,\text{cm}}^{1\%}$ values are 6.67 for bovine serum albumin and 12.5 for trypsin, when both are measured at 279 nm.

Spectrophotometric Titrations Spectrophotometric titrations are based on the ionization of the phenolic chromophore of tyrosyl groups which occurs in the pH range of 9 and above. When a protein is titrated with alkali, the tyrosyl residues are ionized, which results in a new stronger absorption peak at about 293–295 nm. The change in the absorption intensity at 293–295 nm can be used to estimate the tyrosine (and also the tryptophan) content as well as the acid strength of these groups in proteins. Figure 16.6 shows the spectrophotometric titration of bovine serum albumin at ionic strength 0.03 and temperature 25°C, measured at 295 nm.

Difference Spectra When the unperturbed material is taken as the reference and the perturbed material is taken as the sample, the shift in the absorption spectrum as observed along the wavelength axis produces the difference spectrum. Perturbation can be produced by adding a different compound (such as LiCl) or an organic solvent (such as alcohol) to the aqueous protein solution to affect the environment of the tyrosyl residues. The effect of these perturbation materials is not strong enough to change the conformation of the native protein, but is large enough to cause measurable shifts in the spectrum. The purpose of using difference spectra is to study the location of the tyrosyl and tryptophyl

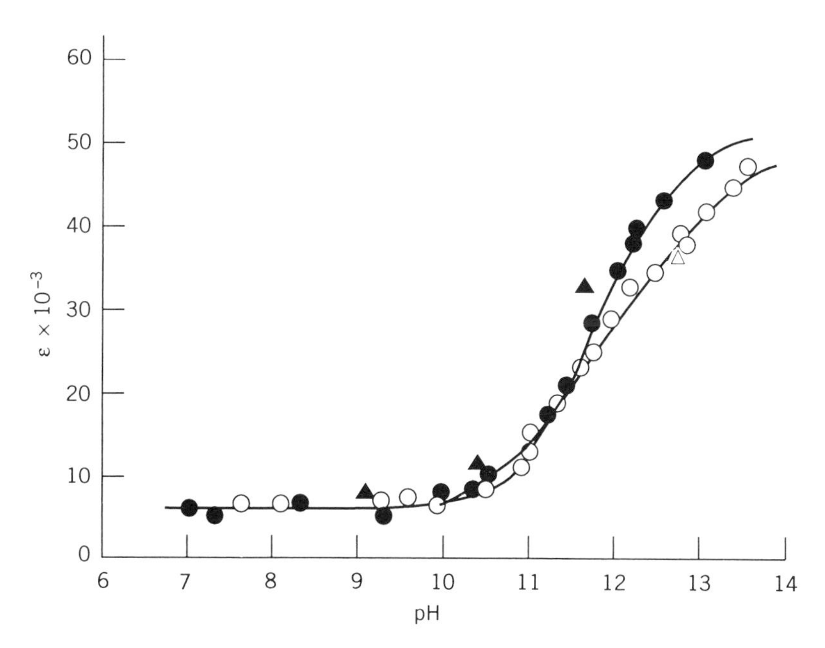

FIGURE 16.6 Spectrophotometric titration data for bovine serum albumin at ionic strength 0.03 and 25°C. ○, in dioxane–waver (1:3, v/v); ●, in water; △, in dioxane–water (1:3, v/v) reversed; ▲, in water reversed. (From Sun (1970). Reproduced with permission from Elsevier Science Publishers B.V.)

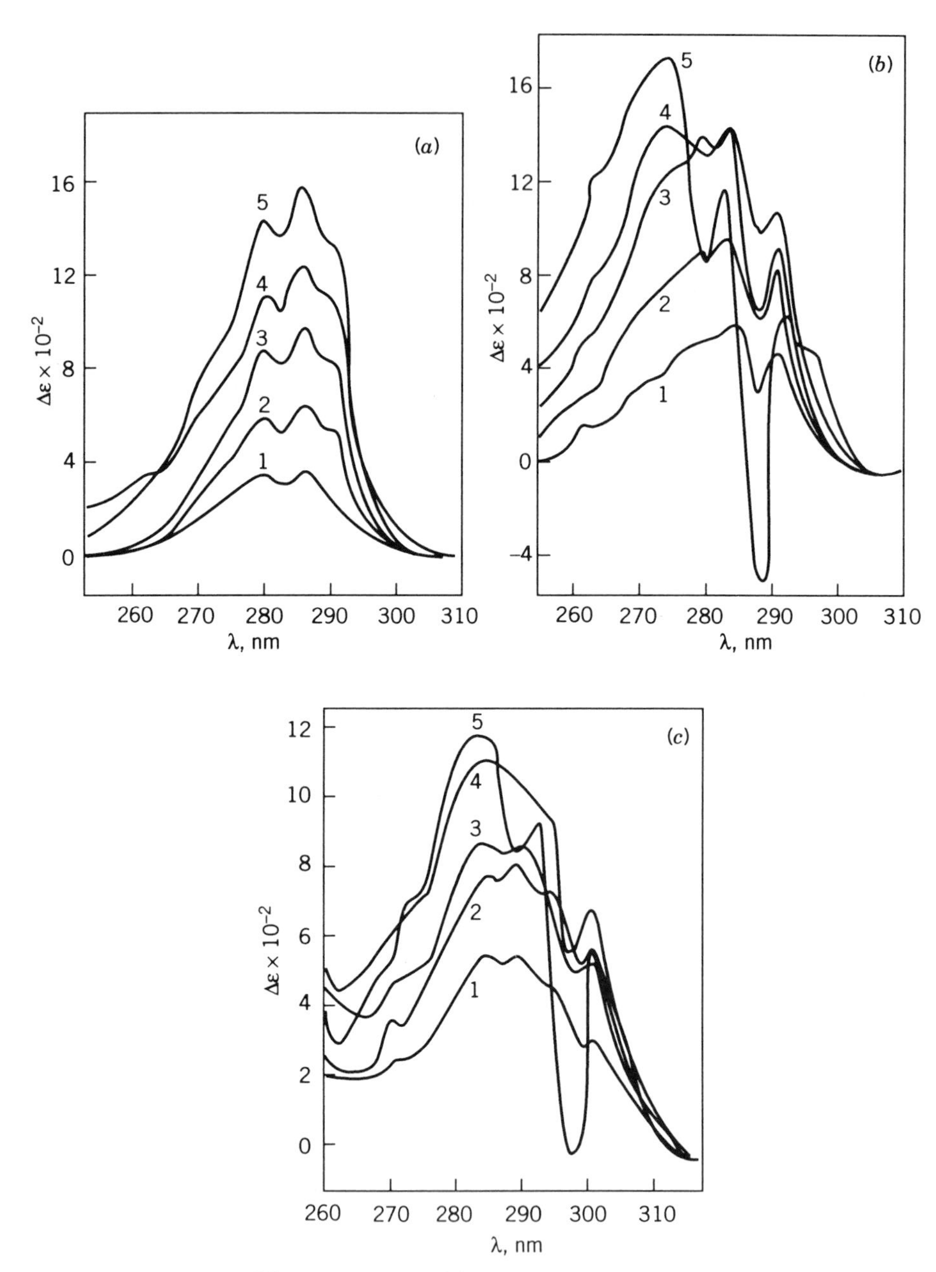

FIGURE 16.7 (a) Difference spectra of bovine serum albumin at pH 5.2–5.4, ionic strength 0.03, and 25°C produced by ethanol. Concentrations of ethanol (v/v): 1, 5%; 2, 10%; 3, 15%; 4, 20%; 5, 25%, (b) Difference spectra of bovine serum albumin at pH 5.2–5.4, ionic strength 0.03, and 25°C produced by *n*-propanol. Concentrations of *n*-propanol (v/v): 1, 5% 2, 10%; 3, 15%; 4, 20%; 5, 25%, (c) Difference spectra of bovine serum albumin at pH 5.2–5.4, ionic strength 0.03, and 25°C produced by dioxane. Concentrations of dioxane (v/v): 1, 5%; 2, 10%; 3, 15%; 4, 20%; 5, 25%. (From Sun et al. (1983).)

residues and their environment in a protein. For example, difference spectra can be used to determine how many tyrosyl or tryptophyl residues lie in the surface of the protein molecule and how many are buried in the protein molecule. Difference spectra can also be used for the titration of protein samples to investigate the change in pK values of the aromatic groups. Figure 16.7 shows the difference spectra of bovine serum albumin produced by ethanol, *n*-propanol, and dioxane, respectively.

16.1.5 Nucleic Acids

The absorption spectra produced by nucleic acids are due to their individual bases. The spectra can be obtained either for the whole molecule or for the individual bases. The absorption peaks for the four bases occur at the following frequencies: adenine, 260.5 nm; thymine, 264.5 nm; guanine, 275.0 nm; and cytosine, 267.0 nm. The ratio A_{250}/A_{280} for the four bases are often used to characterize DNA molecules under particular conditions. The values of these ratios are adenine, 2.00; thymine, 1.26; guanine, 1.63; and cytosine, 0.31. As in protein chemistry, the UV absorption spectra for nucleic acids can be used to determine their concentration. In addition, UV absorption can be used to study denaturation and renaturation of DNA molecules.

Determination of Concentration The amount of any base in a nucleic acid can be determined by measuring its optical density. The $E_{1\,\text{cm}}^{1\%}$ values for the four bases are adenine, 13.4×10^3; thymine, 7.9×10^3; guanine, 8.1×10^3; and cystosine, 6.1×10^3; all at 260 nm. For example, if the OD_{260} of adenine were 0.80, the molar concentration would be $0.80/(1.34 \times 10^3) = 6.0 \times 10^{-5}$ M. As in the case of proteins, the value of OD_{260} is often used for estimating the concentration of DNA or RNA; for example, $\text{OD}_{260} = 1$ corresponds to 50 μg/mL.

Denaturation and Renaturation When DNA molecules are heated to certain temperature (for example, 100°C), the two polynucleotide strands separate. The transition from the double strand (original form) to the single strand (denatured form) can be observed by the change in optical density at 260 nm. The plot of the optical density versus temperature gives a sinusoidal curve that is similar to an acid–base titration curve. In Figure 16.8 the point T_m, which corresponds to the equivalence point in an acid–base titration, is the hypochromic point and denotes where a mixture of the native and denatured strands occurs.

In contrast to denaturation, renaturation is the process by which the separated DNA strands are brought back together. If the denatured sample is reheated from 0°C to $T_m - 25$°C and maintained at $T_m - 25$°C, then renaturation can be determined by measuring the absorbance at 260 nm as a function of time (Figure 16.9). Both denaturation and renaturation can be caused not only by heat, but also by ionic strength and certain chemicals (for example, alkali salts).

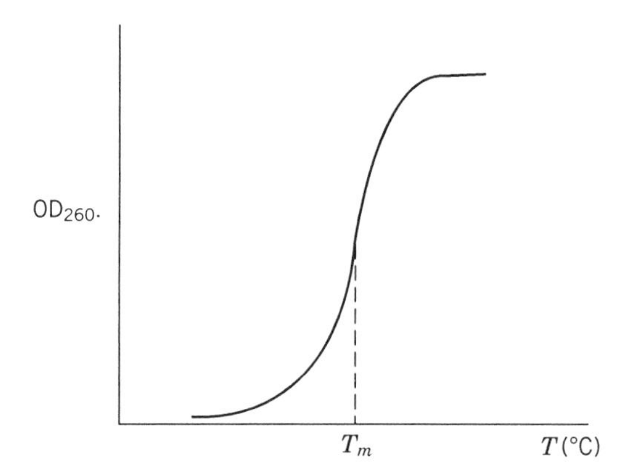

FIGURE 16.8 DNA denaturation process.

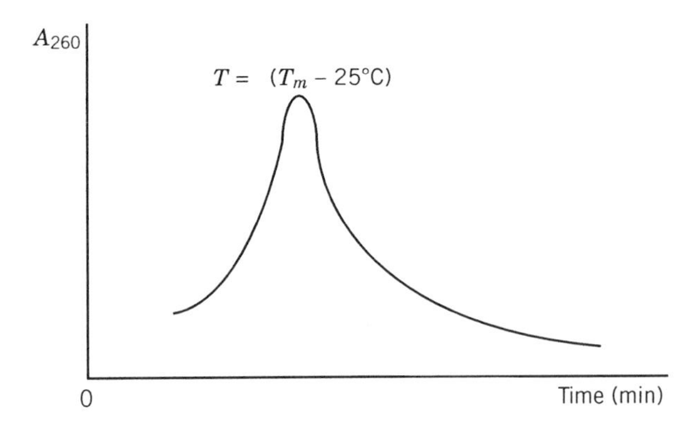

FIGURE 16.9 DNA renaturation process.

16.2 FLUORESCENCE SPECTROSCOPY

16.2.1 The Fluorescence Phenomena

The absorption of radiation energy by a molecule causes an electron to be excited, whereby it jumps from the ground state to a higher energy state. The excited molecule, however, can return to the ground state by emitting photons of the same wavelength as the exciting radiation. The phenomenon is called resonance fluorescence. The excited molecule may also drop to various lower states by emitting photons with wavelengths different from that of the exciting radiation. This is known as fluorescence.

Most molecules have electrons that are spin paired in the ground state,

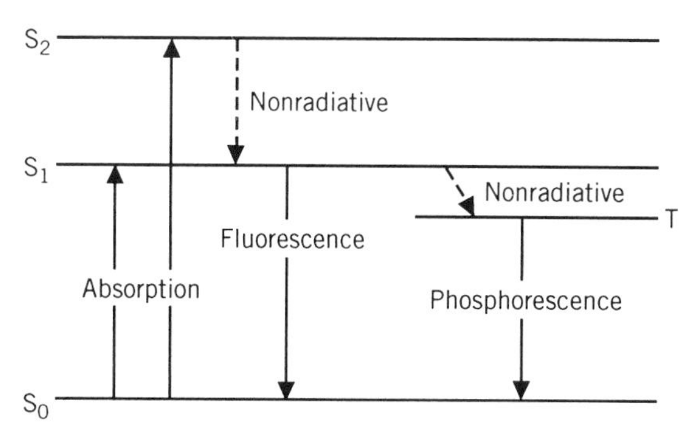

FIGURE 16.10 Fluorescence and phosphorescence: S, singlet; T, triplet.

nanemly ↑↓. The ↑ represents spin up and the ↓ represents spin down. If one of the two electrons is excited to a higher energy state, we say that electron is in a singlet excited state. There are also molecules that have unpaired electrons, namely ↑↑ or ↓↓ or a lone electrons↑ or ↓. When an electrons in these molecules is excited, we say that the electron is in a triplet state. Fluorescence is the emission that results from the return of the paired electron to the lower orbital. The return of the unpaired electron to the lower orbital emits radiation, called phosphorescence. Figure 16.10 shows the energy levels of a molecule that produces both fluorescence and phosphorescence. Since fluorescence consists of both an absorptive excitation and its subsequent emission, a fluorimeter is constructed with two sets of manometers, one to record the excitation and the other to record the emission. Because of their high sensitivity and high specificity, fluorescence measurements have long been used in biological studies of macromolecules. Phosphorescence measurements of biological macromolecules have yet to be explored.

Macromolecules may or may not fluoresce. Those that do are considered to contain intrinsic fluors. The common intrinsic fluors for proteins are tryptophan, tyrosine, and phenylalanine (the same three groups that absorb UV radiation). Macromolecule that have no intrinsic fluors can be made fluorescent by adding an extrinsic fluor to them. This is done by the process of chemical coupling or sample binding. The most common extrinsic fluors for proteins are 1-aniline-8-naphthalene sulfonate, 1-dimethylaminonaphthalene-5-sulfonate, dansyl chloride, 2-*p*-toluidylnaphthalene-6-sulfonate, rhodamine, and fluorescein. The most common extrinsic fluor for nucleic acids are various acridienes (acridine orange, proflavin, acriflavin) and ethidium bromide.

16.2.2 Emission and Excitation Spectra

A fluorescence emission spectrum is a plot of fluorescence intensity versus wavelength (or wave number). It is a description of the distribution of photons in the fluorescence of a molecule excited by radiation. There is another type of spectrum, called the fluorescence excitation spectrum, which is also a plot of

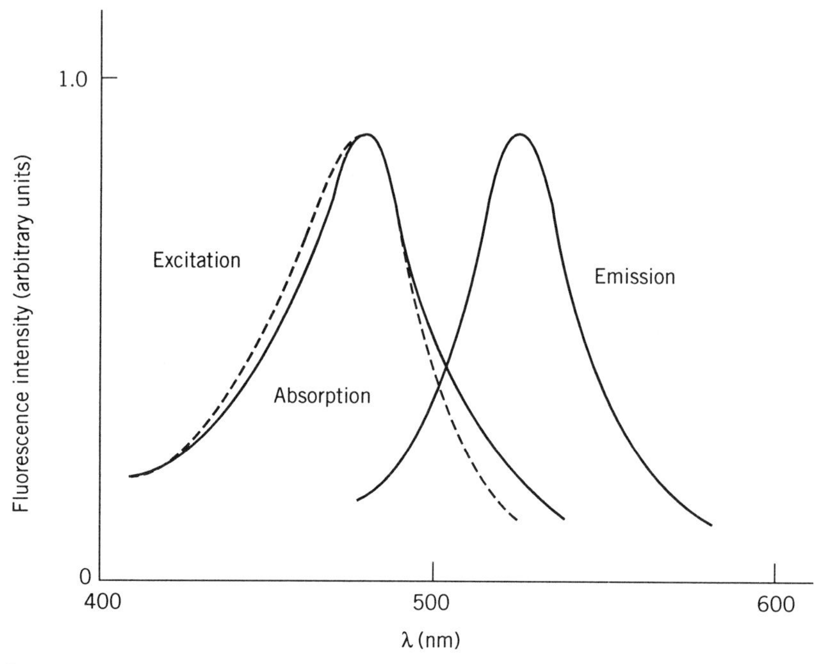

FIGURE 16.11 Fluorescence emission, fluorescence excitation, and visible absorption spectra of a fluorophore such as anthracene. The emission spectrum is obtained by irradiating the sample at one wavelength λ_{max} (absorption) and recording the intensity as a function of wavelength. The excitation spectrum is obtained by plotting the intensity of the fluorescence emission at different wavelengths, but the recording is at one wavelength. Note that the excitation spectrum is virtually identical with the visible absorption spectrum.

fluorescence intensity versus wavelength and which is also a description of the photon distribution. The difference lies in the method of obtaining the spectrum. For a fluorescence emission spectrum, it is a straightforward measurement of the intensity as a function of wavelength. For a fluorescence excitation spectrum, the measurement is carried out by varying the wavelength during the excitation, but the intensity of the emission is recorded at a constant wavelength. The fluorescence excitation spectrum is more similar to an absorption spectrum. Figure 16.11 shows the three spectra. In a UV difference spectrum the tyrosyl groups of a protein are the primary concern as to whether the tyrosyl groups are buried or on the surface of the molecule; in a fluorescence difference spectrum, it is the tryptophyl groups.

16.2.3 Quenching

In a collision with another molecule, Q, or with a surface, an excited molecule, M*, may lose part or all of its excitation energy without any emission or radiation. Such a process is called quenching, and Q, with which the excited molecule M* collides, is called the quencher. Moreover, the excited molecule M* may encounter two competing parallel reactions:

$$M^* \xrightarrow{k_1} M + h\nu \qquad \text{(Fluorescence)}$$

$$M^* + Q \xrightarrow{k_2} M + Q \qquad \text{(Quenching)}$$

The efficiency of fluorescence ϕ_F, which is also called the fluorescence quantum yield, or, simply, quantum yield, can be calculated based on the above two equations:

$$\phi_F = \frac{\text{Number of photons emitted}}{\text{Number of photons absorbed}}$$

$$= \frac{k_1[M^*]}{k_1[M^*] + k_2[M^*][Q]}$$

The concentration term of [M*] can be eliminated by reversing the equation:

$$\frac{1}{\phi_F} = \frac{k_1[M^*] + k_2[M^*][Q]}{k_1[M^*]} = 1 + \frac{k_2}{k_1}[Q]$$

The resulting equation is called the Stern-Volmer equation, which is often used to obtain information about the nature of the local environment around a fluorescence molecule. The equation can be put into a slightly different form which is more easily appreciated from the experimental point of view:

$$\frac{F_0}{F} = 1 + k\tau_0[Q] = 1 + k_D[Q]$$

where F is the fluorescence intensity, the subscript 0 refers to the absence of the quencher, k is the bimolecular quenching constant, τ is the lifetime of the fluorescence, and k_D is the Stern-Volmer constant, defined as $k_D = k\tau_0$. The lifetime τ usually means the lifetime of the excited state. It is the average time the molecule spends in the excited state prior to its return to the ground state. The reciprocal of the fluorescence lifetime τ^{-1} is the fluorescence decay rate. Both F and τ are measurable quantities. They are related by

$$\frac{F_0}{F} = \frac{\tau_0}{\tau}$$

Figure 16.12 shows oxygen quenching of tryptophan as observed by fluorescence lifetimes and yields and is plotted according to the Stern-Volmer equation. Common quenchers are oxygen, xenon, hydrogen peroxide, iodide, and amines.

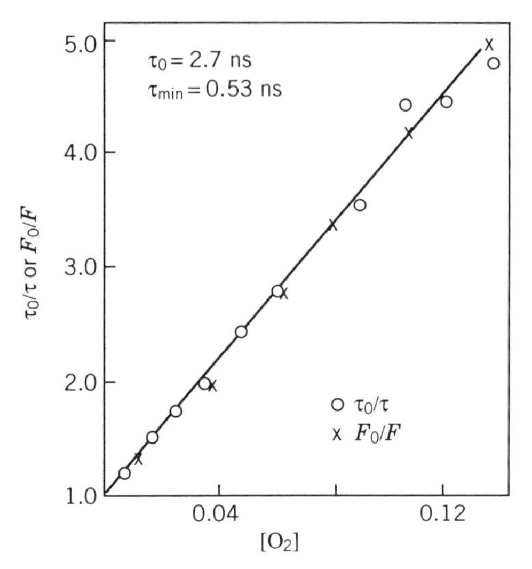

FIGURE 16.12 Oxygen quenching of tryptophan. (From Lakowicz and Weber (1973) with permission of Dr. Weber.)

The constant k is is a second-order rate constant for a diffusion-controlled process. It is related to the translation diffusion coefficients of the fluorescence molecule and the quencher, and hence is related to the radius of encounter.

16.2.4 Energy Transfer

Instead of the usual decay by a radiative or nonradiative process, an excited molecule may decay by another process, energy transfer. The excited molecule, which is called the donor, may transfer the excitation energy to another molecule, called the acceptor. The transfer is primarily a phenomenon of dipole–dipole interactions between the donor and the acceptor. The rate of energy transfer depends on several factors: the overlap of the emission spectrum of the donor with the absorption spectrum of the acceptor (J), the orientation of the donor and the acceptor transition dipoles (κ), and the distance between the donor and the acceptor (r). The overlap J is described by the integral

$$J = \int_0^\infty F_d(\lambda)\varepsilon_a(\lambda)\lambda^4 d\lambda$$

where F_d is the fluorescence intensity of the donor, and ε_a is the extinction coefficient of the acceptor. Both depend on wavelength λ. The orientation factor κ^2 is given by

$$\kappa^2 = (\cos\theta_T - 3\cos\theta_d \cos\theta_a)^2$$

where θ's are the angles, as shown in Figure 16.13.

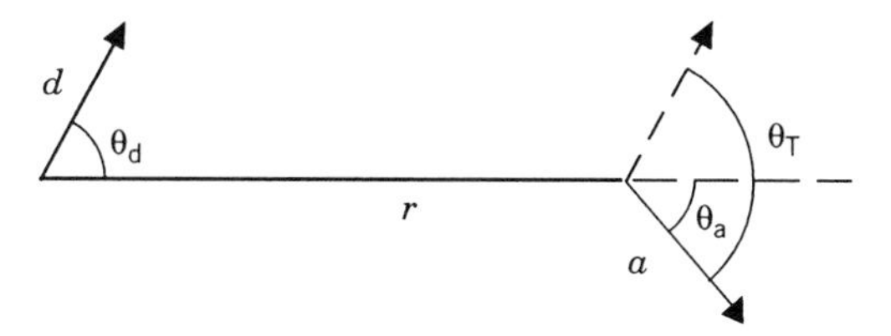

FIGURE 16.13 The orientation factor κ^2.

The distance between the donor and the acceptor r is described by Forster's theory. Forster derived an equation for the rate of energy transfer from a specific donor to a specific acceptor k_T:

$$k_T = \frac{1}{\tau_d}\left(\frac{R_0}{r}\right)^{-6}$$

where τ_d is the lifetime of the donor in the absence of the acceptor, and R_0, which is now called the Forster distance, is defined as the characterized distance

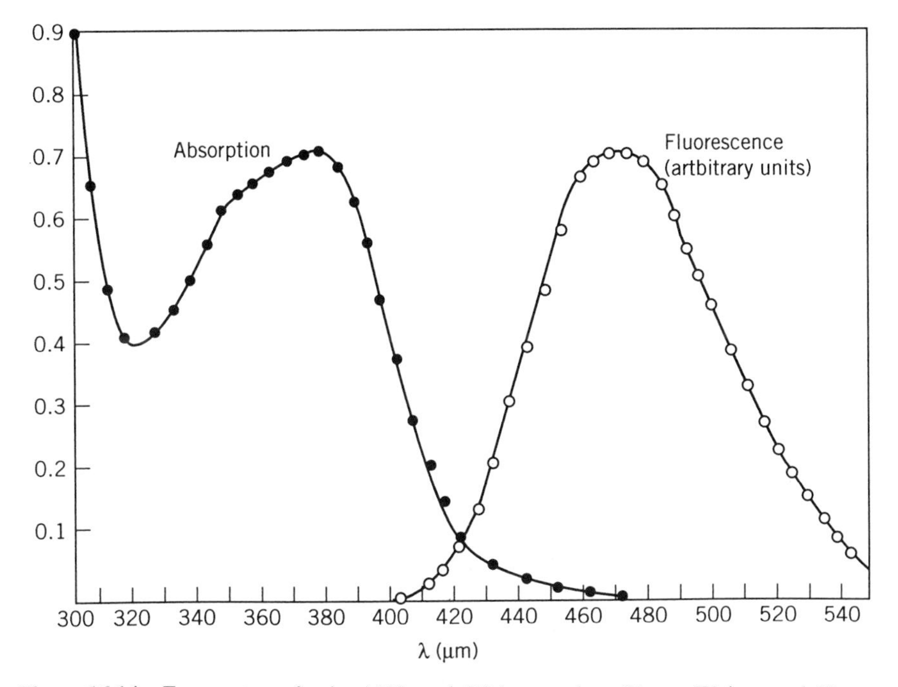

Figure 16.14 Energy transfer in ANS and BSA complex. (From Weber and Young (1964). With permission from Dr. Weber and the American Society for Biochemistry and Molecular Biology.)

at which the transfer rate is equal to the decay rate of the donor, that is,

$$k_{\mathrm{T}} = \tau_{\mathrm{d}}^{-1}$$

At R_0 the efficiency of transfer is 50%. The donor fluorescence quantum yield in the absence of transfer ϕ_{d} and the overlap integral J are both incorporated in the equation of R_0:

$$R_0 = [(8.8 \times 10^{-25})(\kappa^2 n^{-4} \phi_{\mathrm{d}} J)]^{1/6}$$

where n is the refractive index of the medium (for example, water). In experiments, the efficiency of energy transfer (E) can be directly related to the distance r, using

$$E = \frac{R_0^6}{R_0^6 + r^6}$$

Figure 16.14 shows energy transfer in 1-anilinonaphthalene-8-sulfonic acid (ANS) and bovine serum albumin complex (BSA). In this system, r is less than $\frac{3}{2}R_0$, $\tau = 8.0$ ns, $J = 2.0 \times 10^{10}$ cm^3/mM^2, and $R_0 = 24$ Å.

16.2.5 Polarization and Depolarization

Polarization of fluorescence is the result of the relative orientations of the absorption and emission oscillators. Consider Figure 16.15. By manipulating the two polarizers we can measure the fluorescence intensities $I_{\parallel}$ and $I_{\perp}$, respectively. $I_{\parallel}$ is obtained when the electric vector of the linearly polarized light is parallel to an arbitrary axis (such as z), while $I_{\perp}$ is perpendicular to the arbitrary axis (z), as shown in Figure 16.16. The degree of polarization p is defined as

$$p = \frac{I_{\parallel} - I_{\perp}}{I_{\parallel} + I_{\perp}} = \frac{I_{zz} - I_{zy}}{I_{zz} + I_{zy}}$$

If all the molecules are oriented in the same direction, for example, parallel to the z axis, then p is clearly equal to unity. If the molecules are partially oriented, $p \neq 1$. For $p < 1$, we call p fluorescence depolarization.

Depolarization is a manifestation of the intrinsic properties of the macromolecule. It is the result of Brownian motion and it involves the translation of molecules and their rotational movements. Since Brownian motion is affected by temperature, solvent viscosity, and the size and shape of the molecule, these factors also affect the value of p. The angle θ between the original direction of the molecules in a solvent system and their new direction due to some external force is called the orientation angle. The mean value of its cosine, $\overline{\cos\theta}$, is a measure of its orientation. Perrin derived an equation for p that takes all these

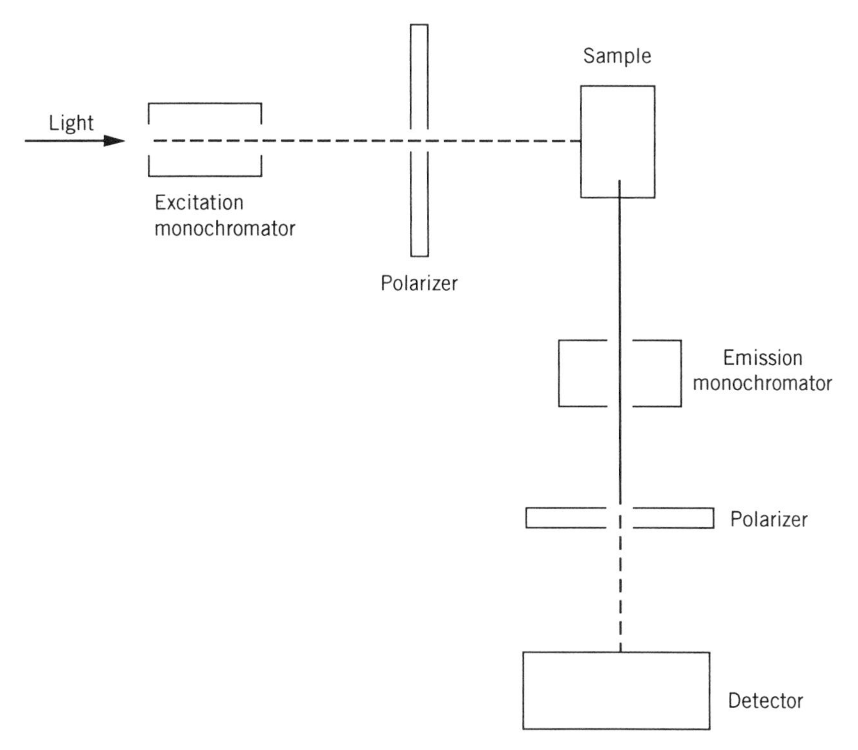

FIGURE 16.15 Diagram for polarization and depolarization measurements.

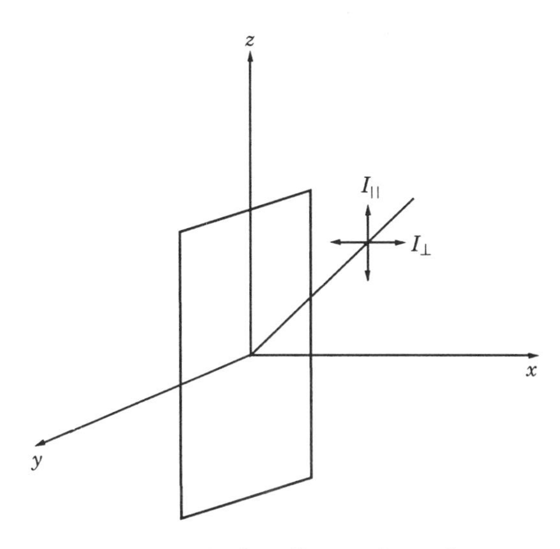

FIGURE 16.16 Coordinates of $I_{\parallel}$ and $I_{\perp}$.

factors into consideration:

$$\frac{1}{p}-\frac{1}{3}=\frac{8}{3}\frac{\overline{\cos^2\theta}}{3\,\overline{\cos^4\theta}-\overline{\cos^2\theta}}\left(\frac{\tau}{\tau_R+1}\right)(3\cos^2\delta-1)^{-1}$$

where τ_R is the rotational relaxation time of the fluorescent molecule, τ is the lifetime of excitation, and δ is the intramolecular vector angle. For molecules in random orientation, the equation becomes

$$\frac{1}{p}-\frac{1}{3}=\left(\frac{1}{p_0}-\frac{1}{3}\right)\left(1+\frac{\tau}{\tau_R}\right)$$

where p_0 is the limiting polarization at infinite viscosity. In addition, if the molecules are spherical, then the frictional coefficient can be expressed in terms of Stoke's law for rotation:

$$\zeta = 8\pi\eta a^3$$

and

$$\tau_R^{-1}=\frac{kT}{V\eta}$$

where V is the molecular volume. The Perrin equation is now in the form

$$\frac{1}{p}-\frac{1}{3}=\left(\frac{1}{p_0}-\frac{1}{3}\right)\left(1+\frac{kT}{V\eta}\tau\right)$$

By plotting $1/p$ or $(1/p_0-\frac{1}{3})$ versus T/η, we can obtain information about V if τ is known, or about τ if V is known from other sources.

16.3 INFRARED SPECTROSCOPY

Infrared (IR) spectroscopy has long been used to determine molecular structure and to identify unknown compounds. IR data have been used to obtain information about the chemical composition, configuration, and crystalinity of polymeric materials. Recently, IR spectroscopy has encountered competition from other techniques, such as NMR and x-ray diffraction. Nevertheless, IR spectroscopy's importance as an experimental technique continues largely because of the rapid development of Fourier transform infrared (FTIR) spectroscopy, which is sensitive to the detailed structure of a molecule.

16.3.1 Basic Theory

Assuming that atoms vibrate in a harmonic potential, the equation of motion that describes IR spectroscopy is the Lagrange equation:

$$Fr + K\ddot{r} = 0$$

where F involves force constants, K involves reduced mass of the molecule, r is the internal coordinate that describes changes in the bond lengths and bond angles, and $\ddot{r}$ is the second derivative of r with respect to time. This equation can be reduced to the well-known eigenvalue equation:

$$GFL = L$$

where $G = K^{-1}$ is the matrix that contains the mass of the atoms and information on the geometry of the molecule, and L is the matrix of elements l. The element l is the vibrational displacement and is defined as

$$r = l_j \cos(\omega_j t + \phi)$$

In the above equation ω is equal to $\nu/2\pi$, t is the time and ϕ is the phase shift. The solution of the above eigenvalue equation is the secular determinant

$$|GF - E\Lambda| = 0$$

where E is a matrix of electric vectors, which is related to the dipole moment, and Λ is a diagonal matrix, with the vibrational frequency defined as

$$\nu_{\text{k}} = \left(\frac{\lambda_{\text{k}}}{4\pi^2 c'^2}\right)^{1/2}$$

where λ_{k} is the wavelength and c' is the velocity of light.

This theoretical calculation is used to idealize a macromolecule and to calculate the fundamental frequency ν_{k}, which is then compared to the experimental frequency. In the calculation of molecular symmetry, periodicity of potential functions and force constants were introduced. If the results of these calculations are in agreement with those from experiment, then the assumed parameters (symmetry, potential function, and force constant) contain information about the actual macromolecule. A detailed calculation is illustrated by the study of polyethylene (see Bower and Maddams, 1989).

16.3.2 Absorption Bands: Stretching and Bending

From the basic theory it is easy to understand that an absorption band appears only when the frequency of the radiation incident upon a compound is the same as the vibrational frequency of that compound. This phenomenon is related to the change of the dipole moment or to the change in polarizability. If it is related to the change of the dipole moment, we observe an infrared band. If it is related to the change in polarizability, we observe a Raman band. Since dipole moment and polarizability are closely related, so are infrared spectroscopy and Raman spectroscopy. These two spectroscopies can supplement the information of each other. The general principles that apply to infrared spectroscopy can also be extended to Raman spectroscopy. In this section we describe infrared spectroscopy only. The absorption range of infrared spectroscopy is sketched in Figure 16.17. The term micro (μ) refers to the wavelength λ and $1\ \mu = 10^{-4}$ cm. In the older literature, prism instruments were used in infrared spectroscopy and the IR spectra were prsented in wavelengths. Today, grating instruments are used in most laboratories and the IR spectra are presented in wave numbers.

Two kinds of fundamental vibration bands are used to describe the vibrational behavior of polymers: stretching and bending. These bands may be further divided into different modes. The six typical modes are shown in Figure 16.18. The terms ν_{as}, ν_s, δ_s, γ_w, γ_t, and γ_r are given under each mode. The arrows show the direction of the vibration, while the plus and minus signs denote the vibrations that are in a direction perpendicular to the page.

In an IR spectrum, there often appear overtones, which have twice the frequency of the normal vibration (for example, the carbonyl group has its fundamental vibration band at 1715 cm^{-1} and an overtone at 3430 cm^{-1}), and combination overtones, which are at frequencies equal to the sum or difference

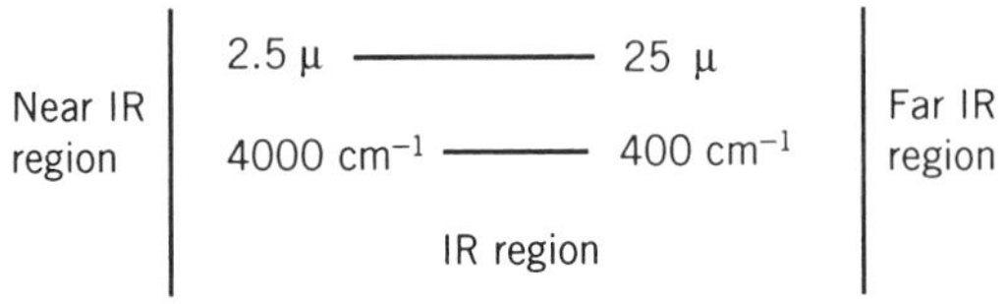

FIGURE 16.17 The absorption range of IR.

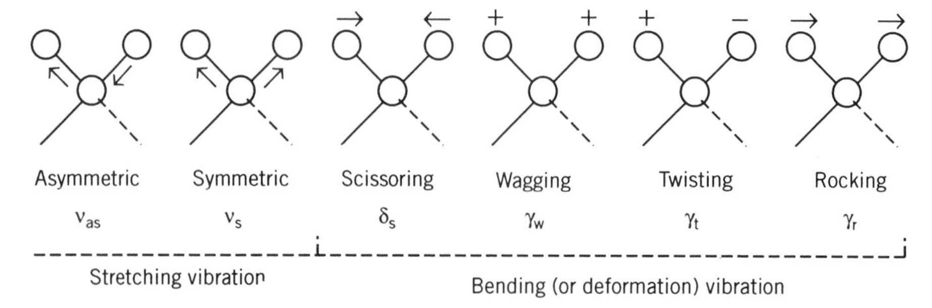

FIGURE 16.18 The six typical modes of vibation.

of two or more fundamental bands. Both overtones and combination overtones appear in the IR spectra as weak bands. They arise from small anharmomicities which tend to couple with the normal vibrations. Detailed information on the structure of the investigated macromolecule can be gained by correlating characteristic bands and band combinations with the aid of knowledge learned from small molecules. The mode of the alcohol group is shown in Figure 16.19 and the well-known IR spectrum of SO_2 is shown in Figure 16.20. In Figure 16.20 we observe fundamentals ν_1 1146, ν_2 519, and ν_3 1360, overtones $2\nu_1$ 2322, and combination bands $\nu_1 + \nu_3$ 2562. The three fundamentals are sketched in Figure 16.21.

The factor that determines the position of the absorption bands is the potential energy of the molecule. For SO_2, the potential energy V can be expressed as

$$V = \tfrac{1}{2}[k_1(Q_1^2 + Q_2^2) + k_\delta \delta^2]$$

where Q_1 and Q_2 are changes in S–O distances (whose equilibrium value is denoted by l), and δ is the change in the bond angle 2α(O–S–O). The quantities k_1 and k_δ/l^2 are force constants (usually in dyn/cm) for stretching and bending motions, respectively.

The stretching frequency ν of a diatomic molecule is, in general, based on the classical mechanics of a harmonic oscillator:

$$\nu = \frac{1}{2\pi c'}\sqrt{\frac{f(m_1 + m_2)}{m_1 m_2}}$$

where c' is the velocity of light, and f is the force constant (bond strength or bond order). Thus, the absorption frequency depends on the bond strength f and the masses of atoms m. As the bond strengths increase from single to double to triple bonds the stretching frequencies also increase from 700–1500 cm^{-1} to 1600–1800 cm^{-1} and to 2000–2500 cm^{-1}. Also, the heavier the atoms are, the lower their frequency. For example, vibration of the O–H bond is at 3600 cm^{-1}, but is lowered to 2630 cm^{-1} in the O–D bond.

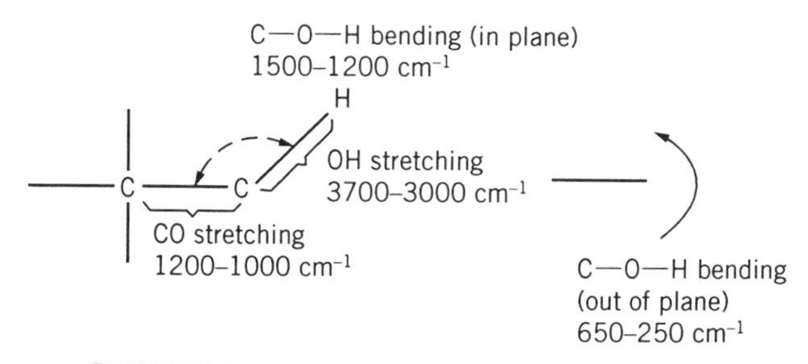

FIGURE 16.19 The mode of the alcohol group.

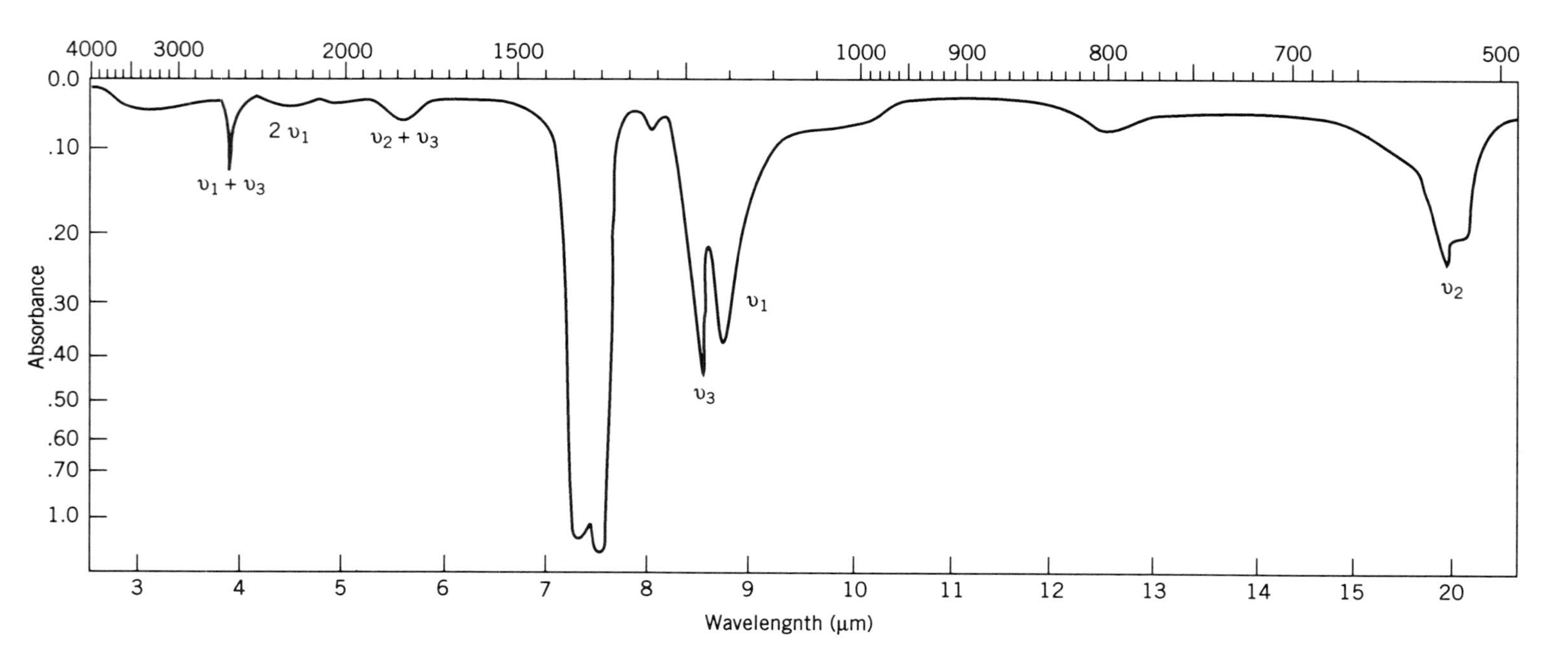

FIGURE 16.20 IR spectrum of SO_2.

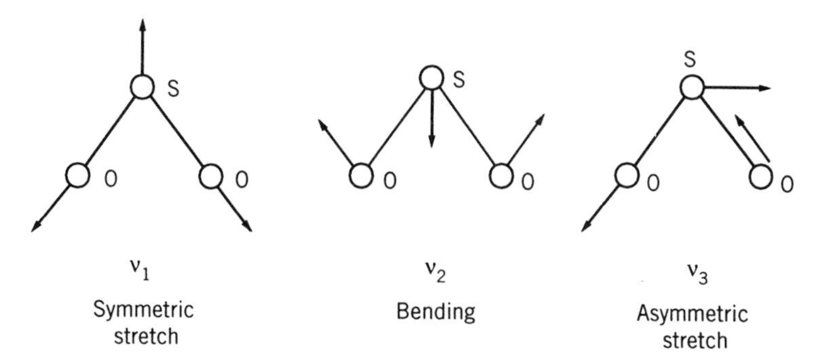

FIGURE 16.21 Three fundamentals of SO_2.

16.3.3 Infrared Spectroscopy of Synthetic Polymers

The infrared absorption spectrum of synthetic polymers is generally simple. In Figure 16.22 we illustrate the polystyrene spectrum, which is fairly representative of aromatic hydrogen polymers. The polystyrene spectrum's three IR bands appear sharply in the spectrum without ambiguity at 2850, 1603, and 906 cm^{-1}. The bands above 3000 cm^{-1} are due to unsaturated C–H stretching vibrations. The stretching vibration of conjugated C═C appears around 1600 cm^{-1}. The presence of the 760 and 700 cm^{-1} bands are attributed to C–H deformations. Monosubstitution is confirmed by the pattern of weak bands from 2000 to 1600 cm^{-1}. Atactic polystyrene has bands at 670, 620, 565 (shoulder), and

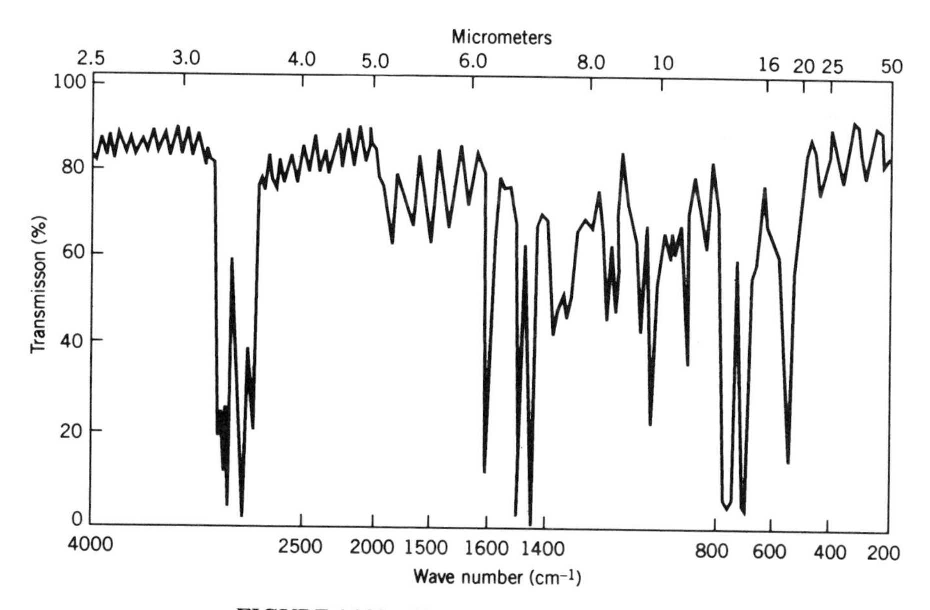

FIGURE 16.22 IR spectrum of polystyrene.

536 cm^{-1}. Isotactic polystyrene does not have any absorption from 550 to 500 cm^{-1} or at 670 cm^{-1}. The interference fringes between 4000 and 3200 cm^{-1} and between 2700 and 2000 cm^{-1} are used to determine the film thickness.

Characteristic Frequency of Functional Groups A large number of characteristic band tables are available. Here, we list the characteristic frequencies for functional groups that are frequently encountered in polymer spectroscopy:

Alkanes	
ν-CH	2950–2850 cm^{-1}
δ-CH	1465–1380 cm^{-1}
Alkene	
ν-CH	> 3000 cm^{-1}
Aromatics	
In-plane bending	1225–950 cm^{-1}
Out-of-plane bending and ring puckering	Below 900 cm^{-1}
Hydroxyl	
ν—OH	3300 cm^{-1}
Ether	1250 cm^{-1}
Carbonyl	1900–1600 cm^{-1}
Triple bond	2300–2000 cm^{-1}
Double bond	1900–1500 cm^{-1}
Single bond	1300–800 cm^{-1}
Typical triple bonds:	
C≡C stretchng	2140–2100 cm^{-1}
C≡N stretching	2260–2200 cm^{-1}
Double bonds of unsaturated polymer chains:	
Vinyl, $R_1CH{=}CH_2$	990–909 cm^{-1}
trans-$R_1CH{=}CH\ R_2$	962 cm^{-1}
Vinylidene, $R_1R_2C{=}CH_2$	885 cm^{-1}
$R_1R_2C{=}CHR_3$	833 cm^{-1}
cis-$R_1CH{=}CHR_2$	704 cm^{-1}
Amides	
Primary —$CONH_2$	1690 cm^{-1}
Secondary —CONH	1700 cm^{-1}

Stereoregularity We have already discussed the stereoregularity of the IR spectrum of polystyrene. Figure 16.23 shows three IR spectra of polypropene. The spectra of atactic, syndiotactic, and isotactic polypropene are virtually the same at wave numbers of 1000 cm^{-1} and above. Remarkable spectral differences are observed, however, in the spectra at wave numbers below 1000 cm^{-1}. This is consistent with our observations on the IR spectra of atactic, syndiotactic, and isotactic of polystyrene as mentioned previously.

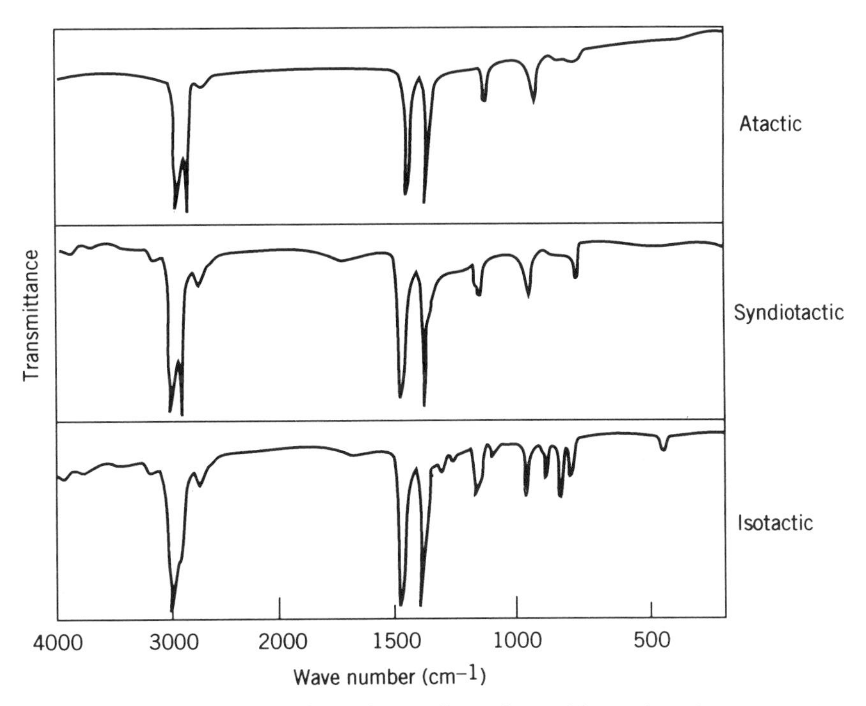

FIGURE 16.23 IR spectra of atactic, syndiotactic, and isotactic polypropene. (From Siesler and Holland-Mørtiz (1980). Permission granted by Marcel Dekker, Inc.)

Crystallinity Crystallinity bands are often assigned for polymer molecules in conjunction with the parameters of other techniques, such as density and x-ray diffraction. They describe the state of order in an investigated polymer molecule. In the spectrum of poly(vinyl alcohol), for example, the absorption band at $1141\,\text{cm}^{-1}$ is assigned to the crystallinity of the molecule. The degree of crystallinity is measured by the intensity of the band at $1141\,\text{cm}^{-1}$, which matches the density of the molecule. Figure 16.24 shows this phenomenon.

16.3.4 Biological Polymers

Because of the complexity of biological polymers, their IR spectra are generally more difficult to analyze than the IR spectra of synthetic polymers. At present, IR spectra have been used basically as an analytic tool to identify certain functional groups of biological polymers. For proteins, the amide I and II bands appear in the region $1700\text{–}1500\,\text{cm}^{-1}$ and are localized in the –CO–NH– group. The frequencies of these bands are not dependent on the neighboring amino acid residue side chains. Three important bands are related to the conformation of proteins: the amide bands at $1650\,\text{cm}^{-1}$ for the α-helix, at $1685\,\text{cm}^{-1}$ for the antiparallel, and at $1637\,\text{cm}^{-1}$ for random structures. The

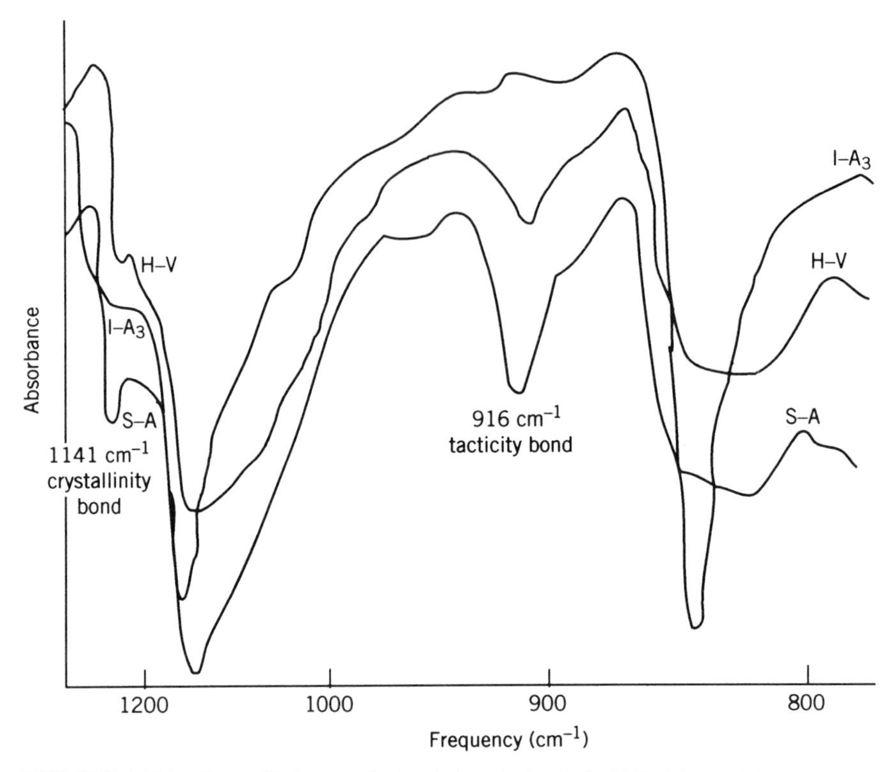

FIGURE 16.24 Crystallinity band of poly(vinyl alcohol). H-V, IA_3, and S-A are three different samples (From Kennedy and Willcockson (1966). Permission was granted by John Wiley & Sons.)

appearance of a carbonyl band in a spectrum for a molecule that contains a hydroxyl group is an indication of its tautomerization. Identification of the number of hydrogen bonds can be determined by dissolving the polymer in D_2O and observing the bands due to the corresponding deuterium. For nucleic acids, since base pairs give bands at different frequencies, the ratio of the base pairs can be determined from the spectrum.

16.3.5 Fourier Transform Infrared Spectroscopy (FTIR)

The revival of interest in a very old instrument, the interferometer (for the study of light beams), together with the development of computer techniques, has given infrared spectroscopy new momentum and sophistication as a probe for investigating the structure of molecules, particularly macromolecules. Before FTIR was available, most spectroscopy instruments used prisms or gratings that could detect only the beams that passed through the slit. Much of the information about the molecular structure that could be obtained from beams other than those that could pass through the slit was completely wasted. The new FTIR device, along with the assistance of a rapid computer program (the Fourier transform algorithm), allows the study of all of the transmitted energy at one time.

Interferometer The interferometer was originally designed by Michelson in 1891. His idea was to divide a beam of light into two paths. The two beams were then recombined after a path difference had been introduced. The device is shown in Figure 16.25. The intensity of the beam can be measured as a function of the path difference by a detector. This optical path difference between the two beams traveling to the fixed and moving mirrors is called retardation and is designated by the symbol δ. When the two beams are in phase, they interfere with each other constructively; when the two beams are not in phase, they interfere with each other destructively. The intensity at any point is equal to the intensity of the source. The quantity measured by interferometer is called the interferogram, and is designated as $I(\delta)$, which is related to the true spectrum $\beta(\bar{\nu})$ by

$$I(\delta) = \beta(\bar{\nu}) \cos 2\pi\bar{\nu}\delta$$

Fourier Transform The interferogram $I(\delta)$ is the Fourier transform of spectrum $\beta(\bar{\nu})$:

$$I(\delta) = \int_{-\infty}^{\infty} \beta(\bar{\nu}) e^{i\pi\nu} \, d\bar{\nu}$$

$$\beta(\bar{\nu}) = \int_{-\infty}^{\infty} I(\delta) e^{-i2\pi\bar{\nu}} \, d\bar{\nu}$$

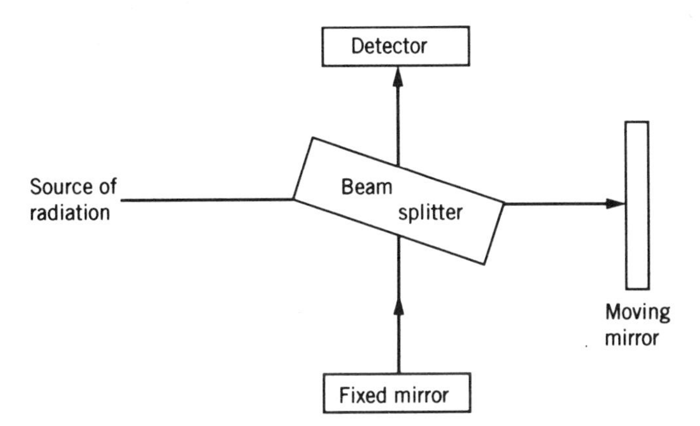

FIGURE 16.25 Diagram of an interferometer.

For numerical computations the spectrum $\beta(\bar{\nu})$ should be changed into the form of a summation instead of an integration:

$$\beta(\bar{\nu}) = \sum_{j=0}^{(N/2)-1} A_j I_j(\delta) \exp(-\mathrm{i} j 2\pi\bar{\nu})$$

where N is the number of spectrum points. Note that here $\mathrm{i} = \sqrt{-1}$ and j is the index of a spectrum point. The values of $I_j(\delta)$ are obtained from the experiment. A_j is a mathematical function called the apodizing function, which is a corrective procedure for modifying the basic Fourier transform integral. The term $\exp(-\mathrm{i} j 2\pi\bar{\nu})$ is usually transformed into sine and cosine functions. One can make a table of cosines or sines and then reduce the argument $2\pi j\bar{\nu}$ to the first period. For each $\beta(\bar{\nu})$, we carry out N multiplications for $A_j I_j \cos 2\pi j\bar{\nu}$ and N multiplications for $A_j I_j \sin 2\pi j\bar{\nu}$. This results in a total of $2N$ multiplications. The adoption of the Cooley–Turkey algorithm in which $N = 2^n$ for some integer n not only makes the computation fast, but also saves computer time, hence, cost.

FTIR Studies of Polymers For the most part, the techniques used in the measurement of FTIR are the same as those applied to conventional samples. Spectral features of a component in a polymer are isolated from the solvent

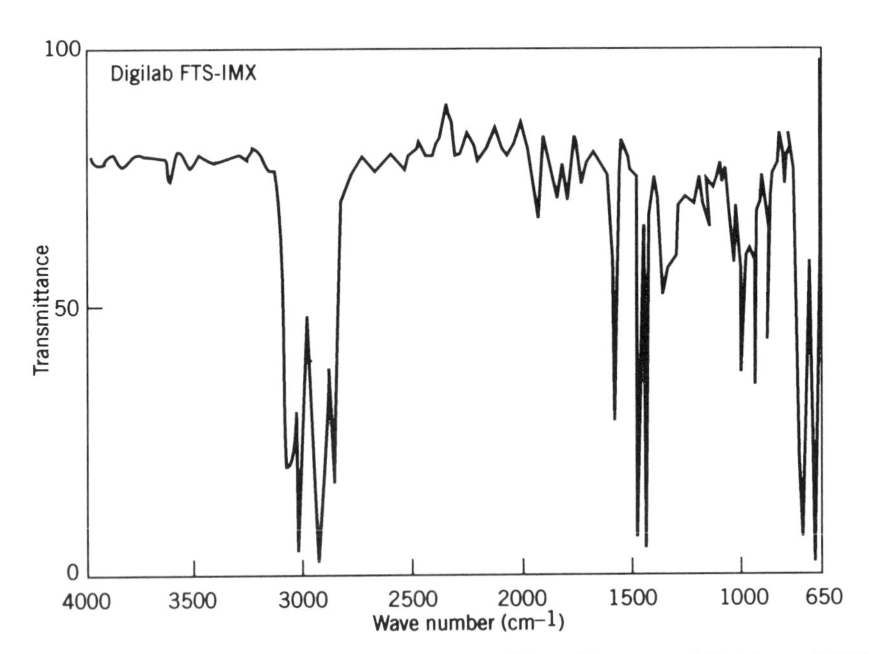

FIGURE 16.26 FT-IR spectrum of polystyrene. (From Ferraro and Krishnan (1990). Permission was granted by Dr. Krishnan and Academic Press, Inc.)

bands and from all other compounds present in the system. Analysis is carried out by computer programs. Figure 16.26 shows the spectrum of polystyrene with an IR microscope through a 10-μm aperture. We see that the base line is flat even below 1000 cm^{-1}. This is very important because the frequency range of the infrared spectrum below 1300 cm^{-1}, especially that below 1000 cm^{-1}, is known as the fingerprint region. Changes in frequency and intensity may be used to determine the microscopic characterization of the polymer to see if there exists any defect due to stress or the environment.

REFERENCES

Alben, J. O., and L. Y. Fager, *Biochemistry* **11**, 842 (1972).

Blout, E., *J. Am. Chem. Soc.* **83**, 712 (1961).

Boaz, H., and G. K. Rollefson, *J. Am. Chem. Soc.* **72**, 3425 (1950).

Bower, D. I., and W. F. Maddams, *The Vibrational Spectroscopy of Polymers.* New York: Cambridge University Press, 1989.

Brand, E. A., *J. Chem. Soc.* 379 (1950).

Briggs, D., *Polymer* 1379, 1984.

Chapoy, L. L., and D. B. DuPré, *Methods Exp Phys.* **16A**, 404 (1980).

Chen, R. F., and R. L. Bowman, *Science* **147**, 729 (1965).

Cooley, J. W., and J. W. Tukey, *Math. Comput.* **19**, 297 (1965).

Duportail, G., D. Froelich, and G. Weil, *Eur. Polym. J.* **7**, 977, 1023 (1971).

Encyclopedia of Polymer Science and Technology, Vol. 13. New York: Wiley, 1970.

Ferraro, J. R., and K. Krishnan (Eds.), *Practical Fourier Transform Spectroscopy.* New York: Academic, 1990, p. 112.

Forster, T., *Ann. Phys. (Leipzig)* **2**, 55 (1948); *Naturforscher* **4**, 321 (1949).

Freifelder, D., *Physical Biochemistry.* San Francisco: Freeman, 1976.

Griffiths, P. R., and J. A. de Haseth, *Fourier Transform Infrared Spectrometry.* New York: Wiley, 1986.

Harrington, W F., P. Johnson, and R. H. Ottewill, *Biochem. J.* **62**, 569 (1956).

Herskovits, T. T., and M. Laskowski, Jr., *J. Biol. Chem.* **235**, 57 (1960).

Herzberg, G., *Molecular Specra and Molecular Structure,* II: *Infrared and Raman Spectra of Polyatomic Molecules.* Princeton, NJ: Van Nostrand, 1945.

Kämpf, G., *Characterization of Plastics by Physical Methods: Experimental Techniques and Practical Application.* Munich: Hauser, 1986.

Kennedy, J. F., and G. W. Willcockson, *J. Polym. Sci. A* **4**, 679 (1966).

Koenig, J. L., *Adv. Polym. Sci.* **54**, 89 (1984); *Spectroscopy of Polymers.* Washington, DC: American Chemical Society, 1992.

Krimm, S., in *Proceedings of the International Symposium on Macromolecules,* E. B. Mano, (Ed.). Amsterdam: Elsevier, 1975.

Krimm, S., *Fortschr. Hoch-Polym-Forschg.* **2**, 51 (1960).

Lakowicz, J. R., *Principles of Fluorescence Spectroscopy.* New York: Plenum, 1983.

Lakowicz, J. R., and G. Weber, *Biochemistry* **12**, 4161 (1973).

Marmur, J., and P. Doty, *J. Mol. Biol.* **5**, 109 (1962).

McClure, W. O., and G. M. Edelman, *Biochemistry* **6**, 559, 567 (1964).

McGovern, J. J., J. M. Grim, and W. C. Teach, *Anal. Chem.* **20**, 312 (1948).

Nakanishi, K., and P. H. Solomon, *Infrared Absorption Spectroscopy*, 2nd ed. San Francisco: Holden-Day, 1977.

Nishijima, Y., *J. Polym. Sci. C* **31**, 353 (1970).

Nishijima, Y., A. Teramoto, M. Yamamoto, and S. Hiratsuda, *J. Polym. Sci. A2* **5**, 23 (1967).

Perrin, F., *J. Phys.* **7**, 390 (1926).

Rao, C. N. R., *Ultra-Violet and Visible Spectroscopy: Chemical Applications*, 3d ed., London: Butterworth, 1975.

Reddi, K. K., *Biochim. Biophys. Acta* **24**, 238 (1957).

Rosenbeck, K., and P. Doty, *Proc. Natl. Acad. Sci. U.S.A.* **47**, 1775 (1961).

Schiller, P. W., *Int. J. Peptide Protein Res.* **15**, 259 (1980).

Siesler, H. W., and U. K. Holland-Mortiz, *Infrared and Raman Spectroscopy of Polymers*. New York: Dekker, 1980.

Snyder, R. G., *J. Chem. Phys.* **47**, 1316 (1967).

Stern, O., and M. Volmer, *Physics 2* **20**, 183 (1919).

Sun, S. F., *Biochim. Biophys. Acta* **200**, 433 (1970).

Sun, S. F., T. S. Chang, and G. M. Lam, *Can. J. Chem.* **61**, 356 (1983).

Thompson, H. W., *Proc. R. London Ser. A* 21 (1945).

Weber, G., *Biochem. J.* **51**, 145, 155 (1952);

Weber, G., *Adv. Protein Chem.* **8**, 415 (1953).

Weber, G., and D. J. R. Lawrence, *Biochem.* J. **56**, 31 (1954).

Weber, G., and L. B. Young, *J. Biol. Chem.* **239**, 1415 (1964).

Wetlaufer, D. B., *Adv. Protein Chem.*, Vol. 17. New York: Academic, 1961.

Wetmur, J. G., and N. Davidson, *J. Mol. Biol.* **31**, 349 (1968).

Yguerabide, J., H. F. Epstein, and L. Stryer, *J. Mol. Biol.* **51**, 573 (1970).

Zerbi, G., L. Piseri, and F. Cabassi, *Mol. Phys.* **22**, 241 (1971).

PROBLEMS

16.1 The ultraviolet absorption extinction coefficient ε for a protein or a nucleic acid often does not represent the true value because of he partial scattering of light. This can be corrected by the equation

$$\varepsilon_{\text{observed}} - \varepsilon_{\text{scattered}} = \varepsilon_{\text{true value}}$$

However, for correction, the extinction coefficient due to scattering, $\varepsilon_{\text{scattered}}$, can be calculated with

$$\varepsilon_{\text{scattered}} = \frac{c}{\lambda^4}$$

where c is the scattering constant, which can be determined by measuring absorption at wavelengths 350 and 400 nm where the protein and nucleic acid have no absorption. Plot the observed and corrected UV absorption spectra for tobacco mosaic virus on the basis of the following experimental data:

λ (nm)	ε
235	19 100
250	15 000
262	16 300
282	14 000
310	3 000
350	0.360
400	0.210

(*Source*: Reddi (1957).)

16.2 The molar extinction coefficient for acetyl esters of tyrosine is 1340 at 282 nm. If absorption was found to be 0.185 at 282 nm for a sample of aqueous solution of acetyl esters of tyrosine, find the concentration of the solution.

16.3 Helical contents of polypeptides can be estimated from UV absorption data at 190 nm (π–π^* transition, which is characteristic of amide) by using the following equation:

$$\left(\frac{\varepsilon_{\text{coil}} - \varepsilon_{\text{peptide}}}{\varepsilon_{\text{coil}} - \varepsilon_{\text{helix}}}\right)(100) = \%\ \text{helicity}$$

where ε is the molar extinction coefficient per residue. The values of $\varepsilon_{\text{coil}}$ and $\varepsilon_{\text{helix}}$ can be obtained from the model compounds (that is, a compound assumed to be 100% helical and one assumed to be 100% coil). For a series of oligopeptide of aspartate derivatives, if we take $\varepsilon_{\text{coil}} = 7.2 \times 10^3$ and $\varepsilon_{\text{helix}} = 3.2 \times 10^3$, what is the helical content in terms of percentage for polypeptides with $\varepsilon_r = 2.5 \times 10^3$ and $\varepsilon_r = 1.8 \times 10^3$, respectively?

16.4 The following data are given for iodide quenching of acridone:

KI (molar)	F_0/F
0.04	4.64
0.10	10.59
0.20	23.0
0.30	37.2
0.50	68.6
0.80	13.7

Construct a Stern-Volmer plot and determine the dynamic (k_D) and static (k_s) quenching constants. Assuming that $\tau_0 = 17.6$ n s, calculate the bimolecular quenching constant *kq*. (*Source*: Lakowicz (1983); Boaz and Rollefson (1950).)

16.5 The following experimental data are given for the reciprocal polarization ($1/p$) as a function of T/η for adsorbates of 1-anilino-naphthalene-8-sulfonic acid on BSA at 25°C in solutions of sucrose in water:

$1/p$	$T/\eta \times 10^{-4}$
3.00	1.23
3.07	1.75
3.16	2.29
3.22	2.75
3.29	3.31

Plot $1/p$ versus T/η to determine $1/p$ and calculate τ, the lifetime of the excited state of the fluorescence of the adsorbate. Assume that ρ_h, the mean harmonic rotational relaxation time, is 118 ns. (*Source*: Weber and Young (1964).)

16.6 A Perrin plot ($1/T$ versus T/η) for fluorescence polarization of β-anthryl conjugate of BSA gave $p_0 = 0.267$ and $t_0 = 4.4 \times 10^{-8}$ s at 25° C. Calculate the rotational relaxation time ρ_h. (*Source*: Chen and Bowman (1965); Harrington et al. (1956).)

16.7 The intramolecular distance r can be estimated by measuring singlet–singlet energy transfer, using the Föster equation:

$$E = \frac{R_0^6}{(r^6 + R_0^6)}$$

where E is the transfer efficiency and R_0 is defined as that donor–acceptor separation. For the angiotensin II analogs H·Tyr-Arg-Val-Phe-Val-His-Pro-Trp·OH in aqueous solution, pH 1.5, the values of Föster parameters were $E = 0.10$ and $R_0 = 11.6$ Å. Calculate the intramolecular tyr–trp distance. (*Source*: Schiller (1980).)

16.8 The oscillater strength f is defined as the number of oscillating units per molecule and, according to the classical theory, f is given by

$$f = 4.32 \times 10^{-9} \int \varepsilon \, d\nu$$

where ε is the extinction coefficient and ν is the frequency of the light. Translating into experimental language, the integral $\int \varepsilon \, d\nu$ is the integrated band area. Approximately, $\int \varepsilon \, d\nu = \varepsilon_{max} \Delta\nu$, where $\Delta\nu$ is the half-width (the width of the bands at point where $\varepsilon = \frac{1}{2}\varepsilon_{max}$). Calculate f for proteins with $\varepsilon_{max} = 1.5 \times 10^3$ and 135×10^3, and $\Delta\nu = 4500$ and $5000\,\text{cm}^{-1}$, respectively. (*Source*: Brande (1950).)

16.9 A characteristic band at $1207\,\text{cm}^{-1}$ of a polymer and its analysis is shown in the following hypothetical spectrum:

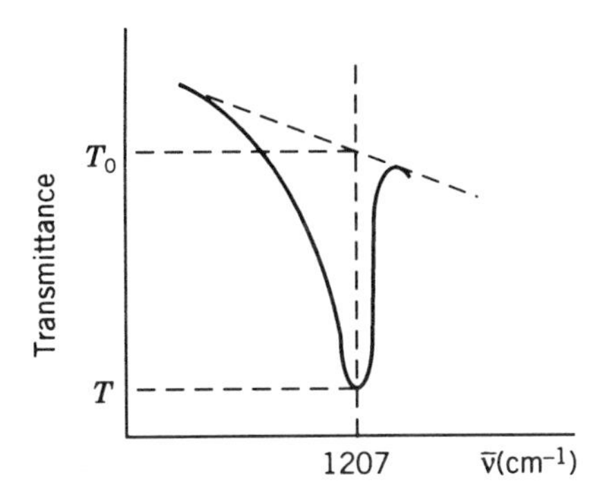

If a set of standard solutions of the monomer that contains the given polymer were available, a calibration curve could be constructed for the transmittance T as a function of the concentration of monomers. Assume that such data are given:

Concentration (% by volume)	10	20	40	60
Transmittance	80	68	45	30

Determine the concentration of a polymer in solution if its IR spectrum gives $T = 54$ and $T_0 = 86$.

16.10 A ligand in aqueous solution shows a characteristic band at $2048\,\text{cm}^{-1}$ in an IR spectrum when a protein is present in the ligand solution. The spectrum consists of two bands: one at $2023\,\text{cm}^{-1}$, which is believed to be the ligand covalently bound to the protein, and the other at $2046\,\text{cm}^{-1}$, which is believed to be the ligand in ionic form. The equilibrium constant can be obtained from $(\text{area})_{2023}/(\text{area})_{2046}$ or A_{2023}/A_{2046}. Calculate thermodynamic constants ΔH°, ΔS, and ΔG on the basis of the following set of data:

A_{2046}	0.019	0.048	0.046	0.044
A_{2023}	0.094	0.084	0.080	0.070
T (°C)	−7	12	25	40

(*Source*: Alben and Fager (1972).)

17

HIGH-PERFORMANCE LIQUID CHROMATOGRAPHY AND ELECTROPHORESIS

17.1 HIGH-PERFORMANCE LIQUID CHROMATOGRAPHY

A chromatograph is a device in which sample components are differentiated into various zones, depending on their velocity of migration through a medium. Control of the migration rates is largely dependent on the device constructed. High-performance liquid chromatography (HPLC), which was first introduced around 1969, utilizes pressure (from 100 to 6000 psi) to push sample mixtures through a column of specific packing materials that are designed to differentiate migration rates. The pressure on the sample is generated by a pumping system. The essential components of a HPLC system are shown in Figure 17.1. The detectors most frequently used in HPLC systems are UV–visible photometric detectors, fluorescence detectors, refractive index detectors, and differential refractometers. Almost all instruments that are capable of identifying a compound can be used as detectors, for example, conductance meter, mass spectrometer, light scattering apparatus, viscosity apparatus, and even molecular beam devices.

17.1.1 Chromatographic Terms and Parameters

Mobile Phase and Stationary Phase *Mobile phase* and *stationary phase* are standard terms used in chromatography including HPLC. The mobile phase is the solvent, and the stationary phase is the sorbent. The stationary phase, which can be a liquid, too, may or may not be supported by a solid. The solid, if used, is called the matrix. The separation of molecular species in HPLC is related to the equilibrium distribution of different compounds between the stationary phase and the mobile phase, while the spreading of the resulting chromatogram

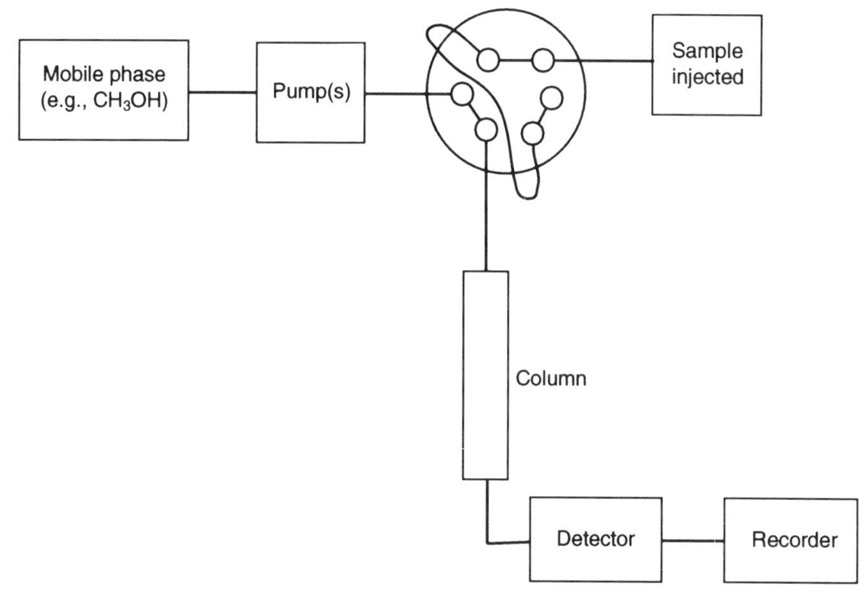

FIGURE 17.1 Diagram of an HPLC system.

is related to the migration rate of the molecules. The various types of HPLC are based on the different substances that are used to construct the stationary phase and the mobile phase.

Retention Retention R is defined as the equilibrium fraction of the sample in the mobile phase; hence, $1 - R$ is the sample equilibrium fraction in the stationary phase. The ratio of R over $1 - R$ is expressed by

$$\frac{R}{1-R} = \frac{c_m V_m}{c_s - V_s} \tag{17.1}$$

where c_m (in the moles per liter) is the concentration of the sample in the mobile phase that is contained in volume V_m (in liters); c_s (in moles per liter) is the concentration of the sample in the stationary phase that is contained in volume V_s (in liters). The product $c_m V_m$ is the number of moles of the sample in the mobile phase, and product $c_s V_s$ is the number of moles of the sample in the stationary phase. If we divide the right-hand expression of Eq. (17.1), both numerator and denominator, by the term c_m, we obtain

$$\frac{R}{1-R} = \frac{V_m}{K V_s} \tag{17.2}$$

where $K = c_s/c_m$ is the well-known constant called the partition or distribution coefficient. On solving Eq. (17.2) for R, we have

$$R = \frac{V_m}{V_m + KV_s} \tag{17.3}$$

This is the classical equation derived by Martin and Synge (1941). Equation (17.3) can be extended to the form

$$R = \frac{V_m}{V_m + \sum K_i V_i}$$

where the retention involves i, the number of mechanisms. Dividing both the numerator and the denominator of the right-hand expression of Eq. (17.3) by the term V_m leads to

$$R = \frac{1}{1 + K(V_s/V_m)} = \frac{1}{1 + k'} \tag{17.4}$$

where $k' = K(V_s/V_m)$, is another constant, called the capacity factor. In practice, k' is the key parameter in describing a given chromatographic system. More specifically, it depends on the chemical nature and temperature of the liquid phases that form the system. It also depends on the surface area of the chromatographic support. Let us rewrite k' in the form

$$k' = \frac{c_s V_s}{c_m V_m}$$

If we let $V_R = c_m V_m$ and $V_R - V_0 = c_s V_s$, the equation of k' is then in the form

$$k' = \frac{V_R - V_0}{V_R}$$

where V_R is the retension volume and V_0 is the unretained volume, also called the void volume. In practice, both the retention R and the capacity factor k' can also be expressed in terms of time t. The retention R in terms of time t is given by

$$R = \frac{t_m}{t_m + t_s}$$

where t_m is the average time that the sample molecules stay in the mobile phase and t_s is the average time that the molecules stay in the stationary phase before they are absorbed. The capacity factor k' in terms of t is given by

$$k' = \frac{t_R - t_0}{t_0}$$

where t_R is the retention time, namely, the time required for a peak to elevate from the time of injection, and t_0 is the time for unretained molecules in the mobile phase to move from one end of the column to the other.

Among the parameters described, the most frequently used in HPLC are V_R, t_R, and k'. The parameters V_R and t_R are further related by a factor F, the flow rate in mL/s of the mobile phase through the column:

$$V_R = t_R F$$

Note that we can also have

$$V_m = t_0 F$$

V_R and V_m themselves are related in the following way:

$$V_R = V_m \frac{t_R}{t_0} = V_m(1 + k') = V_m + KV_s$$

The determination of these parameters is shown in Figure 17.2. We see that the chromatogram is a bell-shaped band or Gaussian curve, which is characterized by the parameter w (in terms of volume) or t_w (in terms of time). Both w and t_w refer to the bandwidth and both can be expressed in terms of the

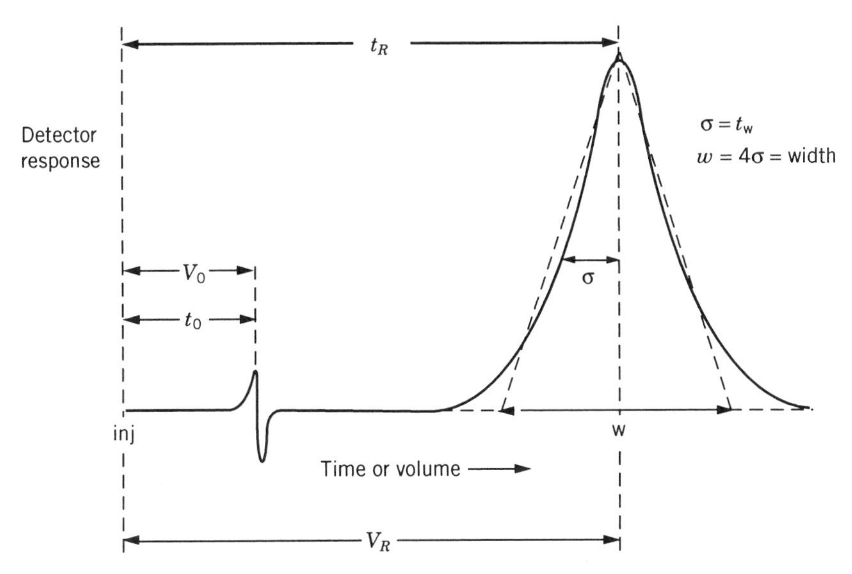

FIGURE 17.2 A typical chromatogram.

standard deviation σ:

$$\sigma = \frac{1}{4} w$$

or

$$\sigma = t_w$$

Resolution The capacity factor k' and retention volume V_R are characteristics of individual chemical species in a given chromatographic system. For two components in the mixture, a and b, in a given chromatographic system, each has its own capacity factor and retention volume, that is, k'_a, k'_b and V_{Ra}, V_{Rb}, as shown in Figure 17.3. The resolution R_s can be expressed by

$$R_s = \frac{2(V_{Rb} + V_{Ra})}{w_a + w_b} = \frac{V_{Ra} - V_{Rb}}{2(\sigma_a + \sigma_b)} \tag{17.5}$$

We can now define a parameter α by

$$\alpha = \frac{k'_b}{k'_a}$$

which is called the selectivity. A separation between a and b in a mixture is possible only if $\alpha \neq 1$. That is, for any separation to be possible, each component must have a different value for the capacity factor and each must be retained to a different degree. The value of the selectivity can be controlled by changing the composition of the mobile and stationary phases, for example, by changing the mobile phase solvent, pH, temperature, and chemical shift.

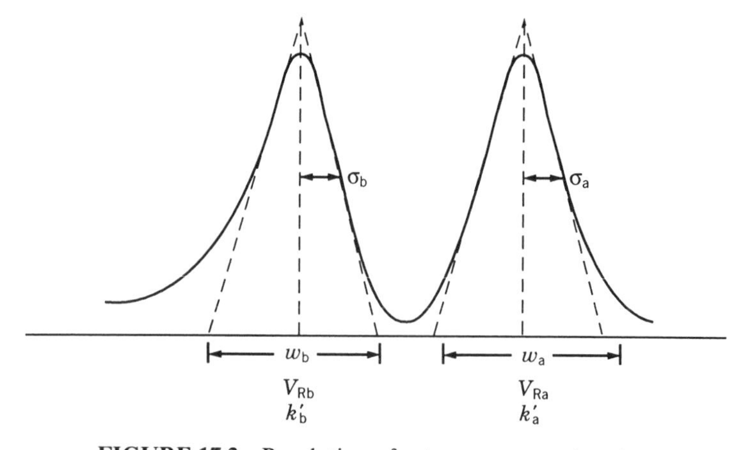

FIGURE 17.3 Resolution of a two-component system.

In addition, the separation depends on the efficiency of a column, N. The term N, which was originally obtained from the efficiency of the column in a fractional distillation, refers to the number of theoretical plates in an HPLC column. N, therefore, is also called the theoretic plate number. In a chromatogram, the narrower the width of an eluting peak, the greater is the efficiency of separating a multicomponent mixture in a column, that is, the greater the value of N. An equation of N is given by

$$N = 16\left(\frac{t_R}{t_w}\right)^2 = \frac{16}{L}\left(\frac{V_R}{w}\right)^2 = \frac{1}{L}\left(\frac{V_R}{\sigma}\right)^2 \tag{17.6}$$

where L is the column length. The three factors (α, N, and k') can be incorporated into one equation:

$$R_s = \frac{1}{4}(\alpha - 1)\sqrt{N}\left(\frac{k'}{k' + 1}\right)$$

17.1.2 Theory of Chromatography

The theory of chromatography is basically concerned with zone spreading. The dispersion of sample zones, which results in chromatographic peaks, causes a distribution of the sample concentration. For a good separation, zones should not overlap. Often a separation is not totally effective because the zones spread into one another.

In 1940, Wilson singled out the nonequilibrium in the local concentration as a major cause of zone spreading. In the following year, Martin and Synge proposed the theory of spreading in terms of flow velocity and the square of the particle diameter. In 1956 Van Deemter et al. related zone spreading to the sorption–desorption kinetics of solute molecules in the column. Giddings in 1958 suggested the random walk model to describe the physical processes involved in zone spreading.

If the eluted chromatographic peak follows a Gaussian distribution, the zone spreading σ^2 is related to the height of the chromatographic curve H, which is equivalent to a theoretical plate (or plate height), in the form

$$H = \frac{\sigma^2}{L}$$

Here σ^2 is the variance in statistics. The quantity H can also be calculated from

$$H = \frac{L}{N}$$

In general, H measures the specific column frequency and N measures the system efficiency.

According to the current chromatography theory, the flow of solute through a column in the presence of a stationary phase and a mobile phase undergoes three processes, which correspond to the three components of H or σ^2. Following are the three physical processes that determine zone spreading (the volume of σ^2) or H:

1. *Translation diffusion* (assuming the random walk model)

$$\sigma_D^2 = 2Dt$$

$$t = \frac{L'}{v}$$

   where L' is the distance of zone migration and v is the solvent velocity.
2. *Eddy diffusion* (due to the inhomogeneity of packing materials)

$$\sigma_E^2 = L'd$$

   where d is the diameter of a zone displaced from the channel.
3. *Sorption and desorption* (due to a local nonequilibrium condition created by the process of sorption and desorption)

$$\sigma_{k_1}^2 = \frac{2R(1-R)vL'}{k_2}$$

   where k_1 is the rate constant of sorption and k_2 is the rate constant of desorption.

   The sum of the three processes constitutes the total zone spreading:

$$\sigma^2 = L'\left(\frac{2D}{v} + d + \frac{2R(1-R)v}{k_2}\right) \tag{17.7a}$$

or

$$H = \frac{2D}{v} + d + \frac{2R(1-R)v}{k_2} \tag{17.7b}$$

Equations (17.7a) and (17.7b) are different forms of the Van Deemter equation:

$$H = A + \frac{B}{v} + Cv \tag{17.8}$$

where

$$A = d$$

$$B = 2D$$

$$C = \frac{2R(1-R)v}{k_2}$$

17.1.3 Types of HPLC

The four major types of HPLC are

1. Ion-exchange chromatography
2. Liquid–solid (adsorption) chromatography
3. Liquid–liquid
   a. Normal-phase chromatography
   b. Reversed-phase chromatography
4. Size exclusion chromatography

They represent four separate mechanisms for the interaction of sample molecules with the stationary phase. Size exclusion chromatography is used for sample molecular weight over 2000, whereas the other types do not have this limit. Usually these other types are used for sample molecular weights below 2000.

The differences between the types of chromatography lie basically among others in the different columns that are used.

Ion-Exchange Chromatography This is a type of chromatography in which the active surface of the column packing carries a charge: An anion exchanger carries a positive charge (for example, quaternary ammonium groups) and a cation exchanger carriers a negative charge (for example, sulfonate groups). The retention of a sample occurs when the ionic sample carries the counter ions:

$$X^- + A^+Y^- \rightleftharpoons Y^- + A^+X^- \qquad \text{(Anion exchange)}$$

$$X^+ + C^-Y^+ \rightleftharpoons Y^+ + C^-X^+ \qquad \text{(Cation exchange)}$$

Ion exchange chromatography is used for amino acid analysis; separations of nucleic acids, nucleoside, nucleotides; identification of sulfa drugs (such as sulfaquanidine, sulfadiazine, sulfisoxazole, sulfamethizide); and food compounds (for example, caffine, ascrobic acid, vanillin). The mobile phase usually contains water or water–organic solvent mixtures. The peak retention can be controlled by pH and ionic strength (salt, concentration). Column packings are made of a polystyrene backbone cross-linked with divinylbenzene attached with ionic functional groups such as $-SO_3^-H^+$ (cation exchanger) and $-N(CH_3)_3^+OH^-$, NH_2^- (anion exchanger).

Liquid–Solid (Adsorption) Chromatography This type of chromatography is based on the competition between the molecules of the sample and the molecules of the mobile phase for adsorbent sites on the active adsorbent surface of the stationary phase. To alter adsorptive activity, two solvents are used in the mobile phase: the principal solvent and the modifying solvent. In most cases, the principal solvent is hexane or dichloromethane and the modifying solvent is water, alcohol, or dimethyl sulfoxide. A modifying solvent is added to the principal solvent to control the absorptive activity of the samples.

Two kinds of adsorbents are used for column packings: polar adsorbents and nonpolar adsorbents. The acidic polar adsorbents, such as silica, are used for aliphatic nitrocompounds or aromatic amines. The basic adsorbents, such as alumina, are used for pyrrole derivatives, phenols, and carboxylic acids. The nonpolar adsorbents, such as charcoal, are used for high molecular weight homologs and aromatic compounds.

Liquid–Liquid (Partition) Chromatography If a third component is dissolved in the immiscible layers of the solvents, there is a distribution of the third component between the two layers. Liquid–liquid (partition) chromatography is based on a multistage distribution of a sample between two solvents within a column. The mobile phase is a liquid. The stationary phase is also a liquid (another liquid), which may be dispersed onto a finely divided inert support. The separation is attributed to the different distribution of the sample compounds between the two liquid phases. There are two kinds of partition chromatography: Normal systems in which the mobile phase is less polar than the stationary phase and reversed-phase systems in which the mobile phase is the more polar liquid. Table 17.1 compares the two different types of partition chromatography.

Reversed-phase chromatography is perhaps the most widely used chromatographic method. It was first developed by Howard and Martin to separate fatty acids by using a polar eluent (mobile phase), and a nonpolar stationary phase that consisted of paraffin oil and octane. Today the stationary phase consists of a liquid that is chemically bonded to a support. For example, the column packings contain octadecylsilyl (C_{18}), octylsilyl (C_8), butylsilyl (C_4), or propylsilyl (C_3), which are bonded to silica supports having various pore sizes (for example, 100, 300, and 500 Å) and particle sizes (for example, 5 and 10 μm). The extent of retention of a molecule depends on the number, size, and stereochemistry of its hydrophobic (for example, alkyl) and hydrophilic (for example ionic) groups.

Reversed-phase chromatography is currently used for the separation as well as identification of substances having a wide range of polarity and molecular

TABLE 17.1 Comparison of the Two Types of Chromatography

Type of Chromatography	Mobile Phase	Stationary Phase
Normal partition	Pentane, hexane, heptane, chloroform dichloromethane	Water, ethylene glycol, polyethylene glycols, trimethylene glycol, acetonitrile
Reversed phase	Methanol/water Acetonitrile/water	Squalane

weight. It is now an indispensable tool in biotechnology. For example, it can be used to separate homologous proteins from different species and synthetic diasteriosomeric peptides. Retention in reversed-phase chromatography has nothing to do with molecular weight, nor does it have anything to do with acidity or basicity. In other chromatographic methods, an isocratic solvent (with a mobile phase of constant composition, such as a single solvent) is usually used. In reversed-phase chromatography gradient elutions (with a mobile phase of different compositions, two or three solvents mixed) are used. Typical examples in reversed-phase chromatography of the mobile phase used in the separation of proteins are acetonitrile for solvent A and aqueous 0.1% trifluoroacetic acid for solvent B. By controlling the percentage of the two or three solvents in the mixture eluted at different time intervals, the sample components (proteins) are separated.

Size-Exclusion Chromatography

General Principle Size-exclusion chromatography relies on the different rates of diffusion or permeation of molecules of different sizes through the pores of packing materials and not on the rates of adsorption and desorption. Size-exclusion chromatography functions as a molecular sieve. The distribution coefficient of molecules in different sizes of pores, K, is defined as

$$K = \frac{V_i}{V_{i0}}$$

where V_i is the pore volume accessible to a molecular species, and V_{i0} is the total pore volume. The retention volume V_R is given in the form

$$V_R = V_0 + K V_{i0}$$

where V_0 is the void of interstitial volume.

Since both the size of the pores and the interstitial volume are fixed, the flow of the sample molecules is limited by their size. The largest molecules that can pass through the column are limited by the size of V_0, while the smallest molecules are limited by the size $V_0 + V_{i0}$. The sample molecule whose size is within the given limit of a column can pass between V_0 and $V_0 + V_{i0}$. Hence, the retention volume V_R for sample molecules is also between V_0 and $V_0 + V_{i0}$.

Two Types of Size-Exclusion Chromatography When an organic solvent is used as the mobile phase, the separation is called gel permeation chromatography (GPC), which is used extensively in polymer chemistry to characterize organic polymers, for example, in the determination of molecular weight. Some of the most commonly used columns are the μ-styragels of Waters and the Micro Pak TSK type H columns in the GPC mode with a mobile phase such as tetrahydrofuran.

If an aqueous mobile phase is used, the separation is called gel filtration chromatography (GFC). GFC is mainly used to separate and characterize biological polymers such as proteins. Some of the most commonly used columns are Water's I-125 and Varian's Micro Pak TSK type SW columns.

Applications of Size-Exclusion Chromatography If a perfect size-exclusion column were available, chromatography could be used for preparative purposes, that is, to separate materials in quantity according to their size. Such an exclusion column is still in development. At present two important applications are used: (1) the determination of molecular weight and molecular weight distribution and (2) the study of the binding of small molecules to macromolecules.

DETERMINATION OF MOLECULAR WEIGHT AND MOLECULAR WEIGHT DISTRIBUTION A universal calibration curve by gel permeation chromatography (GPC) has been suggested in which the logarithm of hydrodynamic volume $[\eta]M$ is plotted against the retention volume V_R. The term $[\eta]$ refers to the intrinsic viscosity and M the molecular weight. Such a plot is shown in Figure 17.4. If the retention volume of a polymer is known, one can read the value of $[\eta]M$ directly from the calibration curve. The value of M can then be determined if $[\eta]$ is also known. This method suffers from two drawbacks: (1) The calibration curve is almost linear, but not really linear; (2) one still has to determine the value of $[\eta]$, in addition to V_R.

A modified method has been proposed in an attempt to improve the universal calibration approach. At a given elution volume, two polymers, 1 and 2, are assumed to have the same hydrodynamic volumes $[\eta]M$:

$$[\eta]_1 M_1 = [\eta]_2 M_2$$

Recall the equation

$$[\eta] = KM^{a}$$

From the above two equations, we obtain

$$\log M_2 = \underbrace{\frac{1}{1+a_2}\log\frac{K_1}{K_2}}_{A} + \underbrace{\frac{1+a_1}{1+a_2}}_{B}\log M_1$$

$$= A + B\log M_1 = A + BCV_R$$

where C is the slope of the calibration curve. Let 1 represent polystyrene standard and 2 any sample. Since K_1 and a_1 are known in literature, and since K_2 and a_2 are either known in literature (for example, poly(methyl methacrylate)) or are obtainable by carrying out a few measurements through intrinsic viscosity and osmotic pressure (or light scattering) techniques, M_2 can be easily determined for any sample of the same species from GPC data V_R. The calibration curve is shown in Figure 17.5, where the slope C can be determined.

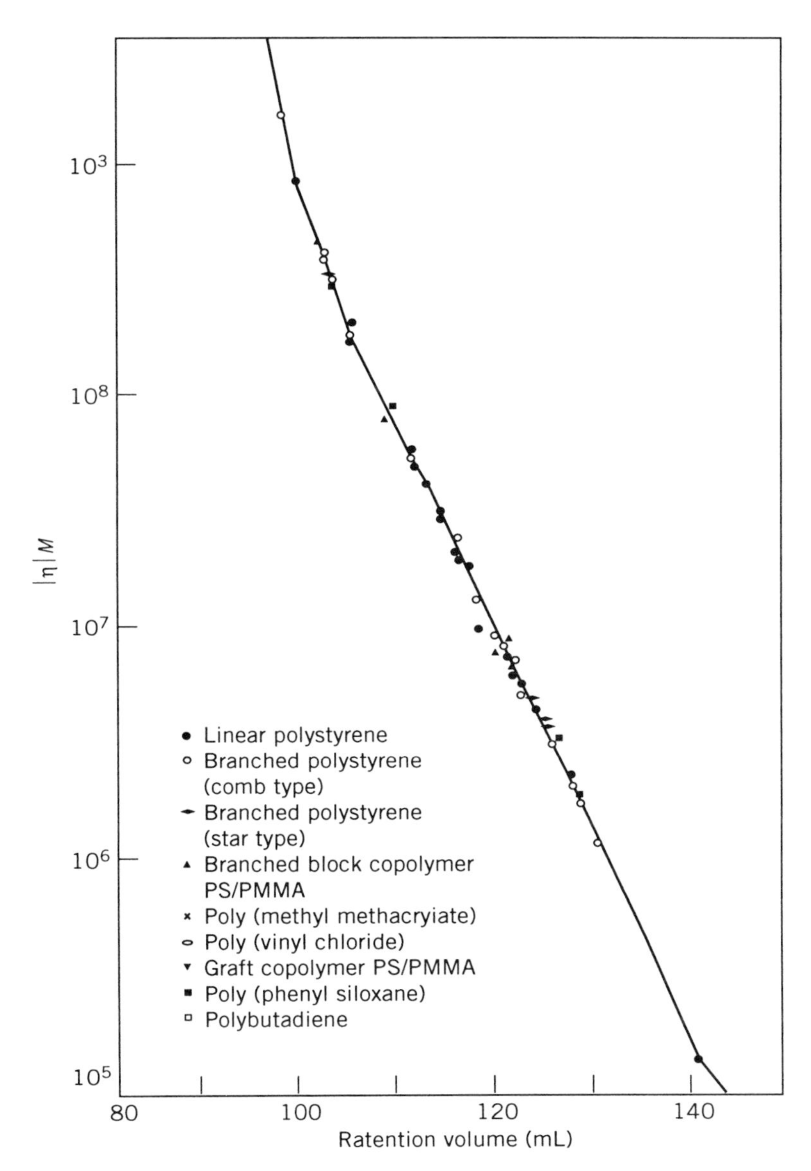

FIGURE 17.4 Calibration curve for GPC. (From Grubistic et al. (1967). Reproduced by permission of John Wiley & Sons.)

The above method still requires viscosity measurement in addition to GPC work. Another method, originally suggested by Waters Associates, enables calculation of molecular weight and molecular weight distribution directly from GPC chromatograms, provided that Figure 17.5 is available.

Consider the chromatograms of poly(methyl methacrylate) shown in Figure

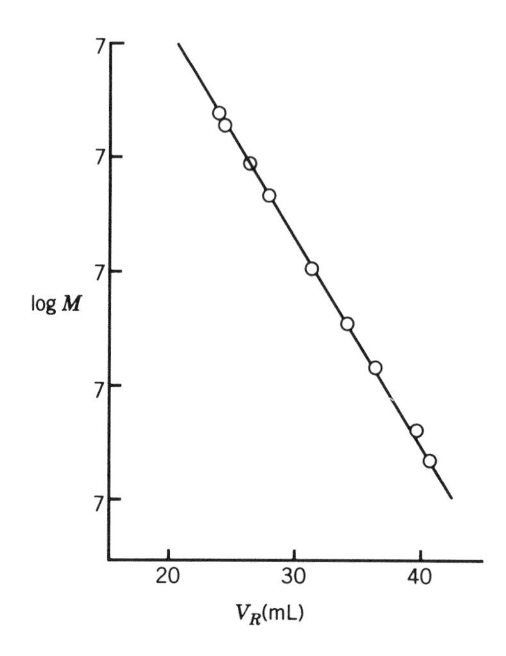

FIGURE 17.5 GPC calibration plot: polystyrene standard in THF at 25°C. The values of viscosity constants used were $K_1 = 1.41 \times 10^{-4}$ dL/g; $a_1 = 0.70$. (From Sun and Wong (1981). Reproduced by permission of Elsevier Science Publishers BV.)

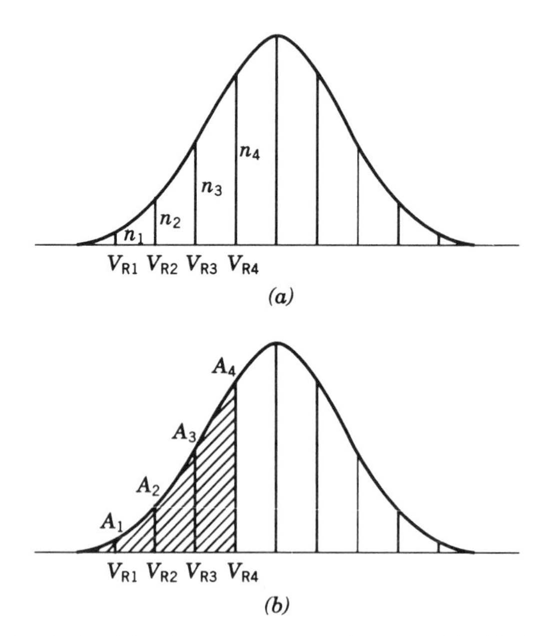

FIGURE 17.6 Analysis of the chromatogram: (a) h_i versus V_{Ri}; (b) A_i versus V_{Ri}. (From Sun and Wong (1981). Reproduced by permission of Elsevier Science Publishers BV.)

17.6. We arbitrarily divide the abscissa of the chromatogram units into s parts and measure either the height of the curve h_i or the area of the curve A_i. There is a one-to-one correspondence between the h_i (or A_i) coordinate and the V_{Ri} coordinate. The value of V_{Ri} can be converted to M_i by reading the calibration curve in Figure 17.5. Hence, we obtain a table in the form h_i versus M_i. The values of $\bar{M}_n$ and $\bar{M}_w$ can now be calculated from the following two sets of equations:

$$\bar{M}_n = \frac{\sum h_i}{\sum (h_i/M_i)}$$

$$\bar{M}_w = \frac{\sum h_i M_i}{\sum h_i}$$

$$\bar{M}_n = \frac{\sum A_i}{\sum (A_i/M_i)}$$

$$\bar{M}_w = \frac{\sum A_i M_i}{\sum A_i}$$

Once $\bar{M}_n$ and $\bar{M}_w$ values are determined, we can calculate the polydispersity $\bar{M}_w/\bar{M}_n$ and the mean average molecular weight $\bar{M}_m$, which is calculated with the following equation:

$$\bar{M}_m = (\bar{M}_n \bar{M}_w)^{1/2}$$

Similar attempts have been made to construct a universal calibration plot for determining the molecular weight of protein polypeptides in gel filtration chromatography (GFC). A straight line of logarithm molecular weight versus retention volume V_R or the capacity factor k' has been reported in the literature. Its application, however, is difficult because (1) protein polypeptides carry charges (positive or negative), while the packing materials for most commonly used columns also carry charges on the surface, making ionic interaction almost unavoidable, and the column cannot function by size exclusion; and (2) since protein polypeptides carry charges, they cannot be viewed as completely random as in the case of most synthetic polypeptides. Thus, in addition to size, we have to consider the shape of the protein polypeptides, particularly under certain environments (pressure and temperature).

STUDY OF THE BINDING OF SMALL MOLECULES TO MACROMOLECULES Consider the binding of a ligand, I, to a macromolecule, M. A mobile phase that contains a buffer and the ligand is suddenly disturbed in the column by injection into the flow system of a small amount of the sample solution (5–100 μL) that contains macromolecules (macromolecules are dissolved in the same buffer or in the mobile phase). Due to the size differences the macromolecules (M) move faster, leaving the small molecules (I) behind. After a short time an equilibrium is reached:

$$M + nI \rightleftarrows MI_n$$

As the macromolecule–ligand complex leaves the column, a void of ligand is created until a new equilibrium is reached. This phenomenon was first observed by Hummel and Dreyer in 1962. In the chromatogram, as shown in Figure 17.7, the peak represents the excess of I (that is, the amount of I in equilibrium with MI_n) whereas the trough represents the deficiency of I (the concentration of I that binds to the macromolecule). The mean number of moles of bound ligands per mole of macromolecules $\bar{r}$ can be evaluated directly from the chromatogram. There are two techniques for the evaluation of $\bar{r}$: external calibration and internal calibration (Sun, 1984, 1985, 1993).

External Calibration A set of experiments is carried out in which a series of mobile phases with known concentrations of ligand are chosen. To each mobile phase two samples are injected one after the other. The first sample contains solvent alone and the second contains the macromolecule in solvent (for example, protein in buffer). The first chromatogram will give one, and only one, peak (negative), the area of which is equivalent to the concentration of ligand in the given mobile phase. Since a series of mobile phases with different concentrations will be used, a calibration curve (a straight line) can be established, as shown in Figure 17.8, in the form of area versus the concentration of ligand.

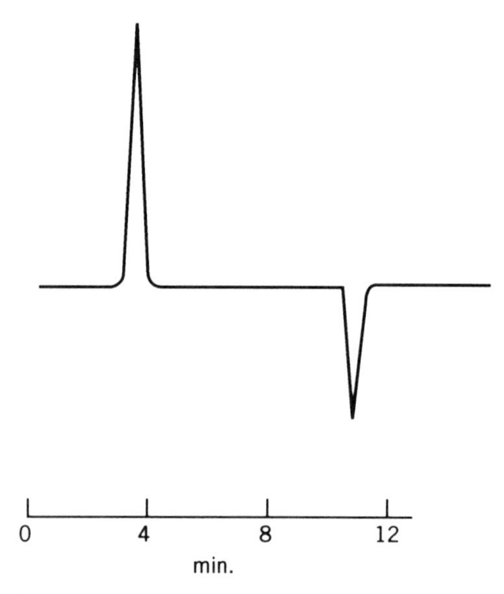

FIGURE 17.7 A Hummel–Dreyer chromatogram: L-tryptophan–BSA binding.

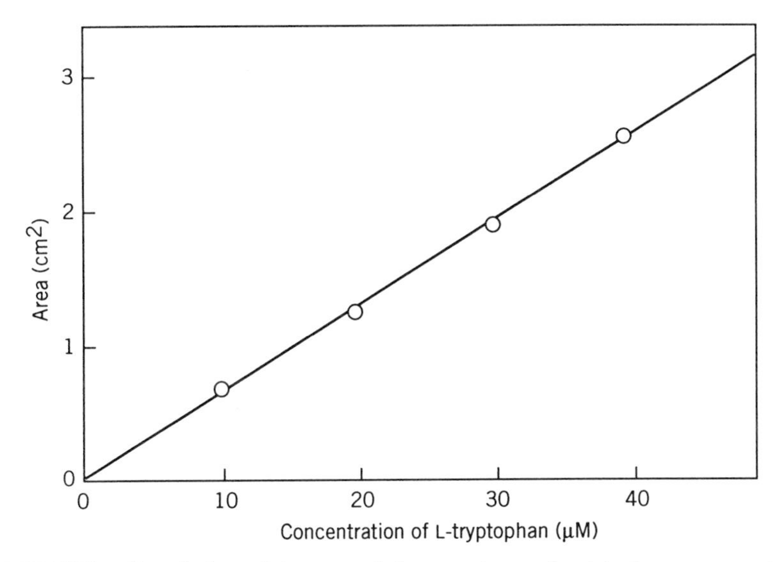

FIGURE 17.8 Correlation of the area of the negative peak with the concentration of L-tryptophan for the external calibration method. (Adopted from Sun and Wong (1985).)

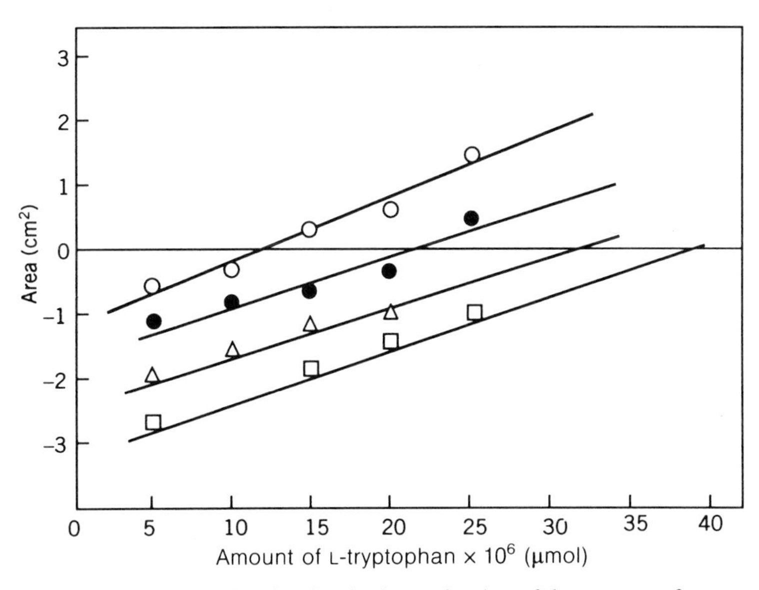

FIGURE 17.9 Internal calibration for the determination of the amount of L-tryptophan that binds to BSA. (Adopted from Sun and Wong, (1985) Reproduction by permission of Friedr. Vieweg & Sohn.)

The second chromatogram will give two peaks, positive and negative, as shown in Figure 17.7. The area of negative peak is equivalent to the concentration of bound ligand, which can directly be read on the calibration curve, that is, area of the negative peak = $[L]_b$. Since the concentration of macromolecules [M] is known at the time the second sample is prepared, we can easily calculate

$$\bar{r} = \frac{[L]_b}{[M]}$$

Internal Calibration Given the same mobile phase (same concentration of ligand in the solution), several samples that contain a known amount of macromolecules and ligand are run. In the samples, the concentration of macromolecules is kept constant, while the amount of the ligand is varied. By plotting the area of the trough (negative peak) versus the amount of ligand in the sample (in moles), the interpolated amount of ligand L_b (where area is equal to zero in Figure 17.9) is the exact amount of ligand bound to the macromolecules. We can thus calculate the value $\bar{r}$:

$$\bar{r} = \frac{L_b}{M}$$

Both the quantities of L_b and M are in moles.

17.2 ELECTROPHORESIS

17.2.1 Basic Theory

The movement of charged particles (ions) in an electric field is called electrophoresis. The basic theory of electrophoresis is related to ionic mobility u, which is also called electrophoretic mobility. When an ion in solution is moving in the direction of a field E, its velocity v depends on three factors: the charge z carried by the ion, the frictional coefficient f arising from the resistance of the solution, and the strength of the field E. The quantity E is defined as

$$E = \frac{d\phi}{dx} = \frac{I}{A\kappa}$$

where ϕ is the potential difference (in volts) of the two electrodes, x is the coordinate, I is the current (in amperes), A is the cross-area (in cm^2) through which the solution passes, and κ is the ionic conductance of the solution (in $\Omega^{-1}\,cm^{-1}$ or $S\,m^{-1}$ (S = siemens)). The product of velocity v and the frictional coefficient f is equal to the force of the field on the charge of the ion:

$$fv = zeE \qquad (17.9)$$

where e is the electrostatic charge in coulombs ($e = 1.602 \times 10^{-19}$ C). Ionic mobility u is defined by Eq. (17.9) as follows:

$$u = \frac{v}{E} = \frac{ze}{f}$$

Its units are in $(\text{m s}^{-1})/(\text{V m}^{-1})$ or $\text{m}^2\,\text{V}^{-1}\,\text{s}^{-1}$. Thus, the mobilities u_+, u_- and the velocities v_+, v_- of ions (+, −) are related by

$$v_+ = Eu_+$$

$$v_- = Eu_-$$

A classical example is the electrophoresis of 0.02 M NaCl. When a current of 1.60 mA was used, the boundary moved 0.020 m in 689 s. The cell is a tube, with an inner radius of 0.188 cm. The specific conductance of the solution was $\kappa = 1.26\,\text{s m}^{-1}$ at 25°C. The electric field strength and the mobility of an ion can then be calculated as follows:

$$\frac{\mathrm{d}\phi}{\mathrm{d}x} = \frac{1.60 \times 10^{-3}\,\text{A}}{[\pi(0.188 \times 10^{-2}\,\text{m})^2](1.26\,\text{S m}^{-1})} = 114\,\text{V m}^{-1}$$

$$u_{\text{Na}^+} = \frac{0.020\,\text{m}}{(689\,\text{s})(114\,\text{V m}^{-1})} = 2.50 \times 10^{-7}\,\text{m}^2\,\text{V}^{-1}\,\text{s}^{-1}$$

Historically, ionic mobility is measured in terms of conductance because there is a linear relationship between the conductance Λ and the ionic mobility u in a dilute solution

$$\Lambda = \mathcal{F}(u^+ + u^-)$$

where $\mathcal{F}$ is the Faraday ($1\mathcal{F} = 96\,487\,\text{C mol}^{-1}$). Therefore, behind the method of electrophoresis there is a rich theory of conductance which has been developed for more than a century. Here we describe three well-known subjects that provide the theoretical background of electrophoresis: ionic atmosphere and mobility, the relaxation time effect, and the zeta potential.

Ionic Atmosphere and Mobility Due to electrostatic forces, an ion is always surrounded by many other ions of opposite charge, which form an ionic atmosphere. The ionic atmosphere can affect the conductance and mobility of the central ion in three ways:

1. *Viscous Effect* Opposing the electrical force that exists between the ion and the field is a frictional viscous drag of the solvent, which, in many

cases, lowers the conductance and the mobility of the ion. The frictional drag is usually expressed by Stokes' law:

$$f = 6\pi a \eta v$$

where a is the radius of the ion, η is the viscosity coefficient of the solvent, and v is the velocity of the ion.

2. *Electrophoretic Effect* While the central ion moves in one direction, the ionic atmosphere which consists of ions of opposite charge move in the opposite direction. Thus, the central ions are forced to move against a stream of solvent. Their velocities are consequently reduced.
3. *Relaxation Time Effect* On the way to the electrode, the central ion leaves the ionic atmosphere behind. As a result, the originally symmetric atmosphere becomes asymmetric. It exerts an electrostatic drag on the ion, thereby reducing its velocity in the direction of the field.

The ionic atmosphere depends on the concentration of both positive and negative ions. On this basis, Onsager derived an equation for the concentration dependence of the equivalent conductance in which all three aforementioned effects are incorporated. The equation is given in the form

$$\Lambda_c = \Lambda_\infty - \left[\frac{29.142(z_+ + |z_-|)}{(\varepsilon T)^{1/2}\eta} + \frac{9.903 \times 10^5}{(\varepsilon T)^{3/2}} \Lambda_\infty \omega\right] 2I^{1/2} \tag{17.10}$$

where I is the ionic strength

$$I = \tfrac{1}{2}(c_+ z_+^2 + c_- z_-^2)^{1/2}$$

z_+ and z_- are the valence numbers of the cation and anion, respectively, ε is the dielectric constant of the bulk solvent, Λ_∞ is the equivalent conductance at infinite dilution, T is the absolute temperature, and ω is defined as

$$\omega = (z_+ |z_-|) \frac{2q}{1 + q^{1/2}}$$

in which

$$q = \frac{(z_+ |z_-|)}{z_+ + |z_-|} \times \frac{\lambda_\infty^+ + \lambda_\infty^-}{z_+ \lambda_\infty^- + |z_-| \lambda_\infty^+}$$

In a uni–uni valent electrolyte, $q = 0.5$. According to Eq. (17.10), the plot of $\Lambda_c - \Lambda_\infty$ versus $\sqrt{I}$ or $\sqrt{c}$ should be linear: Eq. (17.10) is valid only for dilute solutions, $c < 0.01$.

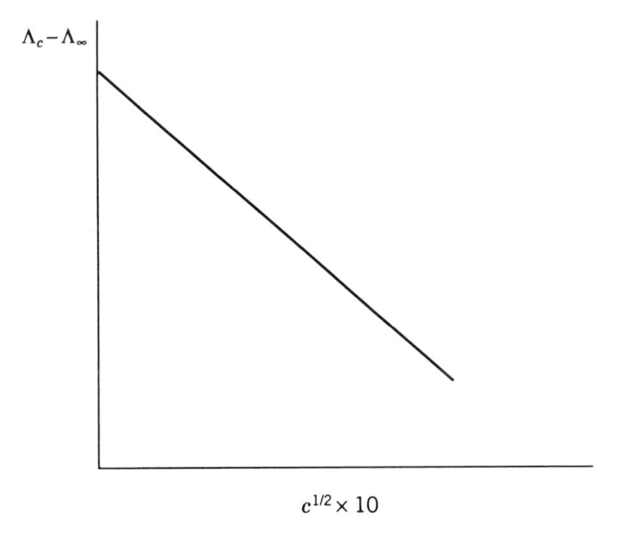

Zeta Potential Zeta (ζ) potential is a parameter used to describe the electrophoretic mobility of colloidal particles. Charged colloidal particles are slightly different from ions in that colloidal particles are surrounded by an electric double layer, which is similar, but not identical to the ionic atmosphere. The inner part of the double layer moves as a unit in transport experiments. The ζ potential is the surface potential of the inner part of the double layer, as shown in Figure 17.10. It is defined as

$$\zeta = \phi_{r=a}$$

Von Smoluchowski derived an equation to describe the electrophoretic mobility

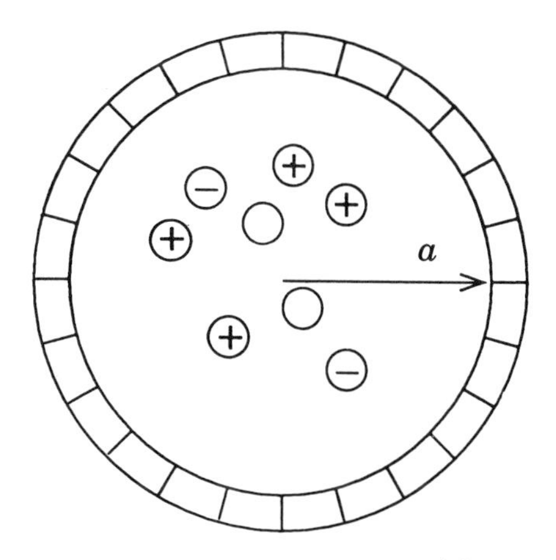

FIGURE 17.10 Zeta potential.

of charged colloid particles u':

$$u' = \frac{\varepsilon\zeta}{\eta} E$$

This equation is valid for relatively thin double layers, $\kappa' a \gg 1$ (κ' being the Debye-Huckel parameter). For high potentials an additional correction is required for the relaxation effect, which is similar to the situation described above.

The Moving Boundary Moving-boundary electrophoresis is performed with the substance in free solution. Historically, this was the first form of electrophoresis. Although it is no longer as widely used as before, it does illustrate the important role that diffusion plays in electrophoresis. The diffusion phenomenon is shown in Figure 17.11.

The moving boundary spreading in the electrophoresis of protein is usually measured in terms of refractive index gradient dn/dx, where n is the refractive index. The results are often recorded as shown in Figure 17.12. The formation of the boundary, its spreading, and its separation (if any) are functions of diffusion. The equation of moving boundary is given in the form

$$\frac{\partial n}{\partial x} = \frac{\Delta n}{2\sqrt{\pi D t_D}} \int_{-\infty}^{\infty} q(u) \exp\left[\frac{-(x - uEt_E)^2}{4Dt_D}\right] du \tag{17.11}$$

where t_D is the time recorded after the formation of the boundary, t_E is the time after application of the electric field, $q(u)$ is the distribution of mobilities, D is the diffusion coefficient, which is assumed to be a constant for all the proteins, and $\Delta n = n_1 - n_2$ is the difference in refractive index of the solution and the solvent.

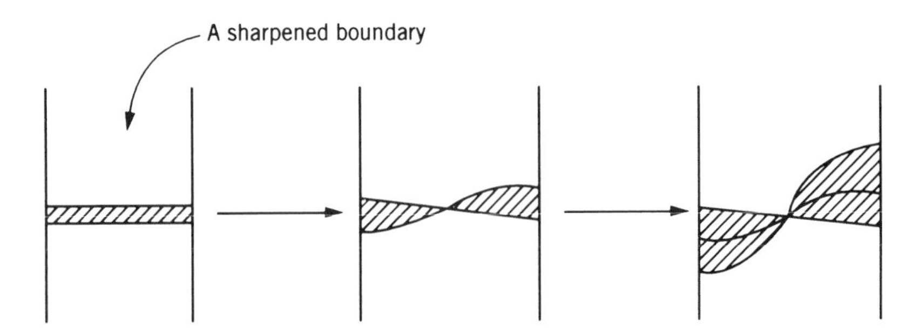

FIGURE 17.11 The diffusion phenomena. One may use a capillary tube to siphon a sharpened boundary.

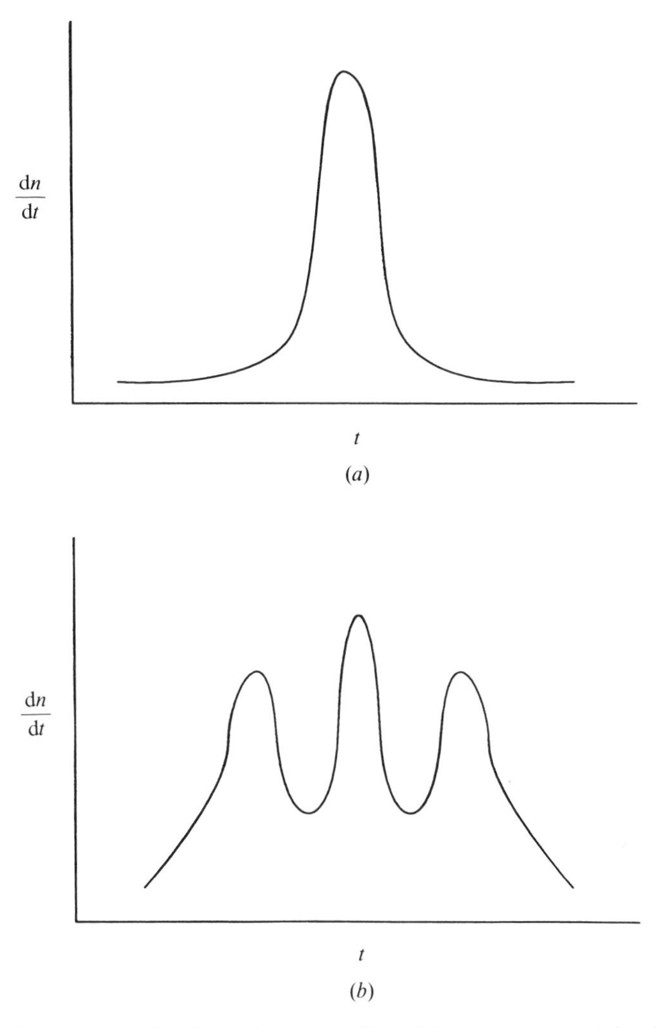

FIGURE 17.12 The moving boundary spreading: (a) one component in the system (sample); (b) three components in the system (sample).

For a first approximation let $q(u)$ be a Gaussian distribution:

$$q(u) = \frac{1}{h\sqrt{2\pi}} e^{-u^2/(2h^2)}$$

where h is the standard deviation of the mobility distribution. Then

$$\frac{\partial n}{\partial x} = \frac{\Delta n}{\sqrt{2\pi}\sigma} e^{-x^2/(2\sigma^2)}$$

where

$$\sigma = \sqrt{E^2h^2t_E^2 + 2Dt_D}$$

The apparent diffusion coefficient(s) can now be calculated from the gradient curves by standard methods, including the measurement of the ratio of height/area from the Gaussian graph.

17.2.2 General Techniques of Modern Electrophoresis

Techniques of modern electrophoresis applied to biological polymers are similar to those of modern liquid chromatography. An apparatus of modern liquid chromatography consists of three basic parts: the eluent (mobile phase), the column (stationary phase), and the detector. Likewise, an apparatus of electrophoresis consists of three basic parts: the electrolyte buffer, the supporting medium, and the mode of detectors. The classification of electrophoresis is in general based on the choice of the three fundamental parts, particularly the supporting medium.

The Electrolyte Buffer The selection of a proper electrolyte buffer is an important step in the successful run of an electrophoresis. The two major factors in the selection of a proper electrolyte buffer are pH and ionic strength. The pH directly influences the mobility of the molecule, and the ionic strength affects the electrokinetic potential, which, in turn, affects the rate of migration. Low ionic strengths increase the rate of migration and high ionic strengths decrease the rate of migration. As shown in the Onsager equation, if the ionic strength of the buffer is increased, the conductivity increases. The greater conductivity generates a great amount of heat, which often poses a series problem in the formation of a temperature gradient. The Onsager equation also indicates that the temperature increase affects the viscosity of the medium, which, in turn, affects the frictional coefficient, and so forth. All these factors suggest that ionic strength should not be kept high. On the other hand, high ionic strength sharpens the boundary, which is desirable in obtaining good resolution. Thus, ionic strength of the system should be neither too high nor too low.

Detectors As in HPLC, almost all physical or chemical methods that can be used for the identification of chemical compounds can be used as detectors. The following are among the most popular:

UV absorption This is often used for monitoring the migration of ionic species.

Autoradiography If the compound initially in a mixture is labeled with radioactive elements, such as ^{14}C, ^{3}H, ^{35}S, or ^{131}I, the separated compound can be detected by radioactive measurements, such as autoradiography.

Staining and destaining This is used in gel electrophoresis where the separation can be visibly detected by dyes such as coomassie blue and bromophenol blue.

Immunodiffusion When a sliced gel that contains the chemical species of interest is separated from the mixture and placed in a medium near another chemical compound, the interaction causes the migration of the chemical species. This is particularly useful for the study of immunomaterials in relation to antigens. Hence, this process has a special name: immunoelectrophoresis.

The Supporting Media The supporting media can be paper, cellulose acetate membranes, agarose, starch, polyacrylamide gels, or nothing, that is, free solutions. Running electrophoresis in free solutions has a serious drawback, however, because the resulting sample components not only will move in the direction of the electric field, but will also diffuse in various directions, as has been discussed in the previous section on the moving boundary. This would make data analysis unnecessarily complicated. Today electrophoresis is often run in supporting medium. In the literature, the apparatus is frequently named after the supporting medium used, for example, paper electrophoresis and starch gel electrophoresis. This supporting media are usually cast in glass tubes or plates (except paper), as shown in Figure 17.13.

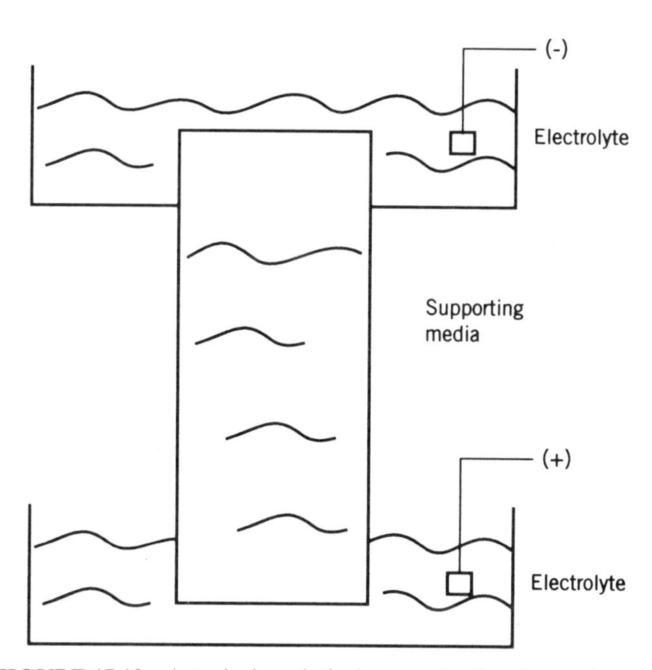

FIGURE 17.13 A typical analytical apparatus for electrophoresis.

There are many types of electrophoresis, such as

Micro-gel electrophoresis
Two-dimensional macromolecular maps
Thin-layer electrophoresis
Affinity electrophoresis
Paper electrophoresis
Starch electrophoresis
Agrose gel electrophoresis
Polyacrylamide gel electrophoresis
Isofocusing and isotachophoresis
SDS-protein capillary electrophoresis

Paper electrophoresis and starch gel electrophoresis were commonly used fifteen years ago. Now, they are seldom used. Agarose gel electrophoresis and polyacrylamide gel electrophorese, particularly the latter, continue to be widely used, and it seems that they are becoming even more important in studies of proteins and nucleic acids. It would not be an exaggeration to say that no research laboratory in biochemistry could perform the separation and identification of gene sequences without a polyacrylamide gel electrophoresis apparatus. In what follows we describe (1) agarose gel electrophoresis and polyacrylamide gel electrophoresis, (2) isofocusing and isotachophoresis, and (3) DNA sequencing (which, of course, can apply equally well to RNA sequencing).

17.2.3 Agarose Gel Electrophoresis and Polyacrylamide Gel Electrophoresis

Agarose Gel Electrophoresis Agarose is a polysaccharide of galactose and 3,6-anhydrogalactose. It dissolves in boiling water, and when it is cooled to room temperature it forms a gel. The pore size of agarose gel is relatively large (500 Å in a 2.5% gel). The separation of the macromolecules, however, is basically due to the gel's charge density and not its pore size. The higher the charge, the faster the macromolecule moves. Agarose gel is widely used in immunoelectrophoretic procedures where immunodiffusion is a major step in the separation of the largest molecules and supramolecular complexes, such as viruses, enzyme complexes, lipoproteins, and nucleic acids.

Polyacrylamide Gel Electrophoresis Polyacrylamide gels are prepared by the radical polymerization of acrylamide monomers ($CH_2{=}CH{-}CO{-}NH_2$) with the cross-linking comonomer *N,N′*-methylbisacrylamide ($CH{=}CH{-}CO{-}NH{-}CH_2{-}NH{-}CO{-}CH{=}CH_2$). The pore size is controlled by the concentration of monomers and cross-linking reagents. It is possible to prepare gels with a variety of pore sizes, and separation is basically due to the size of the macromolecules. Hence, polyacrylamide gel electrophoresis is a molecular sieving apparatus. The largest pores of the polyacrylamide gel, however, are still smaller than those of agarose gel.

Polyacrylamide gels are used for separating and identifying small fragments of DNA and RNA molecules. They are also used for monitoring the process of a preparative scheme by identifying the presence of impurities. In many instances, polyacrylamide gel is joined with agrose gel to form composite polyacrylamide–agrose gel, which is used to separate larger segments. However, for separating intact DNA molecules, agarose gels alone must be used.

SDS–Polyacrylamide Gel Electrophoresis Polyacrylamide gel electrophoresis is often used simultaneously for both separating and estimating the molecular weight of proteins that are solubilized with the detergent sodium dodecyl sulfate (SDS). This special kind of electrophoresis is called SDS–polyacrylamide gel electrophoresis or SDS–PAGE.

SDS binds to proteins to form complexes that carry negative charges in the form of rods. When subjected to PAGE, the molecules migrate at different rates. Larger molecules migrate slowly, whereas small molecules migrate quickly. There is a correlation between migration rate and molecular weight (molecular size). Recall the definition of mobility:

$$u = \frac{d}{vt}$$

where d is the distance of migration, t is the time, and v is the voltage difference. If two compounds are migrating in the gel with the same v and t, then the mobility of the two compounds have a ratio equal to the ratio of distance (migration):

$$\frac{u_1}{u_2} = \frac{d_1/vt}{d_2/vt} = \frac{d_1}{d_2}$$

If d_2 is chosen as a standard, for example, a marker dye, then

$$u_1 = \frac{d_1}{d_2} u_2$$

Arbitrarily we set $u_2 = 1$ mobility unit. Then, we have $u_1 = d_1/d_2$. For the correction of gel length before and after staining due to possible further swelling, we introduce the term l/l':

$$u_1 = \left(\frac{d_1}{d_2}\right)\left(\frac{l}{l'}\right)$$

where l is the length of gel before staining and l' is the length after staining of the chemical compound of interest (for example, protein polypeptide). Since the separation of the mixture is basically due to the sieve effect, the mobility of a compound is proportional to its size, which, in turn, is proportional to its molecular weight. It was empirically found that the plot of ln(molecular weight) versus mobility is linear. Thus, a universal calibration plot can be constructed so that for any unknown compound, if its mobility in electrophoresis is determined, its molecular weight can be estimated.

17.2.4 Southern Blot, Northern Blot, and Western Blot

Blotting analysis is a technique in which the resolved DNA, RNA or protein on the gel electrophoresis is extracted from the gel matrix to a blotting membrane (or filter) such as nitrocellulose. The DNA, RNA or prolein fragments that stick to the blotting membrane (or filter) are then further isolated, purified, and analyzed. These techniques are extensively used in molecular biology to study gene structure, and particularly to define the presence of a gene-related sequence in a genome.

Southern Blot Southern blot is a technique used on a fractionation of DNA fragments. Agrose gel electrophoresis and the nitrocellulose filter are the major apparatus used. DNA from a tissue or cell is isolated, purified, and cleaved with a specific restriction endonuclease into defined fragments. The fragments are separated through agrose gel electrophoresis. After fractionation, they are transferred from the gel to a nitrocellulose filter which is called a blot. The blot is hybridized with a probe that is specific for the gene under study. Autoradiography of the blot enables us to identify the restriction fragments which form complimentary base pairs.

Northern Blot Northern blot is used to separate single-stranded RNA. Formaldehyde/agarose gel is used for the electrophoresis. The fragments of RNA are then transferred to a cellulose filter from the gel, and are hybridized with a specific radiolabeled probe as in the case of the Southern blot.

Western Blot Western blot is a technique used to investigate proteins. Here polyacrylamide gel electrophoresis is used in conjunction with a nitrocellulose filter. Protein products are fractionated by size on polyacrylamide gel electrophoresis, and then are transferred to nitrocellulose for identification with a primary antibody. The bound primary antibody may be further detected with a second species, such as ^{125}I-protein A or biotimylated goat anti-Iq G.

17.2.5 Sequencing DNA Fragments

Gel electrophoresis can be used to sequence DNA (or RNA) fragments, that is, to identify the sequence of bases in DNA molecules. The process involves chemical modification and cleavage of specific nucleotides, followed by electrophoresis on high-resolution denaturing acrylamide gels. In the first step, a base is chemically modified (for example, methylation of guanine), and then is removed from its sugar by cleaving the DNA strand at its sugar. The cleavage is accomplished by a restriction endonuclease. Each base cleaved is contingent on the one that precedes it. DNA is cleaved only at the sugar attached to the modified base. The second step is to end-label the fragments and to extract DNA from the polyacrylamide gel electrophoresis. Sequencing DNA fragments is one of the most important techniques in modern biotechnology.

17.2.6 Isoelectric Focusing and Isotachophoresis

Isoelectric Focusing Isoelectric focusing electrophoresis is based on a pH gradient from anode to cathode. Such a gradient, however, is not created by the buffers of different pH. It is produced and maintained by the electric field on the synthesized materials, known as *carrier ampholytes*, such as Ampholines (a commerical product of LKB) or Biolytes (Bio-Rad Laboratories). Ampholine has the following structure:

$$\begin{array}{ccccc} -CH_2- & N & -(CH_2)_n- & N & -CH_2- \\ & | & & | & \\ & (CH_2)_n & & R & \end{array}$$

where $n = 2$ or 3, R is H or $-(CH_2)_x-COOH$. Its molecular weight is in the range 300–1000. It is water soluble and can produce a pH range of 3.9–9.5 when an electric field is applied. During the electrophoretic process protein molecules migrate anodically (from a basic region) or cationically (from an acidic region), until they lose their net electrical charge and cease to migrate. That particular pH region in which they cease to migrate is the region of an isoelectric point. There are two types of stabilizing media for isoelectric focusing: a density gradient of sucrose or a gel of polyacrylamide. The separation by any of the two types is based on composition rather than size of the protein molecules.

Density Gradient Isoelectric Focusing As an illustration of the general principle of density gradient isoelectric focusing, we chose a U-tube apparatus (Figure 17.14). After the U-tube is filled halfway with a 1% ethanolamine solution containing 40% sucrose (base), nine or more layers are added one by one to form the density gradient. The volume of each layer is approximately 1 mL. Layer 1 contains 1.5% carrier ampholytes in 30% sucrose, layer 2 contains 1.2% carrier ampholytes in 37% sucrose, and so on. The last layer contains 0.5% carrier ampholytes in 5% sucrose. On top of the density gradient are placed 1 or 2 mL of sulfuric acid. The protein sample (1 to 3 mg) is dissolved in an intermediate solution(s). The anode is inserted into the sulfuric acid solution and the cathode into the ethanolamine solution. The voltage increases gradually to 500 V. Focusing should be complete in 4–10 h.

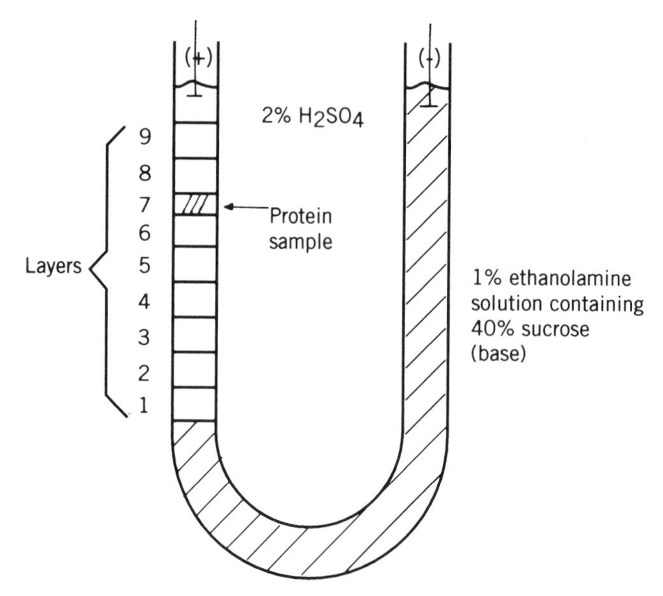

FIGURE 17.14 A U-tube apparatus for density isoelectric focusing.

Polyacrylamide Gel Isoelectric Focusing Two well-established techniques that use polyacrylamide are the gel cylinders and the thin layers. Here we discuss only the gel cylinder technique, because it is simpler and easier to understand. Figure 17.15 shows a disk electrophoresis apparatus for gel isoelectric focusing. The protein sample (approximately 30 μm per component) may be incorporated into the gel, which contains polyacrylamide and carrier ampholytes, or layered on top of the gel under a protecting layer of 2% carrier ampholytes and 5%

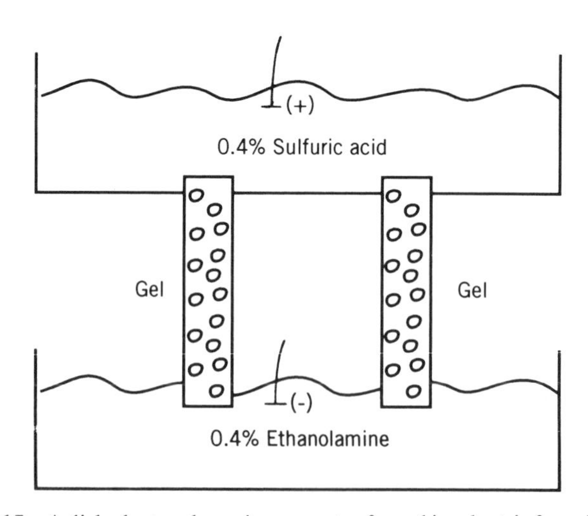

FIGURE 17.15 A disk electrophoresis apparatus for gel isoelectric focusing. ○, sample

sucrose. The voltage is raised gradually to 400 V for 60-mm gels and to 700 V for double-length gels. The following equation should be kept in mind:

$$\underset{\text{Voltage}}{v} \times \underset{\text{Current}}{ma} = \underset{\text{Power}}{\text{Watt}}$$

For protein, v should be high, while milliamperes should be low. If v is high, the velocity (mobility) is high. If milliamperes are low, the temperature is low, and the protein will not be denatured (at 35°C, proteins are usually denatured). For nucleic acid, milliamperes should be high. At high temperatures, the strains of nucleic acid are kept separated.

Two-Dimensional Electrophoresis Two-dimensional electrophoresis is a combination of two different electrophoretic separation procedures. The use of two separation techniques is based on the principle that they have independent parameters; that is, they separate chemical species according to different properties.

Usually isoelectric focusing on a thin gel is chosen for the first dimension separation. Once the ionic species (protein) are focused into narrow bands, they serve as zones for the second dimension analysis. The second dimension can be any of the following: immunoelectrophoresis, a discontinuous SDS–polyacrylamide gel system, or gradient electrophoresis.

Isotachophoresis Isotachophoresis is a steady-state stacking electrophoresis and can be used in capillary tubes, thin-layer equipment, gel rods, slabs, or columns. It consists of a leading ion with high mobility, such as chloride, sulfate, phosphate, or cacodylate, and a trailing ion with low mobility, such as ε-aminocaproic acid, β-alanine, or glycine. In between the leading and trailing ions, sample constituents continue to migrate on the basis of their mobilities. The leading ion and the trailing ion carry the same charge. Once the sample zones in between the leading stack zone and the trailing stack zone are separated, a steady state is reached. According to Kohlrauch's law,

$$\Lambda_0 = \Lambda_0^+ + \Lambda_0^-$$

the conductivity Λ (hence, the mobility u) of an ion is independent of the compound from which it is electrolyzed. On this basis, the concentration of each separated sample zone can be calculated by the concentration of the leading ion zone, which is known by the solution preparation.

Capillary Electrophoresis Because of its remarkable separation capabilities, capillary electrophoresis has been rapidly developed for use in biotechnology, particularly in gene splicing. In this technique a capillary tube of glass is used, with a diameter of 0.3–0.6 mm and a length of 15–100 cm. The voltage is 25–30 kV and the current is several microamperes. The capillary tube is filled first

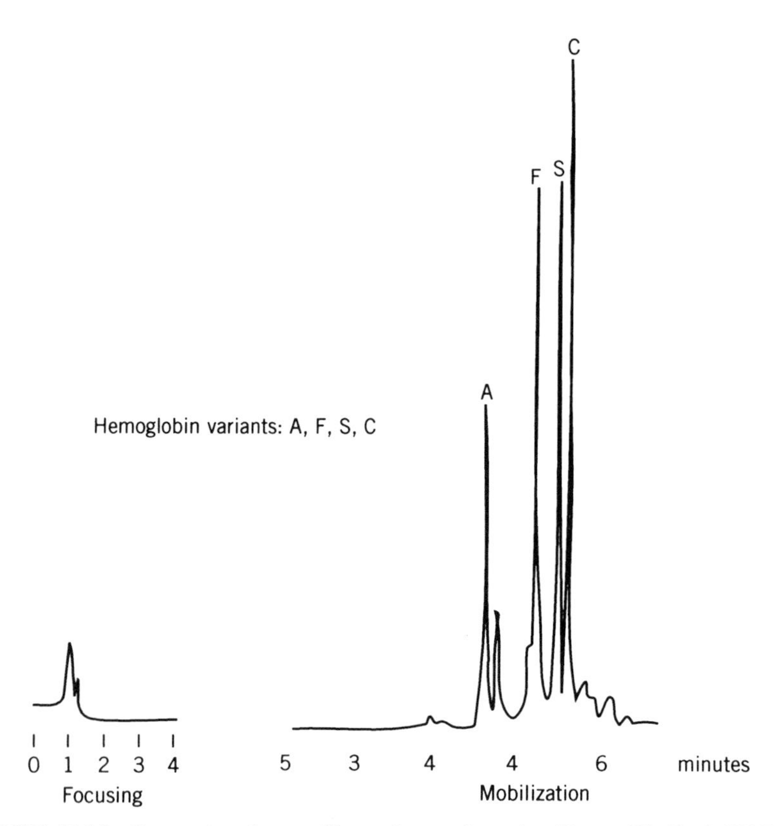

FIGURE 17.16 Separation by capillary electrophoresis. (From Bio-Rad HPE 100 capillary electrophoresis system, Application note 2, 1992. Reproduction by permission of Bio-Rad Laboratories, Hercules, California.)

with leading electrolyte and then with the sample. The terminating electrolyte is filled with after the sample if the capillary electrophoresis is to be used as isotachophoresis. It can be used, however, as isoelectric focusing as well as nongel sieving size-based electrophoresis by adding spacers. Figure 17.16 shows a sample of protein analysis by capillary electrophoresis.

17.3 FIELD-FLOW FRACTIONATION

Field-flow fractionation is a separation method that is similar to both liquid chromatography and electrophoresis. It has only one phase, the mobile phase. It can use an electric field to separate molecules as in electrophoresis, but it is not limited to an electric field and it does not need gels. The method was developed by Giddings and coworkers (1966, 1970, 1993) and is based on the concept of coupling concentration and flow nonuniformities. Separation occurs when an electric field is applied on the channel in which the sample flows. The molecules near the center of the channel have a higher velocity than those near the walls, thereby creating a concentration gradient. Meanwhile, diffusion works in the opposite direction to the driving forces. At equilibrium, the two opposite forces are balanced and the concentrating molecules stay around the equilibrium position. Thus, zones of molecules of different sizes are formed. Figure 17.17 illustrates this concept. The applied field can be centrifugal, gravitational, magnetic, or electric; it can also be a concentration gradient, a temperature gradient, or a chemical potential gradient. Detectors that can be used are similar to those used in liquid chromatography, namely, UV, visible, IR, fluorescence, refractive index, viscosity, density, and osmotic pressure.

The fractionation diagram is similar to a chromatographic design (Figure 17.18). According to Giddings (1970), the retention R can be expressed by

$$R = 6\lambda[\coth(2\lambda)^{-1} - 2\lambda]$$

where $\lambda = l/w$, with l as the ratio of diffusion coefficient D and the vector induced by the applied field V ($l = D/V$), and w as the thickness of the fractionation channel. The value of λ depends on V, which, in turn, depends on the type of the applied field. The following are some common field-flow fractionation methods.

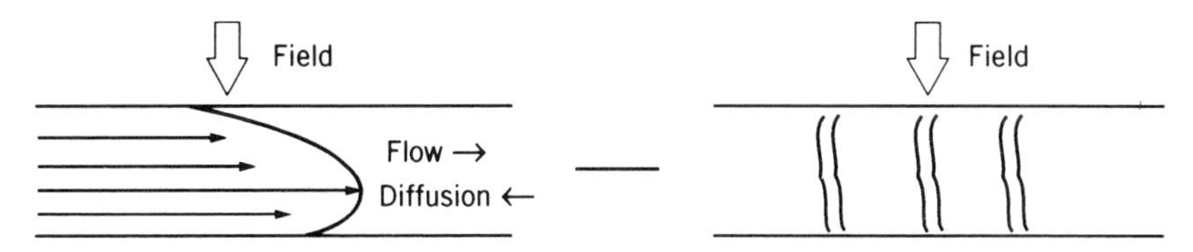

FIGURE 17.17 Field-flow fractionation.

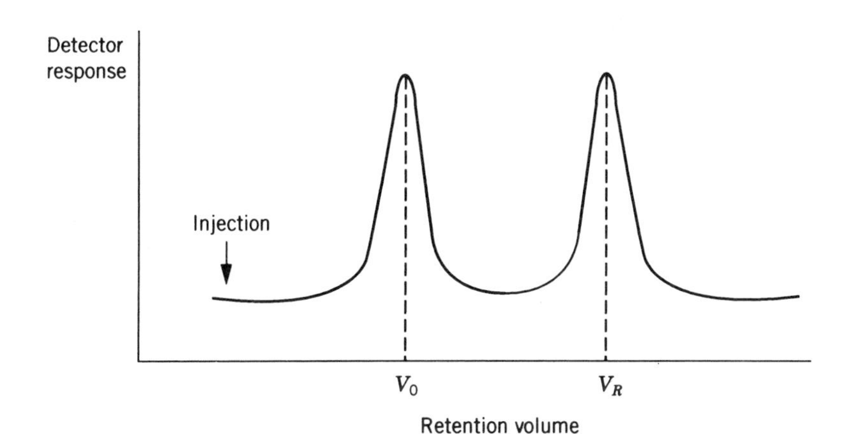

FIGURE 17.18 A fractionation diagram.

Thermal Field-Flow Fractionation This method is based on the principle of thermal diffusion D_T (Figure 17.19). The value of λ may be evaluated by using

$$\lambda = w \frac{D_T}{D} \frac{dT}{dx}$$

where D is the diffusion of the solution and D_T is the thermal diffusion coefficient.

Electric-Field-Flow Fractionation This method is similar to electrophoresis. The electrical field is induced by charging the two parallel plates, as in Figure 17.20. The channel between the two plates is filled with buffer as in the case of electrophoresis. The value of λ is calculated by using

$$\lambda = \frac{D}{uEw}$$

where u is the electrophoretic mobility and E is the electric field strength.

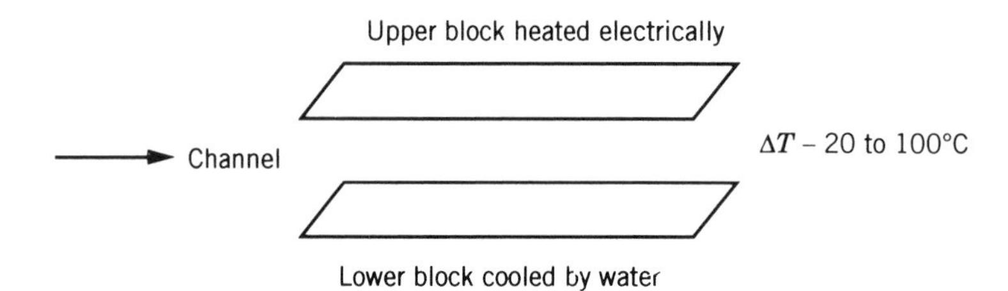

FIGURE 17.19 Thermal field-flow fractionation.

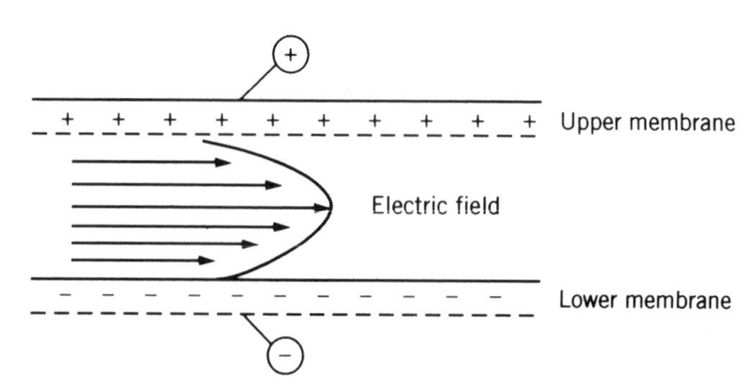

FIGURE 17.20 Electric field-flow fractionation.

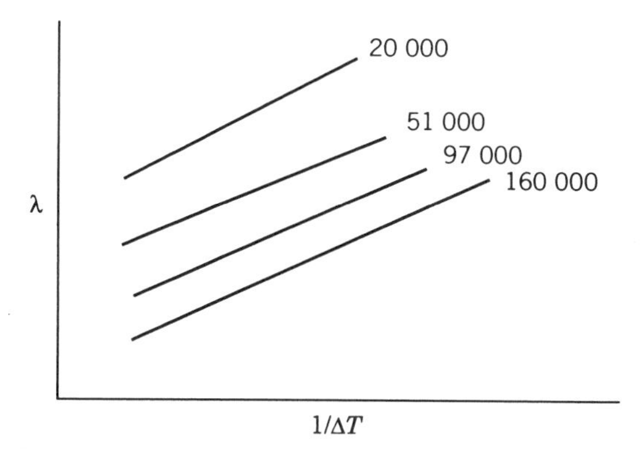

FIGURE 17.21 Separation of polystyrene sample in terms of λ. (Adopted from Giddings (1974).)

Sedimentation Field-Flow Fractionation This method uses the centrifugal field to separate molecules. The value of λ is calculated by using

$$\lambda = \frac{6kT}{\pi a^3 g w \Delta p}$$

where k is the Boltzmann constant, a is the particle diameter, g is the gravitational constant, and Δp is the difference in the densities of the solute and solvent.

Flow Field-Flow Fractionation This method is similar to dialysis or ultrafiltration, with the solvent acting uniformly on all the solutes. The field is generated by the flow of the solvent. The separation is mainly determined by the diffusion coefficient or frictional coefficient. The value of λ is calculated using

$$\lambda = \frac{R' T V_0}{3\pi N_A \eta V_c w^2 a}$$

where R' is the universal gas constant, V_0 is the void volume, V_c is the flow rate of the solvent, η is the viscosity of the solvent, a is the diameter of the particle, and N_A is the the Avogadro's number. Figure 17.21 shows the separation of polystyrene sample in terms of λ.

REFERENCES

Andrews, A. T., *Electrophoresis: Theory, Techniques, and Biochemical and Clinical Applications.* Oxford, UK: Clarendon, 1986.

Clin. Chem. **28**, 737 (1982). The entire issue is devoted to two-dimensional electrophoresis.

Davis, L. G., M. D. Dibner, and J. F. Batley, *Basic Methods in Molecular Biology*. New York: Elsevier Science, 1986.

Deyl, Z. (Ed.), *Electrophoresis: A Survey of Techniques and Applications*; Part A, *Techniques*. Amsterdam: Elsevier, 1979.

Giddings, J. C., and H. Eyring, *J. Phys. Chem.* **59**, 416 (1955).

Giddings, J. C., *Dynamics of Chromatography*, Part 1. New York: Dekker, 1965.

Giddings, J. C., *Sep. Sci.* **1**, 123 (1966); *Anal. Chem.* **42**, 195 (1970); *Science* **260**, 1456 (1993).

Giddings, J. C., and K. D. Caldwell, in *Physical Methods of Chemistry*, ed. by B. W. Rossiten and B. W. Hamition, New York: John Wiley, 1989.

Grubistic, Z., R. Rempp, and H. Benoit, *J. Polym. Sci. B* **5**, 753 (1967).

Hummel, J. P., and W. J. Dreyer, *Biochim. Biophys. Acta*, **63**, 530 (1962).

Heftmann, E. (Ed.), *Chromatography: A Laboratory Handbook of Chromatographic and Electrophoretic Methods*, 3d ed. New York: Van Nostrand Reinhold, 1975.

Howard, G. A., and A. J. P. Martin, *Biochem. J.* **46**, 532 (1950).

James, A. T., and A. J. P. Martin, *Analyst (London)* **77**, 915 (1952); *Biochem. J.* **50**, 679 (1952).

Janca, J., *Field-Flow Fractionation*. New York: Dekker, 1988.

Martin, A. J. P., and R. L. M. Synge, *Biochem. J.* **35**, 1358 (1941).

Maxam, A. M., and W. Gilbert, *Methods Enzymol.* **65**, 499 (1980).

Ravindranath, B., *Principles and Practice of Chromatography*. Chichester, UK: Ellis Horwood, 1989.

Sun, S. F., and E. Wong, *J. Chromatogr.* **208**, 253 (1981).

Sun, S. F., S. W. Kuo, and R. A. Nash, *J. Chromatogr.* **288**, 377 (1984).

Sun, S. F., and F. Wong, *Chromatographia* **20**, 445 (1985).

Sun, S. F., and C. L. Hsiao, *Chromatographia* **37**, 329 (1993); *J. Chromatogr.* **648**, 325 (1993).

Tung, L. H., *J. Appl. Polym. Sci.* **10**, 375 (1966).

Van Deemter, J. J., F. J. Zuiderwig, and A. Klinkenberg, *Chem. Eng. Sci.* **5**, 271 (1956).

Wilson, J. N., *J. Am. Chem. Soc.* **62**, 1583 (1940).

Yau, W. W., J. J. Kirkland, and D. D. Bly, *Modern Size-Exclusion Chromatography*. New York: Wiley, 1979.

PROBLEMS

17.1 Derive the equation for calculating the theoretical plates:

$$N = 16\left(\frac{t_R}{t_w}\right)^2 = \frac{16}{L}\left(\frac{V_R}{w}\right)^2 = \frac{1}{L}\left(\frac{V_R}{\sigma}\right)^2$$

17.2 Following is a chromatogram of the mixture of sample A and sample B:

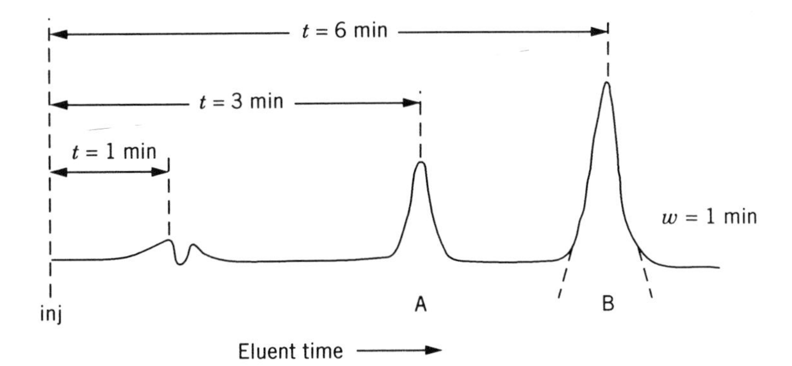

(a) Calculate $N, \alpha, k'_a, k'_b, N_s$ (resolution).

(b) If the column length is 12 cm, what is the plate height?

17.3 In gel permeation chromatography the chromatogram of a compound is in the form of a Gaussian distribution function:

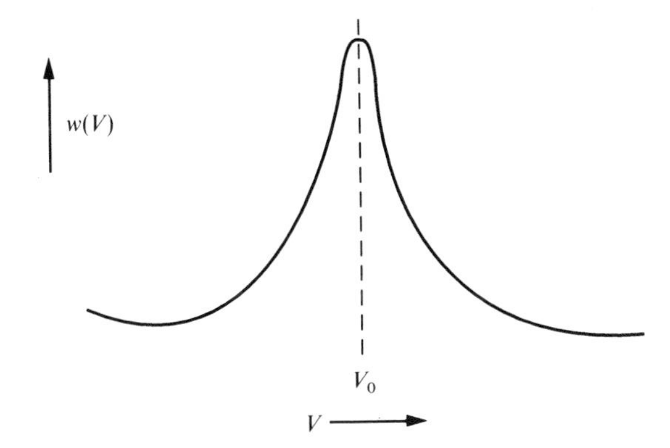

where V is the elusion volume. A relationship exists between the eluent volume and the logarithm of the molecular weight in the expression

$$V = c_1 - c_2 \ln M$$

where c_1 and c_2 are constants.
Show that the log-normal distribution of molecular weight

$$w(M) = \left(\frac{1}{\beta\sqrt{\pi}M}\right)\exp\left[-\frac{1}{\beta^2}\ln^2\left(\frac{M}{M_0}\right)\right]$$

can be translated into the expression of elution volume V in the form

$$w(V) = \left(\frac{c_2}{N\beta\sqrt{\pi}M}\right)\exp\left[-\frac{1}{c_2^2\beta^2}(V - V_0)^2\right]$$

Note that $w(M)$ and $w(V)$ are related by

$$w(M) = \frac{c_2 N w(V)}{M}$$

where N is the normalization constant for $w(V)$; that is,

$$N = \frac{1}{\int_{-\infty}^{\infty} w(V)\mathrm{d}V}$$

(*Source*: Tung (1966).)

17.4 The values of elution volume for some proteins, using same column and mobile phase and under same conditions, are as follows:

Protein	Molecular Weight	V_l (mL)
Sucrose		210
Ribonuclease	13 700	138
Chymotrypsinogen	25 000	135
Ovalbumin	45 000	110
Serum albumin	67 000	98
Thyroglobulin	670 000	68

Sucrose and thyroglobulin are used for references on both ends. For sucrose, $K_\mathrm{d} = 1$ and $V_l = V_0 + V_i$, while for thyroglobulin, $K_\mathrm{d} = 0$ and $V_l = V_0$.
Plot

(a) Mol wt versus V_l
(b) Mol wt verses K_d
(c) Mot wt versus k'

17.5 Because of electroosmosis, the observed mobility u should be corrected by including the gel length l:

$$u = \frac{\mathrm{d}}{tv/l} = \frac{\mathrm{d}l}{tv}$$

Show that if the actual migration length is d' and the gel length after staining is l', then

(a) $u_{\mathrm{act}} = u\left(\frac{l'}{l}\right)^2$

(b) $d' = d\frac{l'}{l}$

17.6 In the isotachophoresis, the concentrations of separated sample fragments can be calculated once the concentration of the leading zone is known. This can be done by using Kohlrausch's equation. Show how this can be done.

APPENDIX

CONVERSION FACTORS

$$
\begin{aligned}
1\ \text{atm} &= 1.013\,25 \times 10^{5}\ \text{N m}^{-2} \\
&= 1.013\,25 \times 10^{5}\ \text{Pa (pascal)} \\
1\ \text{bar} &= 10^{5}\ \text{Pa} = 0.98692\ \text{atm} \\
1\ \text{Å} &= 10^{-8}\ \text{cm} \\
1\ \text{dyn} &= 10^{-5}\ \text{N} \\
1\ \text{cal} &= 4.184\,00\ldots\text{J} \\
1\ \text{eV} &= 1.6022 \times 10^{-19}\ \text{J} \\
1\ \text{poise} &= 0.1\ \text{kg m}^{-1}\ \text{s}^{-1} \\
1\ \text{erg} &= 1\ \text{dyn cm} = 10^{-7}\ \text{J}
\end{aligned}
$$

SI PREFIXES

Multiple or Submultiple	Prefix	Symbol
10^{-3}	milli	m
10^{-6}	micro	μ
10^{-9}	nano	n

FUNDAMENTAL CONSTANTS

Quantity	Symbol	SI Value	egs or Other Value
Avogadro constant	N_A	$6.022\,045 \times 10^{23}$ mol^{-1}	
Boltzmann constant	k	$1.380\,662 \times 10^{-23}$ J K^{-1}	$1.380\,662 \times 10^{-16}$ erg K^{-1}
Speed of light in vacuum	c'	$2.997\,924\,58 \times 10^{8}$ m s^{-1}	$2.997\,924\,58 \times 10^{10}$ cm s^{-1}
Elementary charge	e	$1.602\,189\,2 \times 10^{-19}$ C	$1.602\,1892 \times 10^{-20}$ emu
Planck constant	h	$6.626\,176 \times 10^{-34}$ J s	$6.626\,176 \times 10^{-27}$ erg s
Faraday constant	$\mathscr{F}$	$9.648\,456 \times 10^{4}$ C mol^{-1}	
Permittivity of vacuum	ε_0	$8.854\,1878\,18 \times 10^{-12}$ F m^{-1}	
Nuclear magneton	μ_N	$5.050\,824 \times 10^{-27}$ J T^{-1}	$5.050\,824 \times 10^{-24}$ erg G^{-1}
Gravitational constant	G	6.6720×10^{-11} m^3 s^{-2} kg^{-1}	$6.672\,0 \times 10^{-8}$ cm^3 s^{-2} g^{-1}
Molar gas constant	R'	$8.314\,41$ J mol^{-1} K^{-1}	$8.314\,41 \times 10^{7}$ erg mol^{-1} K^{-1} $82.056\,8$ cm^3 atm mol^{-1} K^{-1} 1.9872 cal mol^{-1} K^{-1}

AUTHOR INDEX

SUBJECT INDEX